NH 19/12

THE 1882–1884 FIELDNOTES OF LUCIEN M. TURNER
TOGETHER WITH INUIT AND INNU KNOWLEDGE

MAMMALS OF UNGAVA & LABRADOR

EDITED BY SCOTT A. HEYES & KRISTOFER M. HELGEN
FOREWORD BY TIM FLANNERY • ILLUSTRATIONS BY BRYONY ANDERSON

A SMITHSONIAN CONTRIBUTION TO KNOWLEDGE

PUBLISHED IN ASSOCIATION WITH THE ARCTIC STUDIES CENTER

Smithsonian Institution Scholarly Press

Arctic Studies Center

Published by the
SMITHSONIAN INSTITUTION SCHOLARLY PRESS
P.O. Box 37012, MRC 957
Washington, D.C. 20013-7012
www.scholarlypress.si.edu

In association with the
ARCTIC STUDIES CENTER
Department of Anthropology
National Museum of Natural History
Smithsonian Institution
P.O. Box 37012, MRC 112
Washington, D.C. 20013-7012
www.mnh.si.edu/arctic

Compilation copyright © 2014 by Smithsonian Institution

Rights reserved. This publication may not be reproduced, stored in a retrieval system, or transmitted in any form or by any means, electronic, mechanical, photocopying, recording, or otherwise, without the prior permission of the publisher.

Text by Lucien M. Turner and the introductions to the Cetaceans, Ungulates, Rodents, Lagomorphs, Insectivorous Mammals, and Carnivores sections are in the public domain. The rights to all other text and images in this publication, including cover and interior designs, are owned either by the Smithsonian Institution, by contributing authors, or by third parties.

Cover image: Woodland Caribou plate by Ernest Thompson Seton. From *Four-Footed Americans and Their Kin* (Wright, 1898).

Library of Congress Cataloging-in-Publication Data

Turner, Lucien M. (Lucien McShan)
  Mammals of Ungava and Labrador : the 1882-1884 fieldnotes of Lucien M. Turner together with Inuit and Innu knowledge / edited by Scott A. Heyes, Kristofer M. Helgen.
      pages cm
  Includes bibliographical references.
  ISBN 978-1-935623-21-2 (hardback)
  1. Inuit—Ethnozoology—Québec (Province)—Ungava Peninsula. 2. Mammals Inuit—Québec (Province)—Ungava Peninsula. 3. Inuit—Québec (Province)—Ungava Peninsula—Social life and customs. 4. Turner, Lucien M. (Lucien McShan)—Travel—Québec (Province)—Ungava Peninsula. 5. Turner, Lucien M. (Lucien McShan)—Diairies. I. Heyes, Scott A. II. Helgen, K. M. (Kristofer M.) III. Title.
  E99.E7T83 2014
  599.09714'111—dc23

2013024908

ISBN: 978-1-935623-21-2 (print)
ISBN: 978-1-935623-28-1 (epub)

Printed in the United States of America

∞ The paper used in this publication meets the minimum requirements of the American National Standard for Permanence of Paper for Printed Library Materials Z39.48-1992.

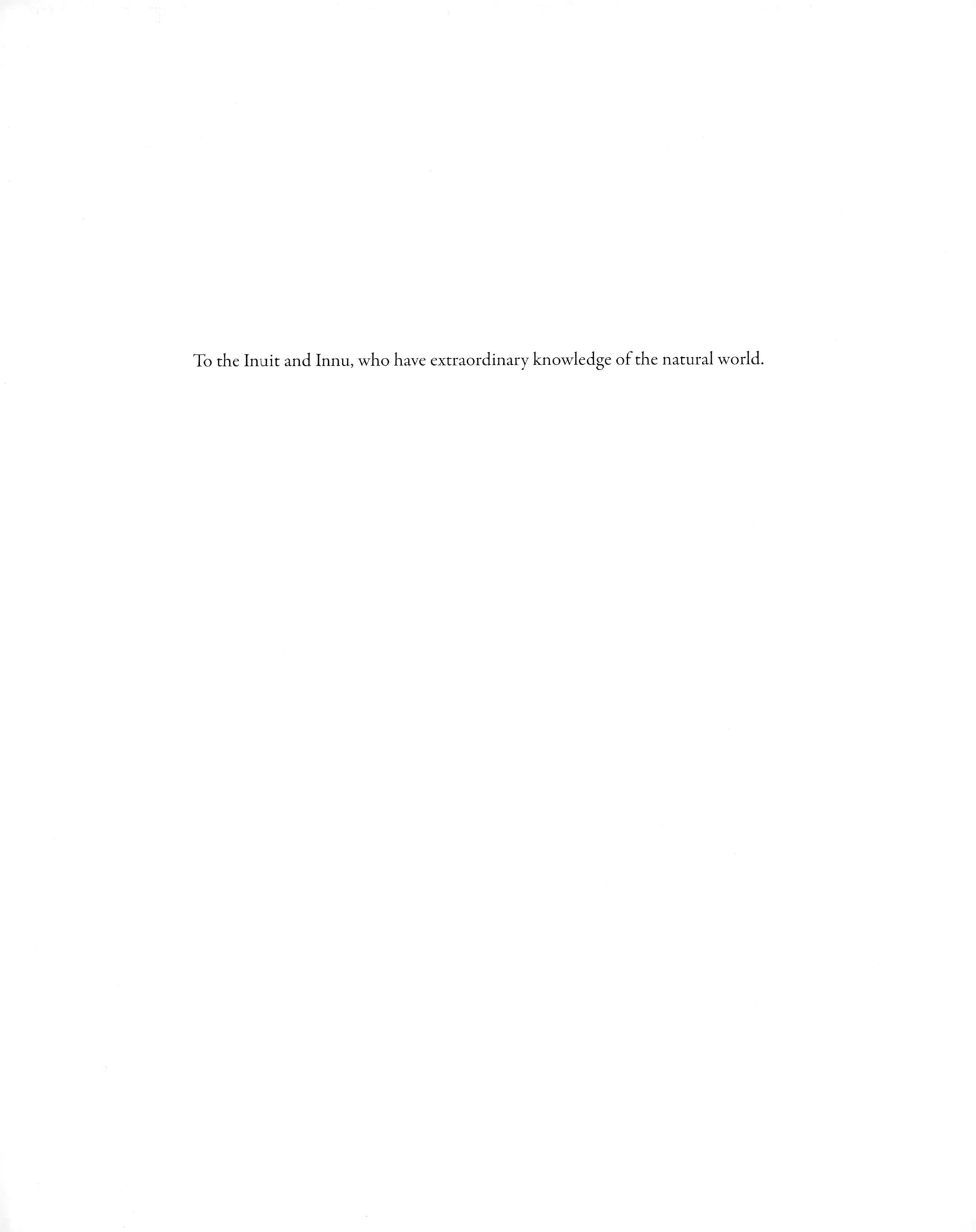

To the Inuit and Innu, who have extraordinary knowledge of the natural world.

Elder Dedication

Sophie Jararuse Keelan

My name is Sophie Jararuse Keelan, Native of Northern Labrador. Since ancient times, when the Inuit, the First Nations people, first inhabited the Earth, they have traveled across the North by dog team in their natural nomadic lifestyle and derived knowledge and wisdom from their way of living. They have always learned through observations.

I believe this book will be an excellent learning tool in schools and communities, especially for our younger generations. The information is cultural, true, and reflects the wisdom of telling stories and legends. It brings genuine meaning to the claim that only our stories of the past can make us real.

To everyone who has an interest in our heritage and in this country's history, I salute you, and those people who share their knowledge in this book–even if it highlights our differences. As people of this harsh country, let us observe, listen and come together.

In these times, when the world is changing, the need for speaking out has come. The knowledge and wisdom of those who have gone and passed on will never be forgotten. They will be treasured and documented.

Sophie Jararuse Keelan

Sophie Jararuse Keelan
Kangiqsualujjuaq, George River, Ungava Bay
12 August 2012

Contents

List of Figures	x
List of Mammal Stories	xvi
Editors' Preface	xix
Acknowledgements	xxi
Foreword	xxiii
An Appreciation of "Something Fiercely Desirable"	xxv
Two Extraordinary Years	xxvii
Legend to the Book	xlvi
A Turner Family Tribute	li

Chapter One — L.M. Turner's Introduction, 1886 — 3

Turner's Letter of Transmittal	5
Turner's Narrative	9

Chapter Two — Mammals of Ungava and Labrador, Canada — 29

Cetaceans — 31

Delphinapterus leucas (Pallas)	White Whale	ᖃᑦᓗᖅ	34
Monodon monoceros Linnaeus	Narwhal	ᐊᑦᓚᖹᐊᖅ	48
Balaena mysticetus Linnaeus	Bowhead Whale	ᐊᕐᕕᖅ	52

Ungulates — 61

Alces alces (Linnaeus)	Moose	ᑐᑦᑐᕙᒃ	64
Rangifer tarandus (Linnaeus)	Woodland Caribou or Reindeer	ᑐᑦᑐ	66

Rodents — 117

Marmota monax (Linnaeus)	Woodchuck or Groundhog	*No Inuit Term*	120
Tamiasciurus hudsonicus (Erxleben)	Hudsonian Squirrel	ᓇᕝᕋᑐᒋᐅᑦ	122
Glaucomys sabrinus (Shaw)	Flying Squirrel	*No Inuit Term*	130

Castor canadensis Kuhl	Beaver	ᑭᒋᐊᖅ	134
Ondatra zibethicus (Linnaeus)	Muskrat	ᑭᗱᓗᒃ	140
Dicrostonyx hudsonius (Pallas) / *Synaptomys borealis* (Richardson)	Lemming	ᐊᕕᙳᑕᖅ	144
Myodes gapperi (Vigors)	Evotomys	ᓄᓂᐋᕝᒃᖅ	148
Microtus pennsylvanicus (Ord) / *Phenacomys ungava* Merriam	Arvicole	ᓄᓂᐋᕝᒃᖅ	152
Erethizon dorsatum (Linnaeus)	Canadian Porcupine	ᐃᓀᖅᑐᑦᖅ	158

Lagomorphs 165

Lepus arcticus Ross	Hare	ᐅᑳᓕᖅ	166
Lepus americanus Erxleben	Rabbit	ᐅᑳᓕᐊᑦᕐᖅ	178

Insectivorous Mammals 183

Sorex cinereus Kerr / *Sorex hoyi* Baird	Shrew	ᐅᑦᕐᓇᖅ / ᐅᕐᕐᓇᖅ	186
Condylura cristata (Linnaeus)	Mole	*No Inuit Term*	188

Carnivores 191

Phoca vitulina Linnaeus	Harbor Seal or Fresh-water Seal	ᖃᓯᒋᐊᖅ	194
Pagophilus groenlandicus (Erxleben)	Harp Seal	ᖃᐃᕐᒡᑦᒃ	198
Pusa hispida (Schreber)	Ringed Seal	ᓇᑦᓯᖅ	200
Erignathus barbatus (Erxleben)	Bearded Seal or Square-Flipper Seal	ᐅᑦᔪᒃ / ᐅᕐᔪᒃ	224
Halichoerus grypus (Fabricius)	Gray Seal	*No Inuit Term*	236
Cystophora cristata (Erxleben)	Hooded Seal	ᓇᑦᓯᕙᒃ	238
Odobenus rosmarus (Linnaeus)	Walruses	ᐊᐃᕕᖅ	244
Ursus americanus Pallas	Black Bear	ᐊᑦᖯᖅ / ᐊᒃᖯᖅ	256
Ursus arctos Linnaeus	Barren Ground Bear	*No Inuit Term*	260
Ursus maritimus Phipps	Polar Bear	ᓇᓄᖅ	264
Procyon lotor (Linnaeus)	Raccoons	*No Inuit Term*	276
Lontra canadensis (Schreber)	Otter	ᐸᒥᐅᑦᒃ / ᐸᒥᐅᖅᒍᖅ	278
Mephitis mephitis (Schreber)	Skunk	ᓂᐅᑦᓗᒃ	282
Gulo gulo (Linnaeus)	Wolverene	ᖃᕝᕕᒃ / ᖃᕝᕕᒃᑕᕐ	284
Neovison vison (Schreber)	Mink	ᐃᖅᓗᒡᓯᔨᑎ / ᑯᑦᓯᑎ / ᙳᒡᓯᔨᑦ	300
Mustela erminea Linnaeus / *Mustela nivalis* Linnaeus	Stoat or Ermine	ᑎᕆᐊᖅ	302

Mammals of Ungava and Labrador

Martes americana (Turton)	Marten	ᖃᐅᕝᕕᐊ(ᕐ)ᕈᖅ / ᖃᐅᕝᕕᐊ(ᕐ)ᕈᒃ	308
Vulpes vulpes (Linnaeus)	Red Fox	ᑲᔪᕐᑐᖅ / ᑲᔪᖅᑐᖅ	312
Vulpes vulpes (Linnaeus)	Black or Silver Fox	ᕐᑭᕐᓂᑕᖅ	318
Vulpes vulpes (Linnaeus)	Cross Fox	ᐊᑯᓐᓇᑐᖅ	320
Vulpes lagopus (Linnaeus)	White, Arctic, Blue, or Stone Fox	ᕐᑯᑦᕐᑕᕐᔪᖅ	322
Canis lupus Linnaeus	Labrador Wolf	ᐊᒪᕉᖅ	328
Canis familiaris Linnaeus	Eskimo Dog	ᕐᑭᒪᕐᖅ	340
Lynx canadensis Kerr	Canadian Lynx	ᐱᖅᑐᕐᓯᖅ	356

Appendix: Turner Collections at the National Museum of Natural History   362

Glossary   374

Bibliography   378

About the Authors   384

List of Figures

Archival Research on Turner	xix
Turner's Manuscript	xx
Innu Campground at Ft. Chimo, Ungava Bay (1909)	xxiii
Turner at His Desk at Ft. Chimo (c. 1884)	xxiv
Lucien McShan Turner (1874)	xxix
Woodland Caribou Plate by E.T. Seton	xxx
Field and Museum Tags	xxxiii
Ledger Book (1884)	xxxiv
Turner's Family (1886)	xxxvii
Entrance to the George River, Ungava Bay (2012)	xxxix
Letter from Turner to Robert Ridgway (1884)	xl
Hudson's Bay Company (HBC) Dwellings at Ft. Chimo (1909)	xli
Caribou Fish Hook and Thimble Used by Inuit	xliv
Lucian Wayne Turner (2010)	1
Garment Paint Brush	1
Lucien McShan Turner (c. 1884-89)	2
Turner's Original Letter of Transmittal	6
Manuscript Cover	7
Labrador Sailing Ship	8
Montagnais Hunters Beside Canoe at Esquimaux Bay (c. 1882)	10
Group of Esquimaux (Inuit), Labrador (c. 1882)	11
Rigolet, Labrador	14
Views of Davis Inlet, Labrador (c. 1882)	15
Mount Ukasiksalik, Labrador (c. 1882)	16

Atlantic Cod, *Gadus morhua*	16
Views of Nachvach (Nachvak) Trading Post and Fiord (c. 1882)	17
Pack Ice, Ungava Bay, Labrador (c. 1882)	18
Polar Bear Skull	19
The S.S. *Labrador* on a Shallow Reef (1882)	21
Innu Camp near Ft. Chimo, April (c. 1883)	22
Inuit Family beside an HBC Dwelling at Ft. Chimo	22
View of Ft. Chimo (1882)	23
View of Rapids on the Koksoak River (1882)	24
View of the Koksoak River (c. 1882)	24
Innu Birchbark Canoe (c. 1883)	25
The HBC Steamer S.S. *Diana*	26
Naskapi and Inuit at Ft. Chimo (c. 1882)	27
Caribou on Fall Migration in October 2006	28
Pod of Belugas	33
Beluga Whale Skull	38
Beluga Whale Fetus	39
Fluid Specimen Storage at the Smithsonian	40
Inuit Hunting Customs	41
Whale Harpoon	42
Accession Card Describing Some Mammals Collected by Turner	43
Harpoon Head for Hunting Whales and Seals	44
Inuit Carving	50
Pod of Narwhal	51
Eskimo Violin with Bow Made from Whale Baleen (1883)	54
Bowhead Whale	56
Hunting Bowhead Whales	57
Herd of Caribou during Fall 2006 Migration	63
Caribou Antlers Collected by Turner	69
Caribou Antlers Collected by Turner	70
Stack of Caribou Antlers	71
Caribou Skin and Sealskin Clothing	72

Innu Skin Tent	73
Caribou Collection	76
Caribou Skulls Collected by Turner	77
Accession Card Notes Describing Caribou Meat Collected by Turner	78
Caribou Skull Collected by Turner	79
Caribou Skin Sack Collected by Turner	82
Inuit Girls in Caribou Clothing	83
Innu Child's Caribou Skin Coat	86
Blade of a Caribou Antler Bearing Field Data in Turner's Handwriting	87
Ice-Bailing Scoop Made from Caribou Antler and Wood	88
Model of a Drying Frame by Inuit	89
Bird Net Made by the Innu from Caribou Leather	90
Inuit Artist Working with Caribou Antler	90
Caribou Skulls Collected by Turner	91
Caribou Skull and Antlers	92
Bone Scrapers Collected by Turner	93
Caribou Shoes	94
Pair of Caribou Shoes	95
Caribou Skin Sleeping Bag (1884)	97
Caribou Roaming the Tundra in Northern Labrador	98
Back Scratcher	101
Sealskin Boots and Skin Clothing Worn by Inuit	102
Garments Worn by Inuit Men and Women (c. 1882)	103
Inuit Caribou Skin Coat	104
Amauti, an Inuit Woman's Parka	104
Pair of White Fur Mittens	105
Caribou Mittens	106
Toy Bow, Bow Case, Quiver, and Arrows	107
Strong Man and the Caribou	111
Inuit Carving	112
Hunting Caribou	113
Porcupine	119
Red Squirrel Skulls Collected by Turner	124

Red Squirrels	126
Squirrel Arrows	127
Red Squirrel Skins	128
Red Squirrel Mentioned by Turner in Manuscript	129
Flying Squirrel Skin	132
Flying Squirrel Skull	133
Beaver Lower Jaws	139
Muskrat	142
Muskrat Skin	143
Female Lemming Skull	146
Fluid Specimens of Lemmings	147
Myodes (Evotomys) Skin and Skull, Type Specimen	150
Myodes Habitat	151
Phenacomys Skin and Skull, Type Specimen	154
Type Specimens of Voles	155
Vole Skulls in Glass Vials	156
Vole Skin and Skulls	156
Vole Skull, Type Specimen	157
Porcupine Skull Collected by Turner at Ft. Chimo	161
Porcupine Brush	162
Porcupine Skin Collected by Turner at Ft. Chimo	163
Arctic Hare Skull	168
Cup and Ball Game Used by Ungava Inuit	171
Arctic Hare Skin	172
Underfur of Snowshoe Hare	175
Arctic Hare Skulls Collected by Turner	177
Articulated Hare ("Rabbit") Skull	180
Shipping Manifest Showing Turner's Ungava/Labrador Collection	181
Little Brown Bat	184
Northern Long-Eared Bat	185
Seal Hunting in Ungava Bay	196
Inuit Cap	197
Inuit Students Celebrate a Hunt	203

Models of Traditional Hunting Implements Used to Procure Marine Mammals	204
Kayak Under Construction (c. 1919)	205
Inuit Family Making Kayak (c. 1919)	206
Sealskin Kayaks	207
Hunter Beside Kayak (c. 1919)	207
Kayak at Ft. Chimo (1882)	208
Hunting from a Kayak	209
Kayak Model, Handheld Size	210
Sealskin Kayak	211
Seal Intestines	213
Harpooning a Seal	215
Seal Intestine Parka	217
The Missing Seals	218
First Seal Hunt	219
Mitilik Creatures	221
Illustration of a Seal Giving Birth near Nachvak Fiord, Labrador	222
Netting and Cutting Implements	223
Seal Hunting	226
Preparing Sealskins	227
Sealskin Tent (1882)	228
Sealskin Tent (1896)	229
Umiaq Model	231
Umiaq Construction (1960)	232
The Little People	235
Inuit Carving Depicting Seal Hunting	243
Harpoon Tip	246
Sealskin Float Used for Walrus Hunting	247
Ivory Ornaments	248
Deer Skin Woman's Outer Dress, Ornamented	248
Adult Male Walrus Skull	251
Walrus Carving	253
Walruses on an Ice-floe	254
Walrus Tusks Collected by Turner	255

Bear Canine Teeth	258
Bear Ornament	262
Bear Ornament with Beads	263
Fighting with a Polar Bear	266
Polar Bear Skull	269
The Man Who Wanted to Be Eaten	272
The Spirit Wind	273
Polar Bear Hunting	275
Female Otter Skull	281
Wolverine Skull	287
Wolverine Skin	288
Wolverine Skull	299
Winter Coat of a Stoat or Ermine	306
Least Weasel Skull	307
Male Marten Skull	311
Red Fox Skulls in Box	314
Red Fox Skull	315
Fox Story	317
Arctic Fox Skull	324
Arctic Fox Skulls in Boxes	325
Arctic Fox Skins	326
Wolf Skull	330
Wolf Skull	337
Female Wolf Skull, Type Specimen	339
Inuit Dog	342
Mushing	345
Illustration of a Story about the Man and His Wife Who Had No Dogs	346
Dog Whip	348
Inuit Dogs with Harness at Rigolet, Labrador (c. 1882)	349
Inuit Dog	350
Illustration of a Story about the Evil Dark Cloud	352
Illustration of a Story about the Miraculous River Crossing	353

List of Mammal Stories

▲ Story related by an Inuit elder

■ Story related by an Innu elder

▲ The Jealous Man – *Turner (1887a: 288)*	37
▲ Woman Who Turns into a Beluga Whale – *Tivi Etok (Heyes 2007)*	45
▲ Three Single Women – *Tivi Etok (Heyes 2007)*	55
▲ Raven and the Deerskin Blanket – *Turner (1887a: 28)*	74
■ A Starving Family of Indians – *Turner (1887b: 632-637)*	108
■ The Indian Youth with a Reindeer Wife – *Turner (1887b: 644-652)*	109
▲ Strong Man and the Caribou – *Tivi Etok (Heyes 2007)*	111
▲ Inuksuit and Caribou – *Johnny George Annanack (Heyes 2007)*	112
▲ Hunting Caribou – *Tivi Etok (Heyes 2007)*	113
■ Origin of the White Hairs on the Under-Eyelid of the Hudsonian Squirrel – *Turner (1887b: 625-626)*	128
■ The Indian and His Beaver Wife – *Turner (1887b: 587-592)*	137
▲ The Child and the Hare – *Turner (1887a: 299)*	173
■ The Venturesome Hare – *Turner (1887b: 582-586)*	173
■ The Greedy Hare and the Lame Frog – *Turner (1887b: 593-600)*	174
▲ Scaring of the Seals – *Turner (1887a: 280-281)*	215
▲ The Missing Seals – *Tivi Etok (Heyes 2007)*	218
▲ First Seal Hunt – *Tivi Etok (Heyes 2007)*	219
▲ Mitilik Creatures – *Tivi Etok (Heyes 2007)*	220
▲ The Little People – *Johnny George Annanack (Heyes 2007)*	235
▲ The Walrus and Grass – *Turner (1887a: 400 [correctly 300])*	248
▲ Origin of the Walrus and White Bear – *Turner (1887a: 282)*	249
▲ Story of the Orphan Boy – *Turner (1887a: 282-291)*	249
▲ Ceremonies Pertaining to Mammals – *Turner (1887a: 65-66)*	250
▲ The Incident at Akpotak Island – *Turner (1887a: 33-36)*	250

- ▲ The Walrus with Small Tusks – *Tivi Etok (Heyes 2007)* — 252
- ▲ Walrus Hunting – *Benjamin Jararuse (Heyes 2007)* — 252
- ▲ Walrus Skull and the Caribou – *Elijah Sam Annanack (Makivik 1984)* — 253
- ■ The Fate of the Bear Who Deceived Her Children – *Turner (1887b: 612-614)* — 259
- ▲ The Giant Polar Bear – *Paul Jararuse (Heyes 2005)* — 270
- ▲ Polar Bear with an Inuk Wife – *Tivi Etok (Heyes 2010)* — 270
- ▲ Rough Ice and Nanuklurk, the Big Polar Bear – *Noah Angnatuk (Makivik 1986)* — 270
- ▲ The Man Who Wanted to Be Eaten – *Tivi Etok (Heyes 2007)* — 272
- ■ Origin of the People on the Earth – *Turner (1887b: 608-609)* — 290
- ■ The Wolverene and the Bear – *Turner (1887b: 615-620)* — 290
- ■ The Wolverene Who Wished to Fly – *Turner (1887b: 621-624)* — 291
- ■ The Wolverene and the Rock – *Turner (1887b: 601-607)* — 293
- ■ A Wolverene Teaches Birds to Dance – *Turner (1887b: 627-631)* — 294
- ■ A Wolverene Goes Begging among Wolves – *Turner (1887a: 675-692)* — 295
- ■ The Origin of the White Spot on the Throat of the Marten – *Turner (1887b: 610-611)* — 311
- ▲ The Man and the Fox – *Turner (1887a: 297-298)* — 314
- ▲ Fox Who Became an Inuk – *Tivi Etok (2010)* — 315
- ▲ Child in the Fox Ear – *Paul Jararuse (Heyes 2007)* — 316
- ■ How Children Became Wolves – *Turner (1887a: 301)* — 331
- ■ The Wolf and Her Otter Husband – *Turner (1887b: 660-674)* — 332
- ▲ Wolf Attack – *Johnny Sam Annanack (Heyes et al. 2003)* — 335
- ▲ Man Who Turns into a Wolf and Other Animals – *Tivi Etok (Heyes et al. 2003)* — 338
- ■ Origin of the White People – *Turner (1887a: 283)* — 351
- ▲ The Evil Dark Cloud – *Tivi Etok (Heyes 2007)* — 352
- ▲ The Miraculous River Crossing – *Tivi Etok (Heyes 2007)* — 353
- ▲ Dogs and the Inugagulliq – *Benjamin Jararuse (Heyes 2007)* — 354

Editors' Preface

This book is a celebration of Inuit and Innu knowledge of mammals of Northern Quebec (Ungava) and Labrador, a region that remains relatively unknown to southern Canadians. Through the writings of Lucien McShan Turner, an accomplished naturalist connected to the Smithsonian, we learn about the rich understanding, knowledge, and respect that the Inuit (formerly Eskimo) and Innu (formerly Naskapi and Montagnais) had for land and sea mammals in the late 1800s. This historical perspective serves to reinforce the profound knowledge that the Inuit and Innu continue to have of mammals today.

Archival Research on Turner
The editors examining Turner manuscripts at the Smithsonian Institution Archives, Washington, DC, November 2011.

In the harsh Arctic and sub-Arctic lands of Ungava, hunters need to know their surrounds and the likely whereabouts of animals for survival. Over the passage of time and through oral tradition, Inuit and Innu knowledge of mammals became a body of wisdom on how to make tools, clothing, shelter, medicines, artwork, and nutritional meals from mammals. Whether using bearded seal skin as a kayak covering or making coats from animal skins and furs, there is no doubt of the ingenuity and resourcefulness of the Inuit and the Innu.

This practical wisdom extended to spiritual beliefs on mammals in which the lines of the natural and spiritual worlds were blurred. During Turner's time in Ungava, when shamanism was still practiced, the Inuit and Innu held superstitions about mammals and believed that the presence or absence of mammals could be attributed to the work of shamans and the spirit world. Presented in this book alongside natural history accounts are stories and legends of mammals that Turner documented in the region. This content includes creation stories, the workings of mammal tricksters, and gender-oriented beliefs about mammals. While most of these forms of belief are no longer held, they remind us of a time when the Inuit and Innu hunted mammals on the basis of many spiritual observances.

This book captures the many modes of knowing mammals by the Ungava Inuit and, to a lesser extent, by the Innu. Turner wrote more about the Inuit largely because he was based in Ungava at Ft. Chimo, where Inuit presence was greater than the Innu. Species by species, we recount Turner's knowledge of more than forty mammals, as told to him by Inuit and Innu hunters, and based on his hunting expeditions – principally around the land and waters of Ft. Chimo. The book follows the format Turner used to describe mammals of the region in his unpublished treatise on the topic.

We have packaged the book to reflect Turner's expansive documentation of mammals, with material sourced from his unpublished manuscripts. For instance, linguistic information relating to Inuit and Innu names that were used to describe each species in the 1880s was extracted from Turner's writings on the Inuktitut and Naskapi languages. These terms appear in the book alongside currently accepted ways of writing and describing mammal species.

The addition of new material to Turner's writings on mammals provides the reader with a sense of how Inuit and Innu knowledge has changed over the course of 130 years, and helps to complete descriptions of mammal species where Turner seemed unsure or lacked information. New materials generated for this project include: introductions to each Order section; line illustrations of species; distribution maps; a map of Turner's travels; wordlists of Inuktitut terms and definitions of mammals; stories and legends associated with mammals as told by Inuit elders today; photos of the region by Turner and others of the period; a glossary; and photographs from Turner's Mammal and Ethnographic Collection at the National Museum of Natural History. Those wishing to work with Turner's Ungava collection will find the inclusion of a list of mammal collection numbers a useful resource.

We consulted with Inuit elders and many other experts on the Ungava region on this project, including linguists, archaeologists, anthropologists, geographers, museum curators, archivists, graphic designers, illustrators, and Turner's direct descendants. We asked Inuit elders from Ungava to provide their perspectives on mammals today and included these in the book along with previously recorded stories on mammals that were documented by Scott Heyes and other researchers. These current day stories and descriptions of mammals remind us that Inuit knowledge of the natural world is constantly being modified and built upon in accordance with a changing environment. We learn that mammals are no longer intricately connected to Inuit belief systems, but they are very much a part of Inuit identity as hunters.

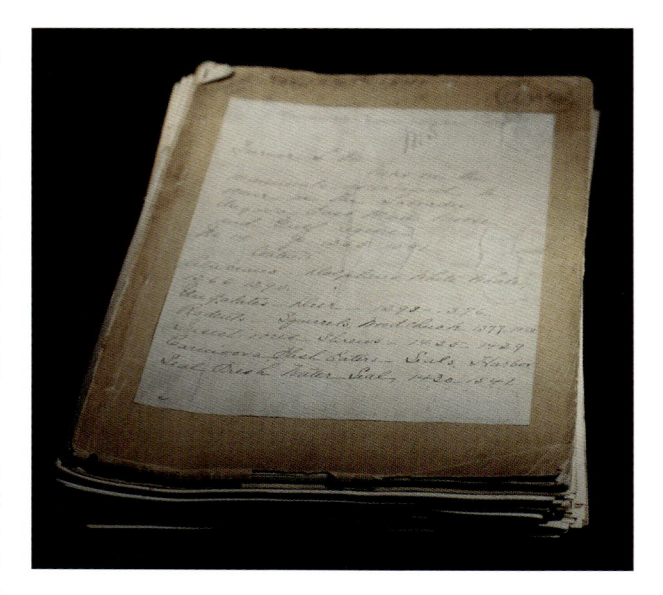

Turner's Manuscript
Photograph of the manuscript by Turner on mammals, titled "Mammals ascertained to occur in Labrador, Ungava, East Main, Moose and Gulf Region." Source: SI Archives, Record Unit 7192, Turner, Lucien M, 1848-1909, Lucien M. Turner Papers.

The book is multidisciplinary in scope and will appeal to a variety of readers, whether they are the Inuit and Innu of the region, Arctic researchers, anthropologists, geographers, mammalogists, or naturalists. The impetus and new content for the project draws from the editors' backgrounds and research in anthropology and mammalogy. The book provides a model for working across different realms of science to produce a work that celebrates Western-scientific knowledge and Indigenous knowledge systems as equal partners.

Though Turner was described as a modest and unassuming scientist, we suspect that he would be satisfied knowing that his notes on mammals from the Ungava and Labrador regions have finally been assembled in one published volume. We hope this book will help to elevate Turner's standing in all realms of science, and particularly in mammalogy and anthropology. This book is a tribute to Turner's achievements, to his discovery of new species, and to the Inuit and Innu people who shared their wisdom of mammals with him.

Acknowledgements

The authors thank the Inuit hunters of Nunavik, Northern Quebec for supporting this project and for sharing their knowledge of mammals with us. We are especially grateful for the time and knowledge that many Inuit elders and other expert hunters from the community of Kangiqsualujjuaq have contributed to the project. In particular, we thank Daniel Annanack, Annie Kajuatsiak, Tommy Kajuatsiak, Tivi Etok, Minnie Etok Morgan, Peter Morgan, Johnny Morgan, Johnny George Annanack, George Don Annanack, Willie Etok, Susie Etok, Sarah P. Annanack, Benjamin Jararuse, Paul Jararuse, Sophie Jararuse Keelan, Johnny Sam Annanack, Mary Sam Annanack, Maggie Lucy Annanack, David Annanack, and Anita Annanack for their hospitality during visits to Ungava Bay, and for giving permission for artwork and their accounts of mammals to be reproduced in this book. Lucina Etok and Jonathan Andersen kindly assisted as interpreters and translators during interviews with elders. We also thank the Northern Municipality of Kangiqsualujjuaq, and the Ulluriaq School, for allowing us free use of their facilities as support towards the project. Thank you to the former Makivik President, Pita Aatami, for arranging airfares for Scott Heyes to travel to the field with First Air and Air Inuit.

The section on Inuktitut terms for mammals in this book was generated with the kind assistance and expertise of the oral historian, John MacDonald, and the Inuktitut linguist specialist, Professor Louis-Jacques Dorais, Département d'anthropologie de l'Université Laval. We thank you both for your comprehensive contribution that you provided freely.

We are grateful for the financial, in-kind, and personnel support that the National Museum of Natural History's Department of Anthropology and its Division of Mammals at the Smithsonian Institution (SI) have provided. In the Department of Anthropology's Arctic Studies Center (ASC), we thank its Director, Dr. William Fitzhugh, for providing publication funds towards the project, and for providing valuable knowledge and material for the book. Dr. Fitzhugh's unwavering support of the project was greatly appreciated. We express much gratitude to the ASC's Arctic Archaeologist, Dr. Stephen Loring, for showing us through the Turner Collection with passion, and for sharing his wealth of notes, opinions, and writings on Turner. Dr. Loring's contribution to the project was invaluable. Scott Heyes wishes to thank the ASC's Dr. Igor Krupnik for guiding the project, and for being a wonderful host. Thank you for your ongoing mentorship and for your ideas on how to bring this and future projects to light. The project has enjoyed the support and assistance of the entire ASC team at various stages of the project including Lauren Marr, Dr. Noel Broadbent, Dr. Beatrix Arendt, and Dr. Chris Wolff. Thank you to the Anthropology staff, Dr. Gwyneira Isaac and her family for kindly hosting Scott Heyes during stays in Washington, DC, and to Dr. Joshua Bell, Dr. Dennis Stanford, and Pegi Jodry, for their friendship, advice, and support. We also thank Felicia Pickering for guiding us through Turner's ethnographic collection at the SI's Museum Support Center (MSC). The archivists, librarians, and supporting staff at the Smithsonian Institution Archives, the John Wesley Powell Library, the National Anthropology Archives at the MSC, and the Library and Archives Canada dedicated much time to helping us locate material on Turner. We would like to especially thank Leanda Gahegan, Stephanie Christensen, Gina Rapport, and Margaret Dittemore for their assistance in locating material.

In the Division of Mammals at the SI, we thank Charley Potter for his expertise on marine mammals and for guiding us through Turner's material in the Whale Collection. A special thank you goes to Sharon Jones for typing out Turner's manuscript material. We appreciate your time and care in deciphering his handwriting. Curator Emeritus, Dr. Don E. Wilson, kindly granted us permission to use his photographic collection of mammals, which were used as a basis to generate illustrations. Darrin Lunde, Craig Ludwig, Renee Regan, and Paige Engelbrektsson kindly helped us track down material in Turner's mammal collection.

The graphical layout of the book was generated with the wonderful assistance of Phil Easson. Thank you to Anna Sundman and Sean Montague for working on draft versions of the layout. The grayscale illustrations and color plates were produced through the talented hand of Bryony Anderson. Thank you as well to the Smithsonian's illustrator, Marcia Bakry, for generating illustrations. We owe thanks to the gifted Smithsonian photographer at the NMNH, Donald Hurlbert, for taking incredible photos of the Collection. We are most grateful of the research assistance that Dr. Angela Frost provided to the project. Angela helped us document the Turner Collection prodigiously and comprehensively. We thank you for your assistance in the production of the book through your talents as a scientist, photographer, artist, and photo-editing expert. Thank you to the Canadian Museum of Civilization, the McCord Museum, and Rare Books and Special Collections at McGill University for granting us permission to publish photos from your archives. We are grateful for the advice that Alaskan Anthropologist, Dr. Ken Pratt, contributed towards the project.

We were truly fortunate to have met Lucian Wayne Turner, Lucien M. Turner's grandson, during the course of generating this book. Thank you, Lucian for providing us with a copy of the book generated on your grandfather's family history, and for permitting us to reproduce material from your family's collection. Your personal study of your grandfather has offered new and valuable pieces of information to the scientific community.

Scott Heyes is especially appreciative of the kind generosity and hospitality rendered by Professors Peter and Ellen Jacobs during stays in Montreal for this project, and for their ongoing mentorship and support. A special thank you also goes to Inez Adams, who kindly accommodated Scott during his first extended stay in Washington, DC.

This project was supported by many funding agencies and institutions, and was initially carried out as part of a research program through the Frost Center's Canadian and Indigenous Studies Program, Trent University, Canada. At the outset of the project, Scott Heyes was the 2010-2011 Roberta Bondar Fellow in Northern and Polar Studies at the Frost Center, and was a Visiting Scholar at the Smithsonian Institution's Arctic Studies Center. Research funding was provided, at various stages of the project, from: the Frost Center's Canadian and Indigenous Studies Program, Trent University; Symonds Trust Fund for Canadian Studies, Trent University; Social Science and Humanities Research Council Award, Trent University; the SI's Arctic Studies Center; the SI Scholarly Press; The University of Melbourne, Australia; Makivik Corporation; Avataq Cultural Institute, Montreal; and The University of Canberra, Australia.

We are most grateful for the commitment that Ginger Strader, Director of the SI Scholarly Press, has maintained for this project, and we are thankful that the Smithsonian Institution has given us this opportunity to bring Lucien Turner's writings on mammals to the attention of science and to the direct descendants of the Inuit and Innu he learned from over 130 years ago.

On the home front, we thank our respective partners, Christine Heyes LaBond, and Lauren E. Helgen, for their encouragement, patience, and support throughout the project.

Scott A. Heyes

Kristofer M. Helgen

Foreword

Tim Flannery

Lucien Turner's notes on Arctic mammals are so vivid – and so full of the distinctive rhythms of the language of his times, that when you read them it seems as if the dead are speaking. Not just the Inuit who told Turner stories and brought him the pelts and skulls of the mammals he was so ardently interested in, but the voice of the man himself, now more than a century in his grave. Indeed Turner's plain, modest narrative is so distinctively '19th Century Yankee' in tone that it recalls an entire, vanished frontier age, an age in which he played a vital role.

Turner's expedition to Ft. Chimo in northern Quebec in 1882 was a journey to the other side of the frontier – a journey into a place where indigenous cultures still flourished largely untouched by the outside world, and where Europeans eked out a perilous existence or did not survive at all. And what an expedition it was! Turner, sent north to make meteorological observations, turned the opportunity into a full-scale scientific reconnaissance of a barely explored region, and as a result brought back a bounty of information and specimens that are unique and uniquely valuable. Moreover, he brought to the task a young man's eye and energy, an unflagging curiosity, and a truly extraordinary ability to record everything – even the seemingly mundane, but which to his modern readers is so fascinating. That he managed to do this in an utterly different land and culture, where language and etiquette were profound barriers to understanding, is a mark of his distinction as a pioneering naturalist. In the end he spent two years studying and documenting the inextricably intertwined lives of the Arctic's humans and other mammals, a labour that resulted in what is arguably the most penetrating account of the subject ever written.

Rarely is Turner a dry recorder of what passed before his eyes. Instead his notes often have an almost uncanny power to evoke the past. When he talks of the Stoat he tamed and which lived in his room at Ft. Chimo, we seem to be there with him:

> *"... it stretched its snake-like neck and with its greenish, glittering eyes which appeared to fairly revolve in their sockets gave the creature a formidable appearance but instead it slowly sank back and resumed its nap."*

Even the jumbled words, unedited and raw as they are, add to the sense of being at Turner's elbow. We are with him too, when he questions reticent Inuit about the ways of beluga and other sea-creatures, and about why they act as they do both during and after a hunt. And as we read we seem to stalk the tundra with him while he hunts hares and other creatures, and seeks understanding of animal ways in the landscape.

Innu Campground at Ft. Chimo, Ungava Bay (1909)
Innu and Inuit hunters would congregate at Ft. Chimo to trade with and work for the Hudson's Bay Company (HBC) in the late 1800s and early 1900s. Note the HBC dwellings in the background. Used with permission of McCord Museum, M2000.113.6.260.

Turner at His Desk at Ft. Chimo (c. 1884).
While stationed at Ft. Chimo, Turner drafted extensive manuscripts on the ethnology and natural history of Ungava, including mammals. 1 cyanotype 4" × 5" mounted on 5" × 8" cyanotype on Stereograph. Source: SPC Arctic Eskimo Labrador NM No ACC # Cat 17548401447800, National Anthropological Archives, Smithsonian Institution.

There is a great beauty, as well as a sadness, in reading of this vanished world. For most of the 130 years that separate Turner's arrival at Ft. Chimo from the publication of this book, the Canadian Arctic must surely have seemed that part of North America safest from the modern world. Perhaps this sense that the Arctic would always be there as it had been led to a complacency that helps explain why Turner's notes were so long neglected. But it's now evident that Turner was documenting a world of snow and ice, and mammals and Inuit, which the economic power of the modern world has all but melted away.

This beautiful book consists of far more than Turner's notes. Astonishing illustrations lavish its pages, as do Inuit stories told more than a century after those recorded by Turner. And then there are the illustrations and photographs of Arctic mammals, and of the many museum specimens Turner collected. There is art in making dry bones look appealing, art that abounds on these pages. And some of the wildlife illustrations, including those of the wolf, are among the finest I've seen.

Scott Heyes and Kristofer Helgen are to be congratulated on the publication of this timely work, and most especially on the resurrection of Turner's notes. It's a work that celebrates the achievements of an extraordinary, pioneering American biologist, and it brings to light a wealth of information which would otherwise have lain in obscurity. It also informs us of how much beauty and diversity is at peril in the modern world.

Tim Flannery
Scientist, author, and the 2007 Australian of the Year.
5 October 2012,
Sydney, Australia.

An Appreciation of "Something Fiercely Desirable"

Stephen Loring

I first encountered Lucien Turner in 1975, having returned from a summer in Labrador as a crew-member on William Fitzhugh's archaeological project on the Central Coast and in the Nain archipelago. That summer irrevocably changed the course of my life, not just because the country was wilderness sublime and the archaeology intoxicating, but also because our travel, in a small 35-ft launch, brought us into direct association with Settlers. It was some measure of my naivety that I was astonished to find the Inuit and Innu so intimately linked to subsistence activities, that my wilderness was their home. And the stories told of fish and birds plentiful beyond description, fierce winters, bears, ice and death, but always a link to generations past and present. This was a place of history, as indeed are all places if you only know how to seek and to listen. These stories, oral histories, observations – call them what you will – sparked my insatiable appetite for all things pertaining to Labrador.

I spent the summer of 1987 closeted in a secluded alcove beneath clerestory-like, vine-encrusted windows, systematically working through Turner's Labrador manuscripts. The SI Archives were then housed in a crumbling corner of the Arts and Industries Building, next door to the Smithsonian's iconic Castle. In addition to his manuscript on mammals, Turner prepared his observations on birds and fishes, altogether providing an invaluable resource on the subsistence options formerly available to the region's Inuit and Innu families. My inquiries put me in the company of a small cadre of northern researchers who had previously recognized the significance of Turner's work, including biologist Francis Harper, biologist and oceanographer Maxwell Dunbar, historian Alan Cooke, and Inuit activist and political leader Zebedee Nungak. Traveling vicariously with Turner through the Ungava District in the late 19th century provided me a much more nuanced appreciation of that distant place and time.

While Smithsonian research interests in Labrador began as early as 1860 with the loan of astronomical and surveying equipment to the U.S. Coast and Geodetic Survey Eclipse Expedition, it wasn't until the arrival of Lucien Turner at Fort Chimo (now Kuujjuaq) in 1882 that Smithsonian research in the Far Northeast can be said to have begun in earnest. Perennially strapped for funds, Smithsonian Secretary Spencer Baird was always on the lookout for opportunities to send his collectors into the field. The participation of the U.S. government in the first International Polar Year (1882-1883) created the opportunity to send researchers Edward Nelson and John Murdoch back to Alaska, and Lucien Turner to the Ungava District of the Hudson Bay Territory.

"...irresistible, something fiercely desirable.... The Labrador peninsula looks blunt and squat on the map; its essence is barbed, and sticks in the heart."
Robert Gathorne-Hardy

There is a strong congruence in the resulting work of the early Smithsonian naturalists-anthropologists that must derive from Spencer Baird's guidance. Baird's death in 1887, after the Alaskan and Arctic fieldwork was completed but before it was fully assembled and published, curtailed the careers of both Lucien Turner and his rival, Edward Nelson, two of the most distinguished and promising naturalists of their generation.

Lucien Turner was perhaps the brightest star in the constellation of extraordinary naturalists with whom Baird surrounded himself. With over six years' experience collecting ethnographic and biological specimens in Alaska and the Aleutian Islands, he had demonstrated an extraordinary affinity as a naturalist. A cornucopia of birds, animals, eggs and bones, minerals and ethnographic objects, Turner's collections fully realized the potential for Arctic science. Turner also possessed an impressive gift for languages, learning Russian as well as enough Unangan and Yup'ik to speak to his Aleutian and Eskimo informants, whose knowledge and stories contributed to the collections' value.

It is a great tragedy that Turner's fieldnotes and journals have, apparently, not survived. The most striking feature of his manuscripts in the SI Archives is his bold, clear cursive script that sweeps elegantly and assuredly across the neatly ruled, unblemished paper, in striking contrast to original documents penned in the dim light and trying conditions of northern camps and cabins. Archaeological and natural history fieldwork proceeds as a series of steps that distance it from its origins; thus observations become fieldnotes, fieldnotes are worked into drafts, and drafts finally codified as publications. In the distillation process, there are always observations unreported and stories left untold. While Turner's manuscript was clearly an arrested work in progress, it retains something of his personality in its weaving together of narrative threads – his own as well as those of Hudson's Bay Company employees and Inuit and Innu informants – that goes beyond a purely descriptive format of biological facts and observations.

Time and curation have a way of enhancing the value and potential of museum collections. As evidenced by this volume, the significance of Turner's observations and the objects he acquired continues to increase. A portion of his manuscript provided a critical reference for establishing the former presence of Grizzly bears in Labrador; the ethnographic collections are a source of inspiration and pride for Inuit and Innu artisans, educators, and community leaders in Labrador and Nunavik, and were featured at the Inuit Circumpolar Conference in Kuujjuaq in 2002; and DNA extracted from hair of the wolverines Turner collected could help confirm their continued existence and viability in the barren lands east of Hudson Bay.

Northern Quebec and adjacent Labrador continue to attract researchers. Situated at the boundary between forest and tundra, the region is especially sensitive to indications of climate change. Northern Quebec/Labrador is a cultural boundary as well; the Inuit and Innu descendants of the people from whom Turner acquired information and collections now reside in villages and, to some extent, still rely on the animals that Turner describes. Much of what now passes as natural history is concerned with loss: the erosion of viable ecosystems, the disappearance of species, the relentless pace of modernization. In bringing forth Lucien Turner's mostly forgotten manuscript on the mammals of Ungava and Labrador, Scott Heyes and Kristofer Helgen have opened a window on an extraordinary place and time, reminding us of what has been lost but also what needs to be saved and valued.

Stephen Loring
Arctic Studies Center, Smithsonian Institution

Two Extraordinary Years

The Editors' Reflections on Lucien Turner's Contributions to Mammalogy and Anthropology in Ungava and Labrador

Scott A. Heyes and Kristofer M. Helgen

Introduction

The Arctic and sub-Arctic regions of Northern Quebec and Labrador were known as the *District of Ungava* when Lucien McShan Turner travelled there in 1882. The most common conceptualization of the placename *Ungava* is that of a historical district, used mostly by the Hudson's Bay Company (HBC), which operated in Northern Quebec and Labrador, to describe the lands and waters stretching across Hudson's Bay, James Bay, Ungava Bay, and the Labrador Sea. It is likely that this spatial descriptor of *Ungava* was derived from a more specific geographical feature known to the Inuit as a "place where there is no river" (Turner 1884: 2846). It referred specifically to a small cove on the Ungava (or Koksoak) River that flows into Ungava Bay. This cove was located near the Hudson's Bay Company trading post of Ft. Chimo, the largest in the Ungava region and the base from which Turner conducted his research. The placenames of Ungava and Ft. Chimo that Turner used so frequently in his fieldnotes may not be recognized by many familiar with the region today. This is because these placenames have been substituted with more specific placenames that better reflect Inuit and Innu occupation of the region and their land tenure arrangements that have been established through modern treaties.

For instance, the Inuit people of the Ungava region now describe the lands they occupy as *Nunavik*, and the Innu people of the region describe the lands they occupy in Ungava as *Nitassinan*. Both terms mean "our land," and each is regarded as the homeland of the Inuit and the Innu, respectively. Just as the term *Ungava* has been replaced as a placename, so too have the terms used to describe the Inuit and Innu themselves. In earlier times, the Inuit were described as *Eskimo*, a term now considered pejorative in Canada. The same term, however, is still used in Alaska to describe indigenous people of the region, and it is used in the study of linguistics to describe the language group of the Inuit people across the Arctic. Today, in Nunavik, the mother tongue is more commonly referred to as *Inuktitut* rather than Eskimo.

The Innu people of the Nitassinan region, whose homeland includes the eastern portion of the Quebec–Labrador Peninsula, were once regarded as the Naskapi and Montagnais people, as well as by several other distinctions based on regional affiliations and Innu language dialects. The language spoken by the Innu today is known as *Innu-aimun*.

The lands and seas upon which the Inuit and Innu have dwelled for thousands of years are harsh and formidable.

"I shall not fail to properly record everything pertaining to the natural history of this place."

Lucien McShan Turner (Letter to Ridgway, 1 Sept 1882)

Survival in this ice- and snow-laden landscape – where bears, wolves, and natural hazards abound – has required Inuit and Innu hunters to develop acute and intimate knowledge of their environments and natural systems. To gain deep, rich, and meaningful knowledge of this landscape, elders say that one must be open to listening and learning from expert knowledge-holders, to learning the local language used to transmit land-based knowledge, and to spending time on the land. The Smithsonian Institution affiliated naturalist Lucien McShan Turner (b. 20 June 1847; d. 7 April 1909) is perhaps one of few non-Indigenous people of his time who attempted to learn and document the land-based knowledge systems of the Inuit and Innu in this fashion. In a modern anthropological context, this mode of study by Turner would be regarded as participant-observation research.

Through interaction with Inuit and Innu hunters, and personal hunting excursions between the summer of 1882 and the fall of 1884, Turner did his best to record their language, customs, traditions, belief systems, stories, and knowledge of animals, birds, plants, fungi, spiders, butterflies, crustaceans, shells, and fish. These accounts of the social and natural history by Turner may not be complete, but they are detailed and informative and were recorded at a time when the Inuit and Innu were still living almost exclusively off the land. His Ungava records are also meaningful and historically significant in that they were written in the context of the first International Polar Year.

Turner's writings on the Inuit and Innu of Ungava are captivating, with his unpublished fieldnotes on mammals being particularly descriptive accounts. These have been contextualized so the reader can appreciate the significance of his research and findings within the fields of mammalogy and anthropology, and how he went about his work with profound respect to, and sensitivity toward, Inuit and Innu knowledge systems.

This is the first time that a succinct account of Turner's Ungava experiences and research on mammals of the region has been prepared. It offers to the Inuit and Innu, who no longer have living memory of Turner being in the region, a window into how their elders from three or four generations ago used and conceptualized mammals.

The Context of Turner's Work on Mammals

"The first important collection of mammals from Labrador was the one made in 1882 by L.M. Turner, while stationed at Fort Chimo, Ungava. Several new forms have been described from this material, which is in the National Museum at Washington."
Outram Bangs (1898: 491)

As noted in this quote by the eminent American zoologist Outram Bangs, Turner's research and findings on mammals of Ungava were significant and well utilized by his peers. With only a handful of non-Indigenous people having ever ventured to the Ungava region – let alone a scientist – Turner's notes on natural and social science were highly anticipated by the academy. Returning to Washington, DC in late 1884, after being stationed at Ft. Chimo, Ungava for just over two years (6 August 1882– 4 September 1884), Turner wasted no time in preparing his fieldnotes for publication.

The task of recording his experiences in the field from an anthropological perspective must have been overwhelming, for he had a backlog of fieldnotes to convert to manuscript from an earlier expedition to Alaska and the adjoining Aleutian Islands, where Turner was based from May 1874 to August 1881. In both Alaska and Ungava, Turner was formally employed as a meteorological observer for the U.S. Army Signal Service, but in reality, he devoted the majority of his time to making observations and collections on natural and social history. And although Turner was engaged by the Signal Service Office, he was largely regarded in the field and by his peers as a contract employee of the Smithsonian. At Ft. Chimo, for instance, he was known as the "Smithsonian man" (Ft. Chimo Journal, 24 May 1883) among the HBC workers. Turner likely secured his position at the Signal Service Office by being nominated by the Smithsonian. It is believed that Robert Ridgway, a friend of Turner's who worked at the Smithsonian, put Turner's name forward as a candidate to Ridgway's superior, Spencer Fullerton Baird. In his capacity as the Secretary of the Smithsonian, Baird was a significant public figure, and his recommendation of Turner to the Signal Service Office would have been seen as a strong reference for the position. Baird provided the following recommendation of Turner:

> *Mr Turner is an accomplished explorer and accustomed to Arctic life, having spent five years at St. Michael's, on Norton Sounds, and, later, several years on the Aleutians. He is of a pleasant, congenial temperament, able to accommodate himself, and of unusual intelligence; and his society is very acceptable to those among he is thrown.*
> (LETTER TO ARMIT, 17 MARCH 1882)

The Smithsonian and the Signal Service Office must have a good working relationship at the time, for when Turner was offered the position, it seems his duties, salary, and field equipment were administered by both organizations.

With an incredible set of information and collected materials about which to write from two disparate Arctic regions, one in the west, the other in the east, and with the hope of securing a permanent position at the Smithsonian through the dissemination of his findings, Turner had to be strategic about his approach to publication. His major manuscript on Alaska, *The Natural History of Alaska*, became a priority upon returning to Washington, DC, for this had been in the hands of the Government Printer for the entire time that he had been in Ungava. He had, in fact, turned this over to the printer in early 1882, prior to leaving for Ungava. Turner became anxious about the delay in the printing process and was concerned that his findings would be stolen and published by Edward W. Nelson, a naturalist based in Alaska from 1887 to 1881: "I hope my report is in the hands of the printer. I do not want Nelson to have access to my manuscript" (Letter to Ridgway, 2 July 1882) and "I hope my report has been published, although I fear someone else will do all he can to prevent it"(Letter to Ridgway 1 Sept, 1882).

Once Turner's *Natural History of Alaska* manuscript was in print it appears that he turned his energy to his Ungava fieldnotes. Working at his desk in the Smithsonian Castle's South Tower (Memo to Baird, Record Unit 305, Accession 13922) on the National Mall between 1884 and 1887, Turner generated six manuscripts on Ungava based on drafts prepared in the field, as well as several short papers on the natural history of the region. The original titles of the six manuscripts, in order of preparation, were:

Lucien McShan Turner (1874)
Turner was 27 years old when he travelled to Alaska. This photo was taken by Bradley & Rulofson in San Francisco prior to his departure. When Turner went to Ungava in 1882, he was just 34 years old. Used with permission of Lucian Wayne Turner, 2010.

1. Language of the "Koksoagmyut" Eskimo at Fort Chimo, Ungava, Labrador Peninsula (1882-1884) (1884);

2. Notes on the Mammals Ascertained to Occur in the Labrador, Ungava, East Main, Moose, and Gulf Region (1886a);

3. Contributions to the Natural History of Labrador and Ungava (1886b);

4. Manuscripts on the Turner Natural History Collections from Labrador (1886c);

5. Descriptive Catalogue of Ethnological Collections Made by Lucien M. Turner in Ungava and Labrador, Hudson Bay Territory, June 24, 1882 to October 1884, Part 1 on Innuit (1887a); and,

6. Descriptive Catalogue of Ethnological Collections Made by Lucien M. Turner in Ungava and Labrador, Hudson Bay Territory, June 24, 1882 to October 1884, Part 2 on Naskapi specimens (1887b).

Of these manuscripts, only the content of the two *"Descriptive Catalogues"* on the Inuit and Naskapi were ever printed, appearing in the Smithsonian's *Eleventh Annual Report of the Bureau of Ethnology, 1894*. The report was authored by Turner and edited with a broad

Woodland Caribou Plate by E.T. Seton
Wildlife artist Ernest Thompson Seton prepared illustrations for the Notes on Mammals *manuscript, but these were unable to be located. This plate of a woodland caribou from Seton, appearing in* Four-Footed Americans and their Kin *(Wright 1898: 276), is indicative of the types of plates he would have produced for Turner. Based on Turner's descriptions, Seton's drawings must have been impressive: "The plates of mammals were prepared under the hand of Mr. E.T. Seton and bespeak for themselves his artistic ability."*

brush by John Murdoch, an Alaskan anthropologist and the Smithsonian's Head Librarian, under the title of *Ethnology of the Ungava District, Hudson Bay Territory*. Although the editing process removed much of Turner's poetic prose and his highly descriptive way of writing, this report has since become an anthropological classic and is widely used today by the Inuit and Innu of the region, as well as Arctic researchers, as an historical reference. Owing to the importance of the report, it was reprinted in 2001 by the Smithsonian Institution Press, with an introduction by Smithsonian archaeologist and Labrador anthropologist Stephen Loring. Rare photographs taken by Turner in Ungava were included and also exhibited in Montreal and Northern Quebec as part of celebrations surrounding reissue of the report.

The photos hinted, for the first time, that there might have been more to Turner's Ungava experiences than his accounts that appeared in the *Ethnology* report. What do we make of Turner's words "I do not find the field of ethnology so attractive and varied as in Alaska"? Did Turner put his energies into other fields of study, perhaps around his deep ornithological and zoological interests? What other material was there in the field that captured his attention? As we discovered in the course of researching material for this book, it appears that Turner's Ungava story is yet to be told in full.

His four unpublished manuscripts, mostly pertaining to language and the natural sciences, and totaling several hundred pages, are not well known by Arctic researchers and became apparent to us only after archival research. It seems that only fragments of these manuscripts have been published since Turner prepared them more than 130 years ago. For instance, Max Dunbar, a zoology professor and oceanographer at McGill University in the mid-1900s, referred to Turner's work on Ungava Bay fish populations and climate studies (see Dunbar 1952, 1954, 1983). In studies of shamanism, Professor Bernard Saladin d'Anglure of Laval University, since the 1960s, has drawn on material relating to Inuit cosmology from Turner's manuscripts (see D'Anglure 1964, 1984). Professor of linguistics Louis-Jacques Dorais, also from Laval, has used Turner's notes on Inuktitut in his research on that language during the last few decades (see Dorais 2010).

The Impetus to Publish This Book

Turner's in-depth knowledge of mammals is apparent in his manuscript titled *Notes on the Mammals Ascertained to Occur in the Labrador, Ungava, East Main, Moose, and Gulf Region*. The 271-page document, held at the SI Archives, shows Turner as a discerning mammalogist, where he reconciles Inuit and Innu knowledge of mammals with his own training and firsthand experiences. His writing is raw and sometimes flowery, but through it we see Turner's thirst for Northern research, his fondness of isolated settings, and his ability to work across two sets of indigenous languages and cultures in his quest to expand knowledge of the mammalian world.

Turner was a humble writer, and he paid due acknowledgement to those who assisted him in preparing the *Notes on the Mammals* manuscript. For instance, Turner recognized the valuable contribution of Clinton Hart Merriam, an influential zoologist and chief of the Division of Economic Ornithology and Mammalogy at the U.S. Department of Agriculture (precursor to the U.S. Fish and Wildlife Service), in the determination of mammals. The absence of wear marks on the manuscript suggests that Turner never had the opportunity to share his knowledge of mammals of Ungava beyond experts such as Merriam. The broader scholarly community never had an opportunity to appreciate Turner's contributions on mammals. The last person to have referenced this manuscript in detail was probably zoologist Francis Harper, who more than fifty years ago consulted Turner's papers to compile his *Land and Fresh Water Mammals of the Ungava Peninsula*, among other works on Ungava (Harper 1961, 1964a, 1964b). The manuscript has sat idle ever since.

Upon reviewing all the unpublished material and associated ethnographic material, visual records, and animal collections by Turner at the Smithsonian Institution (SI), we realized Turner's account of mammals, in particular, deserved to see the light of day. We agreed to embark on this collaborative and multidisciplinary initiative, a project that has been three years in the making.

Turner's Ungava Voyage

Turner was still a young man – only 34 years old – when he departed for Ungava. He travelled to Ft. Chimo without his wife, Mary Elizabeth (nee Lutz), and their sons, who remained in their hometown of Mt. Carmel, Illinois, for the duration of Turner's stay in Ungava.

Turner's introduction provides details about his journey in his own words, though very little about his family. To enhance his vivid descriptions and to help the reader better visualize his journey, we developed a map of his travels and included historical photos of locations he wrote about along the way. Many photos were taken by Turner himself. The other photos of the period are credited to either "J.R.H.," an unknown historical figure, and/or George Washington Wilson (G.W.W.), a famous Scottish photographer. It is unclear whether Wilson travelled on the same vessel as Turner in 1882. It is more likely that J.R.H. was a photographer employed by Wilson, as he is said to have commissioned others to take photographs for his large photographic company. However, there is also a strong possibility that Turner took some of these photos, which were then credited to J.R.H. or Wilson. We suspect that Turner might have been the photographer of the photo of the shipwrecked S.S. *Labrador*, for instance, which ran aground on a tidal reef in Ungava Bay on August 5, 1882 (p. 8).

Turner had camera equipment with him on the S.S. *Labrador,* and based on his descriptions, the vessel was stuck on the reef only for the duration of one tidal cycle, or about six hours. A small vessel that was operating in Ungava Bay came to their aid, but it is highly unlikely that another photographer was on board that vessel. When the S.S. *Labrador* freed itself from the reef and steamed to Ft. Chimo two days later, the HBC postmaster noted the names of those on board. Journals kept by HBC postmasters were precise; the absence of notes about the arrival of J.R.H. or George Washington Wilson at Ft. Chimo provides further evidence that Turner might have been the photographer.

The Ungava set of photos credited to J.R.H. and G.W.W. are held by the McGill University Rare Books and Special Collections Library and the McCord Museum in Montreal. Turner's photos of Ungava are held by the Smithsonian's National Anthropological Archives (NAA).

Turner's voyage to Ft. Chimo, via train from Washington, DC to Montreal and Quebec City, then by ship on the HBC merchant vessels *Tropic* and S.S. *Labrador,* took more than 60 days. Departing Quebec City on June 8, 1882, the first portion of his seabound voyage took a course northeast through the St. Lawrence River, around Anticosti Island, and then northwest along the Labrador Coast to Rigolet (then Rigoulette), an HBC Trading Post within the sheltered waters of Hamilton Inlet. Between June 24 and July 8 in Rigolet, Turner was afforded his first opportunity to collect natural history specimens while waiting for the cargo aboard the *Tropic* to be transferred to the S.S. *Labrador*, a large steamship that would take Turner as a passenger to Ft. Chimo. During the fourteen days in Rigolet, Turner explored the nearby region, collecting birds, plants, fish, mammals, and other natural history specimens. Knowing his time was limited, he traded with local workers, known as the "Planters," to secure specimens that he was not able to find during his short stay. For example, Turner purchased a flying squirrel from them (p. 132), which he recounts as having been collected between the HBC stations of Rigolet and Northwest River.

After departing Rigolet, the next overnight stop along the way to Ft. Chimo was Davis Inlet (185 miles north of Rigolet), where the S.S. *Labrador* discharged supplies to the HBC Trading Post. Based there for three days, from July 16 to 19, Turner again made excursions around the post to collect specimens, making special mention of the physical character of the countryside in his fieldnotes:

> *"...the country in the vicinity of the station is exceedingly rough. Bold and rugged spurs break in every direction, seemingly to have no special arrangement...The tops of the hills are often quite flat and frequently embrace very extensive areas. The sides are mostly abrupt and often precipitous."*
> (Turner 1886b: 14-15)

On the next portion of the voyage, destined for the HBC Trading Post at Nachvak Fiord, the S.S. *Labrador* encountered thick fog, making passage to the shore difficult and dangerous. Deciding that the approach would endanger the crew and vessel, the captain pushed on and made way for the Button Islands in the Hudson Strait at the tip of the Ungava Peninsula, a most treacherous waterway to navigate. The waters of the Labrador Sea

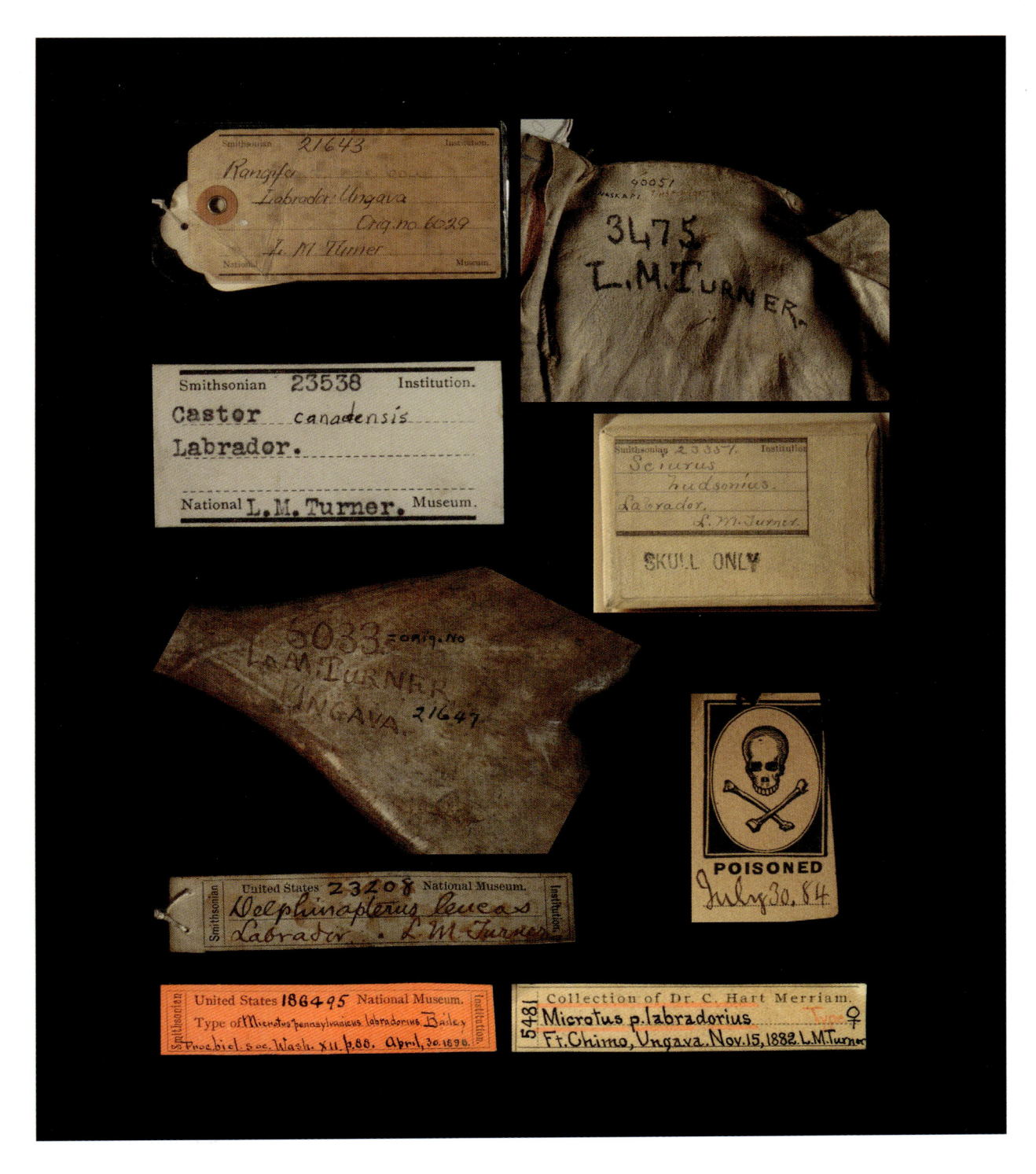

Field and Museum Tags
Some of these original tags bear Turner's writing, while other tags were prepared in the museum after his fieldwork. Photo and arrangement by Scott Heyes, 2012.

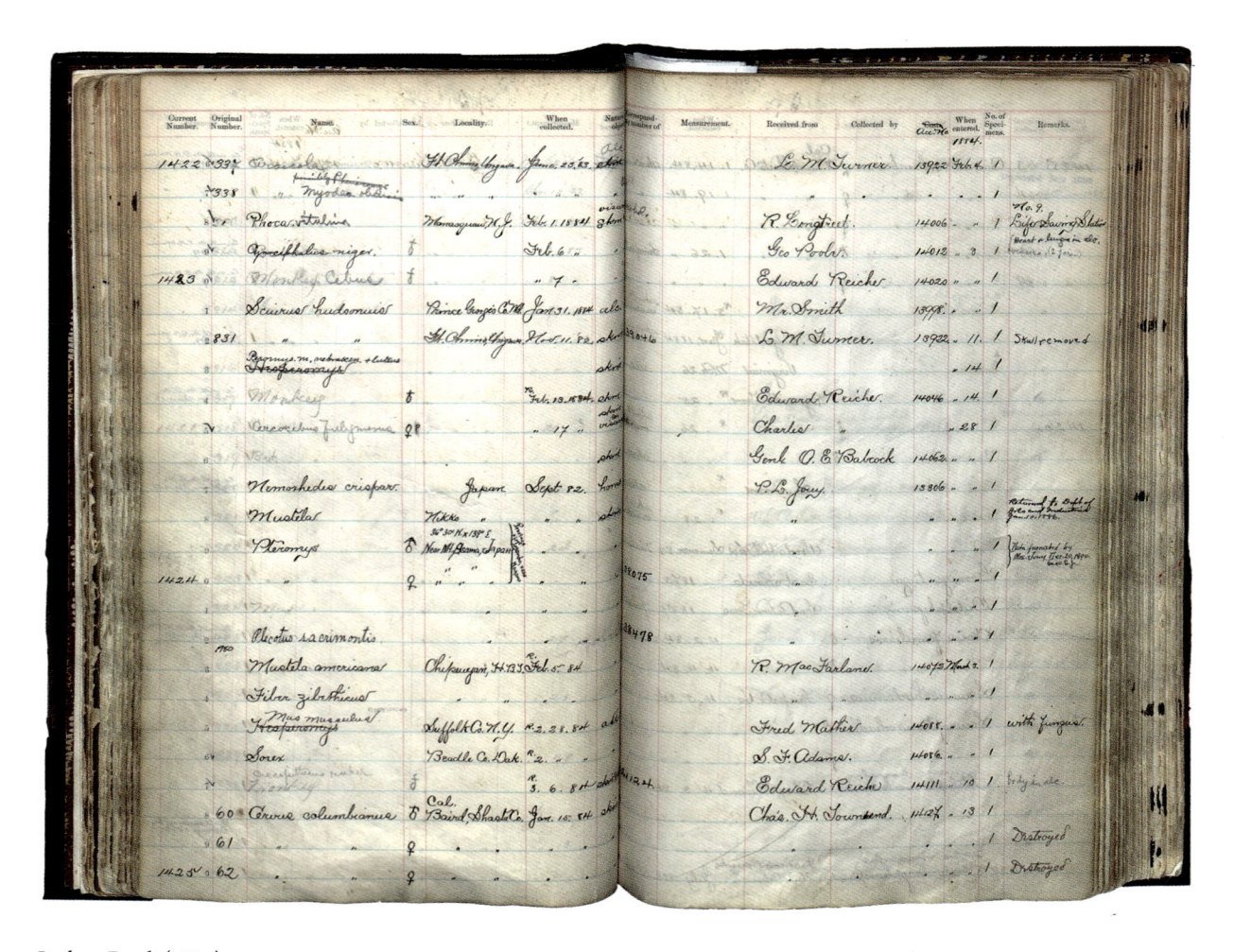

Ledger Book (1884)

A typical ledger book showing an inventory of mammals collected by various Smithsonian scientists. The first two entries indicate that Turner accessioned two specimens of Arvicola and Myodes in the Smithsonian's Division of Mammals. The ledger indicates that he collected these at Ft.Chimo, Quebec on 23 June 1883. Every mammal collected by Turner during his time in Ungava/Labrador from 1882–84 was documented in such ledger books. These books were divided into two sets: "Skins" and "Bones." Photo by Scott Heyes, 2011. Division of Mammals Archives, NMNH, Smithsonian Institution.

and Ungava Bay converge at this location, generating whirlpools and fast-flowing tidal currents. Just after noon on 25 July 1882, the S.S. *Labrador* passed through this confluence after fog lifted to reveal the way. Turner described the tense situation at this location:

> "A course was steered between [Cape Chidley and the "Buttons"], although rarely attempted by a sailing vessel, and the tide being favourable we swept through with great rapidity. The weather changed and we found that the bad conditions existed only outside of the Cape. Everyone felt a relief from the constant strain and anxiety attendant upon a dangerous coast."
> (Turner 1886b: 20)

Beyond the Button Islands, the S.S. *Labrador* took a southwest course, following the eastern coastline of Ungava Bay toward Ft. Chimo. It would be 12 days steaming before reaching Ft. Chimo; a journey that could have otherwise been achieved in a few days had the S.S. *Labrador* not struck walls of pack ice, a shallow reef, and severe fog conditions. On this last leg of the journey, Turner explains that inclement weather cleared sufficiently to allow the crew to disembark near the mouth of the George River on July 30, 1882 for a day trip, to hunt caribou. A caribou hunted by the captain of the S.S. *Labrador,* some six miles inland, was divided into portions, with Turner taking the horns and head for his collection and the rest of the crew carrying the meat. Turner described the long walk back to the vessel carrying his load:

*"It proved a heavy and awkward burden on my shoulders as I trudged along that hot valley; heated by the afternoon sun that called into life and activity myriads of mosquitoes and scores of hungry 'Bulldog' (*Tabanus*) [flies] each of which doubtless attracted by the scent of fresh blood of the deer..."*
(TURNER 1886B: 25)

Returning to the vessel late in the day, which was moored at Gull (Beacon) Island, Turner had the fortune of finding a polar bear skull. This specimen was photographed for this book. It is covered with orange moss and lichen that has remained since Turner picked it up on the island. Inuit hunters in the region indicate that polar bears still frequent Beacon Island in search of seals and other food sources; it is considered an unsafe place to make camp.

With Turner already building his mammal natural history collection along the journey, he was probably buoyed by the fortuitous shell discoveries he made as a result of the S.S. *Labrador* striking a flat-topped reef, which occurred five days after departing Beacon Island for Ft. Chimo. The southern portion of Ungava Bay is riddled with shallow reefs, which are difficult to discern. The area was not mapped on nautical charts. And, with prevailing fog conditions, and tides ranging up to 16 meteres (53 ft) in these waters, it took a watchful eye to avoid being shipwrecked. Turner described the harrowing experience:

"I heard a noise like the rattling of a chain when the anchor is let go. In an instant thereafter I was pitched from my berth and precipitated through a door to the middle of the cabin. The vessel careened so that standing was impossible... The Captain on the bridge gave the order 'Call all hands. The vessel is ashore.'"
(TURNER 1886B: 32)

Waiting out the arrival of the rising tide to free the vessel, which was undamaged, Turner saw an opportunity to explore the rock pools that remained on the exposed reef. He wrote: "When the top of the reef was dry I occupied my time in collecting marine specimens and was fortunate enough to secure some of great rarity, among them a species of shell new to science." William Healey Dall, the Smithsonian's Honorary Curator of Mollusks between 1880 and 1927, studied Turner's shell collection from this reef and enthusiastically reported a new species, *Aquilonaria turneri*, named after Turner (Turner 1886c: 676). The reef on which the S.S. *Labrador* was stranded, and upon which Turner fossicked, was named by the captain as Labrador Reef, after the vessel. Labrador Reef was described by Turner as being 20 miles northwest of the mouth of the Koksoak River. The exact location is still unknown, as modern-day nautical charts do not feature the reef. The photos of the reef and the vessel (pp. 8, 21) – a remarkable find in the SI Archives – enrich this narrative, and this is the first time photos of the S.S. *Labrador* have appeared in print.

From Labrador Reef, the S.S. *Labrador* took another seven days to reach Ft. Chimo, the final destination for Turner and his home for the next two years. Upon arrival, and after enduring thick fog and days stuck in pack ice, Turner wrote: "At last I am on land... I am heartily wearied of the sea. Nothing but icebergs and ice." (Letter to Ridgway, 1 Sept 1882) Turner soon regained his land legs, for he describes in a letter written 25 days after landing at Ft. Chimo that "walking is just the very best." (Ibid) In this time, he collected more than 200 birds and secured the assistance of a local Innu hunter known as "Ghost" or Thomas (*Nastomushjun*) as a guide (Ibid).

Turner quickly settled into life at Ft. Chimo. He was immediately assigned quarters by the HBC postmaster, Mr Keith McKenzie, and unpacked his equipment, which included items such as small traps, arsenic for preserving specimens, a ruby lantern, several dozen photo plates, two letter-copying books, and sheets of

Two Extraordinary Years

Bristol board, presumably for illustrating specimens or pressing flowers (SI Archives, Record Unit 305, Accession 15388).

Turner was impressed with his initial treatment by the HBC employees and the location of his laboratory: "…the white men here form a pleasant contrast to the drunken scamps of Alaska." (Letter to Ridgway, 1 Sept 1882) "I am pleasantly situated and trust I shall thoroughly enjoy my stay at this place. It gives me much pleasure to acknowledge the uniform courtesy received from the gentleman connected with the company, whom I have met." (Letter to Parson, 6 Aug 1882)

Ft. Chimo, located some 23 miles from the mouth of the Koksoak or Ungava River, was a small post that opened in 1830, some 50 years before Turner's arrival. Turner described the setting: "The city of Ft. Chimo contains three dwellings and several other buildings, situated a few yards from the bank of the mighty Ungava and twenty-three miles from its mouth." (Letter to Ridgway, 1 Sept 1882) Inuit from Hudson Bay, Hudson Strait, Ungava Bay and Labrador would travel to Ft. Chimo to trade, while some lived more permanently near Ft. Chimo to work for the Company during the beluga whale and salmon-fishing seasons over the summer. Turner (1887a: 44) noted the Inuit population of the Ungava region, spanning from Leaf River to Cape Chidley, was "not more than one hundred and twenty-six people." The majority of this population likely remained near Ft. Chimo during the summer months. Innu from the interior of the Ungava Plateau also travelled to Ft. Chimo to trade and work, making it a cosmopolitan place, so to speak. English and Scottish fisherman would steam the refrigerated fishing vessel S.S. *Diana* from London to Ft. Chimo during the salmon run each year, which raised the summer population of Ft. Chimo, and added to the variety of languages and accents spoken at the post. Turner took full advantage of Ft. Chimo's status as the center of trade in the vast region, buying where he could afford, and without impinging on the HBC's trading practices or indigenous belief systems, ethnological items and natural history specimens from the Inuit and Innu.

On collecting a beluga whale, for instance, Turner learned that it was important to respect Inuit customs and the protocols of the HBC; contempt for these would have cast him in a bad light, and he would otherwise have been made to feel unwelcome:

"A young one was obtained, from the mother, having a length of less than seven inches and a perfect whale in its exterior outlines. As the Eskimo prize this mammal very highly and are very reluctant to have the young carried away it was a very difficult matter to undertake investigations to determine certain facts without incurring their displeasure and have them quit the field; which, if done, would seriously interfere with the affairs of the Company." (Turner 1886a: 1282)

Turner's friend, Maggie Brown, collected specimens on his behalf in circumstances when he was unsure of the protocols associated with the capture of certain mammals." (Letter to Ridgway, 28 Aug 1883) Upon inspection of captions of photos taken of Ft. Chimo by Turner, it appears that he also employed Inuit and Innu children to collect ethnographic and natural history specimens for him. On the reverse of a photograph of four children, for instance, was a note by Turner: "Kakiki, Harriet, Kittie Gordon and Jennie Brown, my bug catchers at Ft. Chimo." (NAA, Negative 3208)

While Turner was tasked with taking meteorological records at Ft. Chimo, he soon recognized the substantial contributions he could make to the fields of natural history very early in his stay, writing "I think I shall have a very interesting report on this part of the globe." (Letter to Ridgway, 1 Sept 1882) Turner's observations were very much confined to the immediate vicinity of Ft. Chimo, with most mammal specimens taken at, or between, three critical sites that were within reasonable walking distance, or that could be accessed by a small boat the HBC lent him each summer. These sites were around the places called *Whale Head*, "3.5 miles below the HBC Post," a large single rock feature of the Koksoak River (South River or Ungava River) narrows where many beluga whales were netted commercially by the Company; *Forks*, the heavily wooded area and confluence of the Koksoak River and the Larch River (North River); and *Rapids,* a pronounced change in elevation of the Koksoak River, where fresh water met salt water, about 35 miles upstream from Ft. Chimo. These were neither official placenames nor Inuit or Innu names, but rather the HBC vernacular for the region. Historic maps and Turner's descriptions of these places were used to determine their exact location. Other locations near Ft. Chimo that appear on the field tags of mammals collected by Turner – but are no longer placenames used today – are *Juniper Islands*, a line of islands near Ft. Chimo; *Hawk's Head*, which marks the opposite side of Whale Head at the narrows of the Koksoak; and *Whitefish Lake*, some three miles east of Ft. Chimo.

While Turner often travelled around Ft. Chimo on his own to collect natural history specimens – typically through trapping or hunting, a favorite pastime – he also made occasional journeys with others beyond the immediate vicinity. On June 16, 1883, for instance, he went on an egg hunting expedition with two Innu hunters, *Sescumounagan* and *Kanawabano* (from nearby Whale River), presumably to islands along Ungava Bay where many migratory birds are known to nest. (Ft. Chimo Journal, 16 June 1883) Turner also travelled to the Whale River by boat on June 24, 1883, staying there until at least July 5. Records show that he also went caribou hunting on several occasions in the depths of winter in 1884 with his friend John Ford and "three Esquimaux [Inuit]." (Ft. Chimo Journal, 10 Feb 1884) The caribou were plentiful in the region. As Turner noted: "I have killed many Reindeer since I came; and, have seen thousands." (Letter to Ridgway, 28 Aug 1883) "In a month or so I shall have fine sport shooting deer. The deer were so plentiful in this village [Ft. Chimo] in June last [1881/1882] that the women had to drive off the deer with sticks to prevent the animals from knocking down the tents of the people." (Letter to Ridgway, 1 Sept 1883)

In notes by Turner (1887a: 88) about Inuit camps along the Ungava Bay coast, he mentions, in passing, that he visited and made himself known to the inhabitants of several places, including: *Aupálik,* a locality about 150 km (93 miles) west of Ft. Chimo, where two Inuit families dwelled; a community on the Koksoak River; and, "two communities between the mouth of George's River (*Káñukthluáluksoak*) and Cape Chidley (*Kilínik*)."

It seems that the varnish of Turner's stay at Ft. Chimo began to wear off toward the end of the summer of 1883,

Turner's Family (1886)
Turner left behind his wife and two boys to travel to Ft. Chimo. He occasionally wrote letters to his son, Jesse while in the field, often explaining to him how to use guns, identify birds and mammals, and to correctly prepare specimens. Pictured are Mary Elizabeth Turner, Eugene S. Turner, 8 years, and Jesse J. Turner, 14 years. Photo taken at Mt Carmel, Illinois. Used with permission of Lucian Wayne Turner, 2010.

almost one year after being in Ungava. Relations with the HBC postmaster, Keith MacKenzie, soured. He wrote: "The man in charge here is not an educated man and is not at all in sympathy with my work... This is a fine place as regards surroundings but there are other things, which make it undesirable in a very great degree." (Letter to Ridgway, 28 Aug 1883) The HBC workers, many of whom were Turner's friends, must have been in conflict with MacKenzie: "The greater number of the white men are going home this year. They will not stay with this man under no conditions of pay, which I assure you are not equal to a common farm-labor at home." (Ibid) It is believed that Turner's marriage to

Mary Elizabeth also broke down during mid-1883, which might have affected his attitude toward all manner of things. Despite these tensions in the small community and coping with a divorce from afar, Turner stayed on for another year, leaving via the S.S. *Labrador* on September 4, 1884, with all of his packages and contents aboard. He had 26 boxes shipped, "a great number of... immense size" (Letter to Ridgway, 3 Sept 1884), which contained items such as salted shark skin, plants, manuscripts, photo plates, and specimens either stuffed, salted, or in alcohol (SI Archives, Unit 305, Accession 15388). This was in addition to a shipment of 27 boxes of similar effects, including 1100 birds, that Turner had shipped to the Smithsonian on the S.S. *Labrador*, when it passed by Ft. Chimo on its annual route in 1883. (Letter to Ridgway, 28 Aug 1883) The 53 boxes sent to the Smithsonian by Turner went via the S.S. *Labrador* from Rigolet to London then back to New York on an HBC merchant vessel, and finally by train to Washington, DC. Turner likely arrived back in Washington before his last packages did.

On the journey home, the S.S. *Labrador* stopped at the HBC posts at George River, Killiniq, Nachvak, Davis Inlet, and Rigolet, which gave Turner further opportunities to collect natural history specimens along the way, particularly birds. At Davis Inlet, Turner noted that a buck caribou was being held in captivity and being fed crackers, oatmeal, and its natural food, and was to be shipped aboard the S.S. *Labrador*, to be received by the London Zoological Gardens. He mentioned that it was quite a gentle animal, was in excellent condition, and was taken in a small stream while attempting to swim across it. Noting Turner's awareness of the cultural sensitivities associated with the collection of certain mammals, he remarked on the circumstances that must have surrounded the capture of the caribou, suggesting that the Inuit must have been engaged:

> "*Nothing will induce some tribes of Indians to deliver a living reindeer to a whiteman lest the guardian spirit of the deer be offended and cause all the remaining deer to forsake the country. The Eskimo, on the contrary, have no hesitancy in disposing of the living deer as their shamans are able, at their own pleasure, to call the spirits of the deer, already slain, to the earth where they immediately assume a material form ready to be again slain and as often resume the earthly form.*"
> (Turner 1886a: 1363)

At Rigolet, he disembarked the S.S. *Labrador* and boarded another HBC vessel destined for Montreal, which included at least one stop along the way in St. John's, Newfoundland. At this location, on October 10, 1884, thirty-four days after leaving Ft. Chimo, he wrote: "I have arrived this far and will leave soon for Montreal. I shipped all specimens via London except one box of manuscript and photographic apparatus, which I now have sent to care of Adams Express Co. New York per S.S. 'City of Mexico', just sailing." (Letter to Baird, 10 Oct 1884) Turner was obviously concerned about losing his manuscripts if they were to be sent to London. He likely used the services of Adams Express Company because of their reputation in securely handling packages for patrons. Despite this precaution, it is unclear whether all of Turner's original fieldnotes made it to Washington, DC. The location of these particular notes is undetermined, as the SI Archives only possesses Turner's near-finished manuscripts, which must have been compiled from his notes. The maps and illustrations of places he likely prepared at Ft. Chimo, as he did in Alaska, have never been uncovered. And we know firsthand from Turner's descendants that they do not have these materials in their possession. It appears they will never be located.

Turner's collection was accessioned by the Smithsonian, as boxes arrived, from late 1883 to early 1885. Curators of botany, geology, mammalogy, ethnology, fishes, entomology, and other fields at the Smithsonian excitedly awaited the arrival of new shipments. As noted in the 1884 Report of the National Museum (True 1885: 130), Turner's collection of mammals was significant: "From British America the most important accession is the collection of skins and skeletons made by Mr Lucien M. Turner in the vicinity of Hudson's Bay." A newspaper article also commented on the breadth and depth of Turner's collection in 1885:

> "*The accessions received from this specialist by the Smithsonian Institute and the national Museum exploring the regions of "perpetual snow" are said to be greater than ever before "turned in" by any*

Entrance to the George River, Ungava Bay (2012)
The crew of the S.S. Labrador, including Turner, disembarked near this location 131 years ago (30 July 1882) for a day's excursion to hunt caribou. Beacon Island is in the background. Turner collected natural history specimens from the banks of a large stream six miles inland, which was probably the stream branching from a lake called Ammaluttuq *(Big Lake). According to Turner, Captain Gray of the S.S. Labrador* "secured a huge buck in excellent condition of flesh" *near this location. Photo by Scott Heyes, 2012.*

scientist and collector. Mr Turner, because of his indomitable will and indefatigable work in the studio and field, especially the latter, has justly won for himself a rank high among naturalist and scientist of this and foreign countries…"
(The Daily Republican, 15 June 1885)

The vast array of ethnographic materials in Turner's boxes represented the first chance for the New World to see the rich dimensions of Inuit and Innu culture from the Ungava region. While there is a range of material represented from both cultures in Turner's collection, it is interesting to speculate how much larger it might have been had weather conditions allowed for Turner to trade in the winter of 1884. He wrote: "I expected a lot of ethnological material from the S.W. (W coast of Hudson Bay; and, also from the S. side of western end of Hudson Strait) and N.W., but the natives failed to reach here before the disappearance of the snow." (Letter to Baird, 3 Sept 1884)

While we champion Turner's work with respect to his knowledge of mammals in this book, his talents as an anthropologist with a strong command of linguistics must not be understated. From his fieldnotes, it appears Turner was familiar with several languages and had command of

Letter from Turner to Robert Ridgway (1884)
Turner notes his arrival at Ft. Chimo and being shipwrecked in this September 1, 1882 letter to his friend at the Smithsonian, Robert Ridgway. Such letters to his friends, family, and Smithsonian colleagues provide rich details about Turner's life and activities in the North. Photograph of the original. Source: Robert Ridgway Collection, McGill University Rare Books and Special Collections, Montreal. Used with permission.

Inuktitut (known in the 19th century as Eskimo) prior to being stationed in Ungava. In an 1881 letter to Spencer Fullerton Baird, then Secretary of the Smithsonian, Turner wrote: "I speak Russian and Eskimo [Inuktitut] fluently and have knowledge of two Aleutian languages." (Letter to Baird 5 Dec 1881) His fieldnotes reveal that he compiled wordlists in Unalet, Malemut, Nulato, Ingalet, and Aluet. And, in correspondence letters with the Smithsonian, we also know that Turner knew some Spanish, French, German and Latin. His knowledge of Inuktitut and Latin no doubt served him well when he prepared notes on Inuit and non-Inuit taxonomy of mammals while stationed in Ungava. A newspaper article in 1885 reported on Turner's time in Ungava, making special mention of his knowledge of Inuktitut: "…from his proficiency in the Eskimo philology, his speciality, it is universally concluded by contemporaries that he is deservedly recognised the foremost American authority on this most intricate branch of the linguistic science." (*The Daily Republican*, 15 June 1885) Few can question Turner's commitment to recording the Inuit language of Ungava.

In the introduction to his *Language of the "Koksoagmyut"* [Inuit] unpublished manuscript, Turner stated his determination to record Inuktitut to the best of his ability:

> *"[The notes and words] are the results of many hours of hard and patient labor. Everything here pertaining to the language has been repeatedly verified and if given (by others) the sounds I have attempted to imitate they will be found correct in every instance. The exact shade of meaning has been given in each definition and in no instance has a word been perverted in order to swell the number."*
> (TURNER 1884: 2128)

An entry by Turner on seals in his *Language of the "Koksoagmyut"* (1884: entry 959) exemplifies his care in recording orthography, meaning, and application of Inuktitut terms: *pú yi a kák pok-kó ki ut*; "the seal (sank) went under because", *i m'ig nák si u ¢l'û mût*, "the air was so still that the report (of the gun) could be (was) heard from afar." In another entry by Turner (1884: entry 2226), we realize that *i kóg lûk*, an Inuktitut term relating to caribou, has geographical associations. It means "a lake lying in the path of migrating reindeer, and through which they swim to continue their way." In many instances, Turner even distinguished between the use and meaning of Inuktitut terms by Ungava Inuit and Labrador Inuit. The squirrel serves as an example. In Ungava, Turner mentioned that it was known as *na pák ta śyut;* in Labrador the term was *sĭ'k sĭk*.

When Turner returned from Ungava, he reflected on his knowledge of Inuktitut and believed that a further fieldtrip to the region would enable him to write a more complete account on the subject. Lacking funding to continue linguistic work – and likely a tactic to return to the field to study birds, his favorite pastime – Turner wrote to the U.S. Geological Survey and Bureau

Hudson's Bay Company (HBC) Dwellings at Ft. Chimo (1909)
View of the three HBC dwellings described by Turner near the beach of the Koksoak River, Ft. Chimo. The long days of summer permitted crops to be grown to produce hay for the Company's work horses, animals not well adapted to the cold winters experienced at Ft. Chimo. Used with permission of McCord Museum, M2000.113.6.230.

of Ethnology director Major J. W. Powell in the hope that his proposal might be approved. In a letter dated March 6, 1886, he wrote: "I have, at present, over 1000 pages of MS [manuscript] on the subject, but find many things [are] yet necessary to complete it. Would it be possible for the Bureau of Ethnology, under your direction, to send me there for a period of time, extending from May 1, 1886 to October 1887?" (Letter to Powell, 6 March 1886) His proposal to return to Ungava was not granted, meaning that Turner never had another opportunity to follow up on the language notes he made from 1882 to 1884. Turner might have been disappointed, but the detailed works he produced on the subject suggest that his one and only trip to Ungava furnished him with more knowledge of Inuktitut, and other branches of the natural and social sciences, than he realized.

Contributions to This Book

We have drawn upon our respective areas of expertise as well as firsthand experiences to develop this book in a way that combines Turner's original observations and collected specimens and stories with modern-day contributions from the descendants of the indigenous groups with which he engaged.

From the Editors

It is unknown why the original manuscript was not printed, especially given that it was in print-ready form, with notes to the publisher and annotations relating to where illustrations were to be inserted. Perhaps it was due to budgetary constraints, or perhaps the Smithsonian had other priorities. Whatever the case, we have chosen to publish Turner's entire *Notes on the Mammals* manuscript in verbatim form so future generations will have ready access to this historical material, knowing that what they read is from Turner's hand. The reader should note that Turner spelled words using both British and American English. His original text has not been edited by us, so the rawness of Turner's observations can be fully appreciated. The original page numbers are also provided so one can return to the manuscript in the SI Archives and locate passages with relative ease.

We were unable to locate the illustrations of caribou that were originally produced for the *Notes on the Mammals*, which, according to Turner, were prepared by Ernest Thompson Seton, a gifted wildlife artist. A caribou illustration by Seton (p. xxx) was produced for another publication around the same time he prepared plates for *Notes on the Mammals*; it hints at the type of illustrations that would have been featured had the manuscript been printed. To pay homage to Seton's style, and being mindful of the artistic representation of scientific works in the late 1800s, we opted for line illustrations for each species that Turner described and some watercolors, all beautifully produced by artist Bryony Anderson.

When Turner accessioned a mammal at the Smithsonian, its name, original field number, and date and place it was collected were recorded in ledger books and on accession cards. We consulted these ledgers and cards to verify the spelling of each mammal species by Turner and, in some instances, to verify the location where a species was obtained. In so doing, we discovered some anomalies. For example, Turner accessioned a seal, *Phoca* (NMNH Ledger Book, entry 14159), yet this specimen could not be located in the Division of Mammals, and he made no mention of collecting seals in his manuscripts. We also noticed a booking error, where many species were incorrectly labeled as being collected at Northwest River in Labrador. Had we not referred to the ledgers and cards in great detail, it is highly likely that this oversight might have remained undetected.

The book provides exceptional historical information on Inuit names for mammals, including a wordlist of Turner's original terms and orthography. With the assistance of linguistic experts John MacDonald and Louis-Jacques Dorais, and by consulting historical dictionaries on Inuktitut, we were able to generate a wordlist of the modern-day equivalents to Turner's terms. In some instances we could not determine an equivalent term in use today, which points to the ever-changing nature of language; some terms may no longer be relevant, some are replaced with new words, and some are forgotten over time. The wordlists provide some interesting linguistic insights, such as the term Turner documented for porpoise, *Nisak*, which may be derived from the Norse word, *Nisa*. This example raises topics for future research that anthropologists and linguists might explore in relation to Norse and Inuit interactions.

To place Turner's accounts of mammals in a contemporary context, we include an introduction to each of his sections, using these headings: "Cetaceans," "Ungulates," "Rodents," "Lagomorphs," "Insectivorous Mammals," and "Carnivores." These include discussion of certain specimens collected and described by Turner with reference to how they were used by other scientists and how certain specimens have become the standard benchmark, on the world stage, for distinguishing certain types of mammals. The introductions also remark on some of Turner's descriptions where he was unsure about certain aspects of behavior, which have since become known to science, including why beluga whales enter fresh water in large numbers at certain times. We also use the introductions as an opportunity to present other mammals in Ungava that Turner did not locate or describe but which were certainly present in the region, such as certain species of whales, bats, mice, and the carnivorous Fisher.

Each species described by Turner has a dedicated entry page. Near the top of each species entry page is Turner's original handwriting of the scientific name for each mammal. Each entry page also includes a taxonomic table. The tables conform to scientific convention and correspond with original field tags provided by Turner. Each table features the modern-Inuktitut name for the mammal species in syllabics, along with the English common name that Turner identified. Below this is the Inuktitut term (in the Nunavik Inuit dialect, *Nunavim-*

miutitut) for the species in Roman Inuktitut, another writing system used by the Inuit today, especially by younger generations and Labrador Inuit.

In addition to studying Inuktitut, we know from Turner's notes that he embarked on a project to record the Innu (Naskapi) language and customs while based in Ft. Chimo. This effort was halted when his friend and tutor of Innu language and customs, Mrs. Margaret (Maggie) Brown, died on 30 May 1883 from tuberculosis (Ft. Chimo Journal, 30 May 1883). The handful of Innu words relating to mammals that Maggie Brown relayed to Turner were extracted from his various manuscripts and, where known, placed in the taxonomic tables.

When reviewing the taxonomic tables and wordlists, it is important to recognize that the orthography and syllabary that Turner used is no longer the accepted standard for writing Inuktitut. And we should not fault him for misspelled terms or inaccurate definitions, for he was one of the first anthropologists to record the Inuktitut language. It was still only a spoken language at the time.

At the bottom of each species entry page is a distribution map.

Further guidance about the different components of the book is in the "Legend to the Book" (pp. xlvi-xlix). The legend helps readers understand where information was sourced and how it is depicted.

A significant meeting with an important group of visitors to the Smithsonian in 2010 gave us the final touch we needed to proceed toward publication. Turner's grandson, Lucian Wayne Turner, accompanied by his family, visited the Smithsonian in May 2010 to explore his grandfather's Alaskan and Ungava anthropological and archaeological collections. The visit included a tour of Turner's bird collection and a visit to the SI Archives to view Turner's manuscripts and photos. After some discussion with the Turner family of this book project, Lucian Wayne Turner, born after his grandfather had passed away, kindly provided a copy of a book he had prepared in 2008 about Lucien Turner's family history. It contained information about Turner's genealogy, accomplishments, rare portraits, and correspondence from Turner to his family. Thus the book provided information about Turner's life that would otherwise be unknown to us. With the blessing of Turner's family, we have incorporated some of those details in this book as well as a tribute by Lucian Wayne Turner to his grandfather.

From Contemporary Inuit Elders and Hunters

Photos of a number of the ethnographic materials that were made from mammals by Inuit and Innu feature in this book. They appear in the entries for the species from which they were principally made or relate to, in terms of function (e.g., a float used for hunting whales made from inflated sealskin in the "White Whale" entry or a children's game made from a hare's head in the "Hare" entry). These and other photos show the time, effort, and ingenuity that goes into making tools, clothing, games, equipment, shelter, and forms of transport from various mammal parts, often with painstaking attention to detail.

Photos of amulets and carvings collected by Turner and made by the Innu and Inuit from mammal parts as well as modern-day carvings appear throughout the book. A photo shows a charm made from what Turner described as the lip of a "barren ground bear" (grizzly bear), which was present in Ungava in Turner's time but is now extinct in eastern Canada. Carvings made from mammals serve two important purposes to the Inuit: to pass on stories to younger generations and to engender amongst younger generations the importance of using every part of an animal that has been hunted. To the Inuit, the use of every part of a mammal symbolizes respect and honor for its life.

The modern-day carvings of mammals were produced by the Kangiqsualujjuaq hunters Daniel Annanack and Johnny Mike Morgan, as well as by an elder, storyteller, and well-known polar bear hunter from the village of Quaqtaq, David Okpik.

Anthropological content relating to contemporary Inuit use and knowledge of mammals was made possible through the contributions of a number of elders and hunters, particularly from the community of Kangiqsualujjuaq (or George River). Located on the eastern shores of Ungava Bay, this community was

Caribou Fish Hook and Thimble Used by Inuit
Example of ethnographic objects collected by Turner at Ft. Chimo that relate to mammals. Left to Right: fish hook made from Caribou bone, E89935; thimble, E90093; small hook E90300; large hook with line (probably caribou sinew); E89972. Museum Support Center, NMNH. Photo by Angela Frost, 2011.

established in the 1960s and is comprised of Inuit who originate from hunting camps that were situated across the Ungava region. Many of the 750 Inuit residents of Kangiqsualujjuaq regard their ancestral home as being near the Ft. Chimo region that Turner was concerned with in the 1880s. It is highly likely that the elders and other important knowledge holders whom we consulted for this book are direct descendants, some three or four generations past, of those that Turner consulted for his material on mammals. We feature "recollections" and drawings of mammals by Inuit elder Tivi Etok, a well-known print maker and storyteller whose ancestors originate from Ungava Bay and the Labrador Peninsula. Other stories of mammals are featured from other elders (some that trace their origins to eastern Ungava, western Ungava, and the Labrador Coast): Benjamin Jararuse, Noah Angnatuk, Elijah Sam Annanack, Paul Jararuse, Johnny Sam Annanack, and Johnny George Annanack.

The drawings, carvings, and stories from the Inuit residents of Ungava highlight how important mammals are in a contemporary context. These modes of artistic and cultural expression demonstrate the strong link, respect, and understanding that the Inuit maintain with the mammalian world.

Turner's Legacy

The fauna of eastern Canada's tundra and northern forests has changed in many ways since Turner's visit, and his observations and collections are an invaluable resource in documenting the nature and extent of these changes. Since Turner's time, grizzly bears have become extinct in the Ungava Peninsula. Two other species, wolverines and belugas, have been reduced to extremely small populations, and at times have been thought to be extinct in the region. Caribou in Ungava have experienced enormous fluctuations in population size over the decades. Some species, especially the moose, have expanded their range drastically in the Ungava Peninsula with a warming climate over the past century, and the muskox has been introduced to Ungava Bay in recent decades, with an expanding population.

In terms of Turner's contribution as an anthropologist, we should note his commitment to recording traditional Inuit and Innu stories about mammals at a time when Christianity was supplanting Shamanism. Turner was well aware that he was in Ft. Chimo during a transition phase, where the mercantile ways of the HBC, and the customs and habits of the non-Indigenous employees of the Company were beginning to influence and change Inuit and Innu cultural practices. Knowing that the "old" ways of the Inuit and Innu were rapidly undergoing change, Turner paid special attention to recording the Inuktitut language and the customs and ways of the Inuit and Innu, especially their knowledge of mammals, which Turner knew was susceptible to loss due to acculturation.

In discussing Turner's legacy as a respected mammalogist and anthropologist, we must not forget that while he regarded some things as "discoveries" or "findings" new to science, many of his observations were already well known to the Inuit and Innu of Ungava. Turner collected material in a colonial context but in an era when Inuit and Innu were quite familiar with trade practices and the ways of "whitemen." Consistent with good anthropological practice, it seems Turner collected material in an overt fashion and that he formed genuine and long-lasting friendships with Inuit and Innu hunters, who subsequently taught him their customs and shared their land-based knowledge. By all accounts, he was a respected individual and a scientist who paved the way for honoring and acknowledging Indigenous knowledge and contributions in his works. He was one of the first scientists, for instance, to refer to the *Innuit* (Turner's spelling) as opposed to the descriptor "Eskimo people," which remained in use by some Arctic anthropologists well into the 20th century.

This book provides a window into Turner's world, and presents to the Inuit and Innu of the region a celebration of their ancestors' knowledge of mammals. It honors the ingenuity and enterprising nature of previous and current generations of Inuit and Innu who envisaged ways to transform mammal parts into practical objects: some purely for functional purposes and some that expressed spiritual connections to the land and the animals. Future generations of Inuit and Innu might turn to this book to see photos of material objects that their elders made from mammals. This book returns knowledge of previous generations of Inuit and Innu elders to their respective Northern communities, and it highlights the wealth of Indigenous knowledge that can be rekindled and relearned from Arctic collections.

Turner's two years in Ungava were extraordinary. We hope this book exhibits to the world the rich knowledge of the Inuit and Innu, and that it ensures the inclusion of Lucien Turner among the major scientific contributors to the study of Canadian natural history.

Legend to the Book

The legend provides guidance on the elements of the book that originate from Turner's manuscript, titled "Notes on the Mammals Ascertained to Occur in the Labrador, Ungava, East Main, Moose, and Gulf Region" (1886a). New elements added by the editors are also described.

1 | The modern Inuktitut writing system used in Canada known as syllabics. This information did not appear in Turner's original manuscript.

2 | The original term for each mammal that Turner recorded in his "Notes on the Mammals..." manuscript. The original spelling has been retained.

3 | The modern orthography for each mammal in Roman Inuktitut, a writing system used by the Inuit today.

4 | Turner's original handwriting of the scientific name for mammals from his manuscript. This appeared at the top of each of his entries on mammals. The current scientific name for each mammal is provided in the table below the handwriting.

5 | The original page number in which the mammal entry appeared in Turner's manuscript.

6 | The original order name for each mammal provided by Turner. His spelling has been retained.

7 | The current order name for each mammal. All current scientific names follow Wilson and Reeder (2005).

8 | The original family name for each mammal provided by Turner. His spelling has been retained.

9 | The current family name for each mammal.

10 | The original scientific name for each mammal provided by Turner. His spelling has been retained.

11 | The current scientific name for each mammal.

12 | Some mammals are known by more than one Inuktitut name. The syllabic transcription of only the most commonly used term is provided here. This information did not appear in Turner's original manuscript.

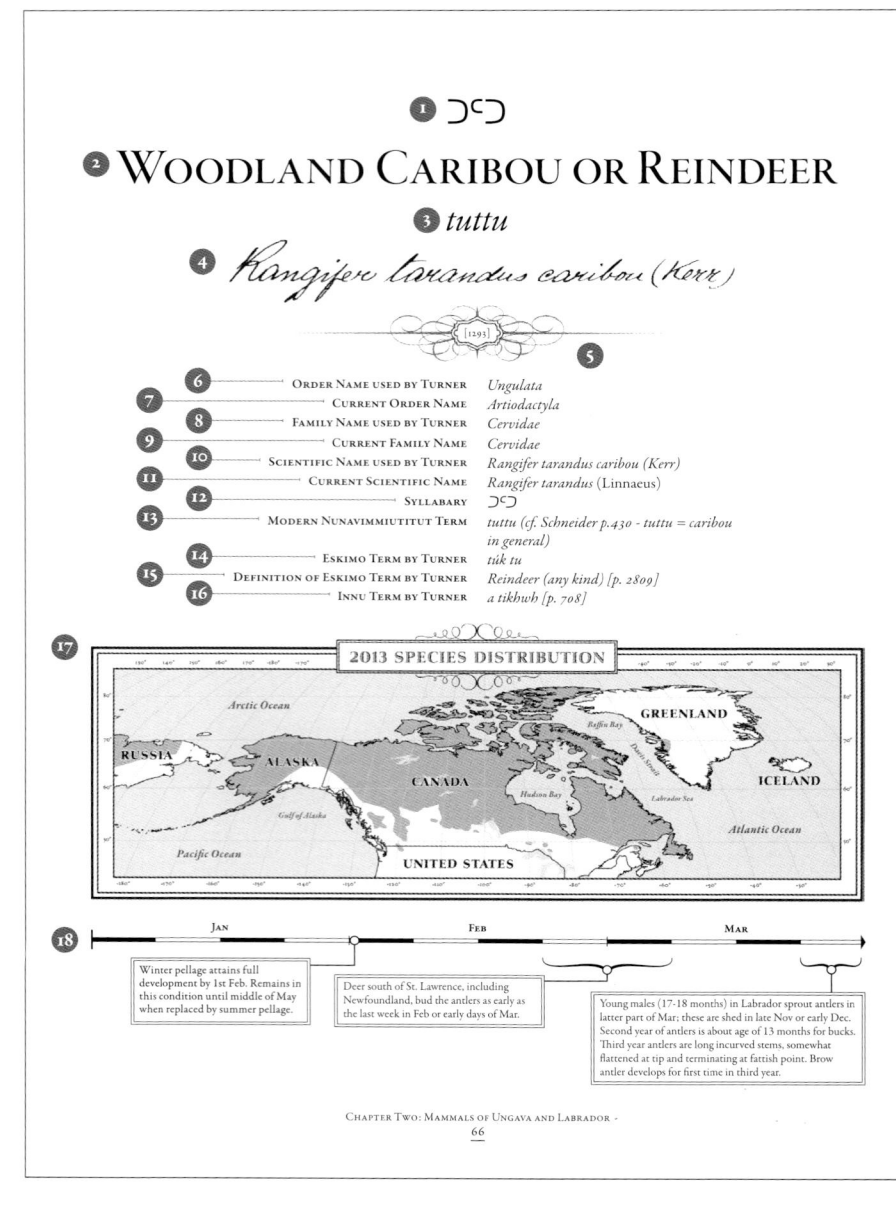

xlvi

Descriptions by Turner

The Reindeer is an animal essentially belonging to the colder regions of both hemispheres. Southern range of its present distribution is confined to the extreme northeastern portions lying south of the St. Lawrence Gulf. It is found sparingly in the extreme northern parts of Maine, although if we may rely upon the statement of DeKay in New York Fauna, page 122, part 1, 1842, this animal occurred, according to Professor Emmons, only few years previously in the northern portions of Vermont and New Hampshire. In the Canadian provinces it occurs in New Brunswick and Nova Scotia; formerly on Prince Edward's Island, but now extirpated. Few are found on Cape Breton, and very abundant on Newfoundland. On the island of Anticosti, they appear to have never

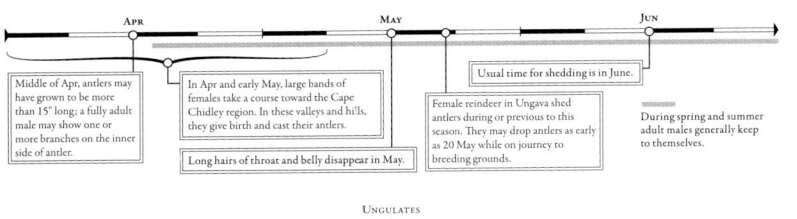

Apr — Middle of Apr, antlers may have grown to be more than 15" long; a fully adult male may show one or more branches on the inner side of antler.

In Apr and early May, large bands of females take a course toward the Cape Chidley region. In these valleys and hi'ls, they give birth and cast their antlers.

Long hairs of throat and belly disappear in May.

May — Usual time for shedding is in June.

Female reindeer in Ungava shed antlers during or previous to this season. They may drop antlers as early as 20 May while on journey to breeding grounds.

Jun — During spring and summer adult males generally keep to themselves.

Ungulates
67

---

13 | The modern Inuktitut orthography and definitions for mammal terms. These are relevant to the Ungava and Labrador dialect known as Nunavimmiutitut. The page number and reference of sources has been provided.

14 | The original Eskimo (Inuktitut) spelling and orthography for mammals that Turner described in his various manuscripts from Ungava. The original diacritics have been preserved. The references for the page number ranges are: 1267-1541 (Turner 1886a); 001-304 & 693-706 (Turner 1887a); 2128+ (Turner 1884).

15 | Definition of Eskimo term by Turner.

16 | The original Naskapi (Innu) spelling and orthography for mammals that Turner described in his manuscripts (Turner 1887b: 305-693; 707-710; 1829-1841). Note that the orthography for these mammals differs today.

17 | Map of the 2013 distribution of the species: adapted from range maps in the current IUCN Red List of Threatened Species (www.iucnredlist.org).

18 | Seasonal diagrams generated from the original manuscript, where possible, to illustrate mammal lifecycles and behavioral patterns.

19 | Artist and naturalist Ernest Thompson Seton prepared illustrations for Turner's original manuscript, but these can no longer be found. Australian artist Bryony Anderson has prepared original line drawings that accompany each species entry.

20 | The original text of Turner's manuscript. Turner's original spelling and arrangement of the manuscript have been retained.

1 | Provides insights by the authors about mammals that appear or are described in the Turner Collection.

2 | Material on mammals extracted from a manuscript by Turner titled, "Descriptive Catalogue of Ethnological Collections...." This was divided by Turner into two parts: 1) Inuit (1887a); and 2) Indians (Innu) (1887b). This manuscript formed the basis for Turner's 1894 publication, entitled: "Ethnography of the Ungava District...." The original manuscript material was significantly altered for publication in 1894. Only the original material is presented here for readers to gain an insight into Turner's observations and descriptions of Inuit and Innu use of mammals.

3 | Stories of mammals that Turner documented during his Ungava and Labrador Expedition from 1882-1884. The stories have been extracted from various manuscripts he produced on the region. The stories are presented here without modification.

---

① Collection Notes

Turner did not collect any barren ground bear or grizzly bear specimens while in Ungava and Labrador. The species is considered locally extinct.

---

② Ringed Seal Ethnography
Turner

THE SEASONAL ROUND (1887A: 102-109)
It may be early in the season and the waters yet afford but scanty sustenance to the eager crowd but when the first seal of the year is secured a feast is held as soon as the party can reach the shore and prepare it. An anxious throng gathers about the carcass and when the skin is stripped from the body the flesh is quickly devoured by the people so long held in expectancy of fresh food.

The flesh is divided according to rank and station, the choicest morsels falling to the favorite children. Ere many hours not a vestage remains except the hide so greatly coveted to repair the soles of their footwear now to undergo roughest usage over the sharp-edged gravel and sand and rocks.

Each day visibly decreases the quantity of ice and the migratory birds appear amongst the open spaces. Soon the earlier arrivals begin to lay and the eggs of all the waterfowl are sought as delicious food by the Innuit who gives no concern to the stage of development of the embryo within the shell. The islands whose barren tops scarcely rise above the swash of the waves are scanned for the nests of the gulls; the higher ones whom summit are clad with grass and weeds are searched for the nests of the eiders; while the ragged shore lines of the islands are peered among for the nests of the surf ducks, sea pigeons and puffins. The higher parts of the marshy tracts of the neighboring mainland are watched to discover the breeding places of the Canada goose or the freshwater ducks and loons.

---

③ How Children Became Wolves
An Inuit story, Turner (1887a: 301)

A poor woman had so many children she was unable to restrain their cries for food. They were changed into wolves and to this day their mother may be heard endeavouring to console her clamouring children with the hope she will find a nice, fat deer for them.

Inuksuit and Caribou
An Inuit story, Johnny George Annanack (Heyes 2007)

The Inuksuit [pl.; Inuksuk is singular] are used to know where there are caribou, or where there are fish around the lakes. When you go caribou hunting, the Inuksuit are not so close to one another so the hunters would know which trail the caribou would take and the hunters would wait ahead of their trails and the Inuksuit would have a shoulder blade on some of them. That's what I used to hear about the caribou trails. The Inuksuit has a lot of meaning and they'd be used to play with and my parents used to fix them when they used to play. They're still standing there. If you could understand their meanings, they're very useful.

Recollections on Inuit Use of Wolves
Tivi Etok (Heyes 2010)

There are different kinds of wolves, just as there are different kinds of dogs. Some are slow. The male ones are generally not hunters. The females are generally the hunters. They run and hunt caribou. Females generally have thick fur and are faster runners than the males. When caribou are killed they make howling noises. The female howls to inform the males and her babies to come along out from their hiding places so that they can eat. The wolf would make their dens in hard sand, like foxes. After they make babies they would move to another place. Like humans, they would move around. Wolves are not like dogs. We used to have a dog and a wolf on a dogteam. We tried to train it like a dog, but they were not good at it; they couldn't have a rope around their head; they were not as strong. Dogs are stronger. Even when the *qamatik* (sled) is very heavy, a dog can pull it; even if an Inuit cannot move the sled.

Caribou – Terms and Definitions

Syllabary	Modern Nunavimmiutitut Term and Definition	Term by Turner	Definition recorded by Turner
ᐊᕐᓇᓗᒃ	**arnaluk** (cf. Schneider p.40. gives "female bird"; *arnalukak* = female whale; walrus)	*ag ná luk*	Doe, female deer [p. 2137]
ᐃᑳᕐᓗᒃ	**ikaarluk** (?): from *ikaar-* (to cross) with the same –(l)luk ending as in *nalluk* (caribou crossing place in a river or lake)? *ikaarpuq* (?) (cf. Schneider p. 58 - *ikaarpuq* = he crosses over to the other side [of a river, lake,... etc]). Johnny Sam Annanack, Kangiqsualujjuaq (personal communication) suggests *ikaarvik*.	*i kóg lûk*	A lake lying in the path of migrating reindeer, and through which they swim to continue their way [p. 2226]

4 | Stories of mammals that were told to the authors by Inuit elders. The stories presented here originate from the same regions that Turner travelled to during his Ungava and Labrador Expedition from 1882-1884. Most of the stories pertain to Kangiqsualujjuaq, Killiniq, Abloviak Fiord (in Nunavik, Quebec), and Nackvak (in Northern Labrador). The graphic symbols and lines that feature above and below the sections on stories and recollections of mammals are based on patterns that appear on Inuit clothing from Ungava Bay. These patterns are principally featured in the trimmings of parkas.

5 | Information on mammals by Inuit elder Tivi Etok of Kangiqsualujjuaq. Etok was interviewed in 2010 by Scott Heyes on Inuit knowledge of mammals.

6 | This section has been produced with the assistance of Inuktitut linguistic experts. It provides the Eskimo terms that Turner used to name and define mammals alongside the modern terms and orthography used by Inuit today. The diacritics and orthography used by Turner have been preserved. The references for the page number ranges are: 1267-1541 (Turner 1886a); 001-304 & 693-706 (Turner 1887a); 2128+ (Turner 1884).

Lucian Wayne Turner (2010)
Lucien Turner's grandson, Lucian Wayne Turner, from Milton-Freewater, Oregon, visited the Smithsonian Institution in Washington, DC with his family to explore his grandfather's collections in 2010. He holds a drum that his grandfather collected from the Innu in 1883 from Ft. Chimo in Ungava, Quebec. His grandfather passed away before he was born. Photo by Scott Heyes.

A Turner Family Tribute

Lucian Wayne Turner

On behalf of Lucien McShan Turner's family, it is my pleasure to write a tribute to my grandfather. I learned about him in my early childhood, but because of the sixty year age difference between my father and me, little information was passed along as to Grandfather's many endeavors and accomplishments. I was only sixteen when Father passed away in 1949 and my grandfather died in 1909, twenty four years prior to my birth.

As our family grew up and I left home, I was given three apple boxes of memorabilia that my grandmother, Mary E. (Lutz) Turner, had kept of family history. It included articles that Lucien had sent to local newspapers about his northern travels, and letters he had sent home to his sons.

It was not until I started a genealogical study of my grandfather that my eyes were opened to this remarkably accomplished man who had an unending thirst for different cultures, habitats, customs and languages. This included documenting the Inuit and Innu of Quebec and Labrador, and the Eskimo of Alaska, as well as the surrounding plant, animal and marine life of these regions.

In early 2000, I started an Internet search to find out more about Lucien, and was amazed at the amount of information I found. To gather more information on my grandfather, I did what he had done early in his life, and made contact with the Smithsonian Institution. I also enlisted the help of family members and universities from Newfoundland and Quebec, as well as several universities in the U.S., including the University of Alaska. With contributions from museums and historical societies all over the United States I was able to gather enough information to write a book for my family in 2008 entitled, *Lucien M. Turner: Genealogy, Biography, and Journals of his work in Alaska and Labrador.*

Through continued contact with the Smithsonian Institution I was introduced to Dr. Stephen Loring, an anthropologist at the Smithsonian's Arctic Studies Center who was well aware of Grandfather's work. After years of correspondence, Dr. Loring invited my family and me to visit the Smithsonian where my grandfather's works are stored. We accepted his invitation, and in May of 2010, we were treated to two days of viewing a remarkable collection of over 4000 Eskimo, Inuit and Innu artifacts from Alaska and Labrador.

I am proud to say there are several published works in relation to my grandfather, Lucien. In 2001, the Smithsonian updated a copy of an earlier publication about Lucien's work in Labrador and republished it as *Ethnology of the Ungava District, Hudson Bay Territory,* with an introduction by Dr. Loring. Earlier works of Lucien's stay in Alaska from 1874-1877 were published under the title *Contributions to the Natural History of Alaska,* by Lucien M. Turner, and published in 1886 by the Smithsonian, though few copies now exist. Mr. Raymond L. Hudson authored a book of Lucien's work in Alaska titled *An Aleutian Ethnography* which I highly recommend reading. Other unpublished manuscripts of Lucien's work are now being brought to life by others such as Dr. Scott Heyes from the University of Canberra, who we met on our visit to the Smithsonian Institution in 2010.

Though Lucien collaborated with the Smithsonian Institution for several years, by 1885 the work had finished. He ended up working part time for them from 1886 to 1887 writing manuscripts of his past work. Lucien stayed around the Washington, DC area for an unknown period of time. He was elected to the American Ornithologist Union in New York and the Microscopical Society of West Chester, Pennsylvania and entertained them with lectures of his past work in Alaska, Quebec, and Labrador. He also wrote articles for several newspapers including: *The Daily Republican* of West Chester, Pennsylvania, in 1885; *The Post* of Washington, DC in 1887; *The Evening Star* of Washington, DC in 1887; *The Register* of Mt. Carmel, Illinois, in 1883 and 1894. He also wrote comments on *The Commonwealth of Pennsylvania Act (109)* in 1886.

Raymond L. Hudson, in his book, *An Aleutian Ethnography*, refers to Lucien as, "A poorly paid member of the United States Signal Service for whom natural history and ethnography were passions." Grandfather had to join the U.S. Army Signal Corps on a U.S. Army Private's pay to do the work for the Smithsonian. I believe it is only fair to say that Grandfather's continuous work from 1874 through 1884 brought a lot of hardships to his wife and remaining children, Jesse and Eugene. He returned home about every year and a half, and only for short periods of time. This ultimately brought about a divorce in the fall of 1884.

Grandfather continued to look for work in his field for the next several years, but the Smithsonian had nothing more to offer. From 1894 to 1898 he worked in the Seattle and Yakima areas of Washington State, collecting specimens which he sold to the Smithsonian and other museums to maintain a salary.

In the early 1900s Grandfather moved to San Francisco, California. It is not known what he did during this time; however, he was there during the 1906 earthquake and ultimately passed away there on April 7, 1909.

In May of 2012 Dr. Scott Heyes wrote and asked if I would write a tribute about my grandfather for this book, *Mammals of Ungava and Labrador: The 1882-1884 field notes of Lucien M. Turner together with Inuit and Innu Knowledge.* Scott enclosed a draft of the book and from the first pages I found myself amazed at the amount of detailed, unpublished information my grandfather had sent to the Smithsonian Institution on the mammals in the Ungava and Labrador regions.

It is more than 130 years since Grandfather Turner studied and documented this rich natural history, and just as many years that it has been in storage at the Smithsonian. Dr. Heyes and Dr. Helgen have gathered a wealth of Grandfather's papers regarding mammals, as presented in this book. It is a book that my family and others will be able to read with a keen sense of who Lucien M. Turner was and the importance he placed on the surroundings, the local inhabitants, and the animal life.

Grandfather's never-ending desire to not waste a single minute in his search for, and record information about, the inhabitants and animals of this area was recorded in the following quote from a letter Grandfather wrote to his childhood and lifelong friend, Mr. Robert Ridgway (then Curator of the Smithsonian Institution's Bird Collection) on 28 August 1883 saying,

"I have written so much that from April 10th to July 20th, I could not use my hand to write as I had the writer's cramp so bad that motion of my wrist was agonizing torture. I contracted it in Feb, Mar, & April while writing stories as related to me, together with writing vocabularies."

It is our desire that Grandfather's work and Heyes and Helgen's book be used to further the studies on these animals so they will continue to live and propagate for the people of this world to enjoy.

Thank you
Lucian Wayne Turner
Grandson of Lucien McShan Turner
30 May 2012
Milton-Freewater, Oregon

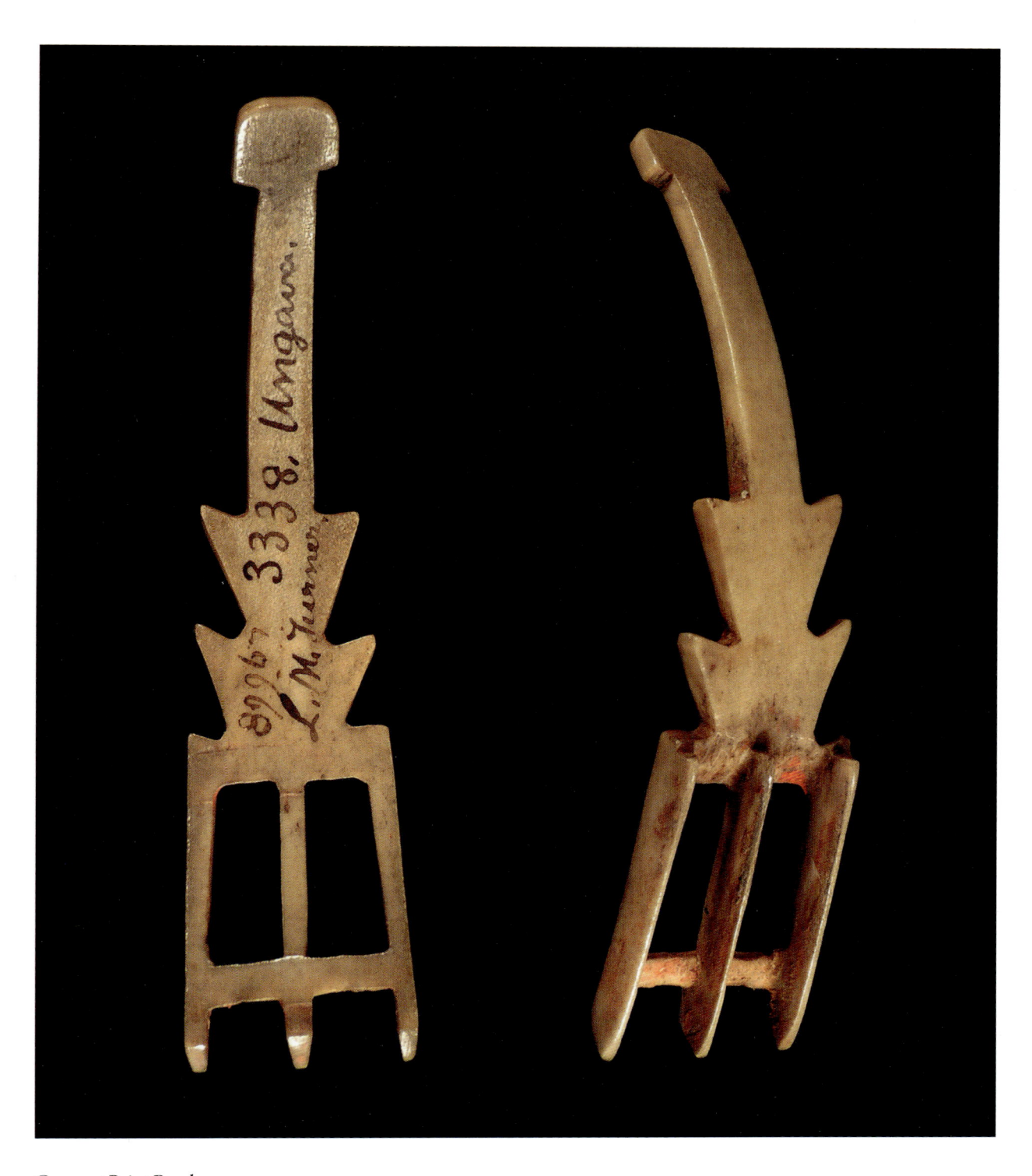

Garment Paint Brush
Three-pronged bone implement for painting robes collected by Turner in Ungava Bay, 1884. USNM E89967-0, Museum Support Center, NMNH. Photo by Angela Frost, 2011.

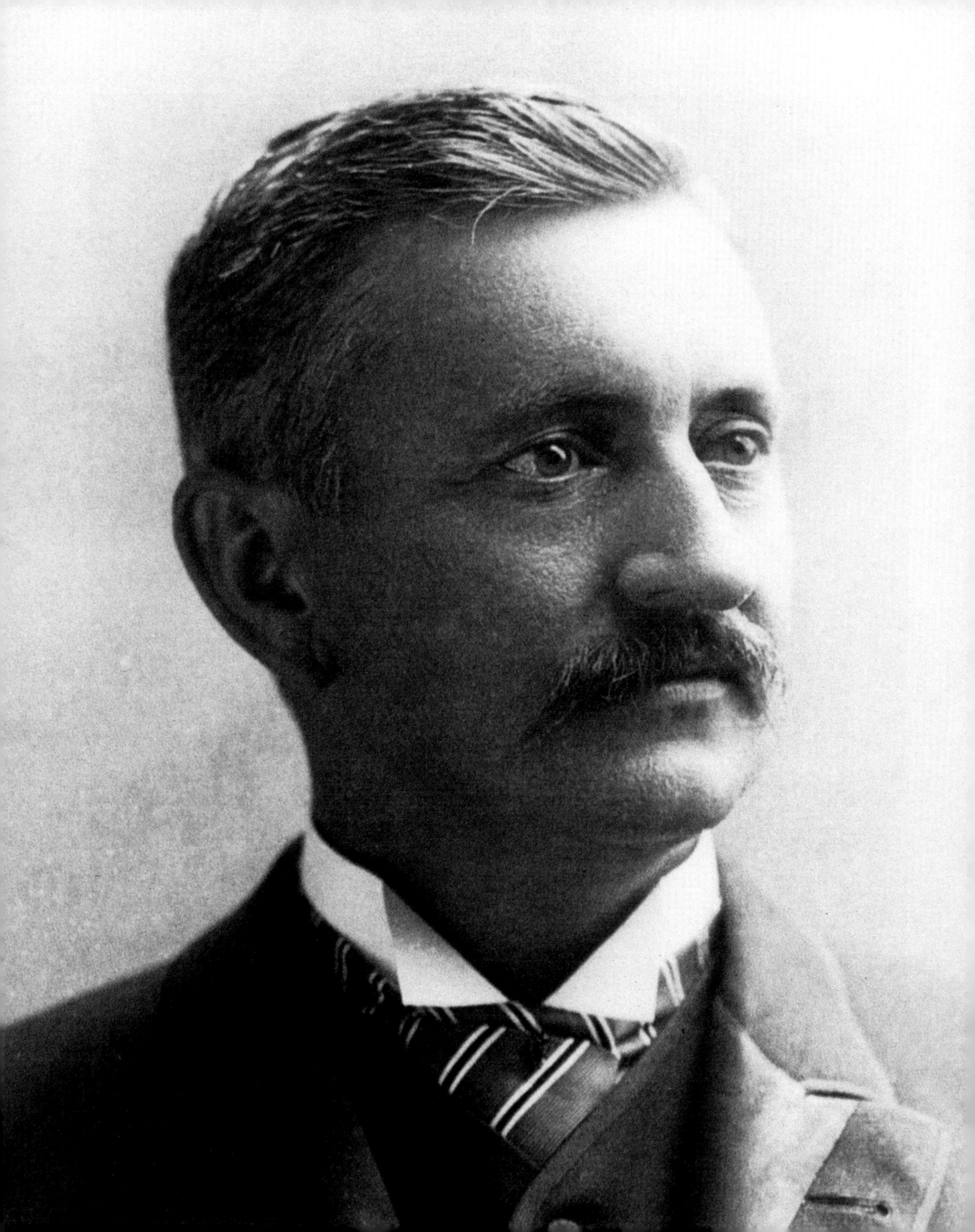

Chapter One

L.M. Turner's Introduction, 1886

This chapter incorporates the front matter material that originally appeared in Turner's manuscript, "Contributions to the Natural History of Labrador and Ungava, Hudson's Bay Territory, May 15, 1886." This manuscript was broken into several smaller manuscripts, including the one upon which this book is based called "Mammals ascertained to occur in Labrador, Ungava, East Main, Moose and Gulf Region." The original spelling and punctuation have been retained, with the original page numbers in square brakets. The Narrative pages 34-38 are missing from the archives, and we were unable to locate them.

Lucien McShan Turner (c. 1884-89)
Born 20 June 1847 in Mainville, Ohio. Died 7 April 1909 in San Francisco, California. Date of photo unknown, but probably 1884-1889 in Washington, DC. Photograph taken by C.M. Bell, Washington, DC. Image used with permission of L. Wayne Turner, 2010.

Turner's Letter of Transmittal

Sir:- I have the honor to present, herewith, for your consideration and publication the results of the observations made by me in Labrador and Ungava. The investigations were carried on by myself under the auspices of the Office Signal Officer; and in the leisure time, from the arduous duties pertaining to the official work connected with meteorology, I was to devote myself to securing all other objects which would serve to elucidate the history of that region. Much attention was given to the study of the faunae and to the collection of ethnological material and a comprehensive vocabulary of the language spoken by the natives. To the study of linguistics I devoted the long winter evenings; the special object being to collate words and sentences of the Innuit; while to the Indians but little time was, however, spared.

The results of my labours are elaborated as far as was practicable and consistent with the very short time given me in which to prepare them for publication. The objects of natural history have been donated to the Smithsonian Institution; while the meteorological records have been turned over to your Office.

I desire to express my warmest thanks to Professor S.F. Baird for his characteristic liberality in permitting me to have access to such material as was necessary for comparison with that obtained by me. To Mr. Robert Ridgway, Curator of Ornithology, and to Dr. L. Stejneger, Assistant Curator of that department, my thanks are specially due for many favors and kind suggestions. To Dr. C. Hart Merriam, Agent of Economic Ornithology, Agricultural Department, I am under special obligations for valuable aid in determination of the mammals. To Dr. T.H. Bean, Curator of Ichthyology, I am under great obligation for the determination of the fishes and also for many valuable suggestions. To Major J.W. Powell, Director of the U.S. Geological Survey and Bureau of Ethnology, I owe a special debt of gratitude for his characteristic liberality so freely extended to me. To many gentleman, employees of the Hudson Bay Company, among them Capt. A. Gray, master of the S.S. "Labrador"; Mr. Robert Gray, at Ft. Chimo; Messrs. Irvine, Miller, Brown, Saunders and Davie, my thanks are due for the many favors extended to me while in that region. [4] To Mrs. Maggie Brown and Mr. John Ford for their painstaking care in the compilation of the linguistics and stories that no errors should creep into the perplexing study of the language my sincere thanks are specially due.

The original drawings of the plates of the birds were prepared under the immediate supervision of Mr. Robert Ridgway and to him should be accredited full merit for the admirable manner of executing that difficult work. The plates of mammals were prepared under the hand of Mr. E.E.T. Seton and bespeak for themselves his artistic ability.

The illustrations of the ethnological material was prepared at the Bureau of Ethnology under the personal supervision of the renowned artist Mr. H.W. Holmes and to him and assistants special credit is due for the faithful delineations of those subjects. The plates of Reindeer antlers are reproduced from photographs taken by myself from subjects forming a part of the collection made at Ft. Chimo. The arrangement of the subjects has been for my personal convenience rather than an attempt at a natural sequence.

In the preparation of the results, herewith presented, I have endeavored to leave out all discussion and confine myself to actual field observations and for this reason have used the personal pronoun I as a warrant of individual responsibility for all assertions made by myself.

I am, sir, &c.
To the Chief Signal Officer, U.S. Army
Washington, D.C.

Lucien M. Turner.
Smithsonian Institution
May 15 - 1886.

Turner's Original Letter of Transmittal
Front page of Turner's Letter of Transmittal, which featured in his manuscript, titled "Contributions to the Natural History of Labrador and Ungava, Hudson's Bay Territory, 15 May 1886."

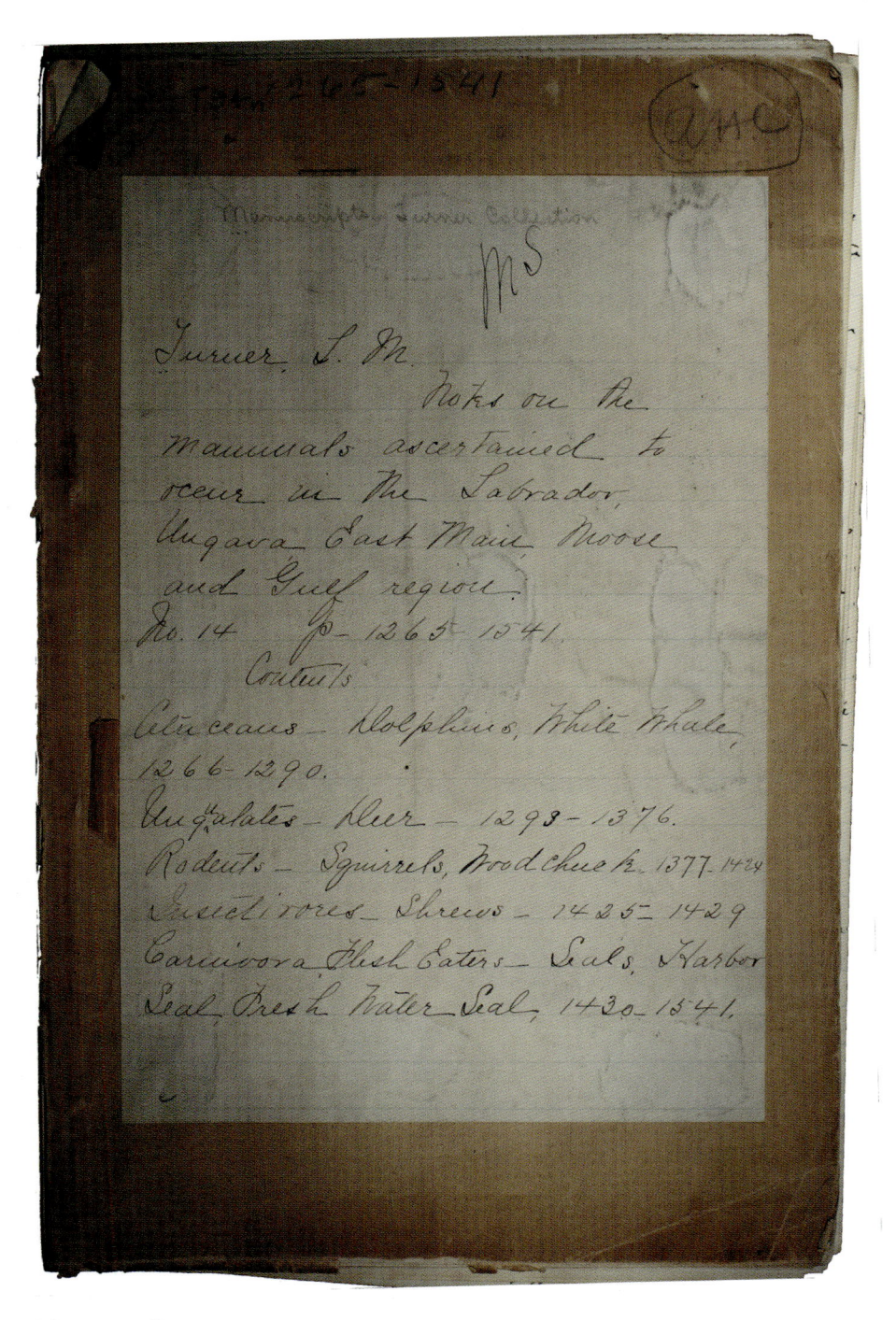

Manuscript Cover

Front page of the manuscript by Turner on mammals, titled "Notes on the Mammals Ascertained to Occur in Labrador, Ungava, East Main, Moose and Gulf region." Source: SI Archives, Record Unit 7192, Turner, Lucien M, 1848-1909, Lucien M. Turner Papers.

***Labrador* Sailing Ship**
The Hudson's Bay Company brig, Labrador, *captained by Alexander "Sandy" Gray. Turner was aboard this ship when it struck a shallow reef near southern Ungava Bay, 1882. The* Labrador *was undamaged. Used with permission of McCord Museum MP-0000.1269.1.*

Turner's Narrative

In compliance with Special Orders, No. 60, from the Office of the Chief Signal Officer, dated May 27, 1882 at Washington, D.C., I proceeded to Ft. Chimo, Ungava by way of Montreal. I arrived at Montreal on the evening of June 2, 1882, and put myself in communication with the Hudson Bay Company upon whose vessel I was to take passage for Ft. Chimo. The Company's schooner "Tropic" left Montreal for Quebec on the 4th of June to receive the remainder of her cargo at the latter place.

On the evening of the 6th I started for Quebec, arriving there the next morning, finding the "Tropic" would sail the next (8th) day. In the afternoon of the 8th we were towed into the stream and under the direction of the river pilot accompanying us we slowly made progress to Bic Island where we were met by a pilot vessel which relieved us of the pilot we had on board.

The "Tropic" now pursued a course to the southward of Anticosti Island, [2] lying in the northern portion of the Gulf of St. Lawrence. Light winds prevailed and our progress was necessarily very slow. Calms of tedious length held the vessel between that Island and the Strait of Belle Isle. As we neared Cape Blanc Sablon, on the north shore, lightest winds and densest fogs rendered greatest caution necessary. For four days we were completely enshrouded in such dense fog that of the scores of small fishing craft cruising here and there only about half a dozen could be seen; although, the voices of the crews could be heard in every direction and our usual diversion was to climb the rigging to look over the sea of fog to observe the dozens of mask-tops and parts of sails which appeared, in every direction, like phantoms and as magically vanishing.

On the morning of the 21st of June we pushed through the Strait and were soon abreast of Greely Island.

Light winds fell and not until late morning of the 23rd did we approach George's Island, the scene of unfortunate wrecking of the small schooner "Walrus" several years before. On the morning of the 24th we entered Hamilton Inlet (or *Ivuktok*). At 2 p.m. we came to anchor at the Hudson Bay Company's station of Rigolet. The character of the weather had been such that but little of the shore had been seen and that only when baffling winds allowed us to drift in dangerous proximity and on such occasions our most strenuous endeavors were to get farther from a coast studded with unseen dangers.

The officer in command of the "Tropic" was Captain Alexander Main as hearty and whole-souled a Scotchman as ever walked the quarter of a vessel. To Captain Main I am deeply indebted for many acts of courtesy and full well know that under the limited means at his disposal he was personally obliged to forgo many comforts in order to render the trip to me not only advantageous but pleasant. The only other passenger was a Roman Catholic missionary priest, Father Fafard of the Quebec diocese. I found this gentleman to be an agreeable companion [4] and believed him to be sincere in his desire to promote the teachings of his Church among the Indians of Labrador and Ungava.

Father Fafard was well acquainted with the various localities along the northern side of the Gulf and gave me much interesting information concerning them. At Rigolet we parted company, he visiting the station at Northwest River where a church had been constructed for the assembling of the people who chose to listen to him when they came to that place to barter the products of the chase obtained during the previous year.

Montagnais Hunters Beside Canoe at Esquimaux Bay (c. 1882)
Turner passed through Hamilton Inlet, also known as Ivuktok or Esquimaux Bay, on his voyage to Ft. Chimo. Photo by J.R.H. Used with permission of McGill University Rare Books and Special Collections, 120623.

It was the intention of the missionary to travel overland from that place to Ft. Chimo in the Ungava district; a trip which had been accomplished but few times, and then under most discouraging circumstances. The failure of the Reverend Fafard to continue the project was never fully understood and soon after his return to his superiors in Quebec he became ill and died.

The religious discussions carried on between Captain Main and the Reverend Fafard were always a source of diversion [5] to me. The Caption was ever agreeable and when about to bring some weight to his forcible arguments it was most amusing to hear the missionary plead inability to understand the English language.

At Rigolet I found the Hudson Bay Company's steamship "Labrador" lying at anchor, having arrived on the 17th of June and was now awaiting our arrival. A small steam-tug to be taken by the "Labrador" to Ft. Chimo and which she now carried on her decks was the cause of a change having been made in the usual programme of the "Labrador's" sailing.

The "Tropic" discharged her cargo, mostly directly on board of the "Labrador" and immediately departed for Quebec, leaving Rigolet on the 28th of June. As I was to take passage on the "Labrador" from Rigolet to Ft. Chimo I amused myself on the shore until the vessel should be ready to depart for that place.

The Hudson Bay Company's affairs at Rigolet are ably managed by Chief Trader M. Fortescue, a gentleman, of rare attainments and fine social qualities [6], who endeavored to render my stay at that station as agreeable as was in his power to do. Mr Fortescue is overwhelmed with titles and distinctions, principally due to his situation in oriental Labrador, being chief magistrate, collector of customs, bailiff and a multi-

Group of Esquimaux (Inuit), Labrador (c. 1882)
Probably aboard the Labrador. *Photo by J.R.H. Used with permission of McGill University Rare Books and Special Collections, 121239.*

plicity of other favors conferred upon him, mostly, I suspect because he receives no salary for his services in those capacities.

Mr Fortescue's renown as a physician and surgeon extends throughout the land and the courteous manner displayed by him in administering drugs or bandaging a wound is sufficient to command the gratitude of the numerous applicants for his skill. I understand that the Newfoundland authorities supply that place with a quantity of medicines to be administered to the poor fishermen repairing thither during the summer fishing season.

Assistance in the Company's affairs is rendered by Mr. E. Adams and Wilson, clerks in the employ of the company. [7] The station of Rigolet is situated about fifty miles within Hamilton Inlet and on the north shore, nearly opposite a constructed portion of that water locally known as the "Narrows". The buildings are one large dwelling occupied by the agent, one house used as a general office and sleeping apartments of the clerks. Another dwelling occupied by the cooper and his family; a large house for the men-servants. Two or three shops, two large warehouses and a retail store. Several other smaller structures comprise the entire number of houses at the place. The conformation of the point of land upon which the station is situated prevents any attempt to a regular disposition of the buildings as each one is placed wherever it is most convenient. A small jetty projects into the water and vessels of little tonnage and draft may lie alongside of it.

The stations presented an attractive sight when viewed from the approach, showing that even in such a desolate region comfort may be had with but little expense and care. [8] The surroundings are quite hilly; to the northward a high ridge of granite of more than two hundred feet elevation runs parallel with the inlet.

Turner's Journey to Labrador & Ungava

1 | Montreal
6 June 1882 – Departed Montreal for Quebec City on train. Arrived morning of 7 June.

2 | Quebec City
8 June 1882 – Boarded HBC schooner "Tropic," taking a course to the south of Anticosti Island.

3 | Bic Island
8 June 1882 – Brief stop at Bic Island.

4 | Cape Blanc Sablon
16-20 June 1882 – Stuck near the Cape for four days due to dense fog and light winds.

5 | Strait of Belle Isle
21 June 1882 – Pushed through the Strait on this day.

6 | Hamilton Inlet
24 June 1882 – Entered Hamilton Inlet.

7 | Rigolet
24 June 1882 – Arrived aboard the "Tropic." Anchored at the HBC Station at Rigolet.

24 June to 8 July 1882 – Turner collected natural history specimens around Rigolet

8 July 1882 – Departed on "Labrador" northward for Davis Inlet.

8 | Indian Harbour
11 July 1882 – Until this date, fog prevented "Labrador" from steaming northward beyond Indian Harbour.

9 | Davis Inlet
16 July 1882 – Arrived opposite HBC station in evening. Large fields of ice closed off entrance due to NE wind.

19 July 1882 – Departed on "Labrador" for Nakvak.

10 | Offshore near Nakvak (Nachvak) Fiord
23 July 1882 – Momentary glimpse of immense mountains of Nakvak through thick fog. Too dangerous to stop at HBC Trading Post. Labrador continues to steam north toward Ft Chimo.

11 | Hudson Strait
25 July 1882 – At noon the fog lifted to reveal entrance between Button Island and Cape Chidley. A course was made between these places. The "Labrador" swept through with great rapidity with the tide.

12 | Entry to Ungava Bay
26 July 1882 – Vessel cautiously advanced. At 3 pm "Labrador" was surrounded by acres of ice. Vessel unable to release itself from pack for four days. Dense fog prevailed. Vessel drifted up to 55 miles with a single tide.

13 | Beacon Island
30 July 1882 – Vessel anchored at western side. *1 August 1882* – Departed for Ft. Chimo on a course towards the mouth of the Koksoak River.

14 | Day Excursion
1 August 1882 – Party of "Labrador" goes ashore. A large caribou is hunted and many more seen. Turner takes natural history specimens from banks of a large stream.

15 | Ungava Bay
1-4 August 1882 – Vessel closed in by ice by noon on 1st August. Some ice pieces up to three miles long were observed. Course was made for the Koksoak River when the ice cleared on 4 August.

16 | Labrador Reef
5 August 1882 – "Labrador" strikes shallow reef at low tide at 2:45 am. No damage done. Vessel re-floated on rising tide at 10:10 am on 6th August. Turner collects fish and molluscs on reef. Reef given name of "Labrador" that day based on the name of the vessel that struck it.

17 | Ft. Chimo (Kuujjuaq)
6 August 1882 – Arrived at HBC Trading Post, Ft. Chimo, in the evening in thick fog. The party went ashore on 7 August 1882 at daylight. Turner stayed in Ft. Chimo until 4 September 1884, a period of 2 years and 27 days in the village.

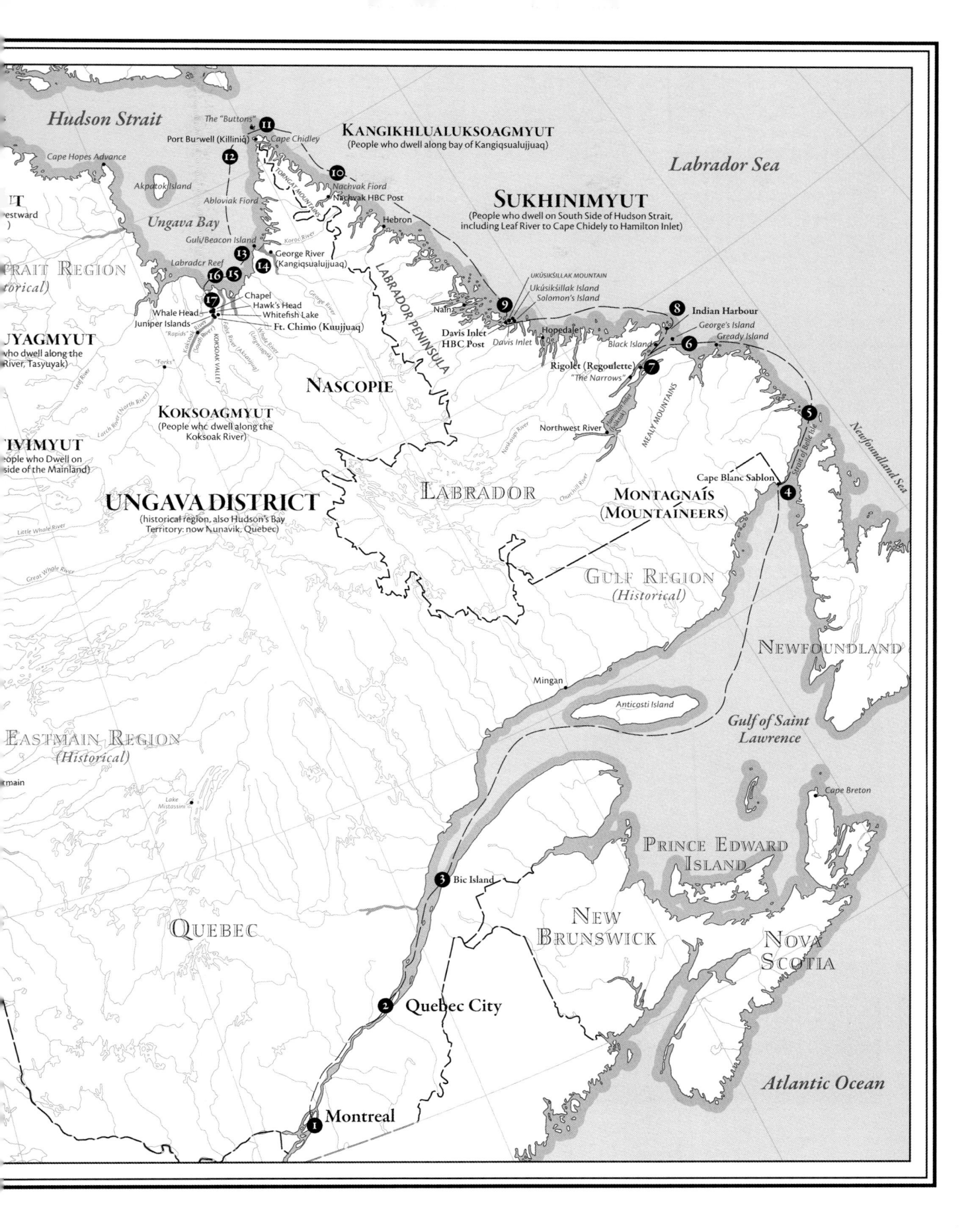

Hudson Strait

The "Buttons"
Port Burwell (Killiniq) — Cape Chidley
KANGIKHLUALUKSOAGMYUT
(People who dwell along bay of Kangiqsualujjuaq)

Labrador Sea

Cape Hopes Advance
Akpatok Island
Ablovik Fiord
Ungava Bay
Gulf/Beacon Island
Labrador Reef
Whale Head
Juniper Islands
Chapel
Hawk's Head
Whitefish Lake
Ft. Chimo (Kuujjuaq)

Torngat Mountains
Nachvak Fiord
Nachvak HBC Post
Hebron

SUKHINIMYUT
(People who dwell on South Side of Hudson Strait, including Leaf River to Cape Chidely to Hamilton Inlet)

STRAIT REGION (historical)

UYAGMYUT
(who dwell along the River, Tasyuyak)

KOKSOAGMYUT
(People who dwell along the Koksoak River)

Korok River
George River (Kangiqsualujjuaq)

LABRADOR PENINSULA

Nain
UKÚSIKSILLAK MOUNTAIN
Ukúsiksillak Island
Solomon's Island

Davis Inlet HBC Post
Davis Inlet
Hopedale
Black Island

8 Indian Harbour
George's Island
Gready Island

7 Rigolet (Regoulette)
"The Narrows"

IVIMYUT
(People who Dwell on side of the Mainland)

Little Whale River
Great Whale River

NASCOPIE

Northwest River
Naskaupi River
Churchill River

MEALY MOUNTAINS

5

UNGAVA DISTRICT
(historical region, also Hudson's Bay Territory; now Nunavik, Quebec)

LABRADOR

MONTAGNAIS (MOUNTAINEERS)

Cape Blanc Sablon
4
Strait of Belle Isle

Newfoundland Sea

NEWFOUNDLAND

EASTMAIN REGION (Historical)

Lake Mistassini

Mingan
Anticosti Island

GULF REGION (Historical)

Gulf of Saint Lawrence

Cape Breton

PRINCE EDWARD ISLAND

QUEBEC

3 Bic Island

NEW BRUNSWICK

NOVA SCOTIA

Atlantic Ocean

2 Quebec City

1 Montreal

But few rods back of the station the ground rises to irregular heighths and in the rear of this the general aspect of the surface is very broken with ridges and spurs, valleys and slopes extending in every direction. On the lower tracts quite large lakes are not uncommon. Some thirty-miles inland begin the foothills of the Mealy Mountains, which attain heights of from 2400 to 4000 feet; their summits clad with perpetual snow.

The country in the vicinity of Rigolet is well covered with timber; the large species are Black Spruce, *Picea nigra*; White Spruce, *Abies balsamea*; White Birch, *Betula papyracea*; Larch, *Larix americana*; Balsam Poplar, *Populus balsamifera*, and several species of *Salix* and the ever present *Alnus*. It may be well to state that the spruces and other trees have local names such as "Pines" or "Firs" for the Spruces and "Juniper" for the Larch or Tamarac.

[9] The first four species mentioned embrace, in that vicinity, the most important kinds; and, from them is obtained the fuel used and the timber for building and other purposes. The size obtained along the coast is rarely half or one third that which obtains thirty to fifty miles interior. Along the headwaters of Hamilton Inlet the timber attains such size and heighth that a party cut quite a number of trees well adapted for spars and masts of vessels. The industry was pushed with a vigor characterizing an impulsive enterprise undertaken in a region not fully explored for the purpose.

It was finally abandoned as hastily as begun; although there was an abundance of material for the purpose. While at Rigolet I saw a small vessel discharge a cargo of staves for barrels. The character of the staves was excellent and proved that with little exertion the timber of the country might furnish such material in abundance. [10] The weather was simply execrable at Rigolet from the 24th of June until I parted on the 8th of July; but during that time I occupied myself with collecting specimens of natural history; each which will be noted under the proper heading of those subjects. The lateness of the season prevented me from procuring more than I did as the greater number of birds were breeding and the fishing season had not yet begun; in fact, the first salmon (*Salmo salar*) was taken the day previous to our departure.

Rigolet, Labrador
Turner stopped at Rigolet for 15 days on his outbound journey to Ft. Chimo in 1882. He collected natural history specimens during this period. Top: Regoulette (Rigolet), Labrador, c. 1882. Photo by J.R.H. Used with permission of McCord Museum, MP-0000.638.1. Left: The Rigolet harbor, c. 1882. The photo description reads: "Store-houses. Regoulette, Labrador." Photo by J.R.H. Used with permission of McGill University Rare Books and Special Collections, 120551.

Chapter One: L.M. Turner's Introduction, 1886

The returns of furs and other products of the country had been, as usual, rather limited in quantity. There were so many purchases for furs and fish that by the time each procured his share the quantity dwindled to but a small number for each. In late July and August of each year a number of Grampuses frequent Hamilton Inlet, and were to be purchased with but little trouble. Some arrangements [11] were subsequently made for their capture but with what success was not learned. The species was suspected to be the Blackfish (Pilot Whale, *Globiocephalus melas*, Traill) and is usually termed "Black Grampus" by the "Planters".

On the morning of the 8th of July the Hudson Bay Company's steamer "Labrador" weighed anchor for Davis Inlet about one hundred and eighty five miles farther north on the Labrador coast. Captain Alexander ["Sandy"] Gray had had command of the "Labrador" for twelve consecutive seasons and under his ability and experience the vessel steamed rapidly toward the mouth of the Inlet. Opposite Black Island a dense wall of fog rendered farther progress impossible. Not until the morning of the 11th did we succeed in getting beyond the northern heads of the group of islands on which is situated the fishing location of Indian Harbor. At this place a boat was dispatched to deliver letters for the regular [12] Newfoundland and Labrador mail-service vessel pledging from St John's to Hopedale during the codfishing season to allow those engaged in that industry to have communication with the civilized world.

The fishing season had begun and was now being vigorously prosecuted with very fair prospects of abundance. A more dreary spot I had not beheld on the face of the earth. The small islands were smooth-sided, worn by the ice and tides of countless ages, surging with a force scarcely to be resisted with less stable foundations than those conglomerate quartz, granite and hardest sandstones. The only vegetation growing on them was *Empetrum*, and scantiest grasses; and these only as mantles beyond the reach of the spray and waves dashing constantly upon their lower sides. In the protected lower grounds a few, scanty sedges and a species of hardy willows (*Salix*) struggled with the treacherous sphagnum [13] for its puny growth.

All wood and fuel must be brought from the inner portions of the Inlet by vessels sent for that purpose. After getting beyond the coast bad weather set in and not

Views of Davis Inlet, Labrador (c. 1882)
Top: The Hudson's Bay Company buildings and harbor of Davies (Davis) Inlet, showing the long wooden jetty. Photo by J.R.H. Used with permission of McGill University Rare Books and Special Collections, 120936. Middle: Cape Harrington, looking south from Mount Ukasiksalik, Labrador. Photo by J.R.H. Used with permission of McGill University Rare Books and Special Collections, 120951. Bottom: Winter scene of Davis Inlet, Labrador, about 1890. Used with permission of McCord Museum, MP-0000.637.7.

until the evening of the 16th did we come to anchor opposite the station of Davis' Inlet. Outside of the Inlet great fields of floating ice were met and no sooner were we within the harbor than it closed the entrance and held so as long as the north east wind prevailed. We learned that the ice in the "Run" had broken but three days before our arrival and "went out" on the 15th instant.

Mount Ukasiksalik, Labrador (c. 1882)
A summer scene of Mount Ukasiksalik, Davies (Davis) Inlet, Labrador. Turner noted in his narrative that the Inuit term Ukasiksalik *means a place where soapstone material can be found. The Inuit used soapstone to make seal-oil lamps and pots. Photo by J.R.H. Used with permission of McGill University Rare Books and Special Collections, 120331.*

The station of Davis Inlet was erected for the purpose of carrying on a fishing and fur trade. It is situated on the south side of a rather large but rugged island bearing the Eskimo name of *Ukúsiksillak*, or place where material (soapstone, steatite) for making native cooking utensils is found. The dwellings are two in number connected at their ends with [14] communication from one to the other by means of a doorway. Two large stores and a cooper shop with a few smaller structures comprise this station's buildings. A long wooden jetty extends beyond the shallow beach and forms a convenient stage to land upon. Only small craft can approach the landing at high water.

Mr William Davie was in charge of the station and expected to remain there for several years. On the 17th an ascent was made to the top of the high hill back (north) of the station and found to be 875 feet high as determined by means of an aneroid barometer taken for that purpose. The place had hitherto received no name expect the "Hill". It was now proposed to name it in honor of Captain Alexander Gray of the "Labrador". With the usual solemnities attendant upon such occasions the locality received the name of Mount Gray.

The character of the country in the vicinity of the station is exceedingly [15] rough. Bold and rugged spurs break in every direction, seemingly to have no special arrangement. The higher land attaining an elevation of 1200 to 2000 feet. The tops of the hills are often quite flat and frequently embrace very extensive areas. The sides are mostly abrupt and often precipitous. The valleys are consequently deep and usually narrow, having wild currents of water rushing headlong to the various arms of the sea separating the numerous islands studding this portion of the coast apparently as breaks for the rarely absent fields of sea-ice crushing back and forth along their sides with every change of the wind and tide.

The valleys and protected hillsides are covered with growths of heavy Spruces and Larch. I did not observe any Birch trees, but was informed that small trees of it grow but few miles to the interior. Willows and alders grew in luxuriant profusion on the sheltered hillsides moist from the [16] subsoil water between the roots and probably only a couple of feet to the solid rock of the hills. In the swampy tracts surrounded by trees the alders attain a size not elsewhere seen by me.

Many species of flowering plants were observed to be already in blossom and among those most attractive were *Traxacum* and *Viola*. Few heads of *Kalmia* and *Rhododendron* showed signs of flowers to open in but a few days. Until the 19th instant I occupied my time collecting such specimens of natural history as could be obtained. Several species of fishes, birds and marine invertebrates were secured at this locality.

Several of the neighboring streams are of such size that great numbers of trout (*Salvelinus*) visit them for spawning purposes. Hundreds of barrels of these fish are caught along this portion of the coast and thence shipped to

Atlantic Cod, *Gadus morhua*
Turner wrote about the cod fishery of Labrador in great detail. Illustration by Marcia Bakry, Smithsonian Institution, 2012.

Views of Nachvach (Nachvak) Trading Post and Fiord (c. 1882)
Nachvak was the northernmost Hudson's Bay Company (HBC) Trading Post along the Labrador Coast, operating from 1868 to 1905. Turner was unable to stop at Nachvak on his outbound journey to Ft. Chimo due to inclement weather. Left: The HBC stores at Nachvak. Photo by J.R.H. Used with permission of McGill University Rare Books and Special Collections, 120115. Right: Lion Head, Nachvack Fiord, Labrador. Turner described the fiord as "tortuous and narrow." Photo by J.R.H. Used with permission of McGill University Rare Books and Special Collections, 120718.

the markets of Canada and [17] other provinces. The quantity of the trout taken at this place is certainly very fine. The individuals captured vary from four to fourteen pounds. The inhabitants along this portion of the coast are furnished with barrels and salt, nets etc. for the capture and preservation of these fish and in some seasons obtain an abundance for which they receive a stated sum for each barrel of two hundred pounds. The price varies according to the state of the market in civilization and also, to a very great extent, according to the amount of rivalry existing among the petty traders who sail along the coast with the ostensible object of fishing but not failing to secure the furs and other commodities the "Planter" may have in his possession. It requires a keen watch from the Company's agent lest those to whom supplies have been furnished do not take everything to some other trader offering but slight more [18] advantageous terms.

The runs of trout last about three weeks and as the appearance of these fish is regulated by the proximity of ice the season is very uncertain. Offshore winds and moderate weather are the things most wished for during the fishing season. At times the Codfish is found to be quite numerous and during scarcity of other fish they are taken in moderate quantity. The Codfish obtained in this vicinity are usually small fishes ranging from nine to thirty inches in length without the head. The larger Codfish of course keep farther offshore and are taken by the fisherman from Newfoundland.

On the afternoon of the 19th of July the "Labrador" was again under way for Nakvak. Dismal weather greeted our entrance to the Atlantic. By the next morning (20th of July) a cold rain and headwind with fog coming from the ice fields, which were known to be [19] but few miles outside of us, rendered the next three days the most uncomfortable one could imagine and our progress was necessarily very slow. On the morning of the 23rd we found by a momentary glimpse at the immense mountainside walls that we were then about thirty miles to the north of Nakvak. Captain Gray knew from previous experience that it would be useless to attempt to enter the tortuous and narrow fiord at the head of which is the station Nakvak.

It was then determined to proceed to Ft. Chimo and deliver the goods, destined for Nakvak, on the return trip. Until nearly noon of the 25th of July was spent endeavoring to find the entrance to Hudson Strait. The fog was so impenetrable that only by it lifting a slight degree could it be ascertained that we were between Cape Chidley and the "Buttons". A course was steered between those places, although rarely attempted [20] by a sailing vessel, and the tide being favourable we swept through with great rapidity. The weather changed and we found that the bad conditions existed only outside of the Cape. Everyone felt a relief from the constant strain and anxiety attendant upon a dangerous coast.

On the morning of the 26th ice was discovered ahead. The vessel was cautiously advanced and by three o'clock we were surrounded by a single field of broken pieces ranging in size from a few yards to many acres. The contending tides swept the ice and with it the vessel wheresoever they went. For four days we were powerless to release the vessel from the pack. At times the ice appeared to be running away from the vessel and in five minutes time a solid cake of ice appeared. The swirling tide currents caused the most wonderful changes to be made in a moments' time. [21] Densest fog prevailed with the calms and again the slightest breath of air dispersed it to reveal our situation unchanged. No less than five consecutive times was the vessel diverted past a single island and upon obtaining an observation it was shown that the vessel had drifted fifty-five miles with a single tide.

After several fruitless attempts the "Labrador" was urged toward the mouth of George's River where open water could be discerned and at the mouth of that river a tolerably secure anchorage was afforded. On the evening of the 30th of July the vessel came to anchor near Gull or "Beacon" Island near the shore of the west, or left, bank of that river.

It was determined to remain there until some signs of the ice being carried out were apparent. Plans were laid for a run on the shore which appeared so temptingly near. The next morning a hunting party was organized [22] and each determined to do his best. The party landed and divided into two groups, the one taking to the higher ground while I followed the course of a stream which debouched near by. This stream took its rise from an immense lake some twelve miles inland. It was exceedingly tortuous, cutting a deep bed into the narrow valley into whose bottom had showered many feet depth of fine gravel as the final seas of ice of the glacial period had passed over the hilltops, cutting terraces here and there, scooping out great depressions on the more level land and again throwing purest quartz sand into them so that their outlines are scarcely less apparent today then when last formed. The heighths of the bank of that creek were from ten to thirty feet high and in many places as

Pack Ice, Ungava Bay, Labrador (c. 1882)
Turner described the vast fields of ice within Ungava Bay. This photo, probably taken from the Labrador, *provides a sense of the pack ice that Turner confronted on his journey to Ft. Chimo. His notes reveal he was stuck in the pack ice for four days in Ungava Bay, and that "the contending tides swept the ice and with it the vessel wheresoever they went." Photo by J.R.H. Used with permission of McGill University Rare Books and Special Collections, 120132.*

perpendicular as though cut with an immense blade. The water was very shallow in certain places and [23] unfathomable at other places where extraordinary currents of rushing water had gouged deep holes in the bottom.

About ten o'clock of that forenoon a shot was heard and followed by two more. Signs of reindeer were everywhere abundant; and, although, each one of the party was anxious to kill one of the animals yet it was little expected to find that Captain Gray had secured a huge buck in excellent condition of flesh. The carcass was cut up and apportioned among the three men accompanying us. Other parts were removed for us to carry after we had finished our stroll.

The high hills farther up the river were partially inspected. Even at a heighth of six hundred feet could be found the first terrace formed by the ice of former years having slowly but visibly cut its way down the tortuous valley turning at the lake, nearly to a right angle; and opening [24] to the sea. The terrace varied from ten feet to two hundred yards wide; and as far as the eye could reach to the eastward this same line had been cut out. On the opposite side of the valley could be seen the same evidence.

About eighty feet below this was a narrower terrace and only about thirty feet below this was a third. Then came the irregular, lower sides of the valley, each side of which presented the same character of erosion.

In certain tracts of the upper and middle terraces were deposits of fine sand and gravel that appeared as little changed as through deposited but a couple of years ago. The rocky character of these spots was such that the few lichens growing upon them were of less than an inch in heighth and these so sparsely growing that but little sustenance could be obtained from a soil of purest rock.

At 3 p.m. we turned toward the [25] vessel; while in the distant lake could be heard the clear notes of the Canada Goose, calling to her companions or young, for in that vicinity is one of the noted breeding-places for this species. Descending to the scene of the Captain's good fortune each one shouldered a goodly sized portion of the remains of the reindeer. The horns and head with neck attached fell to me. It proved a heavy and awkward burden on my shoulders as I trudged along that hot valley; heated by the afternoon sun that called into life and activity myriads of mosquitoes and scores of hungry "Bulldogs" (*Tabanus*) each of which doubtless attracted by the scent of fresh blood of the deer and my moving person thronged with an energy seemingly irresistible, impeded as I was. After carrying the load for nearly six miles I was glad to perceive one of the men coming to relieve me.

When we arrived at the beach we [26] found that the tide had left the large boat high above water. In the meantime I secured over a score of Sandpipers (*Tringa fuscicollis*) and *Aegialitis semipalmata*. Among the former were the birds of the year scarcely beyond the downy stage yet fully capable of flight as they flitted from shoal or rock to search each cap of the waves for freshly revealed animal life forming the food of these birds.

After a couple of hours we embarked and visited "Gull" Island behind which we had anchored. It was my good fortune to find here the remains of a Polar Bear; although they had doubtless lain there many years.

Polar Bear Skull
Detail of a polar bear picked up by Turner at Beacon or Gull Island, near the mouth of the George River, 30 July 1882. NMNH, Smithsonian Institution, USNM A 23207. Photo by Scott Heyes.

This island is surmounted by a beacon of piled stones which at a distance greatly resembles an Eskimo woman in skin clothing. The beacon was erected some three years previously by Messrs. Miller and Olsen of the Hudson Bay Company's service. [27] Early on the morning of August 1st the anchor was again weighed and the "Labrador" attempted her course toward the mouth of the Koksoak River. About noon the ice closed in on us and until the early afternoon of the 4th of that month we were taken at the will of the tides. The ice was even of large pieces, one was as wide as the view extended and appeared to be not less than three miles long. Any attempt to force such a mass of ice was futile; hence, to keep along the outside of it was deemed the better course. Breaks occasionally occurred and allowed the vessel to tack and then attempt to force a passage between the narrow openings. The force with which the iron shod prow of the "Labrador" struck some of the cakes, much larger than herself, caused the vessel to recoil with such shock as to prostrate those [28] not attentive. The ice cakes would turn in every possible position while the water around them seethed and foamed. I observed several small fishes which I was desirous of procuring. A dipnet was brought into requisition and over seventy specimens secured. They proved, upon examination to be small Polar Cod (*Boreogadus saida*).

Often, while the vessel was quiet, the persons on her took occasion to run upon the ice. The ice was often measured by computing one-seventh of its heighth to be out of water the thickness was found to be, in many instances, as much as forty-two feet. Thicknesses ranging from ten to twenty-five feet were the usual measurements.

The first ice which we encountered was doubtless from the Atlantic Ocean, and had been drifted into the Strait by the long-continued northeast wind [29] which had prevailed for the past three weeks. This ice had been carried into the Strait as far as the northeast point of Akpatok Island and was there being swirled hither and thither by the contending tide-currents and winds. Fortunately we were at the eastern edge of this mass of ice. The currents had deflected a portion of it down into Ungava Bay and this was the ice that surrounded us.

By the 4th of August the Atlantic ice had been carried out of our course and that which we encountered from the 1st to the 4th was the ice from the shores, coves and rivers of Ungava Bay. It was with a deep feeling of relief when we pushed into the quiet blue water beyond and quickly headed for the mouth of the Koksoak River less than fifty-miles distant; but having a coast line so low that the utmost caution was necessary to approach it. The [30] bottom of the southern portion of Ungava Bay is studded with reefs and inlets exposed only at low water, hence the condition of the tide must be known before navigation may be safely pursued.

Soundings were being constantly made showing a depth ranging from fourteen to seventy fathoms in depth. The character of the mud brought up was bluish clay color becoming much lighter when dry. Several species of animal life clung to the mud proving that sufficient sustenance was afforded in the cold waters to support their life.

Early on the morning of the fifth of August I was aroused by the sound of voices speaking of a light seen in the distance. The name of the steamer "Diana" was frequently mentioned so that I knew the refrigerating steamer "Diana" was not far off.

At 2.45 am of the 5th I heard a noise like the rattling of the [31] chain when the anchor is let go. In an instant thereafter I was pitched from my berth and precipitated through a door to the middle of the cabin. The vessel careened so that standing was impossible. As soon as I could recover my astonishment I went on deck and saw the people gazing around. I looked over the side of the vessel and saw land at a distance of twelve miles. I then inquired what was the matter. At that moment the Captain on the bridge gave the order "Call all hands. The vessel is ashore". I went below and put on extra clothing and packed my guns and ammunition, determining to have means to procure game if necessary to take to the boat.

In the course of a few minutes it was ascertained that the vessel had struck a flat topped reef; and, when the water had ebbed far enough inspection revealed the fact that no damage was done and that the vessel was [32] lying in a position as safe as though in a dock.

The ebbing tide swept with an almost irresistible fury past the vessel and around us appeared a smooth-faced reef of nearly a mile and a half in length by five hundred yards in width. We had struck on the northwest portion about two hundred yards from the north end. Fifteen minutes earlier would have allowed the vessel to pass safely over the top of the reef and ten minutes later would certainly have stove in the bows of the vessel. When the top of the reef was dry I occupied my time in collecting marine specimens and was fortunate enough to secure some of great rarity, among them a species of shell new to science.

The steamer "Diana" was now in full view some five miles to the eastward and from her a boat was coming toward us. By 8 a.m. Captain Riches of the "Diana" and Captain Gray of the "Labrador" had concluded [33] their conference and as the tide was rising rapidly it was determined to have everything in readiness when the tide should float us off the rock.

At 10.10 a.m. of the 6th August the stern of the vessel lifted and in a few minutes we were free. The flooding tide swept us rapidly over the face of the reef and we headed our way to the mouth of the Koksoak River some eighteen miles distant.

We were in fact at the time of striking the reef which was now named "Labrador's" Reef, within the long point extending from the west (or left) bank of the river to the outer portion of Leaf River Bay.

The beacon at the mouth of the river was plainly visible but a dense fog settling caused us to come to anchor in the mouth of the Koksoak River where we had to await the ingoing tide at day light of the next day. At 8 p.m, a boat came alongside, containing Mr. Keith Me...[34][1]

---

1. Pages 34-38 are missing from the original manuscript and were unable to be located.

The S.S. *Labrador* on a Shallow Reef (1882)

Top: The Hudson's Bay Company brig, Labrador *became lodged on a tidal reef system in lower Ungava Bay at 2:45 am on 5 August, 1882. This photo is probably a record of this event. It shows crew members inspecting the state of damage. While waiting for the tide to refloat her, Turner collected a number of fish and shells new to science from the rock pools. The S.S.* Labrador *was released from the reef without damage. The reef was named Labrador Reef after this event. Photo by J.R.H. Used with permission of McGill University Rare Books and Special Collections, 120754.*

Left: View of the S.S. Labrador's *rigging, sails, and mast. The photo description reads: "Fine Weather. Bound North." Photo by J.R.H./George Washington Wilson. Used with permission of McGill University Rare Books and Special Collections, 120222.*

Turner's Narrative

Innu Camp near Ft. Chimo, April (c. 1883)
Turner learned about the natural history of Ungava and Labrador from the Innu and Inuit. Photo by Turner. Source: SI Archives, Record Unit 7192, Turner, Lucien M, 1848-1909, Lucien M. Turner Papers.

Inuit Family beside an HBC Dwelling at Ft. Chimo
It is quite likely that this was Turner's quarters during his stay in the village. Photo by J.R.H. Used with permission of McGill University Rare Books and Special Collections, 120501.

View of Ft. Chimo (1882)
Two photos of Ft. Chimo taken by Turner have been digitally combined to generate this panorama. Source: 1 cyanotype 4" × 5" mounted on 5" × 8", NM No ACC # Cat 175484, National Anthropological Archives, Smithsonian Institution.

[39] ...one used partly as an office and the remainder as a dwelling for the clerk and temporary persons at that place. In this house is a room containing a well-stocked library containing about 400 volumes, beside magazines etc. Most of the books belonged to a former agent, who left them there for the benefit of subsequent readers.

There are three large warehouses for the storage of goods, furs, nets and other articles. These buildings are of good size and afford commodious shelter for many of the boots, tools &c. The next is a retail store in which is done the bartering for furs and other products. The store is rather well-stocked, though principally with staple articles of trade such as cloth of various kinds, ammunition, beads, hardware, tea, sugar, molasses, flour and a kind of hardbread.

The requirements of the trade are so well known that the requisition for supplies each year [40] is scarcely more than a repetition of that of the preceding year. An oil-house, cooper shop, dwelling for the men servants, three other buildings for the occupancy of the men-servants having native wives. A blacksmith shop and three miserable structures occupied by Eskimo complete the number of the larger buildings. A salt house capable of containing a number of tons of that article as a requirement of former years but now serves merely the purpose of storage.

The arrangement of the houses is without design, convenience alone determines the position of each additional structure. The principal ones are so crowded together that a fire breaking out in one is certain to destroy the remainder. On the agent's house is a bell on which is rung the time to go to work at 7 a.m.; a cessation from 12 noon to 1 p.m. and again at 6 p.m. During the short days of winter the bell is rung only for the meal hours. [41] The round of occupation for the men at Ft. Chimo has for each year much sameness; and it is now not difficult to know what will be done next.

As each labor is dependent upon the season there is necessarily much routine about it and renders it somewhat difficult to divide into a starting point. After the season's work is finished two or three men are chosen to go to the forks of the Larch River with that of the Koksoak, of which the former is a tributary, for the purpose of setting traps for foxes and other furbearing animals.

These men are provided with a small boat which is loaded with their provisions for the length of time they are to remain away, together with ammunition, guns, traps and a tent. They are usually ready to start by the 1st of October, and are expected to remain away until the 1st of January or perhaps two weeks later. The condition of [42] the weather causes the trip, of seventy five miles, up the Koksoak River to be made in from five to eight days as the current is swift and several rapids, two of which cause great delay unless the force is able to move the load over the obstacles. A few hundred yards up the Larch River

View of Rapids on the Koksoak River (1882)
Turner collected mammals along the Koksoak River. Photo by Turner. Source: BAE /SI GN 03264 06542900, National Anthropological Archives, Smithsonian Institution.

View of the Koksoak River (c. 1882)
Turner was stationed at Ft. Chimo on the Koksoak River. Photo by Turner. Source: SPC Arctic Eskimo Labrador NM No ACC # Cat 175484 01448100, National Anthropological Archives, Smithsonian Institution.

is a hut, of a single room, which affords shelter to those going there for that purpose. The men set their lines of traps, one down the Koksoak and the other up the Larch. Each line is visited on alternate days and anything procured from them is taken to the house and from there to the post where the number of skins are credited to the captors at fixed rates but slightly above the prices paid to the natives for such furs. The men at the post during this period are engaged in making nets for the following season. By the middle of January, or early February, the more able hands are sent to some locality to cut fuel [43] for the ensuing year. A locality is usually selected as convenient to the post as circumstances will allow. Proximity to the river bank is also considered and when these two fall near each other the woodcutting is but a short task.

A sufficient amount of provision, axes, tenting, cooking utensils and a grindstone completes the outfit. The affairs are taken a day or two ahead, on sledges, and the tent and stove put in order for occupancy. The men follow and the trees are cut down and trimmed. As the day's cutting progresses toward the evening the men then carry the trimmed logs to a place where a pile, formed of sticks stood on end, is erected and this conical shape pile is taken as the unit of measurement, containing little more than a cord. About eighty of these stacks of wood are cut each year and when that amount has been stacked three or four teams of dogs with [44] sledges and native drivers, are sent to haul the wood to the river bank where

it is placed just above the high water mark when the ensuing spring freshet occurs. In late June or early July the Indians are sent to raft it to the post where the Indian women put it in stacks of about five cords each. It remains there until wanted for the next winter for fuel.

During the time the woodcutting had been going on the men have had two or three fox traps set, which they visit on occasions of three or four days apart. By the tenth of March the men are again at the post and the making of salmon and trout nets occupies their time until the warmer days of spring. Several of the laborers have trades such as cooper, blacksmith; these begin such work as their trade may demand in the way of repairing. The cooper has a shop and with rough staves furnished from Canada prepares fifty to [45] one hundred casks for oil and fish. The blacksmith is also the engineer of the small tug "Ft. Chimo" and the repairs and attention to this vessel requires all the time of himself and the steersman until the middle of June while the other men are engaged in "backing" the nets for the salmon season later in the summer.

The number of boats required for the eight or ten fishing stations, have to be overhauled and repaired. The ice in the river is soon expected to break and all repairs, cleaning up &c must be done by the time the river is free from ice. All these occupations consume the time to the middle of June and by that time summer is at hand. A new life is opened for everyone. The spring trade has

Innu Birchbark Canoe (c. 1883)
This canoe was probably used to assist with the rafting of wood to the Ft. Chimo HBC Station. Turner reported that the Innu captured white whales at Ft. Chimo from their birchbark canoes with the "same facility that obtains by the Innuit from his skin-covered kayak" (1887b: 372). Photo by Turner. Source: SI Archives, Record Unit 7192, Turner, Lucien M, 1848-1909, Lucien M. Turner Papers.

been completed by the clerk and the outfits given to the scores of Indians anxious to get away [46] from the post to their hunting grounds, which each group or related families, have, during their stay at the post, selected.

When the river is free from ice a reconnoitering party is sent down the river to observe the movement of the ice in the bay. The Eskimo have been away for a couple of weeks, eager to secure the now well fatted seals that come to bask in the warm sun. They never wait for the breaking of the ice but go to meet the open sea, which is to afford them food for the next several weeks. They have understood that their services will be required when the White Whales begin to come up the river; hence the Eskimo usually designate at what locality they may be found by a certain date. A party of white men is sent for these people, who return by the end of the first week in July.

[47] The ice in the Koksoak River breaks from the 15th of May to the 15th of June and clears up by the end of a week after breaking. Nets for small, river trout (*Salvelinus fontinalis*) are put down and with them are obtained many whitefish (*Coregonus*). This netting is continued until the latter part of July and these fish form the principal article of food until that time.

About the 1st of July preparations are made for the whaling season. The White Whale enters the river in considerable numbers and are taken by the method described under the article relating to the White Whale (*Delphinapterus catodon*). The whaling season continues from two to three weeks and employs every available hand, including white men and Eskimo. The latter repair to the whaling station seven miles below Ft. Chimo, [48] with their families, who are fed by the Company, while engaged in this work.

During this period the Indian men have rafted the wood down to the station and are now ready to scatter over the country to obtain young reindeer skins for their winter clothing. The few, scattering Indians, who remain behind, are employed to remove the blubber from the whale skins and crush the fat. The Indian women remove the mucous coating and dirt from the whale skins which are then salted and barreled. The refrigerating steamer "Diana" makes her appearance about the last week in July, or ten days to two weeks before the Salmon appear in the river. This period is taken to get the nets and other affairs, including provisions, boats and tents, in readiness.

A crew of two or three persons is usually sent to the mouth [49] of the river to put down one or two nets so as to discover the arrival of the fish which are, by the 5th of August, daily expected. As soon as they begin to arrive, the remaining stations, eight or ten all told, are fitted out and in a few days the regular fishing season begins, requiring the attention of nearly every man, woman and child about the place.

The HBC Steamer S.S. *Diana*

Top: This refrigerated steamer was used to process fish, beluga whales and seals by the Hudson's Bay Company in Ungava Bay during Turner's time in the region. The fish were kept frozen and transported to London at the end of the fishing season. Photo description reads: "Salmon cleaning, 'S.S. Diana,' Fort Chimo, Ungava Bay." Photo taken by J.R.H., c. 1882. Used with permission of McCord Museum, MP-0000.1269.3.

Left: The S.S. Diana *stuck in ice in the Hudson Strait off the Quebec–Newfoundland coast, 1887. Used with permission of McCord Museum, MP-0000.636.12.*

The fishing season lasts until the end of the first week in September, although there are so many controlling influences that their arrival and disappearance is subject to as much as four or five weeks variation. The earliest appearance has reached the 20th of July and the latest not until the last week in September. The subject of the salmon fishing is described more fully under the article on *Salmo salar*.

After the fishing season is over [50] the natives are paid off and a rest of a couple of weeks taken. During this time the Eskimos are busy preparing for the fall hunt of reindeer, at the place where they assemble during the rutting season, in order to obtain skins from which to make winter clothing.

The white men are employed in hauling up the boats and chinking the cracks of the log houses which comprise the buildings of Ft. Chimo. This work completes the yearly round and has but little variation in it.

The arrival of the "Diana" in late July and the "Labrador" anytime in August or early September causes the monotony of such routine life to break into a bright gleam for a day or two. These vessels are the only means of communication with the civilized world; hence their arrival is hailed with joy and their departure noted with pleasure.

Naskapi and Inuit at Ft. Chimo (c. 1882)
Top: The photo description reads: "Nascopi squaws in anticipation of something to eat." The Koksoak River is in the background. Photo by J.R.H. Used with permission of McGill University Rare Books and Special Collections, 120255. Bottom: The photo description reads: "Esquimaux from Whale River, Ungava, Labrador." The building was likely an HBC dwelling at Ft. Chimo. Note the polar bear skins drying on the roof and the Inuit clothing made from various skins and furs. Photo by J.R.H. and George Washington Wilson. Used with permission of McGill University Rare Books and Special Collections, 120908.

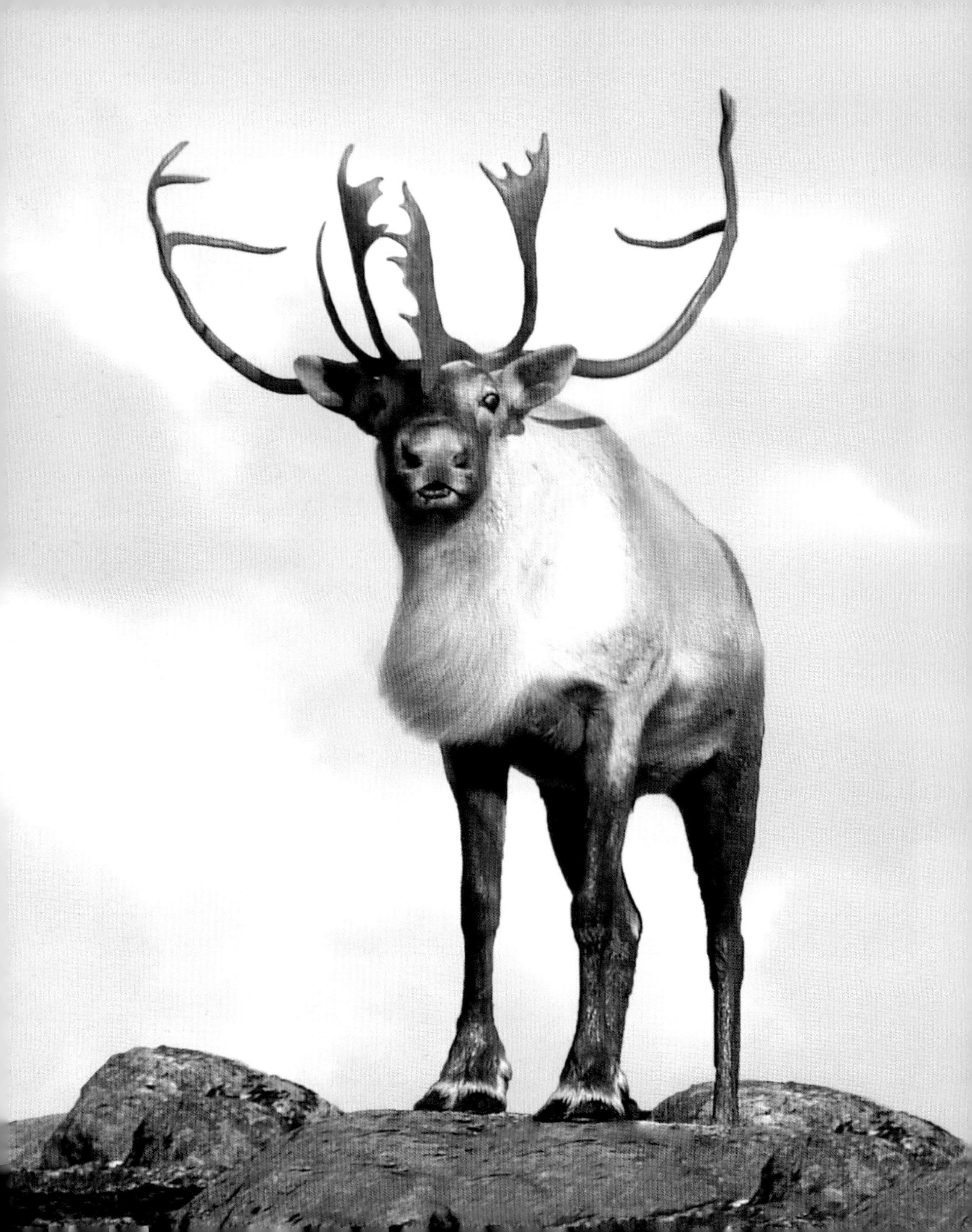

CHAPTER TWO

Mammals of Ungava and Labrador, Canada

This chapter contains the original text by Turner from his 1886 manuscript "Notes on the Mammals Ascertained to Occur in the Labrador, Ungava, East Main, Moose, and Gulf Region." This is the first time the contents of this manuscript have been published. The original manuscript page numbers have been retained and appear in brackets. Photographs of mammals collected by Turner while he was based in eastern Canada appear in this chapter, along with accounts, descriptions, illustrations, and stories of mammals by Inuit and Innu from the region. Historical and contemporary knowledge of mammals by the Inuit, together with accounts by non-Inuit, has been included to support Turner's fieldnotes and to illustrate Inuit and Innu connections with mammals on cosmological and practical levels. Wordlists of the Inuit terms for mammals that Turner documented between 1882 and 1884 have also been included, with the modern Inuit terms and orthography provided for historical reference.

Caribou on Fall Migration in October 2006
Taken at Kamestastin near the Quebec-Labrador Height-of-land border, in the barrenlands east of the George River. Courtesy of Stephen Loring, 2013.

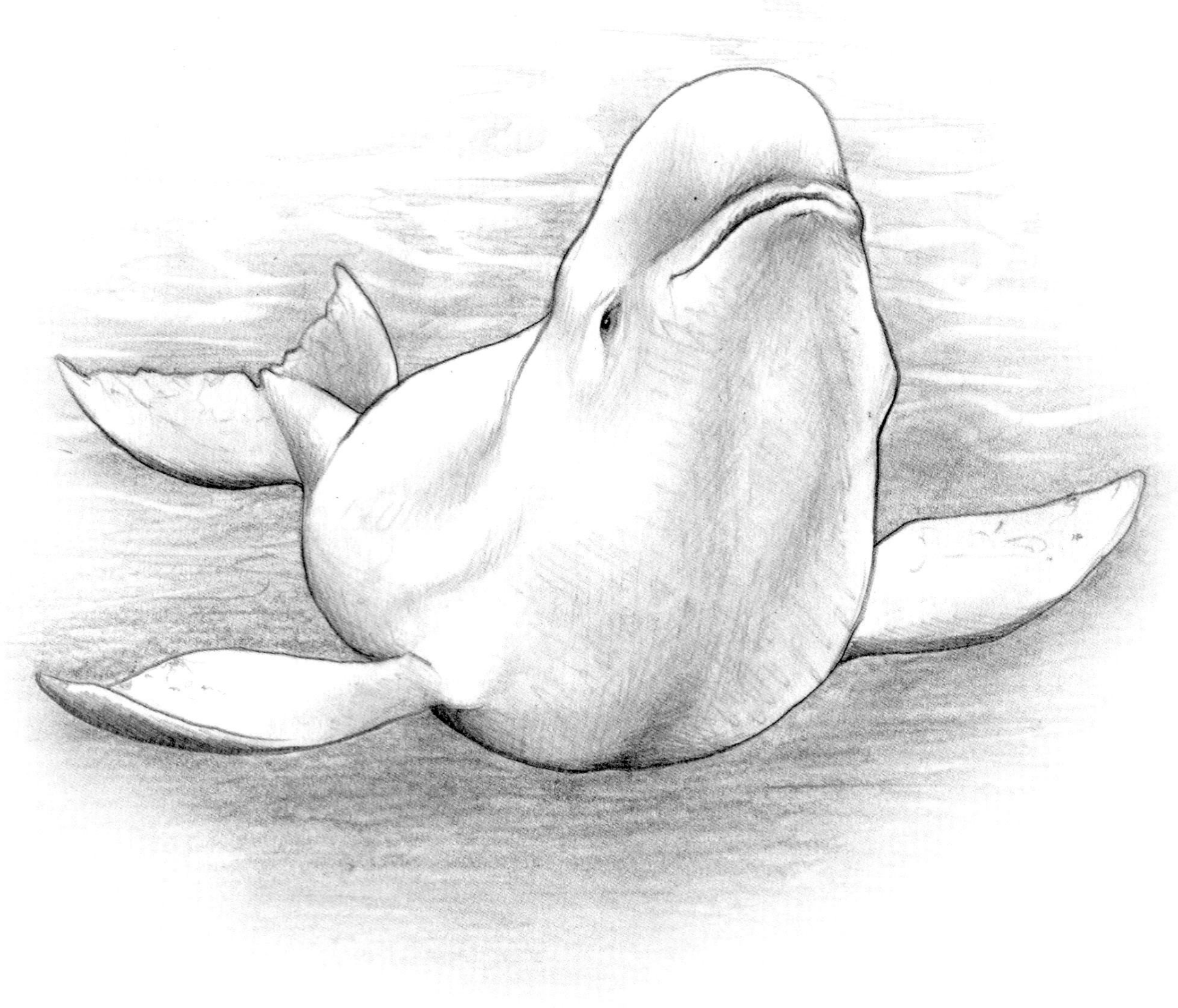

Cetaceans

Introduction to Whales and Dolphins

Turner's accounts of Ungava whales are fascinating. His manuscript provides formal sections devoted to three species, those most heavily utilized by the Inuit – the "white whale" (beluga), narwhal, and bowhead whale. Turner's accompanying notes demonstrate his familiarity with two other species in the region, the "Black Grampus" (the Long-Finned Pilot Whale, *Globicephala melas*) and a small porpoise (the Harbor Porpoise, *Phocoena phocoena*). It is not clear why he did not list separate species accounts in his manuscript for these latter two species.

Many additional cetacean species, not definitively recorded by Turner during his tenure in Canada, are known to occur in the waters off the Ungava Peninsula. A careful observer, Turner wrote in his manuscript, "Other species of whales are known to occur in the waters of Hudson Strait and along the Labrador coast. I do not attempt to determine the species as they were seen under such conditions that would render their identification purely conjecture." Among larger whales, these include the Minke (*Balaenoptera acutorostrata*), Sei (*B. borealis*), Fin (*B. physalus*), Blue (*B. musculus*), Humpback (*Megaptera novaeangliae*), Northern Right (*Eubalaena glacialis*), and Sperm Whale (*Physeter catodon*). Smaller cetaceans occurring off the Ungava shores include the Northern Bottlenose Whale (*Hyperoodon ampullatus*), Orca or Killer Whale (*Orcinus orca*), White-Beaked Dolphin (*Lagenorhynchus albirostris*), and Atlantic White-Sided Dolphin (*Lagenorhynchus acutus*). Turner's writings on the whales of Alaska (Turner 1886: 197-202) vouch for his knowledge of many of these species and their natural history prior to his Ungava travels – especially the Orca, Sperm Whale, and Humpback Whale – so their absence from his Ungava accounts surely stems from a lack of definitive encounters rather than a lack of familiarity on his part. Some of these species are relatively rare in the North Atlantic region, or generally occur far offshore (Turner 1888b), and some of them may have been rarer at the time than they are today due to impacts of 19th century whaling. Apparently only the Minke and Orca occur around the southern shores of Ungava Bay itself (Banfield 1974; Hoyt 1984), the area where Turner made most of his observations (though other species are occasionally recorded in Ungava Bay; Sperm Whale: Reeves et al. 1986; Humpback Whale: Higdon and Ferguson 2011). Of all the species not recorded in the manuscript, it seems most puzzling that Turner obtained no information at all about orcas, seemingly an indication of their rarity during his visit to Labrador and Quebec (Turner 1888b; Mitchell and Reeves 1988; Reeves and Michell 1988a, 1988b; Higdon and Ferguson 2009).

The only whale specimens that Turner collected for the Smithsonian were two belugas, an adult skull and a fetus preserved in alcohol, and most of Turner's notes on whales concern the beluga. Turner's manuscript and ethnographic notes combine into one of the most important 19th century accounts of the beluga: his account of the "white whale" is a rich source of information and a remarkable window into both the natural history and the economic and cultural significance of this animal in the eastern Canadian Arctic 130 years ago. Much of Turner's account concerns the Ungava belugas' summer journey from the waters of Ungava Bay into the Koksoak River, and the means by which these whales were hunted as they transited the river. Turner wrote: "As soon as the ice breaks from the shores of the bays into which large streams flow, the White Whale endeavors to ascend those streams. I could not satisfy myself with a reason for this essentially marine mammal ascending streams of water as far as brackish in some instances and on other occasions, if undisturbed, passing their way to the large lakes forming the source of streams not too long for the Whale to make a safe exit at the proper time." This riddle that puzzled Turner remains incompletely understood even today. As Stewart (1999:292) noted, "The current hypothesis is that belugas enter estuaries to molt. The warmer water allows increased skin temperature, accelerating the molt, and the lower salinity may make the process less irritating. It is after the molt, when the new skin of adults is shiny and white, that beluga are most striking in appearance."

Commercial exploitation of beluga by the Hudson Bay Company took place in Ungava Bay from 1867 to 1911 (Doniol-Valcroze and Hammill 2012). Turner describes how the beluga was hunted and how the skin, fat, oil, and meat of the whale was processed and used, both locally and for export. Turner's account provides a glimpse into the time period where traditional, sustainable hunting of beluga by the Inuit gave way to unsustainable commercial harvesting. This fishery reduced the Ungava Bay beluga population to very small numbers, from which it has never recovered. While some beluga are transient in Ungava Bay, the current resident population of beluga in Ungava Bay (the geographic population discussed by Turner that summers in the bay), thought at times to be extirpated, may now comprise no more than 30-40 animals, a sad change from Turner's day. Ungava Bay beluga are recognized as endangered by the Canadian government, and remaining animals are challenged by changing climate, contamination from pollutants, and ongoing hunting under traditional auspices (e.g. Dewailly et al. 1993; O'Corry-Crowe 2008, Doniol-Valcroze and Hammill 2012). We hold out optimism that protection and careful management of the Ungava beluga herds by all stakeholders will one day bring these beautiful whales away from the brink of local extinction and closer to their numbers and natural behavior in Ungava Bay prior to commercial exploitation.

Pod of Belugas
Taken May 13, 2011 during a spring bowhead abundance survey. Photo credit: Vicki Beaver, NOAA/AFSC/ NMML, Permit No. 14245.

ᖃᓄᓗᒐᖅ

WHITE WHALE

qilalugaq

Delphinapterus catodon (Lin., Gill.)

[1267]

ORDER NAME USED BY TURNER	*Cetaceans*
CURRENT ORDER NAME	*Cetacea*
FAMILY NAME USED BY TURNER	*Not recorded*
CURRENT FAMILY NAME	*Monodontidae*
SCIENTIFIC NAME USED BY TURNER	*Delphinapterus catodon (Lin., Gill.)*
CURRENT SCIENTIFIC NAME	*Delphinapterus leucas* (Pallas)
SYLLABARY	ᖃᓄᓗᒐᖅ
MODERN NUNAVIMMIUTITUT TERM	*qilalugaq (cf. Schneider p.297 - qilalugaq = beluga, white whale [Delphinapterus leucas])*
ESKIMO TERM BY TURNER	*ki l'i luak; kil lil u ak*
DEFINITION OF ESKIMO TERM BY TURNER	*White whale [Beluga] [p. 2298; 1287]*

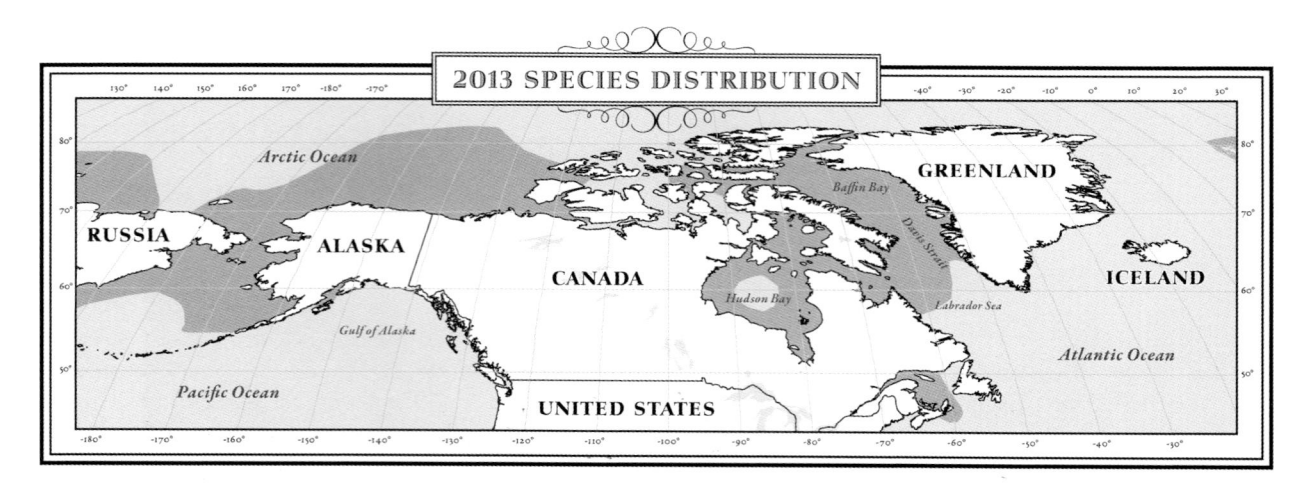

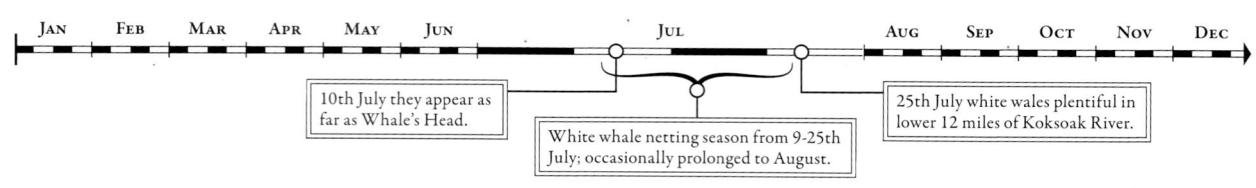

2013 SPECIES DISTRIBUTION

| JAN | FEB | MAR | APR | MAY | JUN | JUL | AUG | SEP | OCT | NOV | DEC |

- 10th July they appear as far as Whale's Head.
- White whale netting season from 9-25th July; occasionally prolonged to August.
- 25th July white wales plentiful in lower 12 miles of Koksoak River.

CHAPTER TWO: MAMMALS OF UNGAVA AND LABRADOR

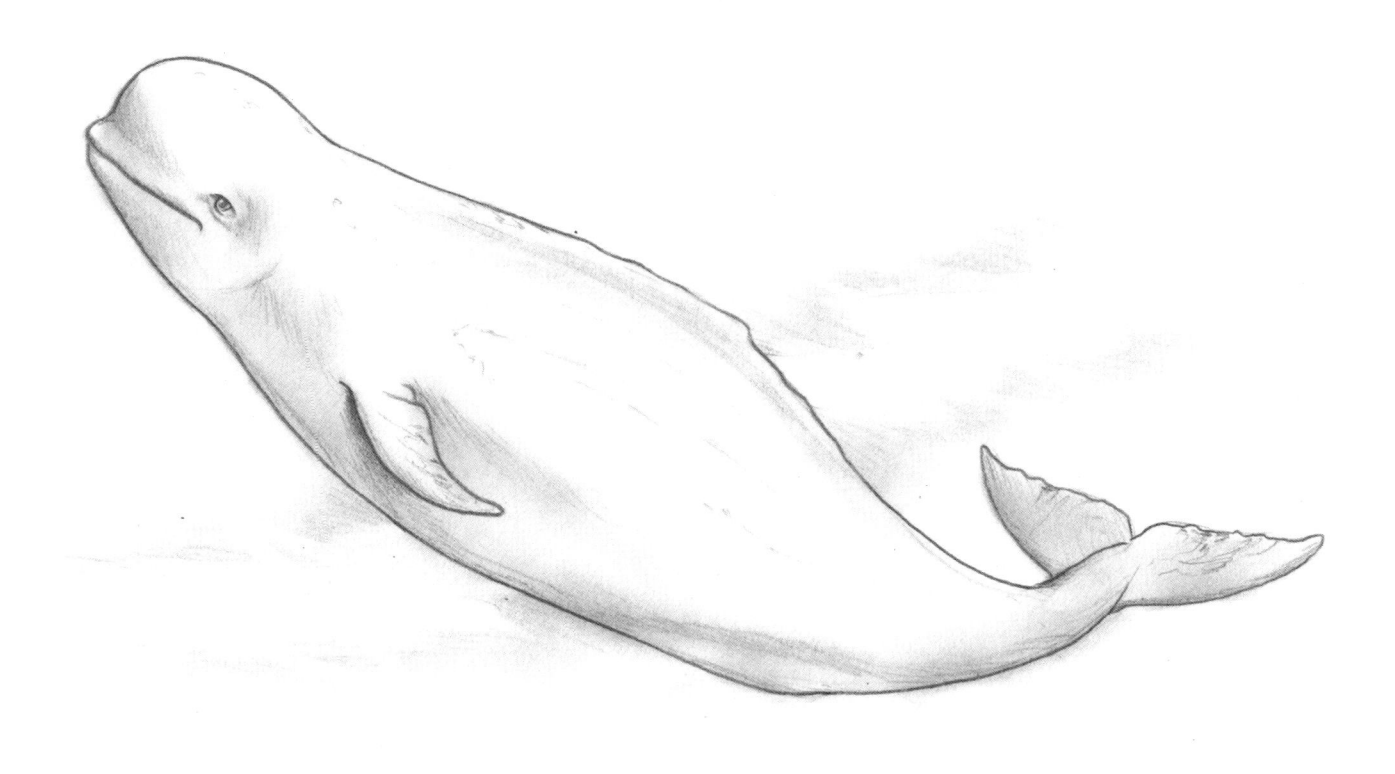

Descriptions by Turner

The White Whale was observed in hundreds along the northern shore of the Gulf of St. Lawrence in the middle of June, 1882. After passing Anticosti Island not a single individual was observed until we entered Hudson Strait. It occurs along the entire coastline of Labrador, but as the vessel in which I ascended that event in June and July encountered so much ice and fog that water was but rarely visible until after entering the Strait, that few opportunities presented themselves to observe the presence or absence of this species.[2] As soon as the ice breaks from the shores of the bays into which large streams flow, the White Whale endeavors to ascend those streams. I could not satisfy myself with a reason for this essentially marine mammal ascending streams of water as far as brackish in some instances and on other occasions, [1268] if undisturbed, passing their way to the large lakes forming the source of streams not too long for the Whale to make a safe exit at the proper time.

---

2. Inuit report the presence of beluga in the George River as far upstream as Helen's Falls. Inuit would hunt for beluga from the shore at a place called Hubbard Point, which was near the George River Hudson's Bay Company Trading Post. Along eastern Ungava Bay, beluga whales frequent the Korac Tunilic, Tuttutuuq, and Mucalic Rivers, Gregson Inlet, Alluviaq Fiord, Cape Kattaktoc, and especially the Killiniq region. Beluga whales move upriver as the tide rises and depart again to the safety of Ungava Bay as the tide recedes. Ungava Inuit would hunt beluga with a harpoon and kayak in August. Hunters say that beluga whales disappeared from their typical migratory paths once motorboats were used in the North. (Makivik 1984).

In the Koksoak River they rarely ascend even as far as Ft. Chimo which is just at the head of brackish water of highest tides and fresh at low tide. Some three and a half miles below the station the river suddenly contracts, having a wide expanse above the narrows and a wider expanse below. The single rock locally known as "Whale Head" forms the western or left side, while "Hawk's Head," forms the eastern or right side of the narrows. These walls of rock, when viewed from certain locations, have the appearances indicative of their respective names. Above this narrow place the Whale rarely ascends and then only when it observes danger below it. Schools of these whales enter the mouth of the river as soon as there [1269] is sufficient water to enable them to swim. It is usual for one or a few of these creatures to appear several days before the general run. The advance guard lingers for several days and returns to the sea. In the course of a few days numbers may be seen in the bay off the mouth of the Koksoak.

By the twenty-fifth of June they are numerous, playing within the lower twelve miles of the stream. Their numbers increase, and by the tenth of July, they appear as far as "Whale Head." Some three miles below "Whale Head" is a house furnished with all appliances for the capture of these mammals. About four hundred yards below it a point of rock juts into the water and is directed upstream. At a distance of near two hundred yards above this tongue and somewhat more out in the stream is a huge boulder or detached rock which has withstood the countless years of surging of the stream of ice several feet thick passing and repassing [1270] with each tide.

About two hundred, or less, yards above it is a similar boulder situated so that it forms a line with the point and the lower boulder. The rocks are something like seventy-five yards from the bank of the river. The peculiar position of these rocks, in the vicinity of the sporting water of these whales, is fortuitous. From the upper to the lower boulder is stretched a heavy net made from

"The whales are like a flock of sheep and where the leader goes the remainder are certain to follow"

Chapter Two: Mammals of Ungava and Labrador

> ### THE JEALOUS MAN
> *An Inuit story, Turner (1887a: 288)*
>
> A man who had two wives was so jealous of them that he would not permit them to look upon another man, and to speak to a man was certain to enrage him beyond measure. The women became tired of the restraint imposed upon them and they resolved to run away. They fled along the coast and in the course of a few days came upon the carcass of a stranded whale. Here they determined to remain as a supply of food was nearby. The man searched in every conceivable place for the truant women but to no avail. A shaman was consulted and after much deliberation the man was directed to journey to a place where he should find the remains of a cast-up whale and to secrete himself in that vicinity to observe signs of the two women. He went and before long had the satisfaction of seeing them approach to obtain some food. The man was eager to overtake them that they saw him approach. They turned and fled but he overtook them and compelled them to return with him and on arrival at his village he put out their eyes to prevent them running away.

stout cotton line, the meshes of the net being fourteen inches. The height of the net is sufficient to touch the bottom at high tide, which at this locality is near forty feet rise and at extreme tides covers the rock several feet in depth. The bottom is quite smooth formed by sand and mud thrown out by the deflected waters swirling through the narrows above. From the second boulder to the point below a similar net is stretched but as the point is longer than the height of the rock [1271] above it becomes necessary to station one or more persons at that place lest the whales make a break for that locality when driven within the pond. Encompassed, in part, by the point is a small cove or pool which becomes dry at low water. From the upper rock to the shore a net termed the "Gate" is stretched.

As the whale is more venturesome and more easily driven during the approach of high tide, several natives (Eskimo) are directed to descend the river to search for whales. If any are to be seen, the Eskimo endeavor to creep along the shore until they are below them. They now carefully approach the school and strive to urge them to ascend the river where, at convenient stations, other natives join the party and by the time the whales are opposite the pond everything was made ready for the drive. The party in the boats (Kaiaks) drive the whales above the mouth of the pond [1272] and at a given signal turn them down and keep them near the left bank of the river. The "gate" has already been prepared by sinking the netting which formed it to the bottom and by tying cords to it with stones attached; the back rope leading to a windlass on shore. The cords holding the net are not so stout but that by the force of the windlass they are easily broken and the buoys attached to the net cause it to rise to the surface. A person is usually stationed near each of the rocks and sometimes one between to prevent the whales from breaking through the net. The utmost quiet must be observed as the animals approach the vicinity for the least noise is certain to cause them to break. Dogs are kept from the place lest their bark at the wrong time drive the whales beyond reach.

The whales are like a flock of sheep and where the leader goes the remainder are certain to follow. If a portion of them enter over the [1273] gate, the remainder often returns and follow those already within the pond. As soon as a sufficient number is within the windlass is started and the gate is soon above the surface, effectually barring egress in that direction. The whales soon discover the trap and make the wildest efforts to break through the netting which if the creature has but a short lunge to make at it, it will withstand his strength. Occasionally one more terrified than others becomes frantic, jumping from side to side of the pond and if those who drove them do not hasten to the sides of the net a rent will be made through which more or less will escape. The tide is now going down and the Eskimo now begin the attack. Their lances similar to those used in spearing reindeer are brought into requisition and the animals speared each time they appear above the surface.

Beluga Whale Skull
A beluga skull collected by Turner, probably from Ungava Bay in 1882. USNM A23208, Division of Mammals, National Museum of Natural History (NMNH). Photo by Angela Frost, 2011.

They are now pursued and the work of killing continues until [1274] the victims cease to struggle. When low water occurs a boat is manned with a good supply of ropes and these are affixed to the bodies of the whales and as soon as the tide begins to rise, the bodies are drawn to the beach and dragged upon shore.

White Whales are very erratic and no one is able to know beforehand how they will act. The appearance of the sky overhead is nothing. The water in which they sport is nothing for they may be driven in rough weather as well as calm, although calm days are best as they allow the least ripple on the surface to be discerned. The whales may be driven to almost the very head of the net and some freak or panic, not discoverable, cause them to rush wildly beyond control. If persecuted they will leave the river for several days and not enter it until their fear has subsided. Their actions are at times unaccountable and often most tantalizing but when attempted to be driven under such circumstances [1275] but little success attends. The result of hours labor often ends in disappointment at a moment when everything appeared most favorable. How they act this year is not how they will drive at the next season.

It not infrequently happens that only a score or two will be taken when hundreds are within sight. Again the nets have been known to enslave over a hundred at a time. The average is perhaps seventy-five for the season, lasting from the 9th to the twenty-fifth of July. Occasionally the season is prolonged into August, but as the salmon season is usually so near at hand, preparation must be made for that work. When a great number of whales have been taken at a single haul, it requires a period of several days to dispose of the carcasses.

In the capture of these whales, there is a two-fold object, the oil and skin. Both of these parts are removed at the same time. The blubber adhering closely to the skin and rather loosely to the flesh. A [1276] longitudinal incision is make from the throat to slightly beyond midway of the caudal peduncle. The skin and blubber are now removed from the slightly adherent dark flesh. The unwieldy animals occasionally requiring turning in order to get at the blubber under the superimposed carcass.

If the animal be an adult the size is so great as to necessitate the skin being cut into two pieces for convenience in handling. The skin is then cut along the back, forming "sides" of skins. The skin of the head, tail and fins is not taken off, those portions forming waste. Each side or skin shows either one or two holes, those made by cutting around the wrist of the pectoral fins. The skins are now dragged to the water and allowed to awash with the tide. At the next rise of the tide, the skins are taken on boats to the station of Ft. Chimo, some seven and a half miles farther up the river. Here the skins are taken from the vessel and [1277] either temporarily stored or else the blubber is immediately removed from the skins. It is cut into strips of various sizes, depending on the thickness of the deposit of fat, and run through a mill for the purpose for breaking down the oil globules. A vessel below receives the oil and crushed blubber. From there it is poured into huge tanks of nearly cubical form, prepared from stout galvanized iron. In these tanks the fat is allowed to remain until the fleshy portions settle to the bottom and the clear oil remains at the top. This oil is dipped from the tanks as occasion requires and put into oak casks of forty to forty-three gallons. It is then known as pale oil and commands a higher value than the remainder which is boiled in large kettles arranged for the purpose. The operation of rendering the oil by heat tinges it a brownish color which causes it to be sold for an inferior price. The rendered pieces of fat are put into a press and the last [1278] drop obtainable is squeezed from it. The casks containing the brown oil are marked in order to distinguish them. A cooper is engaged for a term of years to prepare the casks for the oil and when this task is finished performs the general service exacted from all the servants of the Company.

As it is necessary for the Company to retain a number of Eskimo dogs in their service to perform the various labors for which teams of other animals are employed in other localities, there must be laid by an amount of food for these brutes. The heads, fins, tails, and strips of flesh cut from the carcasses of the whales are taken to Ft. Chimo and stored in barrels placed in pits dug in the ground where it is covered first with brush and then with dirt. The delay often occurring before these two

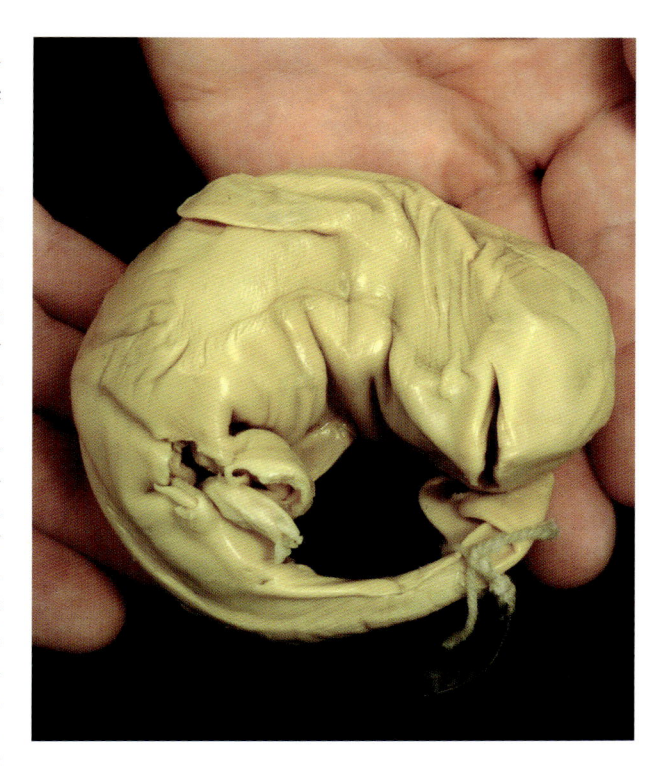

Beluga Whale Fetus
The male beluga fetus was removed from a beluga harvested by Inuit or Hudson's Bay Company employees at "Whale Head" on the Koksoak River in July 1884. Young mammals were often collected in delicate circumstances, as apparent in Turner's manuscript (1886: 1282): "A young one was obtained, from the mother, having a length of less than seven inches and a perfect whale in its exterior outlines. As the Eskimo prize this mammal very highly and are very reluctant to have the young carried away it was a very difficult matter to undertake investigations to determine certain facts without incurring their displeasure and have them quit the field; which, if done, would seriously interfere with the affairs of the Company." The fetus remains well preserved in alcohol. USNM 14867, Division of Mammals, NMNH. Photo by Angela Frost, 2011.

last operations are attended to give the blow fly an excellent opportunity to deposit innumerable eggs upon the meat. In the natural course of events the ground [1279] becomes literally a moving mass of squirming maggots. I was well pleased at this course adopted as that locality was an excellent spot to collect the fall birds which invariably stopped a few days to partake of the fat larvae of those flies. Wagtails, sparrows, Rusty grackles, an occasional raven and gull fell to my lot at this attractive point. The blubber of the Whale varies according to the size and condition of the creature. The thickest deposit measured was but slightly more than four and a half inches and from that thickness down to less than two inches. The thicker the fat the cleaner it is. The color when fresh is very pale cream but becomes pale yellowish in the course of a day or two, the temperature and exposure having much to do with it. To obviate the light discoloring the raw oil in the tanks they are usually shaded if not covered.

In thrusting these whales with the lance a death blow may be given and yet not allow the [1280] animal to bleed thoroughly. The congested blood remaining in the smaller veins have also a tendency to discolor the oil.

The flesh of the whale is a dark color, appearing quite repulsive as an article of food. The Eskimo strip portions of the flesh from the bones and partially along it for food. They, however, do not lay it by for future supply but only for the immediate future and usually have none in two or three weeks after the whaling season is over.

Of the many whales of this species opened by me, I failed to discover any food within the stomach in either an undigested or other form. The contents, if any, consisted entirely of a semi-viscid yellowish brown fluid. The lower intestines never contained fecal matter. I was thus led to conclude that the purpose of their coming into the fresh water was not to obtain food. Schools of these whales certainly remain for several days in the river and do not during that [1281] period go to the sea, proving that they may abstain from food for that length of time.

I did not detect any parasites upon any portion of the body or infesting any of the organs or flesh. The size of the white whale as found in Hudson Strait varies though not so greatly among the fully adult individuals. Males of fourteen to fifteen feet were not rare. The females being but slightly less in size. The circumference measuring as much as nine and a half feet just anterior to the dorsal. The smallest individual measuring less than three feet in length. The weight of an adult male in good condition will reach not less than sixteen hundred pounds while the female will attain as much as fourteen hundred. Every size from the huge male of nearly snow white (ivory white is nearer the color) to the small one of dark lead blue may be [1282] found in the same school.

Fluid Specimen Storage at the Smithsonian
Charley Potter, the Smithsonian Institution's Collection Manager for Marine Mammals at the NMNH locates the beluga fetus at the NMNH Museum Support Center, Suitland, Maryland. Photo by Angela Frost, 2011.

The coming together of the sexes apparently takes place at any time of the year. The female lying with back down to receive the male which glides along her, pausing but an instant in the act. A young one was obtained, from the mother, having a length of less than seven inches and a perfect whale in its exterior outlines. As the Eskimo prize this mammal very highly and are very reluctant to have the young carried away it was a very difficult matter to undertake investigations to determine certain facts without incurring their displeasure and have them quit the field; which, if done, would seriously interfere with the affairs of the Company. Such studies having to be done surreptitiously or in such a careless way as not to attract attention.

A certain individual of a few years before had created great consternation among all the people by his disregard for propriety; and, [1283] as one must, if he desires to obtain the good will of an Eskimo, exercise a due amount of circumspection and discretion in his relations with those people, it was necessary to not be too rash in obtaining information. The young individual is now in the collection of the National Museum. [See USNM A14867.]

To return to the skins which have been placed in barrels to undergo a process of fermentation which reaches the desired condition in about two weeks, although this depends much upon the state of the weather. Cool, murky weather retarding the process and dry clear weather hastening it. When the skins are ready to be cleaned from the blubber they are laid upon the ground with the fatty side up. By means of sharp knives the Indian women remove it. The fat or oil collecting on the skins, undergoing the operation, is from time to time removed and ladled into buckets standing [1284] conveniently by. As much of this fat as can be removed from the skin is taken off and the skin itself then placed into the barrels before mentioned. After a time the skins are inspected and if the scurf collected upon the exterior of the skin is easily detached the skins are now given to Indian women to clean. The process of cleaning these skins is exactly like that of removing the hair from the deerskins mentioned in another connection. The same implement is used for removing the scurf skin from the whale. After each skin has been cleaned, it is washed and salted, put into barrels and shipped to London where the skins are sold at a price averaging thirty shillings for each "side." These skins undergo a special process which converts them into a quality of leather commanding a great value and of surprising durability for footwear.

As much as eighty gallons of oil [1285] has been taken from a single white whale. The value of the oil varies from thirty-five to sixty cents per gallon and is used for an infinite variety of purposes.

The food of the White Whale is unknown to me so far as my own observations extend. What others have seen has no bearing on the subject in this connection. An occasional individual is taken in the nets set in August for salmon. Such captures fall to the lot of the one in charge of the net and to him extra compensation is paid. Where several of these creatures attempt to force a passage, behind some of the runs between a small island and the mainland, blocked by a salmon net, great havoc if not an entire destruction of the net is made as the receding tide warns these whales to hasten to the open area. An occasional individual is found caught, in some one of the numerous basins in those sloughs, by the water flowing out so rapidly as to prevent exit.

Inuit Hunting Customs
Inuk hunter, Elijah Annanack, stops passing boats in Ungava Bay to share his catch of a beluga. The Inuit consider the soft-textured skin to be a great delicacy. One of the customs associated with the hunting of belugas in Labrador was recorded in Moravian journals in 1883. One entry indicated: "According to an old custom, one half of a white whale is divided amongst the men of the village, and the other half belongs to the owner of the net in which it was caught" (PA 1883: 587). Photo by Scott Heyes, summer 2010.

[1286] The Eskimo often secure the whale by creeping stealthily toward a school and selecting one as a victim. The greatest caution is necessary and as the pursuer knows every habit of the creature, he has only to follow it until it is exhausted by keeping it under water and not allowing it time to regain full breath when it comes to the surface for the purpose of exhalation and inspiration. The large seal spear is used and when a convenient

"The Eskimo often secure the whale by creeping stealthily toward a school and selecting one as a victim"

opportunity occurs, the whale is struck and the line with float attached is quickly released and allowed to retard the progress of the whale as much as possible. If this occurs within the river the pursuer endeavors to keep the animal headed upstream or against the current in water that the *ávatuk* or float may the sooner exhaust the victim. When signs of fatigue are shown, the Eskimo redoubles his exertions and by coming alongside the whale, is enabled to plunge the hand spear [1286] or a knife into the body.

If the capture take place near the trading post, the skin and blubber is immediately taken there and sold. Among the distant Eskimo the capture of a whale is considered a valuable acquisition. The fat is placed within bags of sealskin and the meat consumed on the spot. The Eskimo apply the name of *kil lil u ak* to the White Whale. I have often heard them pronounce the word with the accent on ú making the word sound quite differently; and, suspected that they were attempting to imitate the Whiteman who always places the accent on the penult of that word. Ludicrous instances of the mispronunciation of Eskimo words occur and at times they are so perverted as not to resemble in the least, the proper sound.

Whale Harpoon
Illustration of a whale harpoon model with attachments that was used by Ungava Inuit (USNM E90103). Illustration originally produced for Turner's BAE Report (1894, Fig. 138: 314). Mason (1902: 257) described this harpoon: "The shaft is wood, the foreshaft is bone. The base is wedge-shaped, and fits into the slit at the end of the shaft, being held in place by a lashing of sinew cord. On the foreshaft fits the toggle head, with iron blade held fast by two rivets. The body of the toggle head is rectangular in cross section. The line passes through the sides and is not seen on the lower part. The wide barbed end is cut into three or four tooth-shaped parts. The line is of rawhide, plaited. The peculiarity of this harpoon is a board, somewhat circular in form, on the lower end of the shaft, which acts as a drag to the wounded animal, in place of a sealskin float. The line passes between this board and the shaft, and has a handle or toggle fastened at the other end to be held in the hand of the fisherman." Illustration source: SI Archives, Record Unit 7192, Turner, Lucien M, 1848-1909, Lucien M. Turner Papers.

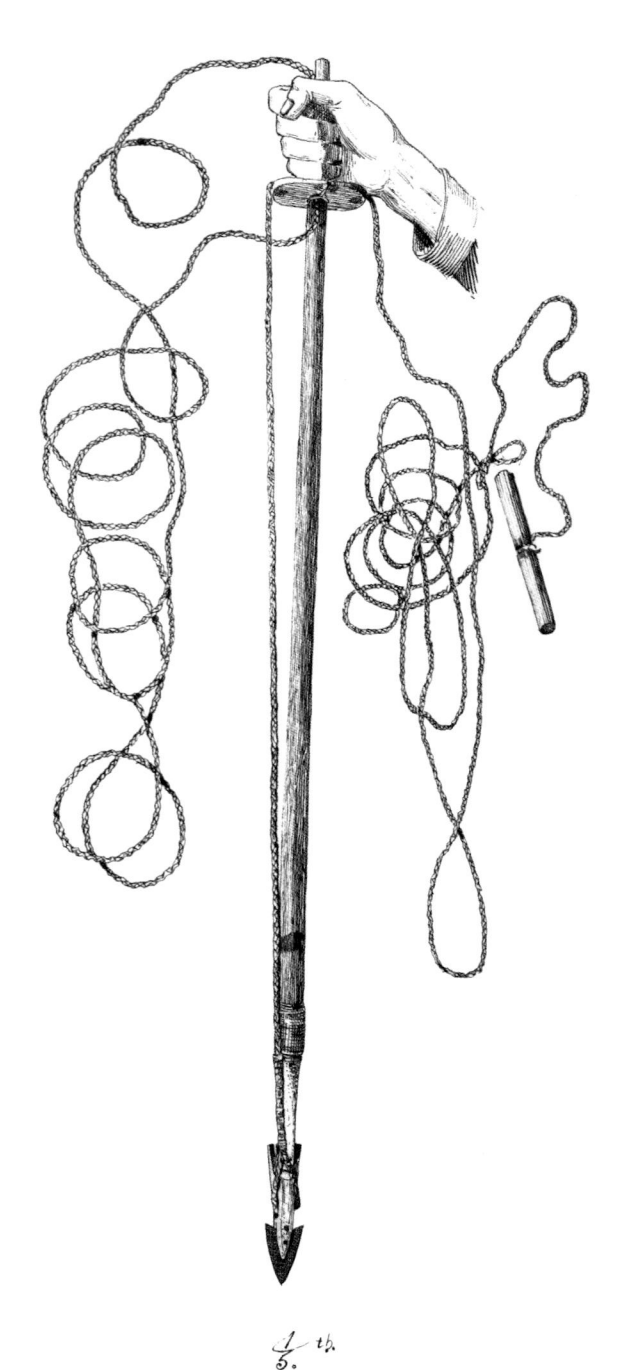

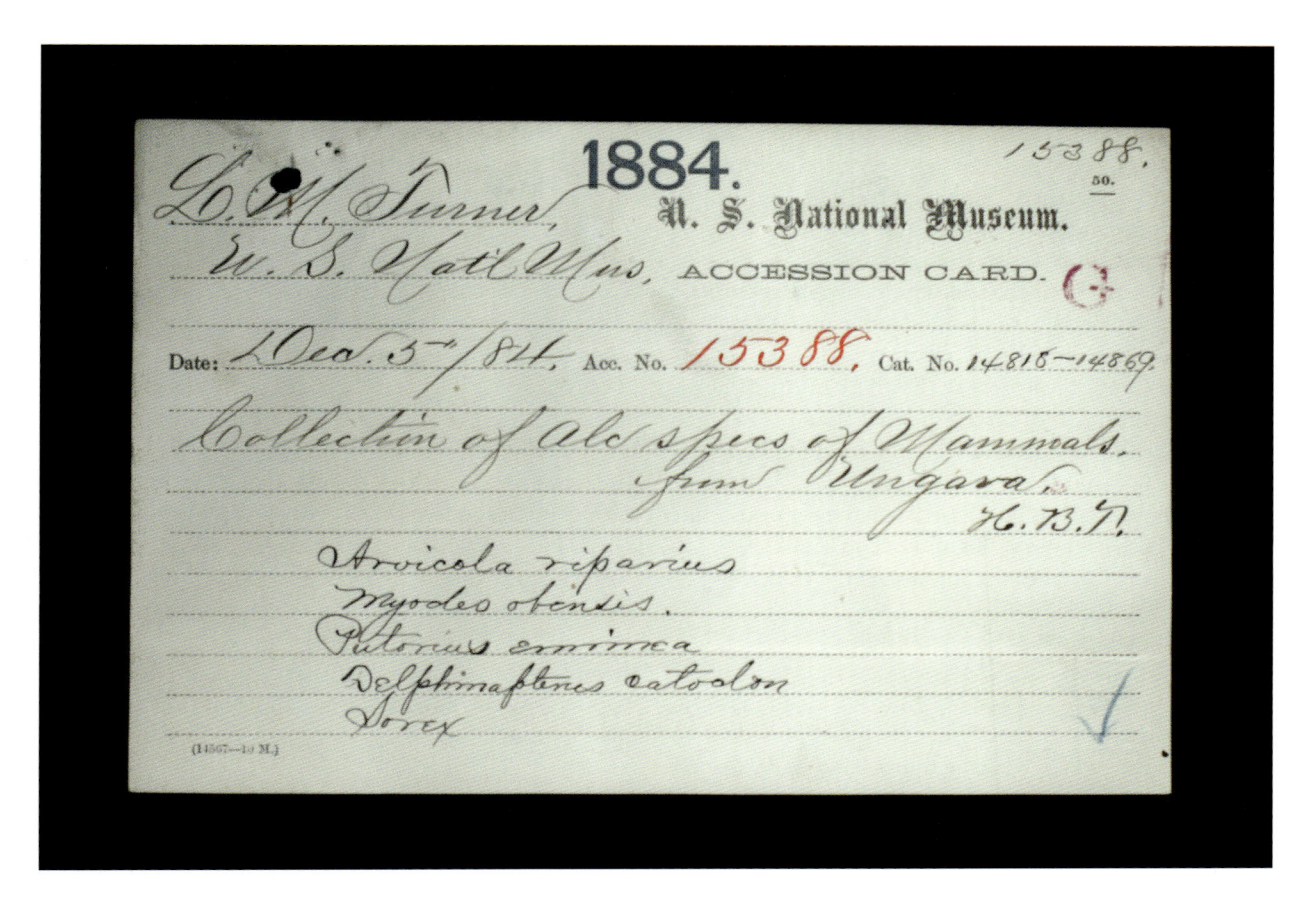

Accession Card Describing Some Mammals Collected by Turner
The card mentions the beluga fetus (see previous page) was received at NMNH on 5 December 1884. The accession card 15388, pictured, indicates that 55 specimens (catalog numbers USNM 14815-14869) collected by Turner were also preserved in alcohol. Source: SI Archives, Record Unit 305, Accession 15388.

Beluga Ethnography
Turner

Beluga Skin (1887a: 106-107)
The sealing season is over by the first of July except such straggling individuals that may be seen at any time during the period of open water. By this time the white whales are showing their cream white sides like a piece of ice lifting with each undulation of the water. The flesh and fat of those mammals are prizes valued highly by the Innuit. The skin makes a soft cover for the khaiak or tops for his boots while the stomach affords a receptacle for the fat taken from its body. These oil bags, like those from the seal, have and represent certain values, depending on the kind of oil and the size of the receptacle.

Beluga Oil (1887a: 107)
The skin of the huge Bearded Seal may contain as much as four hundred pounds of oil while the stomach of the white whale contains seldom more than forty pounds of oil that has a trading value of double in quantity. After the spring season hunt is brought

to a close the bags of oil are stored in places where they will be safe from the ravages of predatory beasts. The mouse and ermine are most destructive, for those diminutive creatures climb all sorts of places to gnaw at the bags and cause the contents to flow out, covering the vicinity with the rank oil. A quantity of this oil is reserved for household purposes and the remainder, if not too remote from the trading station, is sold to that individual for other articles necessary for the Innuit to prosecute the chase of other game, now obtained in different manner than previous to the advent of the trader, who has exerted himself to cause the Innuit to become more or less dependent upon him and thus be necessary to the very existence of the native whose tastes and occupations are being rapidly subverted to the interest of the trader.

BELUGA SPEAR (1887A: 153-154)

[A] short spear [USNM E90164/3194] is used [by the Koksoak Inuit] for stabbing white whales and seals that have been either wounded and slow to die or else when white whales have been driven into a shallow arm of the sea which when the tide ebbs shoals that the backs of the creatures are visible. The Innuit paddle after the whales and with this instrument stab them repeatedly or else sink the point so deep into the flesh that the spear is imbedded and becomes disjointed at the coupling and the struggles of the animal cause the point to twist aside and the head of it to catch under the skin.

A line of variable length is attached to the rear end of the short wooden shaft to prevent the instrument from dropping overboard and also to retard the progress of the creature in the water by the line being attached to some part of the Khaiak or to a small float made of inflated intestine or skin of a seal. The hand spear consists of a wooden shaft, of about sixteen inches in length, having on the anterior end a head of ivory which forms a socket into which the posterior portion of the walrus tusk point of eight inches length, fits and is held in position by means of two-thonged attachment or connectings passing through perforations in the ivory point and into the wooden shaft so as to make a joint capable of lateral motion or unhinging.

The Innuit endeavors to pierce the body of the whale near the arms or pectoral region for here the skin is thinner and the wound soon causes that cavity to fill with blood which results in speedy death.

Harpoon Head for Hunting Whales and Seals
A harpoon head collected by Turner at Ft. Chimo, probably in 1883. He wrote that "The Innuit paddle after the whales and with this instrument stab them repeatedly or else sink the point so deep into the flesh that the spear is imbedded and becomes disjointed at the coupling and the struggles of the animal cause the point to twist aside and the head of it to catch under the skin." USNM E90164/3194, Museum Support Center, NMNH. Photo by Donald Hulbert, 2013

Woman Who Turns into a Beluga Whale
An Inuit story, Tivi Etok (Heyes 2007)

This drawing is about a woman who turns into a beluga whale. A blind Inuk boy was living in an igloo with his mother and two younger sisters. The boy's mother was not always good to him. One day the family was bothered by a large polar bear that tried to knock down their igloo. The mother told her son to kill the bear with his bow and arrow, even though he could not see. He took aim by pointing his arrow through a small hole in the igloo. Using his very good sense of hearing, the boy fired off an arrow to where he thought the bear was standing. His arrow struck the bear, but his mother told him that he had missed and instead killed their dog. The boy was sure he had killed the bear because he heard it slide on the snow when it collapsed. The mother insisted that it was their dog. Just to make her son believe her she secretly killed the dog and asked him to touch it as proof. The boy knew his mother was lying.

The family had been close to starvation up until the moment the bear was killed. The family feasted on the polar bear meat in secret, away from the boy. His mother gave him dog meat instead. The boy's sister was not happy with her mother feeding him dog meat so she would secretly provide him with polar bear meat when the mother was out from the camp. His sisters kept the boy in good health, unbeknownst to their mother.

One day in summer, two loons landed on a lake when the boy was fetching water. The two loons were not regular loons. The two loons spoke to the boy and explained that they would help him to see. The boy was very excited about the prospect of gaining the use of his eyes. He agreed to follow the instructions of the loons. In order to gain his vision the two loons put him under the water. They said to the boy: "When you are starting to run short of air make a sudden movement so we will know when to bring you up to the surface." When he made the sudden movement and they brought him up for air the loons asked the boy what he saw. The boy saw nothing. So, the loons told the boy that he should hold his breath underwater again, but this time when you make a sudden movement we are going to keep you there for a short while longer. After this second dive, the loons asked the boy what he see, and the boy replied, "I see a little bit of land." The loons told the boy to dive for a third time, but explained that when you are struggling to find air we will keep you down a little longer. So the boy dived again. When he surfaced the loons asked him, "What do you see?" The boy replied, "On that far away mountain I see green grass and lemmings holes." The boy's eyesight had become too sharp. So the loons told the boy that if he dived in the water and resurfaced again that his eyesight would become normal.

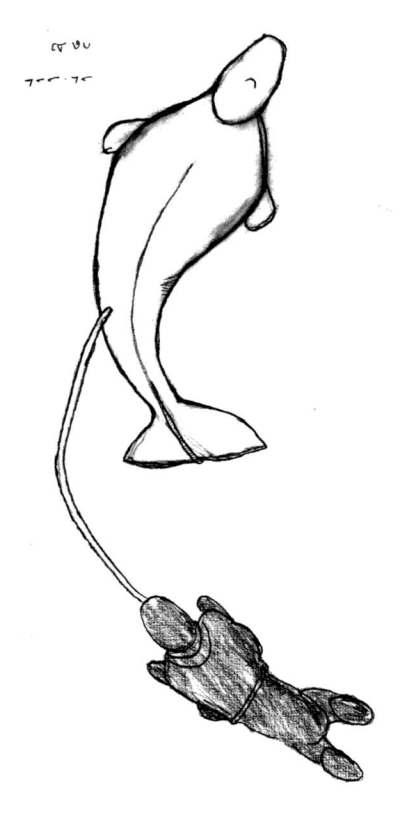

An illustration of the story of a woman who turns into a beluga whale. By Kangiqsualujj-amiut elder Tivi Etok, 23 May 2005. Reduced from 19" × 12" B&W Charcoal on Canson paper. Scott Heyes Private Collection.

So the boy went home pretending to be blind. He saw the polar bear's skin; the bear that he had shot earlier. The boy faked that he was blind but in reality he saw everything. When the mother discovered that her son was no longer blind, the boy took all of her belongings that were of use to her like meat and skins.

In the Bible it says that you should listen to your parents and respect them, yet the boy was not following this principle, he wanted to pay back his mother for the way she had treated him.

The family went beluga hunting not far from the tent. The son tied a rope around his mother telling her that when they harpoon a beluga whale everyone will be pulling at the same time to retrieve the whale. The mother asked the son to kill a baby grey beluga because she said to her son that they are easier to pull and are weaker than a full-grown beluga. But the son told his mother the opposite, "we are going to hunt a white beluga because they are easier to pull and are weaker." The boy was telling a lie to his mother because he was still annoyed that his mother fed him dog meat instead of polar bear meat. So, the son harpooned a white beluga. The mother immediately began running towards the water because she knew that the pull of a white beluga was powerful. The rope was long, but in the picture I have drawn it short. The others did not grab the rope to assist. So, the mother was pulled into the water. Even when the mother was in the water with the beluga, her children did not feel inclined to help her because she was responsible for doing bad things to her children.

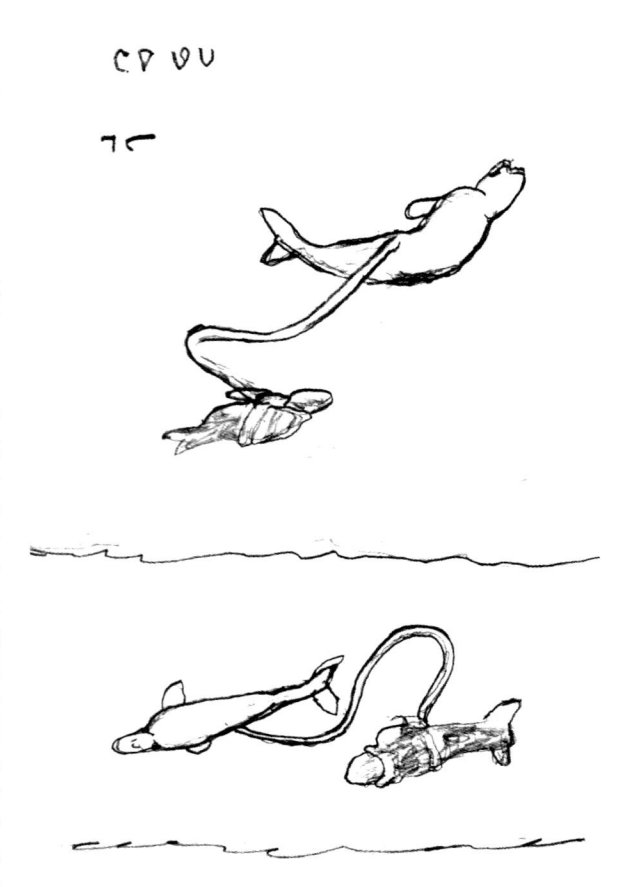

An illustration of the transformation of the woman who turned into a beluga whale. By Kangiqsualujjamiut elder Tivi Etok, 23 May 2005. Reduced from 4 $^{23}/_{32}$" × 7 $^{3}/_{32}$" lead pencil. Scott Heyes Private Collection.

The mother then left with the beluga. When the mother was in the water she would yell out "Lumauk" (pronounced Loo-maa-ook), "Lu-Lu-Lumauk." These words were heard from great distances; the sound travelled a long way. I (Tivi) think that she was saying "my son," because it was her son's actions that caused her to be attached to the beluga. My father told me that he used to hear that noise. My father even saw the relatives of the lady-beluga along the Ungava coast.

This lady was alive for a long time. The rope became beluga fat and that is how she survived and transformed into a half-human, half-beluga creature. The rope never snapped off; it remained like an umbilical cord. She was able to breathe through the rope that was connected to the beluga.

This lady beluga was still around about 60 years ago. I (Tivi) heard a story from someone that they saw the lady-beluga and that people had planned to go and kill it because they felt sorry for a human being in the water. So they killed both her and the beluga and buried her on land like a regular human would like to be buried. I have never seen the grave, however. I believe that she is dead. She must have regretted doing bad things to her son.

But when he does not find her, he thinks she was drowned, turns back, climbs out on the land and becomes a whale's bone again. The two girls, however, from that time on obeyed their parents.

Recollections on Inuit Use of Beluga Whale
Tivi Etok (Heyes 2010)

The tail fins of a beluga whale, in Inuit tradition, is cut up by the hunters and saved for Inuit women and men and shared amongst them.

Inuit used to eat the muscles in the tail part first. The sinew in the tail was saved for threads to make kamiks and mitts. It is very strong sinew. If we ate the sinew in this part of the tail we wouldn't have had any thread. All the meat of the beluga was eaten. The beluga whales stomach was retained and carefully removed so that it was not damaged. The stomach was converted into a bag. Inuit used to place the white meat and the fat inside this bag to save food. Even boiled *muktuk* (fermented whale fat) was placed in the bag. No meat was wasted. If we had buckets available, we also used to place meat in this, too.

The outside of the beluga whale skin (the white section before the fat) was saved and cut to become material for tents, mitts, and kamiks. All the meat of a beluga can be dried and hanged. The meat was preserved and stored for winter supplies and as dog food. The intestines were dried and hanged. There was no *Qallanat* (non-Inuit people) then. This was our only food. The beluga fat was used as a fuel to cook over a lamp. A small piece of fat was put on the soapstone lamp and lit, which produced a small flame. We only used it as fuel in winter; in summer we were able to collect small twigs along the beach for fuel. There are so many useful things about the beluga whale.

The snout of the beluga is used by the whale to break thin ice to make a breathing hole. The seal has nails but the beluga does not have nails to scratch to make a hole. Beluga whales can stay under water like a seal for quite a few minutes. Sometimes the beluga hangs around areas where there are lots of clams.

NARWHAL
allanguaq

Monodon monoceros, Linné.

[1288]

Order Name used by Turner	*Cetaceans*
Current Order Name	*Cetacea*
Family Name used by Turner	*Not recorded*
Current Family Name	*Monodontidae*
Scientific Name used by Turner	*Monodon monoceros, Linné.*
Current Scientific Name	*Monodon monoceros Linnaeus*
Syllabary	ᐊᑦᓚᖑᐊᖅ
Modern Nunavimmiutitut Term	*allanguaq* (cf. Schneider p.19 - allanguaq = narwhal [Monodon monoceros])
Eskimo Term by Turner	*ag laṅg o ak; ag lang wak*
Definition of Eskimo Term by Turner	Narwhal (Monodon monoceros Linne). The word refers to the variegated markings on the body. (Turner 1884: 2138, 1289)

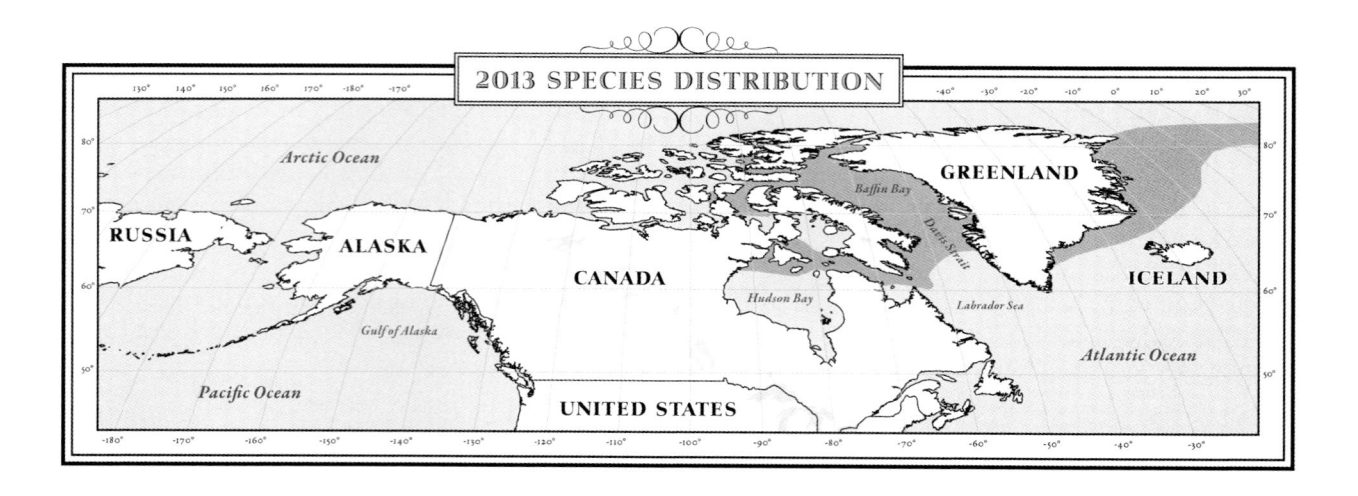

2013 SPECIES DISTRIBUTION

Chapter Two: Mammals of Ungava and Labrador

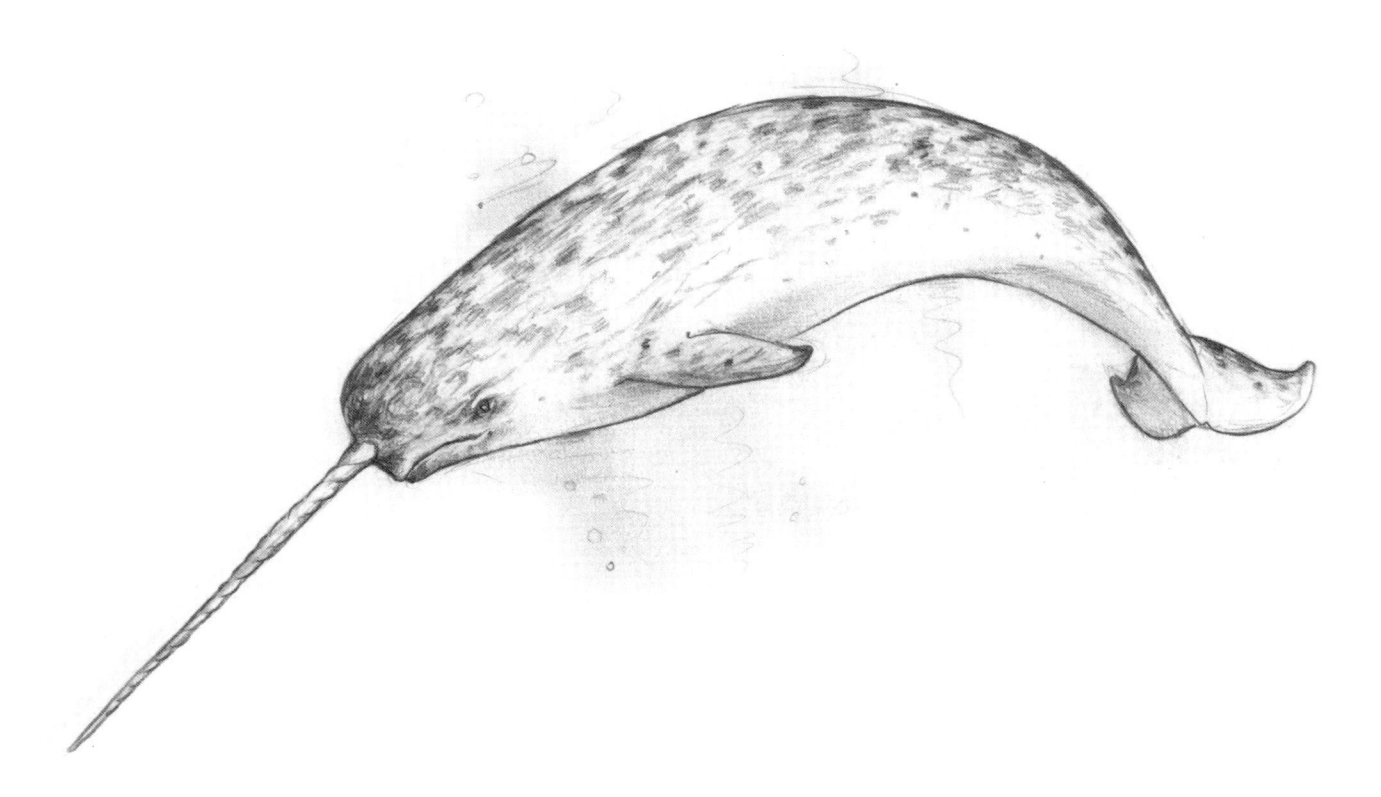

Descriptions by Turner

The narwhal is, so far as I could learn, not common along the Labrador coast. Its southern range was not determined, appearing most plentiful along the northern extremity of that coast. Within Hudson Strait its occurrence is more common, especially off the mouth of George's River, and to the westward of Akpatok Island.[3]

It occurs but rarely in the southern portions of Ungava Bay. The Eskimo dwellings along the south side of the western extremity of Hudson Strait capture a considerable number of these creatures. The Eskimo of Ungava Bay rarely attempting their capture for the simple reason that they fear them. They asserted that the Narwhal immediately swims toward their Kaiak and would doubtless overturn it if the occupant did not paddle away from it. I was led to conclude that this action on the part of the Narwhal was due [1289] to its desire to discover one of its kind for they are somewhat sociable in their nature and rarely found alone unless it be an individual astray or roving. The "Northerners" do not hesitate to attack it, and consume the oil and flesh as quickly as that of the White Whale. I endeavored to procure a specimen but was unable to do so on account of the distance which it would have to be brought. The "Narwhal" is known to the Eskimo under the name of *ag lang wak*, a word referring to the abrupt markings on the body.

[1290] Other species of whales are known to occur in the waters of Hudson Strait and along the Labrador coast. I do not attempt to determine the species as they were seen under such conditions that would render their identification purely conjecture.

---
3. Inuit hunters report that Narwal once frequented the Korac River. The muktuk (outer layer of fat) of a Narwhal is considered an Inuit delicacy and is thicker than beluga skin (Makivik 1984).

Recollections on Inuit Use of Narwhal
Tivi Etok (Heyes 2010)

The name for narwhal is the same word used for pen. It takes its name from all its spots and colors on its body. The body is black, but the spots are white. The tusk is very expensive to buy. One time I noticed a narwhal with no tusk. The female does not have tusks, but the males do. There used to be a narwhal around here, near George River, a long, long time ago. But they are coming back slowly.

Inuit Carving
A carving of narwhal, seals and sea birds made from caribou antler by Kangiqsualujjamiut artist Daniel Annanack, 2005. Narwhal are rare along the east coast of Ungava Bay. Narwhal tusks are prized by Kangiqsualujjamiut artists as a carving medium. Most Inuit carvers elsewhere in the Arctic prefer to use soapstone. Scott Heyes Private Collection.

Pod of Narwhal
Narwhal near Arctic Bay, Nunavut. Courtesy of Teema Palluq.

ᐊᕐᕕᖅ

Bowhead Whale
arviq
Balaena mysticetus

[1290]

Order Name used by Turner	*Cetaceans*
Current Order Name	*Cetacea*
Family Name used by Turner	*Not recorded*
Current Family Name	*Balaenidae*
Scientific Name used by Turner	*Balaena mysticetus* Linné
Current Scientific Name	*Balaena mysticetus* Linnaeus
Syllabary	ᐊᕐᕕᖅ
Modern Nunavimmiutitut Term	*arviq (cf. Spalding 1998, p.12 - arviq = Right or Greenland whale [Eubalaena glacialis]). Also Bowhead Whale.*
Eskimo Term by Turner	*aġ vik*
Definition of Eskimo Term by Turner	*Bowhead Whale [p.1290]*
Innu Term by Turner	*mist a neg [p. 709]*

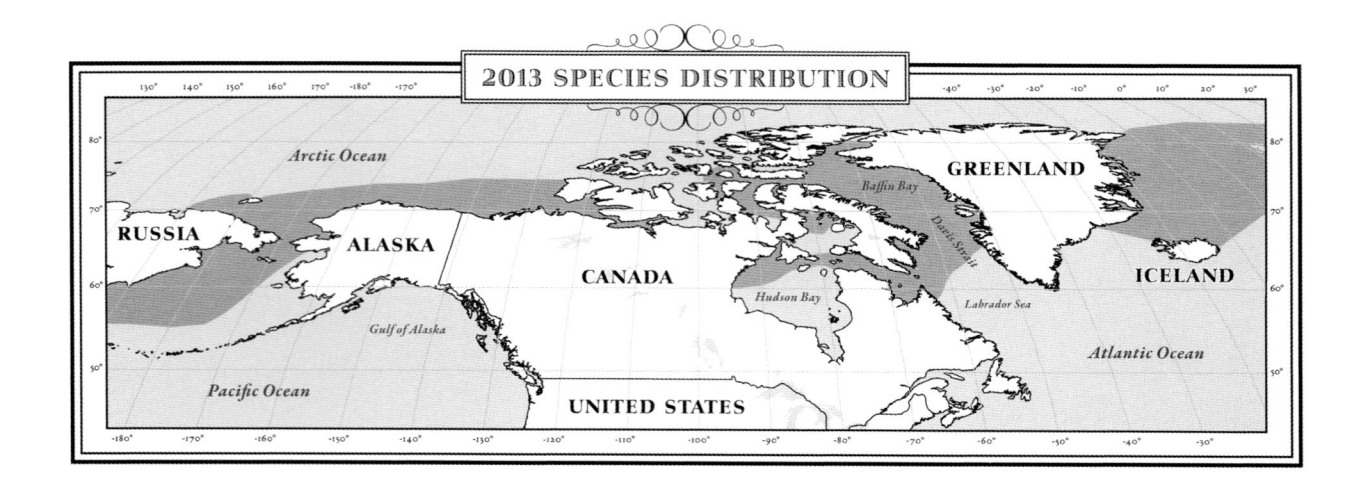

2013 SPECIES DISTRIBUTION

Chapter Two: Mammals of Ungava and Labrador

Descriptions by Turner

The [Bowhead Whale] is known to the Eskimo as *Aġvik*. I know nothing of its habits in those waters.[4]

---
4. Bowhead whales are extremely rare in Ungava Bay, with many hunters having never seen one. Noah Angnatuk, an Inuit elder from the Killiniq region of Ungava reported in a 1981 interview that he had seen bowhead whale bones on several mountain tops - although the reason for them being there was never explained (Makivik 1984).

Bowhead Whale Ethnography
Turner

ESKIMO VIOLIN AND BOW OF THE KOKSOAK INUIT (1887A: 193-195)

The instrument [USNM E93524] here denoted is a very crude imitation of the common violin in use amongst civilized people. It is made of spruce wood, cut in to thin boards held together by means of wooden pegs and a few small, iron nails. It possesses but two strings made of twisted sinew. The bow is simply a piece of bent wood and has a strip of baleen in lieu of the hair of a bow. The whalebone strip is rubbed with dry resin obtained from the spruce or from the larch tree. The strings are put in proper tension by moving the bridge forward or back and the sound produced by scraping the whalebone over the strings while the bottom of the violin rests upon the lap of the performer and the neck of the instrument is held toward and to the left of the body. The performer was able to play quite a number of recognizable native tunes upon the instrument. The facility of movement and freedom of sound gave a surprising entertainment. There was but one woman able to play upon it. She was certainly not less than sixty years of age and created a comical spectacle by her antics as she accompanied herself upon the instrument. All of the Innuit are passionately fond of music; one or two possessing violins which had been procured from the sailors visiting the region.

Jewsharps are in constant demand, especially by the young women who are able to imitate, in a degree, some of the bars of the commoner songs of the day.

Among the Moravian Missions of the Labrador coast the violin forms an essential part of the music of the native choir singing in the chapels erected for their worship as taught by the missionaries of that denomination.

Many of these performers are able to execute most difficult passages upon their violins and the amusing part is to hear the leader of the music direct the different keys in the German language while he is unable to understand a word of that [language] not pertaining to music. The accordion is heard in the land at rare intervals; more often as one descends the Labrador coast. The strains of music have a soothing tendency in the breast of the Innuit; and, in my opinion he has discarded the use of the instruments, drums, &c, employed by his ancestors, simply from the introduction of softer toned music.

Eskimo Violin with Bow Made from Whale Baleen (1883)
An Inuit violin collected by Turner at Ft. Chimo, known as Kiyu-iluli-agiagok to the Ungava Inuit (Turner 1884: 2312). This name derives from: kiyu, *wood;* iluli, *hollow;* agiagok, *filer, from the word to file or scrape (Turner 1884: 2312). The bow, made from Bowhead Whale baleen, was known as* pitikshigak, *and the strings were known as* Ivulu, *the term for sinew (Turner 1884: 2761, 2257). Turner wrote that "...the Innuit are passionately fond of music," and "[a Koksoak Inuit] performer was able to play quite a number of recognizable native tunes upon the instrument." USNM E90201, Museum Support Center, NMNH. Photo by Donald Hulbert, 2013.*

Three Single Women
An Inuit story, Tivi Etok (Heyes 2007)

There were three single women, they pretended they were married to the most unusual elements, one of them was married to a small rock, and one of them had a piece of bowhead whale bone, and the third had a small piece of an eagle bone. They did not let anyone know about it, not even their parents or families. One day, they were caught by their imaginary husbands in a way that they could not imagine. The woman who pretended to have a rock for a husband turned into a stone, and the one with the eagle bone was taken to a high cliff and the one with a piece of whale bone was taken by a whale.

The woman who had a bone of an eagle was taken to a cliff where she could no longer come back down. The one that turned into a stone stayed where she became a stone, and the one with the whale bone was taken to an island by the whales and was living in a whale skin and bones tent which was the small bone that she kept as a pretend husband, the whale would turn into a human and had the woman for his wife.

The parents of the three women were greatly hurt by the occasion and ended up with no daughters, so they ended up living by themselves. The one that turned into a stone could not be taken back she stayed being a stone. But the ones that were taken by the whale and the eagle still could return but were impossible to take back.

The parents tried everything to get their daughters back even though the animals that had taken them were dangerous so they had a very hard time. I'm not sure which one was the first, but the one who was by the eagles would be fed by her eagle husband and she ended up trying to make a string of muscles of the birds that the eagle husband could use in order to climb down from the cliff. When the eagle returned to the woman he would also become a human. The woman would not show her eagle husband what she was making out of the muscles of the birds, one day the eagle had caught a caribou and the woman had braided all the sinew of the caribou in order to make a rope or string to reach her father down the cliff. The father of the woman could talk to his daughter from the bottom of the cliff and she'd often check to see if the string of sinew could reach the bottom.

The eagle husband would be out hunting all day and when he returned to his wife, he'd turn into a man. One day the eagle had arrived unexpectedly and the string she had been braiding was almost seen by her eagle husband, she told him to make a sound before he arrives to her next time. Her string of muscles was able to reach the bottom of the cliff one day and she told her eagle husband to make a hole on the cliff in order for her to tie the braided strings before he left for hunting.

Once the string could reach down to her father, she asked her father to tie his bearded sealskin rope to her braided string, so she was able to use the hole on the rock to take down the string she had been secretly braiding. She was able to go down the cliff with her father's bearded sealskin rope before her eagle husband could return. So she made it back down the cliff to reunite with her parents.

By late evening, the eagle husband returned from his hunting trip and found that his wife was no longer up there, he became furious, and started searching for his wife. He arrived at the camp and landed on their tent where there usually is a hole on the ceiling of the tent and peeked through the hole. The father had poked the eagle through the hole with his harpoon and when he did, he turned into the little piece of bone that the daughter had played for a husband. It fell through the tiny hole into the tent, and he got one of his daughters back.

Then they had to try and get their other daughter who was stuck at the island. They took off on their qajaq to get her. The big whale that had turned into a tent for his wife would keep his wife tied to the tent when she wanted to go out, keeping her unable to leave. When the whale husband wanted her to come into the tent, he'd pull on the string to let her know. She could see her family from a distance and accidentally called out saying "Hey! there's a qajaq out there!" So the whale who was a tent asked "What did you say my wife? My partner in bed?" She apologized and answered "I'm sorry, I saw a duck and a rabbit chasing one another." She tied her leash on a rock and was able to get away without the whale husband could find out.

When the whale husband finally found out that his wife had left, he transformed into a big whale and rolled down to the beach to go after them. I didn't mention that the woman was throwing away all her clothes as the whale was catching up with her, when she threw away a part of her clothes, the whale would fight with the clothing and it slowed him down. She finally had no more clothes to throw except her panties, so she had to throw them out too before she could be caught. The whale had fought with the panties for a long time until it realized it wasn't the woman, he started chasing again, the family was not far from the shore and the whale had decided to go after them before they could reach the shore, but the family had reached the shore just in time, so the big whale had turned back into the little bone that it was when the woman had pretended it was her husband. The daughter that turned into a stone could not be changed back to a person. That story has more details into it but that's all I'm going to say.

Bowhead Whale
Bowhead whale. Taken April 26, 2011 during a spring abundance survey. Photo credit: Vicki Beaver, NOAA/AFSC/NMML, Permit No. 14245.

Recollections on Inuit Use of Bowhead Whale
Tivi Etok (Heyes 2010)

One was seen near here (Kangiqsualujjuaq) last year. I saw one once when I was on a cruise ship. We saw quite a few. Some are white, some were black. The bowhead whale comes in all sorts of colors. All large whales are called arviq by the Inuit.

THE INUIT PROCESS OF HUNTING BOWHEAD WHALE

The early Inuit used to camp in Nachvak Fiord; they were our ancestors. The drawing shows how our ancestors used to hunt big whales like humpbacks. There are three whales; one is heading out to sea, one is heading into the fiord and the other is stationary and facing towards the land. The hunters in those days didn't have guns. They had only knives and harpoons. They hunted using their kayaks. The hunters would not bother hunting the whales heading into or out of the fiord. They were difficult to hunt.

Hunting Bowhead Whales
An illustration of whale hunting in Nachvak Fiord, Labrador by Kangiqsualujjuamiut elder Tivi Etok. Reduced from 11" × 17". Lead pencil on bond paper. 28 May 2005. Scott Heyes Private Collection.

They would wait for a whale to come close to shore, swimming in the direction of the land. The hunters would situate their kayaks either side of the whale. The first hunter would jump onto the whale from his kayak. But the whale moved about, it did not want to be killed by that hunter. The second hunter, on the other side of the whale hopped on top of the whale. The whale didn't move, so the man made a slice into the arch of the back of the whale, deep into the blubber. By cutting in this location, the whale is unable to move his bones; he can't dive. So the whale continues landward. Hunters used to wait in the shallows ready to kill the whale. This is how our ancestors hunted whales. I have never tried hunting whales in this fashion; it would have been good to try it. A big whale would give the Inuit food for one year. This is part of our way of life.

WHALES – TERMS AND DEFINITIONS

SYLLABARY	MODERN NUNAVIMMIUTITUT TERM & DEFINITION	TERM BY TURNER	DEFINITION RECORDED BY TURNER
ᐊᓪᓚᖑᐊᖅ	*allanguaq* (cf. Schneider p.19 -*allanguaq* = narwhal [*Monodon monoceros*])	*ag láng o ak; ag lang wak*	Narwhal (*Monodon monoceros*, Linneaus). The word refers to the variegated markings on the body. [p. 2138; 1289]
ᐊᕐᕕᖅ	*arviq* (cf. Spalding 1998, p.12 - *arviq* = Right or Greenland whale). Also bowhead whale.	*aǵ vik*	Bowhead whale [p. 1290]
ᓃᓴᒃ	*niisak* (known in Labrador, cf. Jeddore p. 81 - *niisaatsuk* [*niisaarsuk*; *nisak*]; (cf. Schultz-Lorentzen p. 166 - *nîsa* = harbor porpoise [*Phocoena phocoena* Linnaeus, 1758] also Peck p. 158 – *nisa*; *nisâk* = "porpoise; sea-hog; dolphin"). Note: a Greenlandic term, possibly of Norse origin (ref ?). Also *niisaarjuk* (ᓃᓵᕐᔪᒃ) and *arluasiak* (ᐊᕐᓗᐊᓯᐊᒃ). Tivi Etok, Kangiqsualujjuaq (Pers. Comm., 2010) suggests *aarluas iak*, which clearly derives from *aarluk* = killer whale. In a 1981 interview, Thomas Etok of Kangiqsualujjuaq described "blue dolphin" as *ardluisiak* (Makivik p. 72).	*ní sak*	A small species of porpoise [the harbor porpoise, *Phocoena phocoena*] [p.2719]
ᐸᒥᐅᓕᒑᕐᔪᒃ	*pamiuligaarjuk* (cf. Schneider p.237). Note: Neither the false killer whale (*Pseudorca*) nor the bottlenose dolphin (*Tursiops*) lives around Ungava. As Turner noted, this is the Black Grampus, or what is now called the long-finned pilot whale, *Globicephala melas* Traill, 1809. Johnny George Annanack, Kangiqsualujjuaq, suggests *Pamiulilatchuk* = white beaked whale (Makivik p. 20).	*pú mi ú li gá chuk*	Name of the Black Grampus [the pilot whale, *Globicephala melas*] [p. 2769]
ᖃᑯᖅᑕᖅ (?)[5] ᖃᑯᖅᓂᖅ (?) ᖃᐅᓪᓗᕆᑦᑕᖅ	*qakurtaq* (?) *qakurniq* (?) (cf. Schneider p. 280 - *qakurtaq* = white; *qakurniq* = whiteness) Note: Turner here seems to be confusing a quality of beluga skin - whitish – with its "proper" name *mattaq* (cf. Schneider p.166; also consider Schneider p. 292 - *qaullurittaq* = a full grown beluga that is very white).	*ká ku*	Fresh whale-skin [p. 2264; 698]

---

5. A question mark against an entry appears for the following reasons: 1) when we were unable to make any sense of Turner's original entry, and 2) when some sense could be made, but various degrees of doubt remained.

WHALES – TERMS AND DEFINITIONS

SYLLABARY	MODERN NUNAVIMMIUTITUT TERM & DEFINITION	TERM BY TURNER	DEFINITION RECORDED BY TURNER
ᖃᑯᕐᑕᖅ (?) ᖃᑯᕐᓂᖅ (?)	*qakurtaq* (?) *qakurniq* (See above); also consider Peck p. 92 -*kárkok* = "dried entrails; intestines"	*ka kú tak*	Boiled and dried whale skin [p. 698; 2264]
ᕿᓚᓗᒐᖅ	*qilalugaq* (cf. Schneider p.297 - *qilalugaq* = beluga, white whale *[Delphinapterus leucas]*)	*ki l'i luak; kil líl u ak*	White whale [beluga] [p. 2298; 1287]
ᕿᓚᓗᒐᑦᓯᐊᖅ	*qilalugatsiaq* (?)	*ki l'i lu g'wa chi ak*	Constellation of stars. White whale is the meaning of the word. [p. 2265]
ᕿᐳᕋᖅ(?); ᕿᐳᒃ(?); ᕿᐳᑕᒃ;	*qipuraq*(?) Past participle (old form) of *qipurpaa*, "scratches it" *qipukak* (?) Possibly from *qiputak* (cf. Dorais, 1983, p. 46 – *qiputak* = scoring [lines scratched into something]; also Schneider p.309 - *qipuqqiniq* = a scratch on the surface of something smooth)	*kĭ pú kak*	Creases or folds in the side of a whale. Name of locality north of Indian Harbor near the entrance to Hamilton Inlet. The people – "Planters" – pronounce it *Kibbekok* [p. 2308].
ᓱᕐᑲᖅ	*surqaq* (cf. Schneider p.380 - *suqqaq* = whale baleen [...])	*chú kak*	Hair (or baleen) of violin bow [p. 2212]
ᐅᙳᓂᐊᕕᒃ	***Ungunniavik*** *ungujuq* / *ungumajuq* (cf. Schneider p.451 - *ungujuq* = [the game chased approaches us] whales are chased into the bays when killer whales are in the vicinity). *Ungniavik*. Note: Whale River: river on the south shore of Ungava Bay, between the Koksoak and George Rivers.	*Uń g ni aǵ vik*	Whale River. The word is derived from *Unguktok* to drive animals such as deer on the land or whales in the water; *vik* a place where. Hence a place where whales are driven i.e. the water [p. 2845]

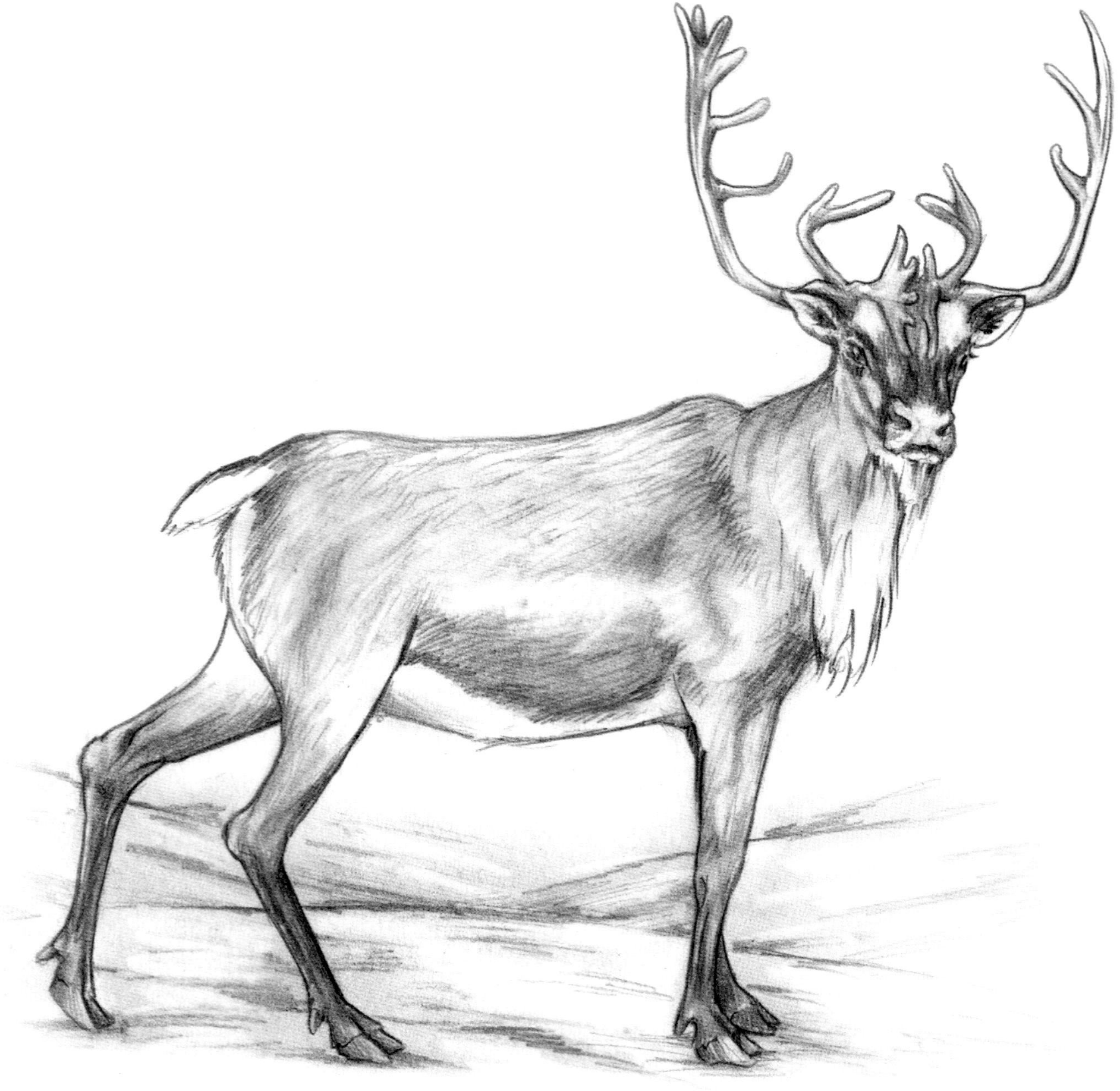

Ungulates

Introduction to Hoofed Mammals

Two species of wild hoofed mammals are found in the Ungava Peninsula, the moose (*Alces alces*) and the caribou (*Rangifer tarandus*). One additional species, the muskox (*Ovibos moschatus*), was not historically present in eastern Canada, but was introduced from Ellesmere Island to the southern shores of Ungava Bay during 1973-1983, where it occurs today with its range apparently expanding (Le Hénaff and Crête 1989; Lent 1999; Chubbs and Brazil 2007).

Prior to 1875, moose were largely absent from Ungava, found only in the far southwest of the region (Peterson 1955; Bergerud et al. 2008). This agrees with Turner's information from 1882-1884, in which the moose accordingly receives little discussion ("The moose is reported to be not common in the southwestern portion of the region. Its range to the eastward has not been satisfactorily determined"). Since the late 1800s the range of the moose has expanded rapidly north and eastward throughout much of Ungava (Peterson 1955; Harper 1961; Mercer and Kitchen 1968), and now apparently reaches the treeline in some areas, including along the shores of Ungava Bay (Franzmann and Schwartz 1997; Bergerud et al. 2008), though it remains rare as far north as Kujjuaq (Ft. Chimo) today. This is one of the more striking changes to the large mammal fauna in Ungava to have unfolded since Turner's visit to the region. The rapid spread of moose northward has been attributed primarily to warming climate (Harper 1961; Bergerud et al. 2008:431).

Fundamental to the natural history and cultural psyche of the Ungava Peninsula is the caribou, and Turner's treatise on caribou in many ways forms the centerpiece of his entire manuscript on Ungava mammals. Turner's manuscript is an extremely important contribution to the biology of the caribou in the 19th century. It is based, like all of his work, on careful firsthand observations of animal behavior, examination (and museum deposition) of specimens, appreciative consideration of indigenous knowledge, and the published observations of other naturalists (in this case especially Samuel Hearne, from whose work [Hearne 1795] Turner quotes liberally). As with the rest of Turner's manuscript on mammals, it is a pity his observations on Ungava caribou were not published, and available to other scholars, during the 19th century.

A longstanding and challenging question in Arctic mammalogy concerns the taxonomy of caribou – how many species are there? Though he considered them a single species, Turner could recognize "two varieties" of caribou: a woodland and a barren-ground form, differing in size, antlers, and behavior. Turner collected a large number of caribou skulls and antlers in Ungava for the Smithsonian, and these have been important to museum zoologists in understanding the physical characteristics of Ungava caribou (e.g., Banfield 1961). For much of the twentieth century, Turner's "varieties" were often recognized as two separate species, *R. arcticus caboti* (or *R. caboti*), the northern, tundra, or barren-ground caribou (living largely above the treeline in summer and undertaking extensive migrations) and *R. caribou*, a southern or woodland form (living largely below the treeline and undertaking short migrations) (e.g., Seton 1927; Harper 1961). Later studies of physical characteristics and more recently molecular genetics (e.g., Banfield 1961; Cronin et al. 2005) have once again led to the recognition of a single species of caribou/reindeer, *R. tarandus,* hugely variable in size and shape across different habitats and climate regimes, that is distributed across most of northern Eurasia and North America (Geist 1998). The caribou of Ungava and Labrador are currently classified as the woodland caribou subspecies *Rangifer tarandus caribou* (Bergerud et al. 2008), though the fine details of caribou taxonomy remain a subject of active investigation (Cronin et al. 2005; Groves and Grubb 2011).

Turner records how indigenous worlds in the northern reaches of Labrador and Ungava were intimately intertwined with the life, fate, and movements of this northernmost of deer. Indeed, the existence of the settlement at Ft. Chimo, where Turner was based, may ultimately derive from predictable seasonal movements of caribou across the Koksoak River in the vicinity (Elton 1942; Bergerud et al. 2008). Turner's detailed notes on the physical characteristics and behavior of Ungava caribou are complemented by his ethnographic observations, compilations of traditional stories, and collections of Inuit and Innu artifacts related to the caribou. His collections of exquisite Innu deerskin coats are beautiful favorites among these remarkable treasures from 1880s Ungava (Burnham 1992).

The most important modern work on the biology of the Ungava caribou is *The Return of the Caribou to Ungava*, a 600-page volume by A.T. Bergerud and colleagues (Bergerud et al. 2008). Bergerud's data-rich book concentrates on the population dynamics and movement ecology of caribou herds in the Ungava Peninsula. It shows that the number of caribou in Ungava has undergone extreme fluctuations both in recent decades and over the past 200 years (with predation, availability of food, and density-dependent effects driving fluctuations), in tandem profoundly affecting the natural history and human ecology of the region. Expanding development, climate change, and the spread of moose and muskox in Ungava will interact in complex ways with these precipitous cycles of rise and fall in the caribou. Simultaneously gazing back to Turner's time and looking ahead to the future, we hope, even as new changes and challenges emerge, that caribou herds will still move across the expanse of Ungava centuries from now.

Herd of Caribou during Fall 2006 Migration
Taken at Kamestastin, near the Quebec–Labrador border. Courtesy of Stephen Loring.

ᑐᑐᕙᒃ

Moose

tuttuvak

Alces malchis (Linné) Gray.

[1292]

Order Name used by Turner	*Ungulata*
Current Order Name	*Artiodactyla*
Family Name used by Turner	*Cervidae*
Current Family Name	*Cervidae*
Scientific Name used by Turner	*Alces malchis* (Linné) Gray.
Current Scientific Name	*Alces alces* (Linnaeus)
Syllabary	ᑐᑐᕙᒃ
Modern Nunavimmiutitut Term	*tuttuvak (cf. Schneider p.430 - tuttuvak = moose, domestic cattle, large mosquito – tuktuujaq p. 417); a crane fly (cf. Spalding 1998, p.167: tuktuujaaq = long-legged long-winged water fly)*
Eskimo Term by Turner	*tuk tú yak*
Definition of Eskimo Term by Turner	*Big reindeer. Name applied to a large Gallinipper fly (Turner 1884: 2810)*

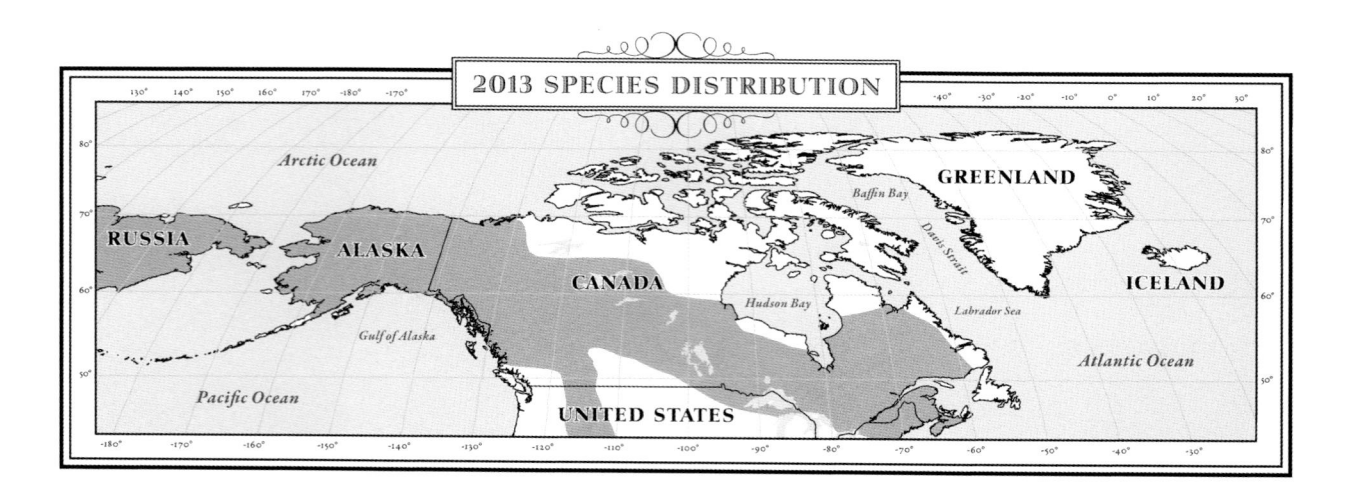

2013 SPECIES DISTRIBUTION

Chapter Two: Mammals of Ungava and Labrador

Descriptions by Turner

The moose is reported to be not common in the southwestern portion of the region.[6] Its range to the eastward has not been satisfactorily determined. The Indians, who visit Ft. Chimo to trade, report that several years ago they saw a great animal, having horns, many times larger than a reindeer, in the forest near the headwaters of George's River. I suspected this animal, which the Indians assert was black, to have been a straggling moose, either persecuted by rivals or seized with a freak to wander. The Indians confessed their fear to attack it.

---

6. Turner did not collect moose while in Ungava and Labrador. The Inuit elders of Kangiqsualujjuaq, in recent interviews on mammals, indicated that moose are not present in the vicinity of Kangiqsualujjuaq or other reaches of Ungava Bay. Some elders report seeing moose tracks near Makkovik in Labrador (Heyes 2010).

Woodland Caribou or Reindeer
tuttu

Rangifer tarandus caribou (Kerr)

[1293]

Order Name used by Turner	*Ungulata*
Current Order Name	*Artiodactyla*
Family Name used by Turner	*Cervidae*
Current Family Name	*Cervidae*
Scientific Name used by Turner	*Rangifer tarandus caribou* (Kerr)
Current Scientific Name	*Rangifer tarandus* (Linnaeus)
Syllabary	ᑐᒃᑐ
Modern Nunavimmiutitut Term	*tuttu* (cf. Schneider p.430 - *tuttu* = caribou in general)
Eskimo Term by Turner	*túk tu*
Definition of Eskimo Term by Turner	Reindeer (any kind) [p. 2809]
Innu Term by Turner	*a tikhwh* [p. 708]

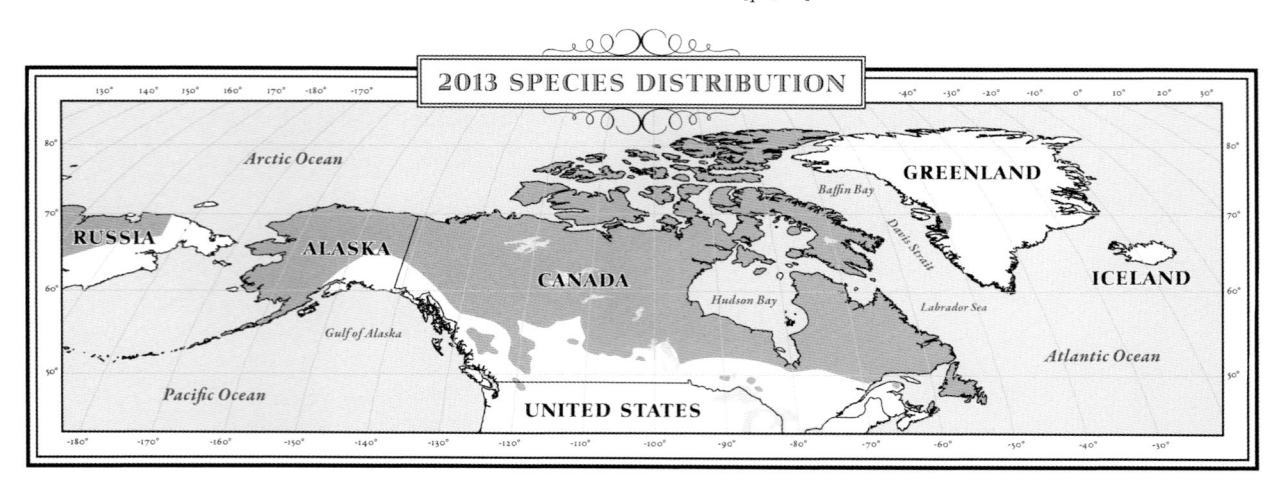

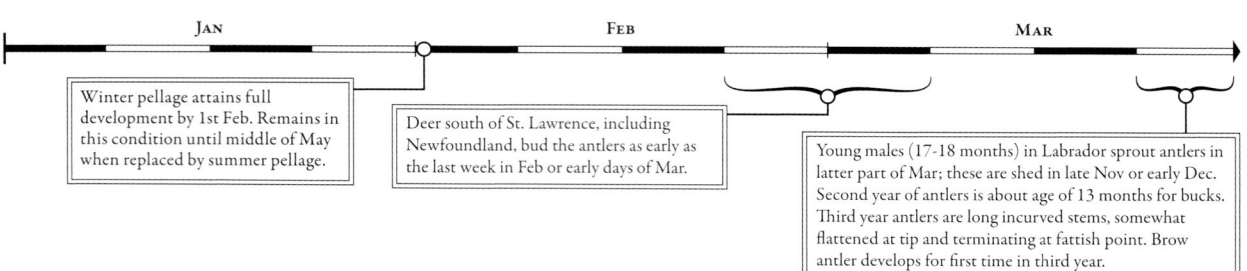

2013 Species Distribution

Jan — Winter pellage attains full development by 1st Feb. Remains in this condition until middle of May when replaced by summer pellage.

Feb — Deer south of St. Lawrence, including Newfoundland, bud the antlers as early as the last week in Feb or early days of Mar.

Mar — Young males (17-18 months) in Labrador sprout antlers in latter part of Mar; these are shed in late Nov or early Dec. Second year of antlers is about age of 13 months for bucks. Third year antlers are long incurved stems, somewhat flattened at tip and terminating at fattish point. Brow antler develops for first time in third year.

Chapter Two: Mammals of Ungava and Labrador

Descriptions by Turner

The Reindeer is an animal essentially belonging to the colder regions of both hemispheres. Southern range of its present distribution is confined to the extreme northeastern portions lying south of the St. Lawrence Gulf. It is found sparingly in the extreme northern parts of Maine, although if we may rely upon the statement of DeKay in New York Fauna, page 122, part 1, 1842, this animal occurred, according to Professor Emmons, only few years previously in the northern portions of Vermont and New Hampshire. In the Canadian provinces it occurs in New Brunswick and Nova Scotia; formerly on Prince Edward's Island, but now extirpated. Few are found on Cape Breton, and very abundant on Newfoundland. On the island of Anticosti, they appear to have never

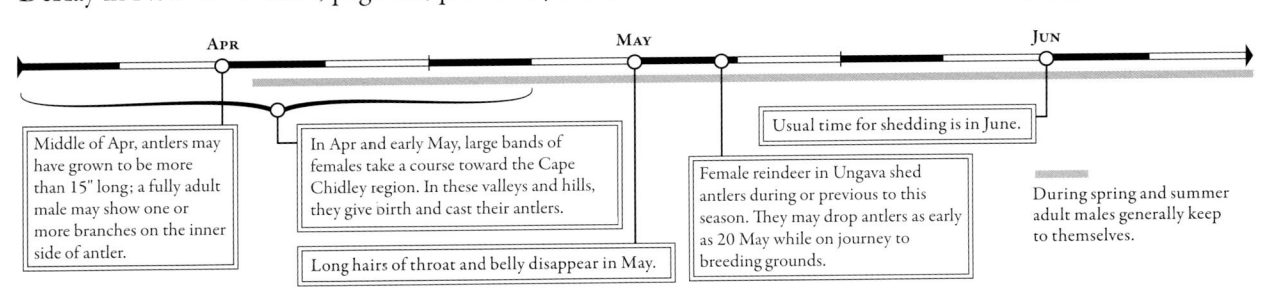

Apr — Middle of Apr, antlers may have grown to be more than 15" long; a fully adult male may show one or more branches on the inner side of antler.

In Apr and early May, large bands of females take a course toward the Cape Chidley region. In these valleys and hills, they give birth and cast their antlers.

Long hairs of throat and belly disappear in May.

May — Female reindeer in Ungava shed antlers during or previous to this season. They may drop antlers as early as 20 May while on journey to breeding grounds.

Usual time for shedding is in June.

Jun — During spring and summer adult males generally keep to themselves.

occurred, or at least there is no reliable record of the occurrence. The peninsula [1294] of Labrador from the Atlantic to the Hudson Bay and from the Hudson Strait to the Gulf of St. Lawrence as far west as the eighty-third degree of west longitude; here the trend of their wanderings ranges in a northwesterly direction to latitude 56°; thence rapidly rising to the fifty-eighth degree of latitude and directed across the valley of the McKenzie River to the Rocky Mountains, after which the trend is to the southwest among the high ridges of the Cascade Mountains and the western slopes of the Rocky Mountains as far south as the fifty-second degree of latitude.

Along the southern shore of Hudson Bay, the woodland reindeer is not numerous but increase as the western shore is ascended in a northerly direction, becoming excessively abundant between York and Churchill Factories, thence to the Arctic Ocean; and, even many of the islands [1295] remote from the mainland bordering on that Ocean. The Sitkan district and the adjacent mainland of Alaska probably contains but few of these animals. From Mount St. Elias to the northward the reindeer become very plentiful and fairly swarm throughout Alaska, excepting the Aleutian Islands. The only island of this chain which has ever been trodden by a wild deer is Oonemak to which the animals repair during exceptionally severe winters when ice blocks up "False" Pass separating it from the Peninsula of Alaska; and from which the deer cross over on the ice to that island. Many persons have, through their ignorance of geography, stated that these animals cross from Bering's Strait to Kamchatka by the Aleutian Islands; a more impossible statement could scarcely be made, as at a [1296] moment's glance at a map will show.

The most southern limit reached in Asia is latitude 50° in Chinese Tartary, thence northerly and westerly to latitude sixty degrees to the western shore of the Scandinavian Peninsula. About one hundred years ago they were successfully introduced into Iceland and have multiplied so greatly as to be a detriment to the inhabitants, on account of the damage these animals create among the grasses and patches of Iceland moss. All Greenland is inhabited by this animal, excepting probably, the extreme southern portion of that island. Probably nowhere in the range of this animal is it so abundant, in the wild condition, as it is in the northern half of the Labrador Peninsula and certain portions of Alaska and the region north of the Hudson Strait. [1297]

From careful considerations of this animal there are two varieties to be recognized in the wild state. The first of these is the inhabitant of the wooded districts as well as the open grounds. The woodland reindeer is the larger of the two, often obtaining a weight of five hundred pounds when undressed, measuring nearly five feet in height from the ground to the shoulder. The females of this variety obtaining a size of about two thirds to three-fifths that of the males. The males of the barren ground variety rarely more than equal in size a medium doe of the woodland variety while the females are much smaller, adult individuals scarcely more than exceeding the height of a large sheep. The weight of the males rarely exceed one hundred pounds when dressed and [1298] females from sixty to seventy pounds dressed. The antlers of the two varieties also present great differences. By certain writers who have never seen a reindeer in the wild state, it is claimed that the barren ground animal has proportionately larger antlers. This is a great error. I have had an abundant opportunity to observe this in the many thousand of these animals which I have seen in Alaska and Labrador. That there is a very great development in the antlers of the individuals of each variety none can deny. I have seen the antlers of an immense buck scarcely larger than a child's chair and in this particular instance I never saw a more perfect set of antlers. All of the branches, prongs and palmations were symmetrical,

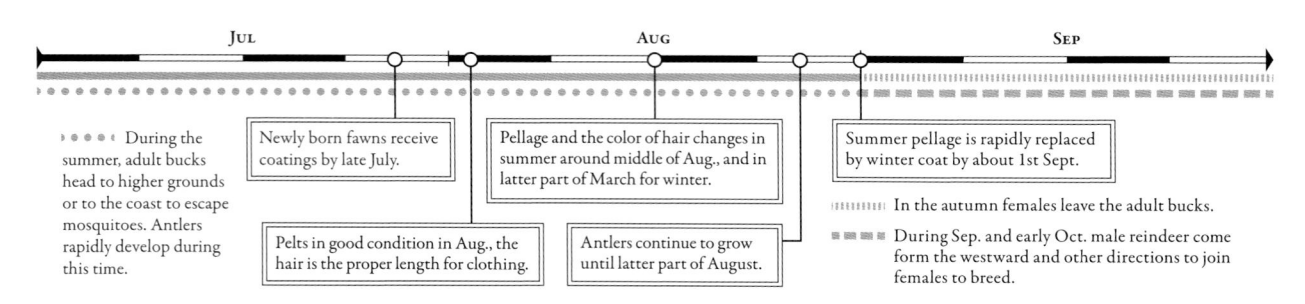

Jul	Aug	Sep

- During the summer, adult bucks head to higher grounds or to the coast to escape mosquitoes. Antlers rapidly develop during this time.
- Newly born fawns receive coatings by late July.
- Pelts in good condition in Aug., the hair is the proper length for clothing.
- Pellage and the color of hair changes in summer around middle of Aug., and in latter part of March for winter.
- Antlers continue to grow until latter part of August.
- Summer pellage is rapidly replaced by winter coat by about 1st Sept.
- In the autumn females leave the adult bucks.
- During Sep. and early Oct. male reindeer come form the westward and other directions to join females to breed.

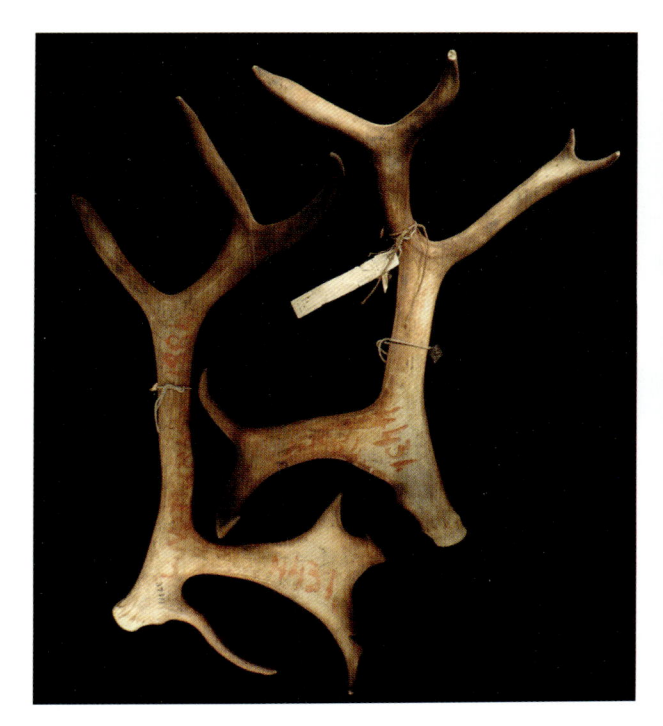

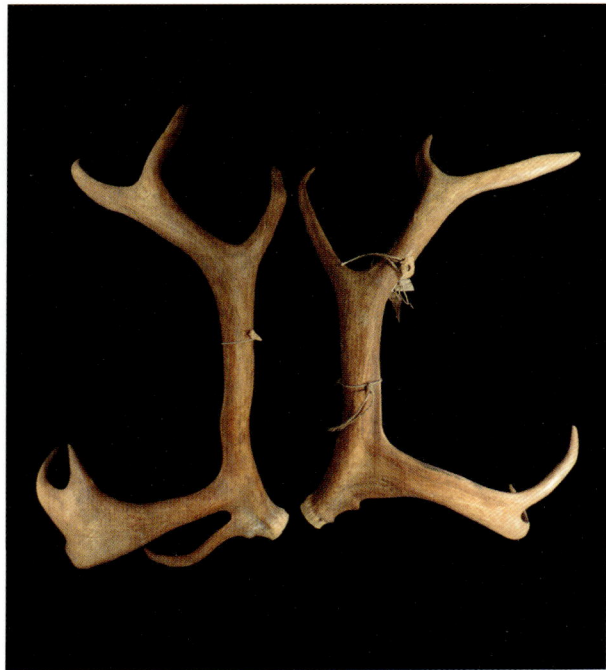

Caribou Antlers Collected by Turner
USNM 14778, Division of Mammals, NMNH. Photo by Angela Frost, 2011.

the brow antlers beautifully pectinate, [1299] the tips of which were gracefully incurved. There were eleven prongs on each of the terminal palmations. It is quite rare to find the horns of the males of either variety to be even somewhat similar. The age of the deer alone determines the size and form of the horns. The young male, in Labrador, sprouts his horns in the latter part of the month of March and they grow as simple stems with one, two, or three prongs irregularly disposed on the stem. These are shed in the following late November or early December. The second growth of antlers is simply an increase upon the size and arrangement of prongs for the first year. It should be understood that the first year as here meant is when the animal has obtained an age of seventeen or eighteen months. The second year of the horns is at an age of about [1300] thirty months for the buck. The third year the horns are long incurved stems, somewhat flattened at the tip and terminating in a flattish point. At this age, the brow antler is developed for the first time. The main stem may or may not have prongs; if these do occur, they are placed mostly on the third next the head and one, rarely two, short prongs having a lengthened base may be found within twelve to four inches from the extreme tip. Exceptionally, the tip may consist of two prongs widely divergent one of which will be longer than the other and somewhat resemble the thumb and index finger spread apart.

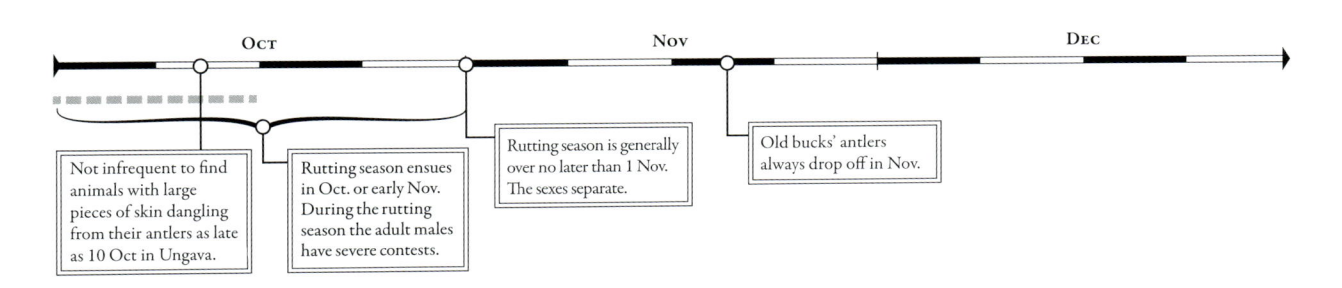

| Oct | Nov | Dec |

Not infrequent to find animals with large pieces of skin dangling from their antlers as late as 10 Oct in Ungava.

Rutting season ensues in Oct. or early Nov. During the rutting season the adult males have severe contests.

Rutting season is generally over no later than 1 Nov. The sexes separate.

Old bucks' antlers always drop off in Nov.

Ungulates

Caribou Antlers Collected by Turner

Turner collected 24 skulls and sets of antlers from caribou during his stay in the region. Caribou were abundant in the region at the time. In a letter to Robert Ridgway, dated 28 August 1883, Ft. Chimo, Turner wrote that he had: "killed many reindeer... and had seen thousands." Turner also documented Inuit beliefs relating to caribou (Turner 1887a: 65): "A deer once killed does not prevent its spirit reappearing in material form at the interoccassion of the shaman. At certain times only may the skins of deer be worked; never at the beginning of the sealing season." Catalog number USNM A21643, Division of Mammals, NMNH. Photo by Angela Frost, 2011.

The fourth year of the horns is a modification of the third and fifth. At the latter time the antler has reached its full development. I have seen Eskimo who professed [1301] to be able to distinguish by the form of the antlers the age of a reindeer up to its seventh year but acknowledged his inability to determine beyond that period. It was perfectly comprehensible to the fifth or sixth year but beyond this I was unable to discern special characters differing from those claimed to be older.

As may be inferred from the foregoing remarks, the horns of different individuals vary so greatly, both as reflects the development of the palmations and the position of the branches along the main stem, that a description of them can be only general. In regard to the sprouting of the new horn the time of its outgrowth varies according to the latitude which the animal inhabits. The reindeer south of the St. Lawrence and including Newfoundland bud the horns as early as the [1302] last week in February or the early days of March. The horn begins as a small lump under the skin and is quite soft to the touch. In the course of a few days, it has increased to an inch or more in height. By the middle of April it may be more than fifteen inches long and, in a fully adult male, may show one or more branches on the inner side of the horn. As the season progresses, the length of the main stem or shaft increases, developing the various branches and palmations. The latter invariably terminate in several, five to fifteen short prongs, rarely more than six inches in length and oftener of only three or four inches. These prongs rarely bifurcate unless it be that they start so closely together that the space between fills up and may then be considered as two prongs having a single, wide base.

[1303] The antler continues to grow until the latter part of August and is during all the period of its growth covered with short, fine hair from a thin skin covering the blood vessels which supply the horn with material for their growth. While in this condition, the horns are spoken of as "in the velvet" from its resemblance to that fabric. As soon as the blood vessels have ceased their functions they shrivel up and the velvety skin cracks and begins to peel off while the tips of the horns may yet be soft and spongy. When the horn has ceased to grow, the animal is seized with an irresistible desire to remove the skin coating from its horns. This is accomplished, in wooded tracts, by rubbing the horns against small trees. So far as my own observation gave evidence the Larch tree is usually selected, probably because of its freedom from branches which might [1304] entangle the horns during the operation. The portion of skin first removed is that at the base of the antlers; the skin from the brow antler and the palmations with their points are the last parts removed. It is not infrequent to find these animals with a large piece of skin dangling from their horns even as late as the 10th of October in the Ungava district of the Labrador peninsula.

The females have horns developed in the same manner as the males, excepting that the brow-antler is never present; this feature forms the principal distinction between the sexes and also that the horns of the females are never as large as those of the males, rarely greater than those of a two-year-old buck. The main stem is rarely greater than two feet in length, the prongs and partial palmations, although rarely numerous, are [1305] closer, forming a less compact arrangement.

The female reindeer in Ungava sheds her antlers during, or sometimes previous to, the fawning season. I have, at Ft. Chimo, seen antlers dropped as early as the twentieth

Stack of Caribou Antlers
Caribou antlers collected by Inuit from the hillsides of an inlet on the east coast of Ungava Bay. Inuit hunters will return to collect these once the antlers turn a deep purple or green with age. These are highly desired by Kangiqsualujjuamiut artists for use as carving material. Photo by Scott Heyes, 2010.

of May from females while crossing the river Koksoak on their journey to the breeding grounds. The usual time there for shedding the antlers is in the month of June. With the female there is only a very short period of rest between the time of shedding and renewal of the horns. This period is about eighteen days after which the antlers sprout and continue to grow until the time when the rutting season ensues in October or early November. At the completion of this season, the velvet is removed from their horns in the same manner as that from the males. It is not uncommon to find patches of small larch trees having all the bark [1306] and limbs removed as far up as the height of the horns of the animals will reach. The females retain their naked horns throughout the winter and until the succeeding May or June. The young does have no horns the first year of their life, but start them just as they enter the second year and continue to shed and renew their antlers each year thereafter, attaining a full development the fourth year; although it differs but little from that of the third year. Among the reindeer north of the St. Lawrence River, all the females have horns while among those south of the St. Lawrence some of the does do not have horns. Many writers assert that none of the does have horns. They would be laughed at by both Indians and Eskimo if those persons made such assertions in the presence of those people and would doubtless inquire [1307] where they had ever seen a reindeer.

Caribou Skin and Sealskin Clothing
Photograph of two Inuit seated in Turner's observatory at Ft. Chimo, April 1884. Photo by Turner. 1 4.5" × 7" photograph. Source: Negative SI 4822, National Anthropological Archives, Smithsonian Institution.

The barren ground reindeer sheds and renews its horns at about the same periods of time as those of the woodland variety. These which inhabit the open grounds removes the skin from its horns by rubbing against rocks, bushes, and banks.

The Eskimo living along the south and western portion of Hudson Strait have often spoken to me of the great difference in size of the reindeer which inhabit that region in comparison with those found along the valley of the Koksoak. Unfortunately, I was unable to procure individuals of the northern animal of that district. The Eskimo informed me that their deer have coarser hair on the body, are lighter colored, and that the largest bucks were not so large as a two-year-old doe of the woodland [1308] variety near Ft. Chimo; and, that the female was quite small, or as they expressed it, a little bit of a thing, about half the size of the buck of their country.

The female reindeer is the only one of the Cervidae whose sex bears horns. This has been accounted for in a number of ways. Some writers affirming in most positive manner that the horns are for the purpose of defending their young from the attacks of the old bucks. To controvert this statement is only necessary to call attention to the fact that the horns of the females are shed at the very time when the fawns are born and in their most defenseless condition. Secondly, the adult bucks are never with the females during the fawning season. Another reason has been given for the use of their horns. Some persons assert that they are for the purpose of [1309] removing the snow from the feeding grounds. Anyone who would study the form of the horns, and their position on the head of the animal, must conclude that the only manner in which they could be used for this purpose would be for the animal to stand on its head and whirl around like a top, or else to lie on its back and scratch with its horns. Either of these positions would be to say the least, something unusual in an animal having four legs.

Innu Skin Tent
Photograph of an Innu Caribou skin tent taken by Turner on the banks of the Koksoak River, August 1883. This photograph (1 cyanotype 4" × 5" mounted on 8" × 5") was illustrated for Turner's BAE report, (1894, Fig. 114: 298). Source: SI Archives, Record Unit 7192, Turner, Lucien M, 1848-1909, Lucien M. Turner Papers.

In respect to the size of the horns, of the two varieties, of the reindeer inhabiting North America, it has been often recorded that the horns of the barren ground animals are larger than those of the one frequenting the woodlands. I have seen and measured the horns of so many individuals of each variety that there is no doubt that while the antlers of the former [1310] may equal, in certain instances, those of the latter, yet, when compared either in weight or bulk the horns of the woodland animal exceed those of the barren ground. The difference in the size of the respective animals has doubtless contributed much to this error. The horns of the woodland reindeer often measure as much as five feet between its outer palmations and four and a half feet width is not at all rare. The length of the main stem to the tip often measures fifty inches. The width of the outer palmation being six inches. The length of the longest branch, usually near the base of the horn, is as much as eighteen inches. The brow antler of the males attains a length of fourteen inches, and a width of twelve and a half inches between the extreme points of the palmation of [1311] this particular branch. It, in fact, attains such length as to interfere with the animal putting its nose to the ground to obtain its food. The main stem of the horn of the woodland reindeer is seldom cylindrical, but flattish, measuring in large examples, one and seven-eighths inches through the longer axis, by one and a third to one and a half inches through the shorter axis.

The horns of the male barren ground reindeer are, for the main stem, rarely more than forty-five inches in length; the diameters of transverse sections are never so great as those of the woodland reindeer. The palmations are not the great sub-triangular forms found in the other variety, for they are more like expansions, to a moderate degree, of the main stem on which much longer tines or prongs grow and have more cylindrical form for these prongs. The horns of the females of each variety are fully as much

> **RAVEN AND THE DEERSKIN BLANKET**
> *An Inuit story, Turner (1887a: 28)*
>
> The raven was a man, who while the others were collecting their household effects preparatory to removing to another locality called to the people that they had forgotten to bring the deerskin bed (the heavy deerskin on which the Innuit sleeps) and kept shouting *Kak, Kak* (Inuit name of the deerskin bed) that the others, although they heard him shouted back *Kak, Kak*. This so angered the man that he changed into a raven and every time the Innuit moves his camp a raven flies near shouting *Kak, Kak*. The Raven's Innuit name is *Tu-lughak* but now speaks the word for a deerskin bed.

[1312] dissimilar. Some of the females are barren, i.e., incapable of producing young. These females are served at the same period as the fertile females. The barren doe is larger than the fertile animal and often equals the size of a buck of three or four years. The horns of a barren female are, when fully developed, exactly like those of a buck of three years, consisting of the long main-shaft with flattish tip somewhat pointed. Branches without palmations are developed so that at a distance the barren females cannot be distinguished from a male. The brow antler is of course not developed in the female of any age or in the male of this age.

Nowhere on the horns of either sex are there to be found any roughness or nodules, except a circlet just at the base, as are to be seen on other cervidae. The horns being quite smooth, excepting the grooves (sulci) along which the circulation of blood was effected; these channels [1313] having a depth of about half the thickness of the veins. (Footnote: I have herewith introduced a few remarks written by Hearne [in quotations], nearly a hundred years ago; and, as they differ in no degree from the present traits of the reindeer on the eastern side of Hudson Bay, and in nearly the same latitude, I have preferred to copy his writing which present that by myself because it thus subserves a double, yet obvious purpose. The writing from his book, Journey to the Arctic, 1771, is sufficiently denoted.)

[1314] "The old buck's horns are very large, with many branches, and always drop off in the month of November, which is about the time they begin to approach the woods. This is doubtless wisely ordered by nature the better to enable them to escape from their enemies through the woods; otherwise they would become an easy prey to wolves and other beasts and be liable to get entangled among the trees, even when ranging for food. The same opinion may be admitted of the southern deer which always reside in the woods; but the northern deer, though by far the smallest in this (Churchill) country have much the larger horns, and at the same time spread so wide that it would make them specially liable to become entangled among the drooping branches of snow-laden trees. The young bucks, in those parts, do not shed their horns as soon as the old ones for even at Christmas time their horns are [1315] still firm on their heads. The does do not shed their horns until the summer; so that when the bucks' horns drop off those of the does are all hairy and have scarcely come to their full growth.["]

That the antler is capable of being reformed even after nearly complete separation, from the effect of an injury, is not to be doubted. I have seen two instances, one of which was a peculiar formed bone, as I supposed, lying on the ground but a few yards from the house at Ft. Chimo. It was glanced at with no thought of what it was until I examined it and found that it was a portion of the antler of a male reindeer. It had been broken short off at the main stem, apparently some twelve inches from the head, this is conjectural only as it was sawed off on each side of the fracture to afford a piece of horn for the handle of some implement. The main shaft was above the union of nearly the same size as below it; the fracture had a series of several small but quite sharp points of one-eighth to [1316] nearly half an inch growing irregularly from the junction or point where the reproduced portions started. A second instance is depicted in one of the plates of antlers herewith presented. The subject of the illustration is in the collection of the U.S. National Museum, and was obtained by me in the Ungava district.

"They wander over the large areas and where this season they may be found in greatest abundance next season not one will be found"

It is not necessary, in this connection to make further reference to it as the delineation speaks for itself.[7]

[1317] The habits of the reindeer are not erratic although with an important study of these animals it would, at first appear that they wander without regard to law or reason. For the present I shall confine my remarks to the reindeer of the Ungava district alone. They wander over the large areas and where this season they may be found in greatest abundance next season not one will be found. When a given locality has been occupied by them for a greater or less period it will be seen that the supply of pasturage has decreased according to the number in the herd or the length of time grazed upon. When this supply becomes exhausted the animals quietly remove to another locality, which instinct has taught them is prepared to afford the necessary subsistence. It is well known that the food of these [1318] creatures consists chiefly of the slowly growing lichens and mosses which require several years to again become of sufficient size to afford attractive pasturage and as there must be such great quantity consumed to afford the individual with flesh and fat it is to acquire the tract is not long in being denuded of its covering for it is not over the entire surface of the ground that it grows. Like other vegetation it requires certain conditions and the surface here in the Ungava district is fully half in area nothing but barren rock exposed through countless ages to the atmospheric changes which have produced but little visible effect upon it. One half of the remainder of the surface is covered with swamps and lakes, the former having but little of the reindeer food growing upon it; only here and there are there small [1319] patches having lichens and mosses growing upon them. Sphagnum moss is not eaten by the reindeer and only few twigs and leaves of other plants are consumed during the winter season.

Small herds may be seen here or there at certain seasons and again during other portions of the year huge droves of reindeer may be seen. There is a certain season in which all the animals of this species come together. During the month of September and early October the male reindeer come from the westward and other directions to join the females which are now in season to receive attention from the opposite sex. Huge bands may now be seen on the hillsides of the rutting grounds. After this season is over, generally not later than the first of November, the sexes separate, the fully adult males [1320] wander off in large straggling bands at first, gradually dwindling in number until but three or four may be seen together. During the latter part of the time thus passing they frequent the thinly wooded tracts and the smaller areas which are successively laid bare or covered by the snow drifting with each change of wind.

When the antlers begin to grow, the latter part of March or early April, rarely more than a single adult male will be found; others may, however, be in the vicinity but only few of them and these separated. During the spring and summer the adult males keep by themselves and not until the latter part of August do they reassemble in about the same way as they scattered during the previous winter and early spring. It should not be inferred that during [1321] the spring and summer that two or even half a dozen bucks may not be seen together, but this will result only from a mere accident bringing them to a focal point whence they separate in a few days.

During the summer, the adult bucks usually repair to the higher grounds or near the seashore; the purpose being to be less exposed to the incessant attacks of the myriads of mosquitoes infesting the interior lower grounds. Here near the seashore or on the hilltops the antlers are rapidly developed and not until the next rutting season are the terminal portions dry and freed from the integument covering them during the process of growth. Now by what instinct the sexes are led to approach a certain, not always the same, locality for the purpose of procreating their own kind is a matter well worthy of consideration. [1322] As the bucks are shedding the velvet from their

---

7. With the view that this manuscript was to be printed shortly after it was written in 1886, Turner made a note here for the printer about including a set of plates depicting caribou antlers. His exact words to the printer were: "[Note to Printer] The four plates of the reindeer antlers should be placed so as to follow in order 1, 2, 3 and 4, the manuscript page 1316. The four plates to succeed each other in that order." These drawings were produced by Ernest Thompson Seton but have not been located.

> *"I have also seen them throw the head to one side and with a smothered snort raise either of the forelegs and strike with the hoof. The blow... I firmly believe... would crush the skull of a man"*

antlers and that skin is removed by a process described in another connection it became a thought with me that as the skin is not at once ready to be removed but requires two to three weeks time before it becomes dry enough to be rubbed or scraped off may it not be that a certain odor remains and thus leads the females to follow the males to this place. It is true that the males approach from nearly every direction; but those arriving with the does (in September) are those which have joined the herds of females as the latter were returning to the locality to meet the bucks.

In the fall, the females leave the adult bucks and with the fawns of that year and the two preceding years gradually disperse into herds or bands, of variable number often numbering [1323] hundreds to as few as half a score. The does and their followers (for it must now be understood that adult bucks or those able to take and maintain a place among the rutting bucks are not allowed to accompany the females after the rutting season is over – not that they are driven off, but it is one of the traits of this animal, repair to the more open areas and there pass the winter. In the bright warming days of April and early May, large bands of females now take a course toward the Cape Chidley region and among the high hills and comparatively safe valleys they bring forth their young and cast their antlers. A month or so previous to this or even about this time (late May or early June) the older bucks (or those termed here as followers), now leave the females and are ready by the ensuing fall [1324] to establish themselves as adults. As soon as the recently delivered fawns are able to run well and make continuous journey, the females pursue on course toward the locality where they will meet the bucks in September and October.

During the rutting season the adult males have severe contests and as I have sat within thirty yards of scores of these huge beasts and saw their fighting so plainly that even the reversed hair and closed eyes of these creatures could be seen as they suddenly came with a clash that reminded one of two large chairs being struck forcibly together. Their manner of fighting is to oppose the horns and push, they rarely rush but seem to calmly put their horns together and shake their heads. It is not unusual to witness a buck dash against the side of another and throw him nearly from his feet. I have also seen them throw the head to one side and with a smothered snort raise either of the forelegs and strike with the hoof. The blow thus delivered is very rapid and I firmly believe it would crush the [1325] skull of a man. It is rare that two bucks, when fighting, become locked together by their horns. The brow antler and the lower branches of the main stem with their broad palmations and often numerous pectinations incurred offer such arrangement that it is surprising that such entanglement does not more often occur. The struggles of the creatures under such circumstances tend only to the more inextricably interlock them. It is difficult to conceive anything but purest accident that would release them from such a predicament.

Caribou Collection
The editors view caribou antlers collected by Turner. Division of Mammals, NMNH. Photo by Angela Frost, 2011.

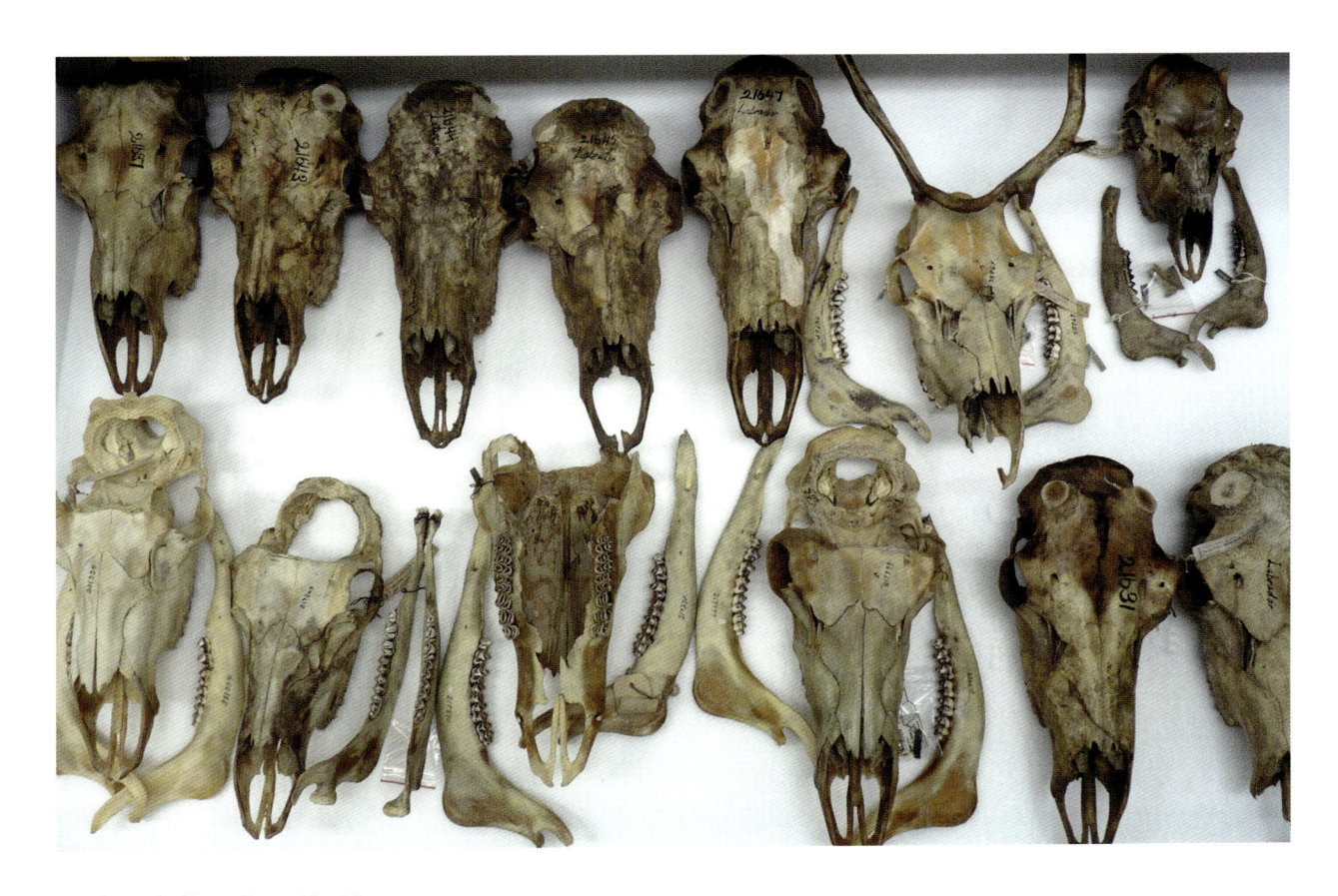

Caribou Skulls Collected by Turner
Division of Mammals, NMNH. Photo by Angela Frost, 2011.

Under such circumstances the creatures must await a slow death for the act of procuring food is nearly impossible and in the course of a few days the creatures starve to death unless relieved by the attack of wolves which are constantly prowling around the outskirts of the herds.

From the above remarks it will be clear that the reindeer are but seldom stationary for more than a month of the year. That once a year there is a center of abundance and that during the remainder of the year larger or smaller bands may be found in proper localities. Each alternate year the females repairing to the breeding grounds pass within few miles of Ft. Chimo; this [1326] has been going on for the past forty years that the station has been established. The females come from the barren grounds lying west of that station. The numbers passing there each even numbered year amount to thousands and the quantity of slain is more than treble that properly made use of; and besides this it should not be forgotten that there are gravid females bearing one or two young so that for each adult female there are at least two lives lost for each one slain.

[1327] The seasonal changes in the pellage and the color of the hairs of the different portions of the body of the reindeer vary greatly. There are two marked seasonal changes of color. The height of these being respectively for the summer about the middle of August and the latter part of March for the winter. There being no special differences between the colors of the males and females in any age the description will apply equally well to either sex.

In the fawn just born there is considerable diversity of the ground color of the hairs of the body, varying from light fulvous-brown to darker, nearly chestnut-brown with always a distinct, darker stripe along the top of neck and extending to the tip of the tail. It is also not unusual to distinguish a more or less evident prolongation of this dorsal color extending at least part way down the fore shoulder; the upper portions of the hams also having a somewhat extended or widened portion of the dorsal stripe spreading on the upper portion of the hind quarters. The abdomen and inside of the legs and flanks are always lighter than the sides, being at times a soiled creamy yellow or even lead gray. The hair of the body is kinky and only on the back and legs is it disposed to lie regularly. The head and ears are about [1328] the same color as the sides, although the face is usually darker yet not so dark as the dorsal stripe. All the legs are of the same color, slightly darker than the sides and lighter than the back. This hair does not appear to fall off, but becomes intermixed with the coating received by the later part of July. It now becomes nearly uniform clove brown on back and upper sides, lighter on neck and nearly the same on face; the sides of the head lighter and whitish on inferior portion of jaw, throat and abdomen, extending upward between the hind legs and to the root of the tail. The leg skin is covered with darker hair and not only on the upper inner parts of the limbs are they lighter than the lower portions. Around the upper edge of the hoof of each leg a whitish band of hair, having a width of one eighth to nearly half an inch, encircles the lower portion of the leg. The posterior portion often has a narrow fringe of longer whitish hairs, extending above the middle of the leg. This fringe and the circle of hairs above the hoof increase in size or width with age and become one of the principal characters of the markings of those members. Between the accessory hoofs and the lower hoofs there is behind a mess of stiffish white hairs forming a sort of tuft which even thrust themselves between the hoofs.

[1329] About the first of September the color of the summer pellage has reached its full growth and is rapidly replaced by the winter coat which is as variable in the adult individuals as to defy description. From between the horns to nearly the entire length of the neck is a narrowing patch of darker hair mixed plentifully with grayish. The back of the neck and front shoulders are grayish, almost whitish in certain individuals, succeeded by a brownish dorsal patch extending nearly to the rump where it becomes lighter, the same color as the sides, neck and hinder parts of the haunches. The sides of the hind quarters and fore-quarters are nearly the same as the patch on the forward portion of the top of the neck. The lower neck now has a long fringe of coarse hairs extending from the middle of the jaw to the part between the forelegs. This fringe is often pure white, usually light grayish, becoming pure white on abdomen and the inside of the legs. The legs are always darker in the adult than any other portion of the body hairs. The white circle of hairs around the hoofs and the fringe at the posterior portion of the legs is nearly pure white. The winter pellage attains its full development by the first of February and remains in this condition until the middle of May when it [1330] is replaced by the summer pellage of nearly pure clove-brown often having a bluish shade beneath the hairs and as often a brownish lighter shade in the tips. The long hairs of the throat and belly disappear in May and are but slightly developed in the summer condition.

The character of the hair of the reindeer is strikingly different than that of other members of the deer family. It is coarse, cylindrical, tapering to a point exteriorly and remarkably fragile. Intermixed with this coarse hair is a quantity of finest hair usually half the length of the coarser hairs. This fur is quite fine and does not change its color or texture with the season, becoming less abundant; however, in the summer condition of pellage and correspondingly more plentiful in winter, forming a thick protection with the coarser hairs sufficient to enable the creatures to withstand the rigors of the Arc-

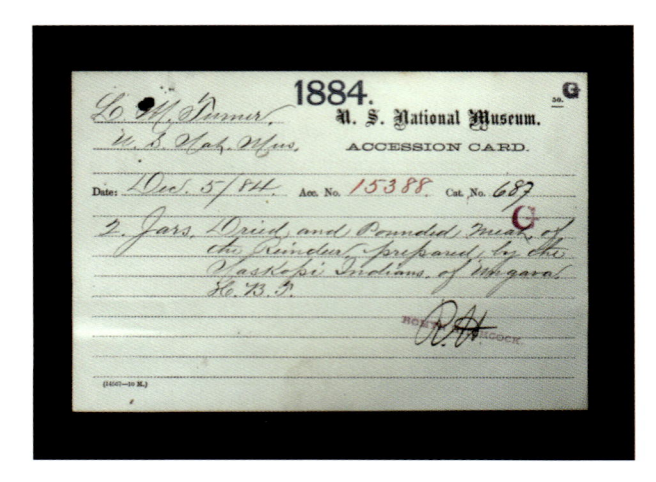

Accession Card Notes Describing Caribou Meat Collected by Turner
Two jars of dry and pounded Caribou meat, sent by Turner from Ungava, were received by the Smithsonian Institution registrar. Turner stated that the meat was prepared by the Naskapi Indians (now Innu) of Ungava Bay. Source: SI Archives, Record Unit 305, Accession 15388.

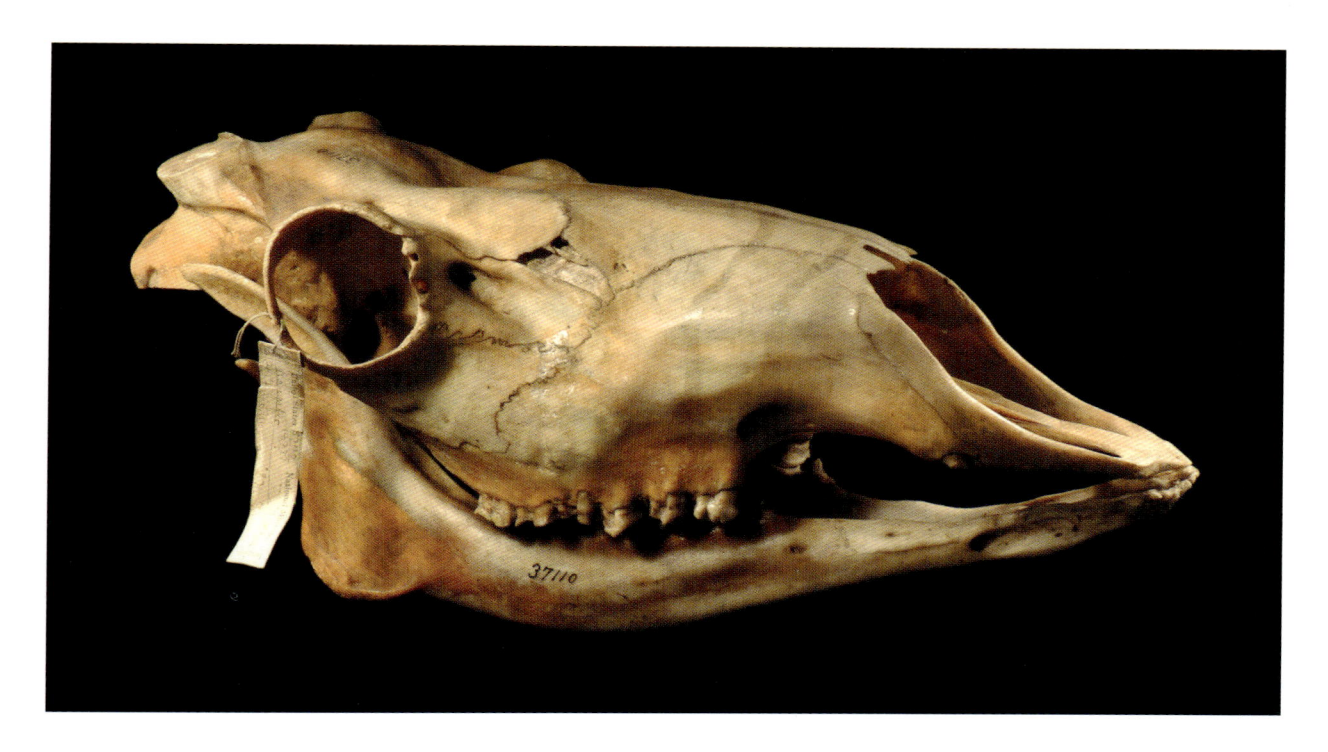

Caribou Skull Collected by Turner
USNM 14777/A37110, Division of Mammals, NMNH. Photo by Angela Frost, 2011.

tic winters with little heed to their severity. Its peculiar adaptability to protect the body from cold causes this animal to be with its flesh and, lightness of its skin, the most valuable of all the Arctic mammals.

After the middle of February the ends of the hairs of the sides become worn off, probably from the animal lying on the snow and the tips freezing to the crust of the heated snow where the [1331] creature has lain. The hairs of various portions of the body vary greatly in texture. The hairs of the legs being stiffish and closely lying to the limbs. The peculiarity of these hairs is that wet snow will not adhere to them and thus do not become as clogged when the creature steps in water as it wanders. This has been observed by the Eskimo and from the skin of its legs of the reindeer he constructs the mittens which are worn when preparing a snow-house or when handling snow. The long fringes on the neck of the deer are so brittle as to be of no service in the economy of the Eskimo household; while the white portions of skin covering the abdomen, flanks, and inside the legs are cut into various shaped strips and used in many ways of ornamentation, often, however, the hairs are trimmed to only half an inch in length and thus present a bristly appearance. The leg-skins are always used for the legs of Arctic boots worn in the country, more especially however in portions of Alaska than in Ungava. These trimmed with the white skin having its hair cut shorter and tastefully ornamented with many colored fringes and tassels of worsted and strips of Wolverene skin make an object well calculated to inspire admiration in the minds of Innuit youths.

[1332] When the reindeer wander in search of food or companionship of its kind, they do not hesitate to take to the water lying in their pathway. Wide and deep lakes or swift streams are crossed with but little apparent exertion. The creature is a bold swimmer and on approaching the margin of the water usually pauses for an instant and walks fearlessly into it. Stragglers have no particular point from which they cross; the larger bands, however; appear to have certain places through which they swim. Of all the reindeer I have seen, I have not come to the conclusion that any particular leader is recognized. It is of course a fact that the stronger ones will keep in the lead and the weaker follow at a greater or less distance.

"On the clear, calm water where the silent hunters strive with noiseless paddle to guide the terrified beasts to a convenient point for the fatal thrust of the spear the horns of the creatures may be heard clashing and the sound borne far over the still water"

The greater amount of swimming across streams is done in the fall when the [1333] sexes are seeking each other for the rutting season. The bucks have at this time the full development and weight of antlers and their flesh in best condition; the layer of backfat having then attained its maximum. The intestines are well covered with a layer of fat and this tends to increase the buoyancy of the creatures. The condition of the pellage also tends to hold the air within and amongst the hairs.

At a distance a band of adult bucks appear, for they swim in a compact group, like some leafless, well-branched tree floating upon the water. The head and several inches of the neck are raised high in the water. About one third of the body appears above the surface. The rump and head are the highest parts above while the foreshoulders are slightly depressed. This is accountable for from the fact that the muscular action necessary to support the weight [1334] of the head and antlers would have the tendency to depress the central portion of the body. The nose and jaws are as dry when emerging from the water as when entered. The head is perfectly free to move in any direction and as the deer is constantly scanning the water for danger or to look back at those following, they often present a peculiar appearance when the antlers are turned differently than the course pursued.

The feet are used exactly as in the act of walking in fact the animal appears to tread its way through the water in the same manner as when upon the land; but when the latter is reached if there be an enemy behind the creature is not slow to urge with heavy lunges its body through the water when its feet touch the ground below. Not a second is paused to free its body from the water but with rapid plunges [1335] it betakes itself to the high grounds beyond.

As previously remarked the members of a group keep closely together while crossing a stream and I have often seen the young of but four or five months directly among the adults of both sexes which at this season are together. The females do not differ at all from the males in the manner of swimming.

It often happens, especially when a canoe of hunters is pursuing the deer through the water, that the males become so panic stricken as to get their horns entangled and one or the other be crowded beneath the water of the frantic efforts of those behind to push forward as rapidly as possible for the hunter guards well the rear that none return whence they came. On the clear, calm water where the silent hunters strive with noiseless paddle to guide the terrified beasts to a convenient point for the [1336] fatal thrust of the spear the horns of the creatures may be heard clashing and the sound borne far over the still water.

It is not unusual, however, that the reindeer has not calculated the currents of the river and as there are numerous, whirling rapids into whose seething waters the strength of the deer is naught, where a dash here or there amongst the jagged rocks soon overcomes its power and the creature drowns and in time cast upon the shore as food for the wolf or other beast.

I could not learn that a reindeer wounded by the agency of man ever takes to the water to escape. They are more often to be found lying under the branches of a tree growing in damp or swampy ground.

[1337] "Deer were very plentiful the whole way; the Indians killed great numbers of them daily, merely for the sake of their skins; and, at this time, August, their pelts are in good condition and the hair of the proper length for clothing."

"The great destruction which is made of the deer in those parts, at this season of the year only, is almost incredible, as they are never known to have more than one young at a time it is wonderful that they do not become scarce: but so far is this from being the case, that the oldest northern Indian in all their tribe will affirm that the deer are as plentiful now as they have ever been. Though they are remarkably scarce some years near Churchill River yet it is said, that they are more plentiful in other parts of the country than they were formerly. The scarcity or abundance of these animals in different parts of the country at the same season is caused, [1338] in great measure, by the winds which prevail for sometime before; for the deer are supposed, by the natives, to walk always in the direction from which the wind blows, except when they migrate from E to W or W to E in search of opposite sex for the purpose of propagating their species."

[1339] "The month of October is the rutting season with the deer in those parts and after their courtship is over the bucks separate from the does; the former proceed to the westward to take shelter in the woods during the winter; the latter keep out on the barren grounds the whole year. This, though a general rule, is not without some exception, for does are to be found also in the woods but their number bears no proportion to that of the males. This rule, therefore only holds good respecting the deer to the north of the Churchill river; for the deer to the south live promiscuously among the woods as well as in the plains and along rivers and lakes the whole year."

[1340] "The deer in those parts are generally in motion from east to west or west to east, according to the season, or prevailing winds; and this accounts for the constant shifting of the northern Indians from one station to another in order to follow the deer. From November to May the bucks continue to the westward, among the woods, when their horns begin to sprout; after which they proceed to the eastward to the barren grounds; and, the does that have been on the barren grounds all the winter are taught by instinct to advance to the westward to meet the bucks. Immediately after the rutting season is over they separate."

Ungulates

Caribou Skin Sack Collected by Turner
USNM E90054, Museum Support Center, NMNH. Photo by Angela Frost, 2011.

[1341] The reindeer does not give birth to young until two years of age. After repeated inquiries I could not learn of a single instance in which a doe was known to bear young before that age.

[1342] It is stated by the Indians, Eskimo, and whitemen, who have had abundant opportunity for observation, that fully one-fifth of the adult female reindeer of the Ungava district have more than one fawn at a birth. I have myself often seen two young taken from a single doe in May and in one instance saw three. This later number is quite rare and it is said that one of the three dies in a short time after birth. Fabricius, Fauna Groenlandica, 1780; p. 27, states "They" the reindeer "assemble about the short days of the winter solstice and deliver their young in May or June; the young mostly single, rarely twins." The exact words are "Congreditus circa brumam, et parit mense Maio vel Junio pullum plerumque solitarium, rarius gemellos."

[1343] "In their summer hunts for the reindeer on the barren grounds, the Copper River Indians pursue the following method of obtaining them. When a herd of deer is desired the hunters study the character of the land surface to discover a narrow defile through which the animals may be driven. The direction of the wind is determined in order to get to the leeward of the deer. A large bundle of long sticks are a part of the summer outfit for deer hunting. These sticks resemble ramrods, which are now arranged in two diverging rows, each stick set about fifteen yards from the other and thus ranged for three or four hundred yards. On the upper end of the stick is placed a piece of skin or cloth or even a bunch of moss hastily snatched from the ground. The nearer end is placed near the mouth of the defile through which the animals are to be driven. The outer ends of the rows [1344] are about one hundred and fifty yards apart. The hunters selected for the purpose station themselves, at the farther end of the defile, behind stones, piles of moss or bushes, if convenient, to await the approach of the animals. The remaining hunters with the women and children separate into two parties and endeavor to form an irregular semicircle to the rear of the animals; the men keeping nearest the hunters so as to be ready to intercept the deer in case they should attempt to break off to one side. They are thus urged on until they are driven within the rows of sticks which they mistake for persons. They now perceive the narrow passage between the hillsides and rush impetuously through the defile where they are assailed by the arrows from the bows of the hunters stationed there. If the animals be numerous two or three fall to each hunter and as many as thirty or more may be procured."

[1345] "Some of the more northern Indian tribes, who hunt the deer on the barren grounds, make long bags of a sufficient number of skins from the legs of the deer and turn the skins so that the ends of the hair will point from the person, who then fills this sack with meat, skins, or other valuables, and drags it over the ground or snow to the edge of the woods where either the old sleds were left behind or where new ones may be constructed. The smoothness of the hair enables the sack to glide over the surface with but little wear upon the receptacle."

[1346] An Eskimo of Northern Sound, Alaska, informed me that he had killed, with a single arrow, two reindeer and so severely wounded a third that it fell under the second arrow. This may appear surprising but anyone who has witnessed an Eskimo shoot an arrow would not doubt his ability to transfix two or more deer if the opportunity presented itself. [1347] The Eskimo dwelling on the extreme Arctic lands often construct the sides of these pounds from rocks, turf, or even snow and ice. The former often enduring for several years' service.

[1348] "It requires the prime parts of the skins of from eight to ten reindeer to make a complete suit of clothing for each grown person during the winter; all of which he killed during the month of August if possible or not later than the middle of September for after this date the hair is too long and so loose that it will readily drop out with but the slightest carelessness on the part of the wearer. Beside these skins which must, for the sake of warmth, be in the hair, he requires several others to be dressed into buckskin for moccasins, stockings and light summer clothing; several more are needed in a parchment state to be cut into thongs and cords for netting their snowshoes, nets, sled-cords and in fact everything where strings are necessary. From this it will be readily seen that each adult requires annually not less than twenty reindeer skins, exclusive of tenting, [1349] bags, and other

Inuit Girls in Caribou Clothing
Photo by Turner, probably taken at his observatory in Ft. Chimo, 1882. Source: SPC Arctic Eskimo Labrador NM No ACC # Cat 175484 01448200, National Anthropological Archives, Smithsonian Institution. 1 cyanotype 4" × 5" mounted on 5" × 8".

domestic articles, for these are usually made of pelts procured at a later date. When the rutting season is over the skins are very thin and of but little service. They are sometimes saved and boiled for food during periods of scarcity and are said to be far from disagreeable in taste."

[1350] The larvae of the deerfly are, in the spring months, carefully collected in a vessel and taken to the camp where they are eaten raw or when boiled. They are consumed alike by the Indians and Eskimos and relished as dainties, asserted to compare in flavor with parasites which infest the bodies of the natives themselves.

[1351] "The grubs and warbles are always eaten raw; and, by some of the Indians, are considered as nice as gooseberries. The children are very fond of them especially when they are full and fat and as large as the first joint of their little finger."

"The reindeer feeds for a few minutes only, and then searches the horizon toward which the wind blows as he trusts to the wind to convey the scent of danger from the direction from which it blows"

[1352] "Blood from the animal is mixed with the half-digested food which is found in the deer's stomach and boiled with sufficient water to render it the consistence of pea soup. Fat and few scraps of meat are mixed with it. To make it more palatable, the blood is alone mixed with the mass and this together with the stomach are placed near the fire for several days until fermentation sets in and gives the contents a slightly acid taste which renders it more agreeable. Some persons would hesitate to partake of this food if they should see the manner in which it is prepared, for most of the fat is chewed by boys and girls who have good teeth in order to break the oil globules; otherwise, the fat would remain in lumps like suet. Old people who have poor teeth and young children who have not clean fingers are not permitted to touch the fat while it is being thus prepared."

"They frequently eat the contents of [1353] the stomach while yet warm from the freshly killed deer in winter, for the summer food of those animals is more varied and contains leaves and twigs which are coarse and indigestible to the human stomach." The young, taken from the dam, are considered delicious food by the Indians and Eskimo of both coasts. Other parts not necessary to particularize are also deemed dainty morsels which must not be cut with a knife but torn with the fingers and teeth. Neither the dogs nor the women are permitted to eat these parts lest ill-success should befall their hunt. The bones are, by the Indians, not permitted to be thrown where their dogs may gnaw them otherwise the deer will forsake their usual haunts.

[1354] In the spring when the nasal recesses of the reindeer become filled with the larvae of the flies, *Oestris* [*Oestrus*], which infest these animals, the sense of smell is rendered dull, but the faculty of sight and of hearing is proportionately more acute (This is doubtless due to the effect produced by the presence of the mass of larvae deadening the one set of nerves while the irritation stimulates the nerves of sight and hearing.) In the months of September and October the sense of smell is doubly acute and the natives assert that by this means alone are the sexes drawn together. At this season also the sight is not clear as I have seen the animals sniffing the air when I was within a hundred yards of the herd and from the action of the animals I believed I was not seen by them.

I was once out hunting geese with an Eskimo of Ungava Bay. The [1355] geese were in a large lake which had to be approached from one side which formed a wall of granite of varying height. The other side was a level tract which rendered approach an impossibility. While crawling along on hands and knees or, at times, prostrate, I had occasion to draw my gun after me but once I thoughtlessly brought the end over end. The native immediately remarked that such an action on my part might do while hunting geese, but would not do if I was hunting deer. He explained that the reindeer feeds for a few minutes only, and then searches the horizon toward which the wind blows as he trusts to the wind to convey the scent of danger from the direction from which it blows and has learned that all danger comes from the opposite direction. Any object moving on the horizon or which was not before detected is the [1356] cause of alarm. The deer watches for that least motion of the object and if discerned, the animal immediately throws its head up and trots a few steps, pauses and again looks.

If satisfied that it is danger it endeavors to get to the leeward of it and by the aid of the sense of smell discovers that presence of its enemy. It does not so readily detect a person or other object below the horizon for the reason that there are so many rocks, boulders, clumps of bushes and other natural objects that the deer does not notice them. It is said that the deer rarely feeds on the ridges for the reason that it chooses to be below the horizon in order that the approach of foes may be the more readily observed. The native when stalking a single deer, which is always more watchful than a herd, approaches only when the deer is feeding and stops the instant [1357] it raises its head. By exercising due caution the single animal may be approached within a few yards. This same native told me that he had gone so close that when he shouted and was seen by the deer it was so terrified as to be incapable of movement, simply spreading its four legs widely apart and thus remaining so long as the hunter is quiet but at the least motion from him away goes the deer like a bird on the wing.

[1358] While in the vicinity of St. Michaels, Alaska, I was repeatedly assured of the Eskimo that the male reindeer will sometimes attain such great age as to be incapable of furnishing sufficient material for the renewal of the horns and that merely a stub with a large knob results. This condition is termed by those Eskimo as Angvót-fuk or very old. When they attain such an age, they are unable to maintain their place with the remainder of the herd and during the rutting season are prevented from approaching the females. They consequently retain the deposit of fat on the back and are always in first rate condition. The Eskimo value the flesh very highly, asserting that it is more tender than that of a young animal and is in good condition throughout the year. These bucks are compelled to hang on the outskirts of the herd for protection from wolves and other enemies.

[1359] Of the many scores of these animals which I have seen dissected, I have never found the gall present in a single individual. The Eskimo assert that this organ is very rare and many contend that it does not occur in the reindeer. It is a fact well known to hunters that the gall is sometimes absent in the common red deer.

[1360] The Eskimo assert that the reindeer have a duct, leading from between the hoofs of the forefeet to the nose, which enables these animals to perceive the scent of an object through the leg. This, they claim, is attested by the fact that when the deer is travelling across the track of an enemy they rarely smell of it but either flee or halt and paw the spot with their hoofs and thus discern the presence or proximity of food or danger. My own observation, in regard to these animals searching for food, showed that the creature does not detect the patches of moss by sense of smell through the nose. The eye is not able to discern the food beneath several inches of snow. The hoofs of the forelegs alone are used in removing the snow from the patches of moss and lichens. Great quantities of snow are certainly eaten with

Illustration by Marcia Bakry, Smithsonian Institution, 2012.

the food procured in this manner. There are, of course, in the northern region perpetual springs of water. I have never heard of the reindeer frequenting such places for water to drink.

[1361] When the fawns are but a few days of age, they evince no faculty to perceive by sense of sight or smell danger even though within but a few feet of them. An incident in proof of this was related to me at Ft. Chimo, Ungava. Up the Koksoak River, nearly 100 miles from its mouth is situated an immense bluff, through which the river has cut its way. The approach to the top of the bluff is along a narrow ridge or neck which suddenly expands and forms a level or table for the top of this bluff and terminates with sides almost perpendicular for nearly a thousand feet. The narrator of the incident ascended to the top and found the plain to be literally moving with does and their young. Retreat for the animals was cut off.

UNGULATES

Innu Child's Caribou Skin Coat

Collected by Turner, probably at Ft. Chimo 1884. Burnham (1992: 241-259) described several Innu caribou skin coats collected by Turner in Ungava, but not this child's coat. Burnham noted that Turner's coats "are documented examples of Naskapi painted coats, something that is very rare." USNM E90051, Museum Support Center, NMNH. Photo by Angela Frost, 2011.

> *"... I saw, at Davis Inlet, a buck reindeer of three years that had been in captivity for many months...It had been fed upon crackers, oatmeal and its natural food. The animal was taken on board the Steamer "Labrador" for shipment to London where it was to be presented to the Zoological Garden of that city"*

The adults fled to the edge of the bluff, deserting their young. The fawns appeared devoid of the elements of fear; and, in a [1362] few moments were approached and even stroked along the back, the little creatures in return licking the hand that caressed them.

On my return down the coast of Labrador in 1884, I saw, at Davis Inlet, a buck reindeer of three years that had been in captivity for many months. It was captured while swimming a small stream. The animal was quite gentle and in excellent condition of flesh. It had been fed upon crackers, oatmeal and its natural food. The animal was taken on board the Steamer "Labrador" for shipment to London where it was to be presented to the Zoological Garden of that city. I have not since learned whether the creature reached its destination alive.

[1363] Nothing will induce some tribes of Indians to deliver a living reindeer to a whiteman lest the guardian spirit of the deer be offended and cause all the remaining deer to forsake the country. The Eskimo, on the contrary, have no hesitancy in disposing of the living deer as their shamans are able, at their own pleasure, to call the spirits of the deer, already slain, to the earth where they immediately assume a material form ready to be again slain and as often resume the earthly form.

[1364] The form of the foot of the reindeer is as peculiar, differing in character from that of other cervids, that a description of it will prove its special adaptability to sustain the body of the creature at any season of the year (for if the ox, hog or horse was suddenly transported to the same localities over which the reindeer travels with the sure-footedness of a goat, the fitness of the deer's hoofs would at once become apparent). The shape of the large toes of the hind and forefeet in the winter season, resemble two large muscle [mussel] shells in which the hinge forms the basal connexion. The color is usually dark bluish-black, more or less glossy. The extreme ends of the toes overlap either the outer over the inner or vice versa, there being no regularity in this for the same individual the hoofs of the different feet may not be disposed in the same manner. The excess [1365] appears that the outer toe overlaps the inner. This would seem to have an important bearing as it would tend to bring the lower edge of the outer hoof more perpendicularly upon the surface and by its sharp edge afford a more secure foothold. The extreme tips of the hoofs being somewhat incurved one or the other must overlap. The undersurface is hollowed out and thus gives the toe the appearance of a shell. The outer edge of the toes (hoofs) are quite thin, varying from one-eighth to one fourth of an inch in thickness. The cavity is of variable depth; usually, about half an inch, somewhat shallower in the fore toes than in those of the hind feet. When the animal is walking, the toes, of course, separate and just as the foot is placed down the ends of the toes resume their respective positions. In certain conditions, the tips of the toes are kept separate and only on level, hard surface will the overlapping be distinguished.

Blade of a Caribou Antler Bearing Field Data in Turner's Handwriting
USNM A21647, Division of Mammals, NMNH. Photo by Angela Frost, 2011.

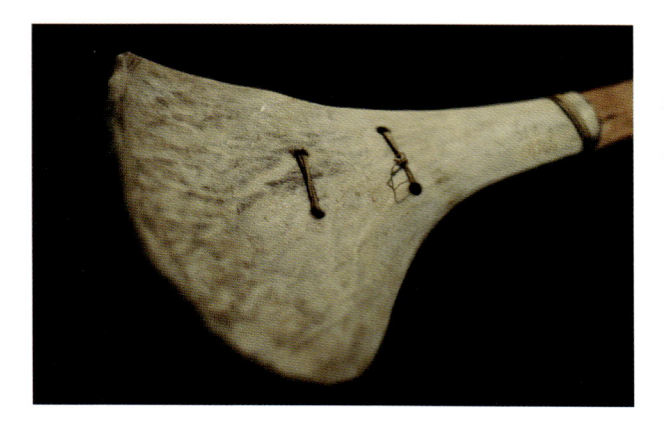

Ice-Bailing Scoop Made from Caribou Antler and Wood
Collected by Turner in Ungava Bay. USNM E90108, Museum Support Center, NMNH. Photo by Angela Frost, 2011.

[1366] When in rapid motion and when on slippery surfaces the toes are then spread widely apart. The size of the track then depends not only upon the gait of the animal but as much upon the character of the ground or surface over which it travels. When in a moderate run, the tips of the toes make a rattling sound like several walnuts placed in a bag and then violently shaken. This sound has a peculiar effect, causing one to imagine that the sinews of the legs were crashing as the creature moves along. The gait of the reindeer may be considered under three characters each differing greatly from the others.

The walk is affected by the foot being drawn up and carried stiffly forward with a swinging motion of the body. The hind foot is brought sometimes behind or in front of the print of the forehoof. The motion when the animal is at ease is usually rapid, faster than a man can walk. This motion [1367] may be accelerated or lessened as the incentive of the creature impels. In a rapid walk, the head is always thrown up, otherwise not usually so. The carriage of the head is a certain index of the next movement of the reindeer for that portion is nearly always placed in position before the animal moves more than three or four steps.

The trot is more difficult and likewise may be fast or slow. In this motion, the forelegs are often lifted high and the knee joint more or less curved. The feet are then thrown forward with considerable force and the edges of the hoofs brought flat to the surface. In the winter season the hoofs are admirable instruments which undergo a change whereby their effectiveness is greatly increased. About the time that the surface of the earth then becomes frozen and this process contracting the yielding, wet grounds causes the water to be forced to the surface where [1368] it is frozen, presenting an area more or less covered with ice. As the season advances, and the dryness of the snow together with the icy surfaces exposed, the hoof then becomes quite changed. The frog of the hoof which in summer consisted of a tough, apparently homogeneous mass of corneous laminae lying flat upon each other now begin to shrivel and part from the interior rim of the hoof. The splitting and separation of the layers of laminae continue to drop from the hoof, doubtless often accelerated by accidental circumstances until by the end of December what was in summer (July and August) a hoof filled on the lower surface with that substance now presents the shell-like form already remarked. The sharp edges now presented on the walking surface are of the firmest or hardest substance. The edges on the ice take firm hold and thus enable the reindeer to progress when without this a special change [1368] would enable its relentless foes to soon destroy the last vestige of its kind. In the spring when the snows are melting and the cutting edges of the hoofs cause the creature to sink its feet like sharp knives into the yielding ice or the rapidly thawing ground nature comes again to the rescue and prepares the summer condition of the hoof.

The dry, hard, but elastic, hoof is now filled on its inferior surface with the tough laminae which by the middle of June or early July present the same appearance usually observed in the ox. The dry rocks and the moist ground tend to keep the frog in such condition that so fast the cavity fills it is abraded by natural causes until the winter season again sets in. The cleft of the hoof is quite deep and the track as produced on substances having a more or less yielding surface are as different that unless one carefully [1370] observe it he will conclude it to be of another individual. On hard surfaces and with the slower paces the track is smaller from the fact that the entire weight of the body is not thrown on that particular member. When, however the reindeer is moving with great rapidity the entire foot is brought to the ground and presents a track of large surface. The third gait is a shambling trot in which the forefeet are thrown outward with a sort of curve reaching far ahead; the hind feet are brought forward with a straight line movement and firmly planted. Upon the hind feet the animal relies for preservation of balance and security from slipping for in the latter case the accessory hoofs of the hind legs may be easily traced upon the smooth ice or snow-crusted rock.

Model of a Drying Frame by Inuit
This frame was placed over a lamp for drying wet boots, mittens, and other articles (Turner 1894: 230-231): "The semi-circular bow has cross strands of sinew or seakskin to form a mesh. On this rests the article to be dried. Under this is a support formed of two sharp-pointed pegs which are stuck into the snow forming the side of the hut. The shape of the support is that of a long staple with square corners. In some instances the pegs form only a wide V-shape and the frame for supporting the articles are laid directly on this. A block of wood hollowed out to receive the convex bottom of the frame is sometimes used to support the latter." USNM E90235, Museum Support Center, NMNH. Photo by Angela Frost, 2011.

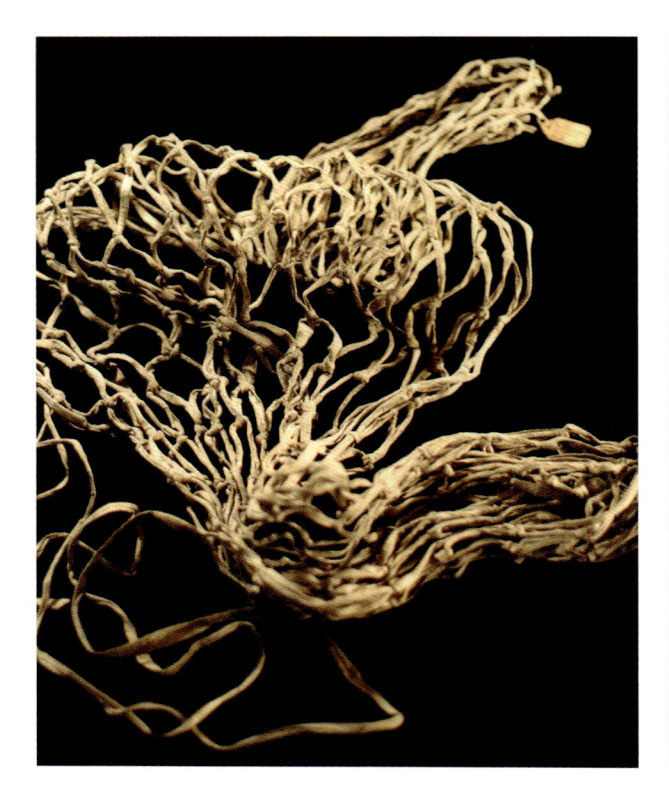

Bird Net Made by the Innu from Caribou Leather
Collected by Turner, 1884. USNM E90029, Museum Support Center, NMNH. Photo by Scott Heyes, 2011.

Inuit Artist Working with Caribou Antler
Kangiqsualujjuamiut artist Daniel Annanack generates a carving from a caribou antler. He carries on a long tradition of carving in his family. Photo by Scott Heyes, 2008.

The accessory hoofs of the forelegs apparently subserve [1371] the purpose of preventing those numbers from sliding outward as those claws are always longer for the outer ones than those on the inner side of the legs.

The track of an adult buck may measure as much as six inches from the tip of the toe to the tip of the dew claw or accessory hoofs. The width of the track depends greatly upon the rate at which the individual moved. The greatest width I have measured was little more than five inches but in this instance the animal was one of the largest I had ever seen and being at that time (October 10th) in best condition of flesh. Considering also that I had put three Winchester rifle balls into his body, it is not surprising he made tracks to get away with the two does whose favors he was then courting.

Another gait of the reindeer should be [1372] considered but as it is only exceptional and that when suddenly seized with alarm the reindeer makes a sudden bound forward, a sort of spring not like the high bound of the Virginia Deer but a plunge of several feet which is continued as a jerking gallop. This motion is rarely continued for more than a score of rods when the swinging trot which is the most rapid of all the progressive movements of this animal carries it quickly beyond reach of its foes. The wolf alone being able and it by its persevering nature to overtake the animal. I have seen misguided yet presumptuous Eskimo dogs attempt to catch a reindeer on the smooth surface of the snow-covered river ice. It was but a minute until the dogs were wondering what became of the reindeer.

The sexes have, of course, hoofs of [1373] variable size according to the age and individual. The following measurements of the hoofs of the fore and hind foot of a medium-sized doe will be sufficient for this purpose. In the absence of an illustration certain parts will require a description in order that it may be understood in the sense or term used.

Chapter Two: Mammals of Ungava and Labrador

	Inches
Left hind foot. Extreme length of hoof, measured from insertion of hairs to tip of the toe[8]	3.00
Width of widest portion of each toe	2.00
Cleft	2.87
Cavity	1.43
Dew claw (accessory hoof)	2.25
(The outer accessory hoof is always larger and longer than the inner.)	
Keel of accessory hoof (the keel is the ridge or recurved inner edge of the accessory hoof	1.62
Height (or width) of keel	0.25
[1374] Greatest length of interdigital tuft of hair, measured on inferior surface of hoof	4.00
Tips of hair on superior surface of hoof reach to within one and five-eighths inches from the tip of the large toes	
Measurement of right forefoot of same individual:	
Length of toe	3.12
Cleft	3.00
Greatest breadth of toe	2.18
Width of cavity	1.87
Length of inferior surface of hoof	3.88
From tip of inter-digital tuft of hair to the tip of the toes	1.12
Length of accessory hoof	2.12
Width of accessory hoof	1.37
Keel of accessory hoof (length)	1.37
Height of keel	0.30
Length of longest hairs of inter-digital tuft	2.00

[1375] The scarcity of reindeer along the eastern shore of James' Bay (the southern extremity of Hudson Bay) has occurred in the past ten years and is due more to the lack of pasturage than any unusual persecution from the Indians, who assert that the deer have joined the herds of the Koksoak valley. This migration of the deer has already caused many of the Indians along that shore to forsake the hunting grounds of their fathers and now blend with the Nascopie Indians who inhabit that valley.

The reindeer forming as it does the staple of life to the Indians of this region and a greater portion of the food of the Eskimo is as frequently referred to in other connections that it is not necessary to present again the various remarks concurring this subject; and with so little material for comparison with that obtained by me it is much to be regretted that a fuller account of the [1376] distinctions of the two (or three) recognized species could not be presented in this connection.

Caribou Skulls Collected by Turner
USNM A21631 and A21632, Division of Mammals, NMNH. Photo by Angela Frost, 2011.

---
8. This table and those on pp. 126, 142, 280, and 289 were prepared and inserted by Turner into the original manuscript.

Caribou Skull and Antlers
An impressive set of caribou antlers on a specimen collected by Turner, probably near Ft. Chimo in 1882. USNM A21651, NMNH Museum Support Center, Suitland, Maryland. Photo by Donald Hurlbert, 2013.

Bone Scrapers Collected by Turner
Skin scraping tools made by Innu hunters from caribou long bones. A portion of the bone was removed for solid purchase on the tool. USNM numbers, from top to bottom: E89922, E89928, E89929, and E89924 (this last flesher contains an iron blade, probably from a wooden plane). Museum Support Center, NMNH. Photo by Angela Frost, 2011.

Caribou Ethnography
Turner

CARIBOU BOOTS AND THE TANNING PROCESS (1887a: 130-139)

The covering for the feet [USNM E90356] here referred to are intended to be worn only about the camp during moderately dry weather. The tanned and smoked reindeer-skin tops have been purchased from the Naskopie Indians, inhabiting the contiguous interior lands. The sealskin tongue and heel strap are prepared from the pelt of either a Harp or a Ringed Seal. The solepiece is also from the hide of either of those amphibians. The process of tanning the sealskin accounts for the difference of color. The dark surface of the tongue-piece is simply the scurf yet remaining on the skin; and, has the property of becoming somewhat glutinous when moist and tends to prevent the continued wet of the exterior from penetrating the follicles of the deep-set hairs on the body of the seal. The white sole has its creamy color due to a combination of processes. The exterior slime or scurf, the adherent epithelium or scarf-skin, is removed by allowing the process of decomposition, necessary to loosen the hair from its follicle, to proceed to such an advanced stage as to permit the scurf to be removed by rubbing it with a dull-edged blade of ivory or horn. The skin is now placed on a stretcher and drawn to its utmost tension. It is now placed in the severest cold, clear air to dry as rapidly as possible. The process has extracted much of the fatter matter within the fibres of the skin thus does not add the brownish color due to partial carbonization of the oleaginous matter, by contact with the oxygen of the atmosphere.

The frost-tanned skins are, of course, not so serviceable and impervious to the water as those from which the scurf has not been removed by scraping or immersion by continued wear in wet weather. It may appear incredible but the wet of melting snow more quickly causes the scurf to peel off than rain would do.

The process of tanning the pelts of the seals differ from the method employed to soften the hides of the reindeer in order to convert them into garments. When the skin of the seal is removed from the body there is a greater or less amount of fat adherent to it. The fat and fleshy particles may be immediately scraped or cut from the pelt by means of a knife and scraper. The knife used is the common butcher knife, purchased from the trader. The scraper is a piece of metal set into a handle; and, has a very close resemblance to the common form of the grocer's scoop, except that the edge is cut more nearly square across to leave a more extensive edge; the size, of course, is not as great as that of the scoop referred to. With this instrument the fat is scraped from the skin to a satisfactory degree, enough however being left to form a kind of soap resulting from one of the liquids used in the process of tanning.

If it be desirable that the hair be taken from the skin the pelts are cast into a corner of the tent or hut and allowed to decompose until the hair is sufficiently loosened to permit it being scraped off with the same scraper that is employed to rid the flesh side of the skin from its fat. The fingers pull out the tufts of hair which are remaining or with the sharp edge of a knife they are shaved from the surface; great care being exercised not to abrade the scarf-skin. The pelt is now worked in every direction to increase its pliability, the very heavy skins, of the largest seals, intended for tenting, boat-cover, boot-soles and the traces of dog-harnesses of course do not undergo the manipulating process for softening them.

The sealskins intended for raiment, boot-legs, mittens and all other purposes, requiring a greater or less degree of pliability, must be thoroughly softened lest they quickly become hard and dry.

After the fat and hair are removed and the skin worked moderately soft it is now immersed in a liquid which has been saved from time to time. After a bath of few hours to as many days the skin is taken from the receptacle and subjected to further rubbing between the hands and softening by means of a stout stick having a sharp edge which is pushed here and there over the flesh side of the skin. In former times a piece of flint, set in the end of a handle, was used in place of the stick mentioned; but, as the introduction of metals has so far replaced the former conveniences of those people it is rare to find them using stone implements for any purpose, although a stone instrument, somewhat resembling a common scythe stone, is used instead of the stick referred to above.

Caribou Shoes
Ice shoes with corrugated soles made by Hudson Strait Inuit. Collected by Turner at Ft. Chimo. USNM E90191, Museum Support Center, NMNH. Photo by Scott Heyes, 2011.

There are certain parts, especially the edges of the skin, that cannot be rendered soft by the friction of the scraper, here the teeth of the operator come well into use. The skin is seized between the teeth and chewed until it is soft. For certain skins, especially those made into boot-soles the teeth are the only means used in softening the parts which the rubbing may not do.

Where a person wears a single pair of boots for two or more consecutive days the sole of the boot becomes inordinately thickened by retention of moisture between the fibres of the skin. When such a boot is laid aside the sole dries and shrinks so much as to be uncomfortable to the foot. The sole must be rendered pliable and is best done by being chewed by the teeth.

The various localities have different patterns or forms of foot gear, so that it is possible to determine at a glance the region whence the boots came.

The sole is cut from the thick hide of a Bearded Seal or from that of the Harp Seal, the former having better lasting properties than the latter. The shape of the sole is peculiar, being nearly circular, slightly longer one way than the other, the center of the toe end having a slight point there and at the heel.

The leg is formed of a single piece and of course has but a single seam; the tongue or portion to cover the instep may or may not be a separate piece; if it is the leg seam comes in front; if the tongue be a portion of the leg-piece the seam comes behind. The insertion of the tongue-piece constitutes one of the distinctive points for determining the locality where made.

The leg is sewed and the tongue fitted and sewed to the leg of the boot. The sole is now placed in proper position and a tacking stitch taken at each side and at each end of the sole to the lower edge of the bootleg.

The attachments of the sole to the leg, by means of sewing, is begun usually at the side, for the boot is turned inside out, and progresses toward the toe and around to the other side tack-stitch; the surplus edge of the sole being drawn in by the successive stitches. The heel is sewed to the upper the same as the toe. Where the boots are long in the leg, termed knee-boots or high-boots, the weight of that portion has a tendency to push down the moist, soft part of the heel of the sole. To obviate this disagreeable consequence the maker of the boot sews several threads of stout sinew either around the heel or else perpendicular seams at that place, slightly drawing the stitches together.

Pair of Caribou Shoes
Caribou shoes that Turner collected at Ft. Chimo, probably in 1883. Turner described the boot making and tanning process in great detail, including how skins were softened: "The skin is seized between the teeth and chewed until it is soft. For certain skins, especially those made into boot-soles the teeth are the only means used in softening the parts which the rubbing may not do." USNM E90356/3253, Museum Support Center, NMNH. Photo by Donald Hurlbert, 2013.

The middle of the side-seam may have a piece of skin inserted, between the sole and upper, to form an eyelet through which a thong may pass and be drawn tight over the foot to prevent the boot from falling from the foot. Around the top of the leg of the knee-boots, a drawstring is inserted in the top seam; this being drawn tight may render the side eyelets unnecessary.

In all the sewing of the boots and articles requiring strong thread, sinew from the reindeer or seal is used. The sinew lies below the superficial muscles of the lumbar region. It is abstracted by making an incision in the flesh near the forward end of the tendon where it disperses amongst the muscular tissue; the finger is inserted and worked under the sinew-layer. A stout pull releases it from its attachments and it is stripped off. In an aged animal it requires greater strength than a man possesses to dislodge it. A stout thong is tied around it at the place where the finger was inserted and by means of a stick the loop of thong twists the sinew from the muscles. It is now washed to free it from blood and a few passings of a knife blade along it rids it of the muscular attachments adhering to it. It is now spread out to dry. The more the blood is washed from it the lighter colored and stronger will be the fibres. When dry it is fit for use. The tendon bifurcates for about one-third of the length posteriorly; here the frayed ends make it convenient for separating the fibres into various sizes and lengths.

A strand is split off by means of the fingernail and of a deft motion between the thumb-nail and finger the strand is freed from knots and irregularities. It is now threaded into the eye of a metal needle and the stitching begun.

The seams are simple, placing the edges of the two pieces of material so that the thread will pass through each and form a common seam. Occasionally it will be a "filled" or a "welted" or even a "whipped" seam. An overlapping or double seam is rare except in heavy sewing of a cover for umiak or Khiak.

Where the Innuit are able to obtain the triangular needles (Glovers needles) they prefer them so they perforate the tough skin more easily than the common needle.

The various articles of course require different kinds of sewing. The western Innuit, inhabiting the south side of the western extremity of Hudson Strait, do not have the abundance of reindeer in winter that their eastern relatives have, so that their food supply is drawn mostly from the sea. This necessitates a great amount of journeying on ice of all conditions; at times so smooth as to render progress next to impossible. To obviate slipping, the ingenuity of the Innuit has devised a method of corrugating the soles of his ice shoes that materially assists him in maintaining an erect position. A number of narrow strips of sealskin are prepared and the end of one is stitched to a piece of skin which when complete will form the sub-sole of the ice shoe. The strip is now looped and pushed against the stitched end, another stitch is taken and the process of alternately stitching and looping takes place until a piece large enough to cover the sole of the foot is made. The sub-sole is now stitched around its edges to the sole of a pair of half-boots and these slipped over the pair of water boots which the native wears.

The manner of sewing and tanning the skins of the reindeer differs somewhat from that here described and will be referred to in another connection.

CARIBOU SLEEPING BAG (1887A: 216-217)

The sleeping-bag [USNM E74466] is, without doubt, an invention of the Hudson Bay employees. That it serves an admirable purpose none will dispute. Many of the Innuit, especially those near the trading-posts have adopted the sleeping bag as one of the necessities among their meager effects. The name is sufficiently indicative of its purpose. It is simply a long bag made of reindeer skins and is usually about two feet longer than the height of the person. If very cold or to be subjected to rough usage a double bag is made, one to slip within the other. Sometimes the outer bag is of sealskin and stands the rough wear and tear better than that made of reindeer skin.

In order to compose oneself within the bag the person must sit down and draw the inner part of the bag over the limbs until the feet touch the bottom. By a deft movement the remaining folds are passed under the body and drawn over the shoulder. The person now reclines and arranges the parts over the head.

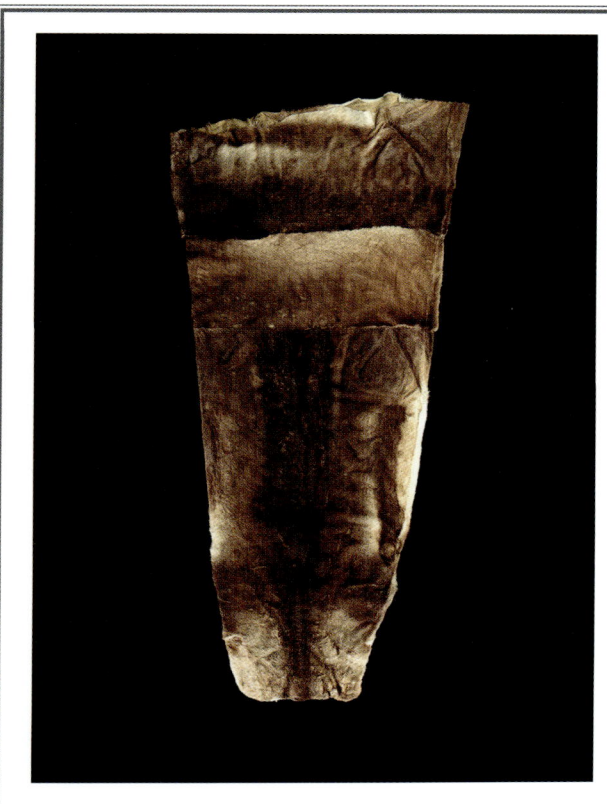

Caribou Skin Sleeping Bag (1884)

A sleeping bag used and collected by Turner at Ft. Chimo. He noted that the sleeping bag was a European concept co-opted by some Inuit. It measures 7′ × 3′ × 8″. Turner wrote about its comfort: "The Arctic traveler who has once indulged in sleep within the soothing folds of a soft reindeer skin sleeping-bag will be loath to expose himself to the vagaries of an uncomfortable bed and shifting blankets." Turner (1887a: 703) noted that pogáluk *was the Inuit word for sleeping bag. USNM E74466, Museum Support Center, NMNH. Photo by Donald Hurlbert, 2013.*

With the skin of a white bear on the hide of a reindeer between the bag and the snow beneath no one need fear to pass the coldest night without other protection. However cold the temperature may be there should be ample space left between the folds so as to permit the exhaled moisture of the breath from congealing about the face or permitting it to be retained within the bag for under such circumstances he will have a cold moist bed for the next night.

On arising the bag should be carefully shaken from the bottom up and then rolled so that the mouth will be the central part of the roll. This insures that no drifting snow on the next day's travel will enter the bag and cause the same unpleasant condition as referred to above. None of the Naskopie Indians have adopted the sleeping-bag as a part of their comforts and here as elsewhere they appear to be less progressive than their neighbors the Innuit. The Arctic traveler who has once indulged in sleep within the soothing folds of a soft reindeer skin sleeping-bag will be loth to expose himself to the vagaries of an uncomfortable bed and shifting blankets.

WRAPPED IN DEERSKIN (1887A: 70-71)

The living [Inuit] mourn their loss with subdued expressions of grief; and when a person, well-respected of the community, dies, great attention is devoted to the burial. The locality determines the disposition of the corpse. It may be exposed on the bare earth at any elevation convenient or it may have stones piled about it to prevent the ravages of beasts and birds. The corpse is wrapped in skins of the reindeer or even the garments worn at time of death. The various belongings such as the Kaiak, gun, ammunition, tobacco and pipe, a cup to drink from, a knife and fire producing means are usually placed beside the dead. If he yet has no gun the spears and other implements of the chase are laid beside him that the spirit of the person may have the spirit of those objects to serve him in the hereafter as was done while the material form existed before decay.

ANTLER USE FOR WEAPONS (1887A: 124)

At the present day the use of the gun has entirely superceded the employment of the bow and arrow to obtain game of any kind by the Innuit in proximity to the trader. The weapon and missile of former times was a bow of wood, usually Larch was preferred and spruce if the former was not obtainable and arrows, having a shaft of wood and tipped with stone or the antler of the reindeer.

At the advent of the White man the substituting of metal points rapidly caused the points of other material to be discarded and are now almost forgotten.

MULTIPLE USES OF DEERSKIN
(1887A: 110-120)

An occasional buck reindeer may be observed in the valley or on the hilltop where he has come to renew his antlers in peace; free from the molestations of the various winged insects that torment almost beyond endurance. His flesh affords an acceptable change to the food that has for weeks been their only diet. The presence of the reindeer remind the Innuit of other important facts.

The salt water and heavy rains have so repeatedly wetted his deerskin garments that they need renewing for they will not, in their present condition, protect against the bitter and merciless storm of the rapidly approaching winter. A few more days of dallying and the leader of the party announces the intention to make for some locality where they will pass the winter. The place chosen is that where they may be convenient to the herds of reindeer that swim the streams in the fall of the year. They anxiously repair to such place and erect their skin tent and await the arrival of the deer which come to meet the opposite sex.

Here the skins for the winter garments are to be obtained; the flesh to be placed in caches of stones until winter. Now the khaiak plays an important part in the struggle.

The umiak is hauled on shore, unloaded; the skin tent is erected and made as comfortable as size and surroundings will permit. The hunters station themselves on the nearer eminences and scan the opposite shore of the stream for the herds of reindeer that approach to cross the water. A herd of few or more reindeer are observed making for the water. The Khaiak is seized and silently placed in the water by the crouching hunter, stooping lest his body be seen above the horizon and the deer take alarm; silent for the ears of the reindeer are always in motion turn-

Caribou Roaming the Tundra in Northern Labrador
These are woodland caribou, with an estimated population of around 400,000. They migrate across vast tracks of land in Quebec and Labrador in search of moss and lichen. Caribou remain a staple to the Inuit. Photo courtesy of William Fitzhugh, 2011.

ing here and there to detect the least unusual sound. Happy is the hunter if the wind blow from the deer toward his camp. That nose possesses nerves so sensitive that odors are sniffed from afar and demands the utmost discretion on the part of the hunter to escape detection by one or the other senses possessed by the timid deer of the far north.

The reindeer boldly takes to the water and when they have swimmed to the middle of the stream the Khaiak and hunter dart from cover and swiftly overtake the deer which discover the presence of their enemy only when it is too late to retreat. As panic seizes them they crowd, surge, attempt to leap upon the others in their frantic struggles, exhausting their strength which soon renders them capable of being driven at the will of the hunter armed with lance having a point soon to pierce their flesh. A prod with the sharp point cause a laggard to lunge forward with such energy that the remainder of the deer toss their heavily antlered heads in such manner that they frequently entangle their branches and remain joined to each other until the fatal moment arrives for they are urged toward the camp and when near a thrust of the lance leaves but sufficient vitality to enable the creature to creep with staggering step to the dry land and there fall.

The huge males are now in best condition; the large mass of fat accumulated over the lumbar region is a delicious morsel relished by all who have tasted it. The heavy skin affords a fine matting of hair to sleep upon and the old crones and haggard men eagerly gaze upon such skins as will be free from the vermin which have tortured their flesh for the past year. The female deer and the yearling fawns afford a skin of thinner texture and finer hair, admirably adapted to protect the human body from the merciless cold of the coming winter.

The skin is stripped from the body in the following manner; an incision is made from chin to the posterior orifice, also along the inside of the limbs. A circular stroke round each hoof joint and the pelt is separated by means of a knife and hand from the body. The head is skinned only on the side of each upper jaw and to between the eyes on the frontal portion of the head.

The carcass is divided by severing the head from the neck and then making an incision from the lower end of the breastbone to between the hind quarters.

The viscera are removed, the blood dipped up and placed upon the pelt of the deer or else the contents of the stomach are thrown out and the blood ladled into it by means of the joined hands. Certain parts of the intestines, the lungs and liver may be reserved for food. The heart is rarely eaten so far as I could observe or determine. The blood being considered, when boiled with fat and water, an especial delicacy; the contents of the stomach mixed with that liquid are relished as a treat indulged at no other time of the year so freely for the reindeer are said to take to the water only when they have filled their stomachs with the mosses constituting their principal food. The spine is now broken about the lumbar region and the limbs cut off at the heel and elbow; the posterior portion is thrust within the thoracic cavity and the entire body left in that position to cool. The valuable sinew is removed from beneath the superficial muscles of the back after the easily detached layer of backfat has been removed from the adult males of the deer (the females have no backfat at this, the fall, season). An incision is made along the back until the sinew is exposed to view; a short transverse cut, at about its anterior attachments to the flesh, and the fingers are thrust under it. A stripping movement disengages it sufficiently to enable the person to pass a stout thong under it. By means of a stick the thong is made to twist the sinew from the flesh. A stout draw pulls it entirely away. It is now washed to free it from blood, and then spread to dry.

Probably only one or two deer may be secured on the course of a day but when thirty to fifty are struck an abundance of labor lies before the party who must strive to save all they may. The flesh is stored under boulders obtained from the river shore and placed upon and about the piles of flesh to protect it from the wolves and the Wolverene hovering in the vicinity and ready to pounce upon the morsels left at the deserted camping site.

The skifts of snow from the sky and the spicules of ice forming in the water round each stone jutting above the surface remind the camp that they would better consider their dwelling site for the winter. When that is selected the party reluctantly gather the more portable of their effects and hasten nearer the coast before the river contains so much ice as to endanger travel in their now heavily laden umiak.

The skins which are required for immediate use are placed with a quantity of the meat within the boat and the party descends to a place where the earlier snows will accumulate and afford the material from which to construct a winter habitation.

While awaiting the fall snow the skin tent is their shelter and here the initial processes of preparing the skins of the deer for clothing begin.

Unlike their neighbors, the Indians, with whom they are more or less familiar in customs yet each people holds to its favorite methods of treating the pelts to render them soft and conformable to the body.

Each member of the family, man and woman, youth and maiden, boy and girl, or the infant in its mother's hood, requires a style of garment fashioned for each sex and stage of life to the adult age, and as the younger children are not able to break in the heavier skins, the pelts of younger reindeer are selected for their wear.

Such pelts as are needed for the garment of the elder males of the community are got in readiness and two or three are worked upon at a time. The flesh side of each is examined and the adherent muscles, left attached in the haste of taking them from the body of the deer, are scraped and broken to remove them and soften those parts. The next process is to dry scrape the flesh side with a scraper made of stone, ivory, or metal, shaped like an adze blade set in or lashed to a short a handle and worked in the manner of a chisel held in the hand and rubbed from the person, shredding, or pushing the fleshy fibres along or aside; the bearing of this implement on the fibres of the skin tends to soften them and render the pelt pliable. The skin is now worked across the grain or transversely to the length of the body. Frequent rubbings between the hands, in a manner similar to the process of rubbing soiled linen, also produces that softness desired. The final process consists in soaking the entire skin in a liquid which has been saved from time to time, in anticipation of that need. When the skins have been immersed for a time, determined upon by the thickness of the hide and coat of hair, it is removed and laid upon the ground or another skin and tramped by the feet, turned and stamped until the liquid has thoroughly permeated every fibre, tending to contract them and thus hold the hair in its follicle and add preservative quality to the leather. It is now hung to partially dry and then again worked, this time with a scoop-shaped scraper of tin plate or other metal in that form. Frequent rubbings of the hand softens the pelt and when it has then received the requisite pliability the skin is deemed fit for shaping into a garment. This process is employed for all skins intended to form clothing. It will be observed that the tanning process differs greatly from that in vogue amongst the Naskopie Indians, who prepare the same character of skins in a greatly superior manner.

The heavier skins for bedding purposes are treated to the scraping and rubbing operations and not to the immersion in the liquid.

When the heavier snows fall the remainder of the skins are brought from the locality where they were obtained and sold to the traders who gives them out to the Indian women to convert into buckskin or parchment.

The season of the umiak and Khaiak has now been brought to a close by the earnestness of the winter. That vessel has been the source of great profit to the hunter if it has borne him to localities affording abundance of seals in the spring, food of various kinds in the summer and in the fall has earned him to and from the swimming places on the river where the reindeer gave him flesh and skin to feed and clothe himself and

Back Scratcher

This back scratcher, Kúmiútik (that which removes lice), was collected by Turner in 1884 and is said to have come from the George River Inuit. Lice were prolific and caused much discomfort to the Inuit. Turner described how the scratcher provided relief: "The back-scratcher…effects a dislodgement of the tormentor in as quick a time as the hand may seize the long handle and thrust the affair along the spine where with an upward movement the sharp edge of the dish-shaped scratcher removes the louse and disperses the itching sensation." USNM E90350, Museum Support Center, NMNH. Photo by Donald Hurlbert, 2013.

dependents. The umiak is now placed upon a stage or scaffold and left to weather the rapidly succeeding storms which drive the Innuit into winter quarters.

REINDEER AND LICE – BACK SCRATCHERS (1887A: 151)

When a person is clad in the thick and heavy skins of the reindeer so conducive to warmth in the coldest weather the lice hold their bloodthirsty revelries and search for yet untouched spots of the human body and by intuition derived from the most ancient times the pest learns that those intangible portions are its safest pasturage.

The raiment hinders one from attaining such localities as between the shoulders where the lice do most congregate and to circumvent the vermin the Innuit has discovered a kind of back-scratcher which effects a dislodgement of the tormentor in as quick a time as the hand may seize the long handle and thrust the affair along the spine where with an upward movement the sharp edge of the dish-shaped scratcher removes the louse and disperses the itching sensation.

The use of this instrument appears to be more common among the eastern Innuit than among those to the westward of the Koksoak River. The Labrador coast Innuit employ it even more often that the people of the vicinity of George's River.

GARMENTS FOR THE UPPER HALF OF BODY OF ADULT MALE AND FEMALE INNUIT, SOUTH OF HUDSON STRAIT AND ALONG EAST SHORE OF HUDSON BAY (1887A: 273-279)

The three divisions of Innuit dwelling from latitude 55 degrees on the east side of Hudson Bay northward and along the south shore of Hudson Strait to Cape Chidley dress in nearly the same style of garment for the upper body. There is, of course, great difference in the clothing for the adult sexes.

The material from which the garments are made consists of the skins of the various species of seal and from the skin of the reindeer; the latter often trimmed with sealskin, though never the reverse.

The coat for the men and boys and for the girls of three years to the age of nearly fifteen is fashioned on the same plan or style. Its length is seldom more than two or three inches below the hip, oftener no longer than to that point. The upper opening is sufficiently large to admit the head through the neck hole and into the hood, which may be thrown back or worn over the head of lieu of a cap. The Innuit of the south side of the western extremity of the strait often cut the coat open in front as far up as the breast, the corners trimmed off like a "cutaway." Around the edges of this garment are

Sealskin Boots and Skin Clothing Worn by Inuit
Inuit man and wife at Ft. Chimo, Quebec, 1884. Photo by Turner. SI Archives, Record Unit 7192, Turner, Lucien M, 1848-1909, Lucien M. Turner Papers.

added fringes cut from a strip of reindeer skin two or three inches wide; or it may have a strand of ivory (walrus tusk, seal tooth, bear tooth, or the teeth from the porpoises and dolphins) pendants. (These objects are referred to in connection with number 90234/3266).

This coat is intended to be worn as an under garment during winter and as a single coat in summer. The winter coat is usually of heavier material than the one just referred to, but is cut on the same pattern. There is no difference in the form of the coat made from either the skin of the deer or seal for the men.

The raiment of the adult women differs in having the hood very ample so as to fall loosely about the head and shoulders. In this hood is carried the child unable to keep pace with its protector. The origin of the peculiar from of the woman's hood is due to a superstition, a decree from an ancient shaman whose spell revealed to him the necessity of woman wearing that from which the stomach of the reindeer gave shape to the pattern. The front of the woman's coat is provided with a short flap while the hinder portion lengthens into a round ended, spatulate form touching the earth. Between the flaps the coat is cut out over the hips so that freedom of the lower limbs may be allowed. The amplitude of the hood does not contribute to great protection from the cold, or when facing a storm.

When sitting the female usually disposes the front flap so that it will lie spread upon the thighs, or with the hand it is pushed between the limbs. The hind flap is either thrown aside or sat upon. As only the females, who have arrived at puberty, wear this style of garment the hind flap is used to indicate a certain condition peculiar to that sex, by turning up the end with the inner side out and attaching the end to the back or side edge about opposite the hips. During that period only is it worn in that manner. Aged females, of course, do not require such notification being given.

It is not unusual for the women to display considerable ornamentation on their garment. Various colors of sealskin having the hair on are used to contrast as well as the steel gray will permit of the Harp Seal and the black of the Harbor Seal will permit. The edges of the hood and sleeves are frequently trimmed with the skin of a young dog of dark color. There may also be edgings of skin from the White Bear, whose long hairs shed the rain better than those of any other mammal.

The pants of the men do not differ in either of the three localities here included. There are two kinds of these nether garments, the one simple like those of the whiteman but not open in front and but barely extending far below the knees. The other style is that of pants and socks combined, or in excessively severe weather the two styles may be worn at the same time. The upper portion of the garment is usually so short

Garments Worn by Inuit Men and Women (c. 1882)
Pictured are Inuit at Ft. Chimo displaying their skin garments. Note the design and hood of the amauti, or woman's parka. Photo by J.R.H. No. 61. Used with permission of McGill University Rare Books and Special Collections, 121041.

that the coat scarcely conceals the body when thrown in certain positions. The seams of those garments are placed as on civilized raiment. The shortness of leg of the pants is due to the fact that if it were longer the thickness around the calf would be so great that the bootleg would have to be increased in size to admit it being thrust within; this condition would seriously impede progress of a person whose knees are thrown together by the continued use of snowshoes which have the tendency to separate the feet and cause the knees to approach each other.

As an additional protection to the lower limbs a long pair of socks are made of thin (short-haired) reindeer skin worn with the hair next the naked person.

The pants worn by the *Itivúk* women consist of two, sometimes three, parts. The upper portion is simply the breeches, the next the leggings which may be the same as those of the men, a sort of combination or else a short legging and the long, detached sock. The breech portion fits the body quite closely; it is also, to use an appropriate but somewhat forcible expression, "cut low in the neck" and with the notch, cut in the hip of the coat, exposes a considerable portion of the naked body if the person bends forward. The upper portion of the leggings are ample in size and remind one of a Spanish bootleg. The short leg of the breeches are thrust within the top of the leggings. The lower part of the leggings are the same as the pants of the men.

UNGULATES

Inuit Caribou Skin Coat
The editors examining an Inuit woman's parka from Ungava collected by Turner. Assisted by Felicia Pickering, Ethnology Collections Specialist, Suitland, Maryland, 2011.

The women display great ingenuity in the fashioning of various colored pieces of reindeer skin so as to form squares, strips and other patterns of white light and dark brown haired skin of the deer to contrast effectively as well as oddly to the beholder. The abdomen and throat of the reindeer are clothed with long white hair during cold weather and these parts either in natural condition or else sheared closely, the latter presenting the erect stiffish ends of the hair. The leggings are supported by means of a thong which is attached to the upper part of the waist of the breeches. The space between the leggings and flesh serves as a convenient depository for all manner of small articles and it is not uncommon to observe a woman reach far down and withdraw a knife, plug of tobacco, scissors, or whatever pertains to a woman's work.

Amauti, an Inuit Woman's Parka
Made by Kangiqsualujjuamiut artists Surra Baron, Surra Annanack, Claire Etook and Ayanaylitok in 1979 for an exhibit on the Inuit of the Eastern Arctic. The parka is made from caribou, seal, and dog fur, and is decorated with glass beads. The patterns and sewing techniques are particular to the Kangiqsualujjuaq region and are based on practices handed down for generations. The parka measures 44 x 192 cm. Used with permission of McCord Museum, M983.184.

The *Taháğmyut* and *Sûhíní myut* do not have such special forms of breeches. Their garments are usually of the same style of lower body covering as the men, except that the ornamentation produced by combination of the various colored portions of skin serve to distinguish the pants of the sexes. These people have, sometimes, a slit cut in the side of the portion covering the thigh and within this they also thrust the small articles of immediate necessity.

In both sexes the nether clothing is supported by a drawstring encircling the top of the waist of the garment.

The men and women in proximity to the trading stations procure clothing of various kinds from the stores and it is not unusual to see a woman wearing a cotton gown over her deerskin undergarments; and if very cold her coat will cover the upper portion of her dress. The men purchase a kind of cloth known as moleskin and from it fashion garments similar in pattern to those of deerskin.

The younger males and young girls dress alike or in the same fashion as the adult men. Infants have clothing of fawnskin or other soft fur. They often are dressed in a single piece like the style known in civilization as "combination." I could never observe a mother dress her child in one of those affairs, but suspected that she doubled the child up like a jack knife and slipped the garment over it.

The younger children indulge in sports or occupations peculiar to the sexes and as the little girl must imitate the mother that gave her birth, she must have had her doll and clothe it in the same style of garment worn by the living. These articles of apparel for the doll are often quite elaborate in ornamentation [...].

Pair of White Fur Mittens
Turner collected these Inuit-made mittens at Ft. Chimo, probably in 1883. Turner noted gender-based belief systems related to the wearing of furs: "No man would debase himself by wearing a particle of the fur of the hare or of the white fox...Either sex may wear the skins of all other mammals, but, at times, under restrictions based upon superstition." USNM E74484, Museum Support Center, NMNH. Photo by Donald Hurlbert, 2013.

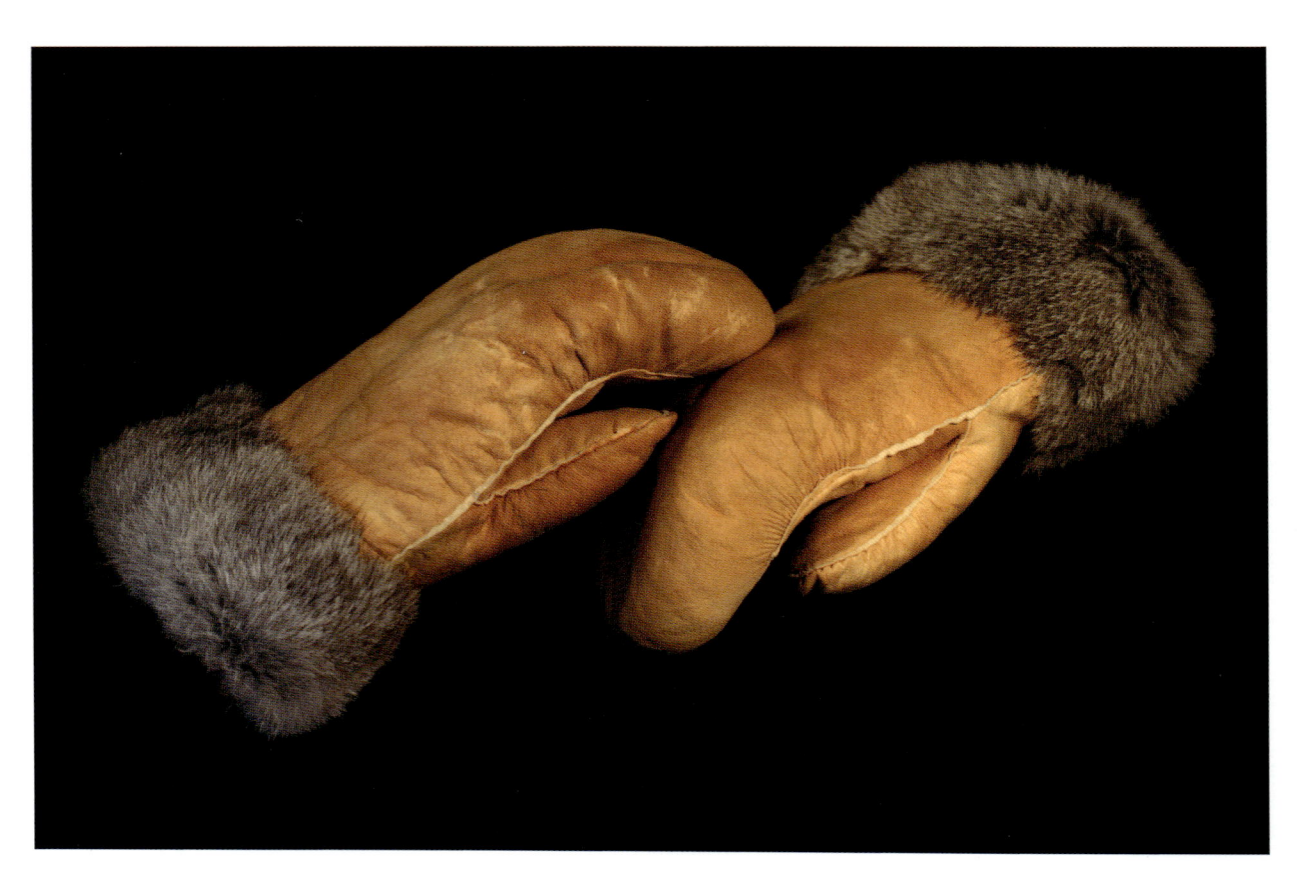

Caribou Mittens
Mittens made from caribou hide and rabbit fur by Kangiqsualujjamiut elder Susie Morgan Etok, 2005. Techniques for making caribou mittens have been passed down for generations. The mittens are waterproof and are smoked using local tamarack wood to cure the leather, which infuses the leather with a strong wooded fragrance. Scott Heyes Private Collection.

BOW, QUIVER AND SEVEN ARROWS (EAST MAIN INNUIT; HUDSON BAY) (1887A: 173)
The material of which the quiver is made is simply the stretched skin of an adult reindeer; the hair having been removed by allowing the skin to decompose to such a degree that the hair was loosened and easily rubbed off. The skin has been further rubbed thin so as to be as light as possible. The loops were for suspending the quiver over the shoulder by means of a strap of skin.

Toy Bow, Bow Case, Quiver, and Arrows
These models of hunting tools, collected by Turner at Ft. Chimo, probably in 1883, were attributed to the East Main Inuit. Turner (2001: 246-247) wrote: "The bow case is made of buckskin and is of sufficient length to contain the bow, excepting the extreme end, which is left projecting for convenience in handling. The case is tied around the bow at the projecting end. The quiver is attached to the bow case and contains two models of arrows for shooting large game. The arrows are tipped with leaf-shaped pieces of tin. They are feathered with portions of feathers apparently taken from the tail of a raven. The mouth of the quiver is also drawn up with a string to prevent the loss of arrows. I have not seen the Eskimo of Hudson Strait use such a cover for their bows and arrows." USNM E90286, Museum Support Center, NMNH. Photo by Donald Hurlbert, 2013.

Ungulates

A Starving Family of Indians
An Innu story, Turner (1887b: 632-637)

A number of Indians had camped so long at a single place that they had exhausted the game from the vicinity; and, with all their endevours to ascertain the whereabouts of the reindeer they were still without food.

A long period of bad weather prevented even the hardiest hunters from going from the tent in search of food. They had eaten all their supplies and starvation was upon them. In vain they appealed to an old man, who had plenty, to give them of his stores. He always refused and bade them begone.

An old man at last applied to him for food, and he gave him a small portion of dried meat. The old man took the meat to his home and when his grandchildren saw it they clamoured for it. He told them to go to the rich man and get some for themselves. They feared to go. Their mother now directed them to go; and the old man, who had the stores of meat, heard her tell them what to do. He said "That is just what I want."

The little boy went to the man's tent and lifted under the doorflap and said he wanted to enter. The old man, whose name was *U s'its kwe n'e po*, said to the boy, what do you want *U sits kwe ne po*? (He now addressed the boy by his own name). The boy answered that he was hungry and wanted some food.

The old man said, Hungry are you? Why do you hunger when the meat is falling from your heavily laden scaffolds? The boy was now directed to sit down. The old man went to fetch a piece of meat for the boy. When he brought the meat he gave it to him and inquired if he had a sister. The boy answered that he had a father, mother and sister. The boy now ate his meat and when he had finished eating the old man told him to come out and see the meat which was upon the stage.

The boy went with him and the old man showed him huge piles of dry meat and large cakes of fat. The old man now told the boy to go home and tell his father that if he would give his sister to the old man that he would get all the meat that the old man had.

The boy went to his tent and told his father to give his sister to the old man, and added, "I shall then have all the meat and we will not starve." The girl's father consented and the boy returned to the old man; telling him that he could have the daughter. The old man took the meat from the stage and prepared three large kettles full of it. He then took into the tent a coat for a man, and two dresses for a woman. He placed one of the dresses alongside of the place where he sat in his own tent; and the other two suits he placed on the side of the tent opposite the door.

The old man now told the boy to invite all the Indians to come to the tent. He said, "There is your father's coat; there is your mother's dress beside it; there, by my side, is the dress for your sister. When they enter the tent tell them to sit by the side of the clothing which is for each of them. The boy now went to tell the Indians and his relations to come to the tent of the old man. His parents and sister came first and each was directed where to sit. The other Indians now began to come in; and they were disposed themselves as best suited them.

The old man awaited the arrival of the last one and when he entered the kettles of cooked meat were brought in and a great feast was held. When they had feasted the old man address the boy, *Usitskwenepo*, why do you not go and get your snares (the old man's snares) and set them for deer? The boy went out and returned with the snares. The old man went with the boy and they soon found and abundance of tracks of the deer in the valley not distant from the camp. They set thirty snares and in the morning all the Indians went out and drove the deer toward the clump of trees from which the snares were suspended. When they came to look at the snares they found that each had caught a deer. The people began to skin the deer and when this was done they had a great feast and an abundance of food to last for some time.

The Indian Youth with a Reindeer Wife
An Innu story, Turner (1887b: 644-652)

An old Indian and his wife had one son, their only child. The fond hopes of the parents had been well-grounded in this child, who endeavoured to fulfil every desire and anticipate every wish of his now aged parents. As the son grew older he became more reserved and appeared contented only when alone or when wondering among the fastnesses of the high hills or when roaming amongst the woods or scanning the plains. Even the streams attracted him and on the bosom of water his canoe would idly float while the occupant seemed absorbed in deep thought. A season of plenty to the home of the parents had passed and was now succeeded by one of want. The son never relaxed his exertions to procure all the food, for his parents, that crossed his pathway. So much was he among the beasts of the field that he was heard muttering strange words to himself. These actions, however, did not alarm the relatives or friends of the youth.

The season was now advanced and the son appeared more buoyant than formerly. One morning he arose from his slumber and went outside of the tent; where he sat down. He remained in that place so long that the father went out; and, perceiving his son pondering over some grave matter, inquired the cause of his apparent perplexity. The son replied that he had had a strange dream and that it distressed him to know how he should reveal it to his father and mother.

After some persuasion the son admitted he had dreamed a reindeer had told him to come and dwell with her among her kind. The father gleefully replied that it was a good sign; that he should kill many deer and thus keep themselves above want. The next day he went to hunt the deer and soon found a number of those creatures quietly feeding.

In a moment the hunter was observed, whereupon a young doe bounded toward him. Just as he was ready to fire an arrow through her body she called to him not to shoot for her father had sent her to him and that then bade him put away the arrow.

As he did so the doe drew nearer and again informed him that her father had sent her to him, asking him to come and live wither them forever. The hunter replied, "How can I dwell with you when it is upon the flesh of the deer that I subsist. I live in a tent and cannot dwell where there is no fire to keep me warm. Here upon the plain there is no water. I cannot live without water to drink." The doe replied, "For fire, meat and water you will never want; your father will never want. You will live forever." The hunter then consented to accompany her to her home. She pointed to a large hill; and said, "There is our home." She requested him to leave his deerskin mantle, snowshoes and arrows behind him; but, she enjoined upon him to retain his bow.

As they were walking along the valley she told her companion they were thereon the path heading to her home. They went under the side of the hill and the hunter saw that the ground was covered with deer. Some of them were greatly frightened when they saw the man in their midst; but, the father of the doe calmed their alarm by stating that the man had come to live with them; and, inquired of the deer "Do you not pity the poor Indians, who have to hunt for a living, while we are surrounded by an abundance of food?" The doe's father addressed the man and inquired if he was hungry. The hunter replied that he was nearly famished. The father gave him a choice piece of meat which was eagerly eaten. When he had finished eating the doe's father asked the young man if his father and mother were hungry. The son answered that they had no food in the tent except what was procured by himself. The doe's father then said that he would send him some nice, fresh meat, adding that hereafter he should not want for food or skins for clothing.

After the son had been absent from his father and mother for one night they became alarmed; and on searching for him discovered nothing but his mantle, snowshoes and the arrows which he had left behind him the day before. Around the place he saw many tracks of the reindeer and among them he found the trail of the bow which the hunter had dragged behind him.

When the father saw this he knew the deer had enticed his son away from him. The sorrowing father returned to his tent and announced to the people that his son had gone to live with the deer. He told the people to prepare a number of snares and they would entrap him on the next day for he cannot run as fast as the deer and in the deep snow will soon be left behind. The next morning the people repaired to the valley and there arranged the snares or nooses among the trees that grew there. Some of the men went to the head of the valley and others stationed themselves along the sides. When the doe's father saw these preparations going on he said to the other deer, "Let us go and give the old man some meat." He told the young man to come with him; but, the hunter replied that he could not run fast enough and would be left behind in a short run.

The father of the doe told him to keep well amongst the other deer and they would help him along. The animals were soon moving down the valley at a swift rate toward the place where the nooses were suspended. The young man kept his place in the midst of the fleeting herd and gave no heed to what was happening. As they plunged through the clump of trees the Indians concealed there raised such an outcry as to throw the terrified deer into great confusion, causing them to run hither and thither.

The Indians behind them closed in and the deer were frightened to the other side of the valley. The young man kept among them, but numbers were caught in the outspread nooses which the hunter saw and avoided. The Indians ran up to the strangled deer and quickly cut their throats. When they came to look for the body of the young man they found he had escaped. The old father was now grieved the more to learn that his son preferred to remain with the deer.

They searched among the tracks and soon found those made of the young man and alongside of his tracks was the trial of the bow. The poor father strengthened himself and said "Let him go. He thinks he can live with the deer. Let him go."

Strong Man and the Caribou
An Inuit story, Tivi Etok (Heyes 2007)

This story originates from the early Inuit. There was small man who had a mother and a wife. Even though the man was small, the man used to carry an enormous kayak. He was a very strong man.

Our ancestors did not have guns. They would kill caribou when they were crossing rivers using spears. Two men, without wives, and their mother came across the small man and his family. The two men stole the small man's wife. All summer long the two men, their mother and kidnapped wife camped close to the small man. When caribou are crossing, big kayaks and paddles are used to hunt them. The small man was capable of doing anything. His mother told him not to do anything to show your strength, just make out that you have ordinary strength. When caribou would cross the small man would let the two men hunt the caribou to make out that he was a weak hunter. The two men told the small man to make the paddle on his canoe smaller, which would make the small man's canoe go slower.

All summer long the small man shot only one small caribou. At the end of August, caribou become fat. The small man wondered whether big caribou bulls would be crossing. The small man wished for the big caribou. The kidnapped wife told the two men that her husband was strong, yet the man made out that he was weak. The two men told their kidnapped wife that the small man was not strong because he could not even hunt caribou. The small man's mother saw that the caribou were getting fatter because she saw that the pollen of small plants was blowing in the wind.

Two big caribou bulls began to cross the river. The two men were the first to leave on their canoe to get the caribou. The small man took his time to get his canoe ready. When the two men got close to the caribou the small man readied himself to paddle fast. It was like the wind took him. He was so fast. The small man even generated a wave from his paddling speed. The small man went in between the two bulls, grabbed them by the antlers and pushed their heads into the water. The bulls drowned and so he successfully killed the caribou. The two men saw all this action and then realized the small man's strength. The two men left by canoe and went back to camp. By the time the small man got to his tent he saw that the two men had left and had also returned his wife to him. The two men were now afraid of the small, strong man.

Strong Man and the Caribou
An illustration of the story of the strong man by Kangiqsualujjuamiut elder Tivi Etok, 28 April 2005. 11″ × 17″. Lead pencil on bond paper. Scott Heyes Private Collection.

Inuksuit and Caribou
An Inuit story, Johnny George Annanack (Heyes 2007)

The Inuksuit [pl.; Inuksuk is singular] are used to know where there are caribou, or where there are fish around the lakes. When you go caribou hunting, the Inuksuit are not so close to one another so the hunters would know which trail the caribou would take and the hunters would wait ahead of their trails and the Inuksuit would have a shoulder blade on some of them. That's what I used to hear about the caribou trails. The Inuksuit has a lot of meaning and they'd be used to play with and my parents used to fix them when they used to play. They're still standing there. If you could understand their meanings, they're very useful.

I will go back to the question you asked, we lived in the *Navvaat* for many years. Almost all the good hunters used to live there; we'd walk a long way to hunt for caribou. There are a lot of people who used to walk long distances to hunt, when game was scarce, we'd be weak and tired but we had no choice but to do it. We could not complain about it, it was the only way in those days, and we were to walk all the way back.

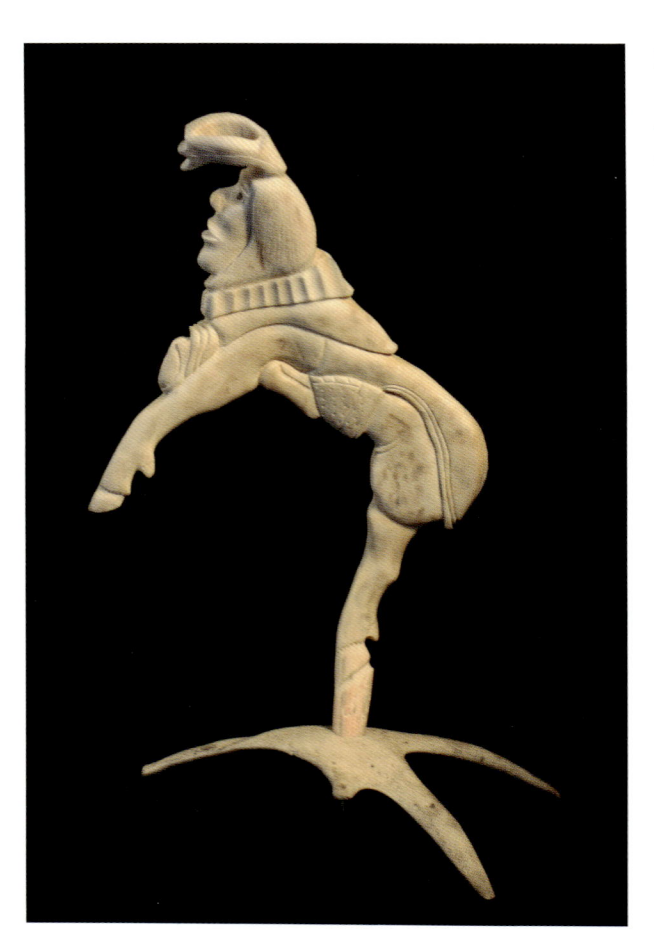

Inuit Carving
Carving about the story of an Inuit woman who becomes a caribou. Made from caribou antler by Kangiqsualujjuamiut artist Daniel Annanack, 2005. Scott Heyes Private Collection.

Chapter Two: Mammals of Ungava and Labrador

Hunting Caribou
An Inuit story, Tivi Etok (Heyes 2007)

This is a picture of my father, a caribou he wounded, and our dog. I am meant to be in the left hand corner of the picture but I didn't have time to finish the drawing. We were hunting beside a lake when we came upon a caribou. The front legs of the caribou were shot off with a single shot .22-calibre rifle by my father. My father intended on killing the caribou with a knife instead of using another bullet, because bullets were precious in the old days. The caribou was kicking and bucking, it even kicked the dog. My father tried to take its antlers and slit its throat. My father ended up killing the caribou by cracking the back of its head. This was the way we hunted in the old days, by preserving our ammunition.

Hunting Caribou
An illustration of a hunting scene by Kangiqsualujjuamiut elder Tivi Etok, 23 May 2005. (Heyes 2007). 19" × 12" Black Charcoal on Cream Canson Paper. Scott Heyes Private Collection.

Caribou – Terms and Definitions

Syllabary	Modern Nunavimmiutitut Term and Definition	Term by Turner	Definition recorded by Turner
ᐊᕐᓇᓗᒃ	*arnaluk* (cf. Schneider p.40. gives "female bird"; *arnalukak* = female whale; walrus)	*ag ná luk*	Doe, female deer [p. 2137]
ᐃᑳᕐᓗᒃ	*ikaarluk* (?): from *ikaar-* (to cross) with the same –(l)luk ending as in *nalluk* (caribou crossing place in a river or lake)? *ikaarpuq* (?) (cf. Schneider p. 58 - *ikaarpuq* = he crosses over to the other side [of a river, lake,... etc]). Johnny Sam Annanack, Kangiqsualujjuaq (personal communication) suggests *ikaarvik*.	*i kóg lúk*	A lake lying in the path of migrating reindeer, and through which they swim to continue their way [p. 2226]
ᐃᓯᕆᑦᓯᒪᔪᖅ	*isiritsimajuq* (cf. Schneider p.98 - *isiritsimajuq* = smoked, filled with smoke); *isiritsisimajuq*. Note: refers particularly to smoked caribou skin used for making mitts.	*í si ghû tsik-si má iok*	Tanned buckskin [p. 2245]
ᐃᑎᒐᒃ; ᐃᑎᒐᐃᑦ	*itigak*; *itigait* (pl.) (cf. Schneider p. 104 - *itigait* = foot, feet [...])	*i tí gok*	Hoof (of deer) [p. 2247]
ᑲᑎᒪᔪᖅ	*katimajuq* (cf. Schneider p.126 - *katimajuq* = who are gathered together [...])	*ka ti mai ok*	Herd of reindeer, or a covey of ptarmigan [p. 272]
ᑯᑭᒃ	*kukik* (cf. Schneider p. 149 - *kukik* = finger, toenail [of a person] hoof, claw [...])	*kú kik*	A Labrador word for the accessory hoofs of a deer. The word is the same as for the nail of the fingers or toes. [p. 2320]
ᓇᕐᔪᐃᑦᑐᖅ	*natjuittuq* (cf. Schneider p. 197 – *natjuk*; *najjuk* = animal horn, caribou antler; cf. Spalding 1998, p. 60 - *nagjuittuq* = hornless male, winter male)	*nûg zu ítok*	Hornless deer [p. 2717]
ᓇᕐᔪᒃ ; ᓇᔾᔪᒃ	*natjuk*; *najjuk* (cf. Schneider p. 197 - *natjuk*; *najjuk* = animal horn, caribou antler)	*nûg zuk*	Antler; horn [p. 2717]
ᓇᕐᔪᒥᒐᖅᐳᖅ	*natjumigarpuq* (cf. Schneider p. 197 -*natjumigarpuq* = he strikes him with his horns; also *natjumigautijuuk* = two creatures that are fighting with their horns)	*nûg zú mi gá ktok*	Fights (deer) [p. 2717]
ᓄᖅᕋᖅ ; ᓄᕐᕋᖅ	*nuqraq*; *nurraq* (cf. Schneider p.226 - *nuqraq* = young caribou less than a year old)	*nó kwak*	Fawn [p. 2694]
ᓄᕗᒃ (?)	*nuvuk* (?) See: *nuvuk* (cf. Schneider p. 231 - *nuvuk* = point, promontory, headland)	*nû lok*	A long point of land extending into the water. It must be resorted to by reindeer before the term may be applied to that portion of land. [p. 2689]
ᐸᖕᒐ ᓕᔾᔪᐃᑦᑐᖅ	*pangalijjuittuq* (lit. "never gallops") (cf. Schneider p.238 -*pangalippuq* = gallops)	*púng a li yu ítok*	A superannuated buck. Literally, one who does not run. [p. 2769]

Caribou – Terms and Definitions

Syllabary	Modern Nunavimmiutitut Term and Definition	Term by Turner	Definition recorded by Turner
ᐸᓂᖅ	*panniq* (cf. Schneider p. 239 - *panniq* = full-grown caribou bull)	*pûn'g nŭk*	Buck (deer) [p. 2769]; Adult, buck reindeer [p. 703]
ᐸᓂᑐᖅ	*pannituqaq* (lit. an "old" caribou bull)	*pûn'gni to'hak*	A very old buck [p. 2769]
ᐱᓗᖅᑐᑦ(?)	*Pilurtuut*(?): personal name (name of the last shaman known in the Kangirsujuaq area ca. 1900); the meaning of the name is not readily understood now; *piluqtuq* (?); cf. Schneider p. 253 - *piluq* = shed caribou hair)	*pi lu'k tut*	Eater of deer hair [p. 2755]
ᖄ; ᖄᖅ	*qaa* from *qaa*; *qaaq* (cf. Schneider p.276 - *qaa* = above, surface, upper side) Note: generally refers to caribou "sleeping" skin used as a ground cover (cf. Peacock p. 127 *qaak* = caribou skin mattress; also in Bourquin Grammar: *kâk* = mattress)	*ka ghá luk* or *khak*	Deerskin bed [p. 2259; 697]
ᖃᑭᒃ (?)	*qakik* (?) Consider confusion with *kakkik* = [nasal] "mucous" (Peck p.85)	*khá kik*	Mass of fat behind eye of deer [p. 2290]
ᓱᓗᕝᕙᐅᑦ	*suluvvaut* (cf. Spalding 1998, p.143 - *suluvvaut* = horn jutting out from caribou's forehead). Note: the ordinary English name for this antler projection is "shovel." A caribou with two such projections, much sought by sports hunters, is called a "double shovel."	*sú lu vaut*	Brow antlers of male reindeer [p. 2777]
ᑎᖓᐃᔪᑦ (?)	*tingaijut* (?) Note: In North Baffin Island the term *tingajuq* refers the thick neck fur of caribou.	*tĭng ái yut*	Long pendant hair on throat of deer in winter [p. 2804]
ᑐᓄᒃ	*tunnuk* (cf. Schneider p. 421 – *tunnuk* = fat, caribou suet [typically for caribou "back fat"])	*tú ńuk*	Backfat of deer [p. 2830]
ᑐᑐ	*tuttu* (cf. Schneider p.430 - *tuttu* = caribou in general)	*túk tu*	Reindeer (any kind) [p. 2809]
ᑐᑦᑐᔭᖅ ᕿᐱᒃ	*tuttujaq qipik* (two words lit. "caribou skin" and "blanket"). See Schneider p. 430 - *tuttujaq* = caribou skin and p.308 - *qipik* = blanket.	*tuk tú yak-kh'i pĭk*	Deerskin blanket [p. 2810]
ᑐᑦᑐᑑᖅ;	*Tuttutuuq*; River, between Kuujjuaq and Kangiqsualujjuaq flowing into Ungava Bay	*Tuk tu' twok*	A small river west of George's River. The word is derived from *Tŭk tu*, reindeers; *twok*, place where; hence a place where deer abound. [p. 2810]
ᑐᑐᕙᒃ	*tuttuvak* (cf. Schneider p.430 - *tuttuvak* = moose, domestic cattle, large mosquito – *tuktuujaq* p. 417); a crane fly (cf. Spalding 1998, p.167: *tuktuujaaq* = long-legged long-winged water fly)	*tuk tú yak*	Big reindeer. Name applied to a large Gallinipper fly [p. 2810]
ᐅᖑᔪᖅ ; ᐅᖑᒪᔪᖅ	*ungujuq* ; *ungumajuq* (cf. Schneider p.451- *ungujuq* = game [or] whales chased into bays)	*ung úk tok*	Drives (wild deer or other mammals) [p. 2845]

Ungulates

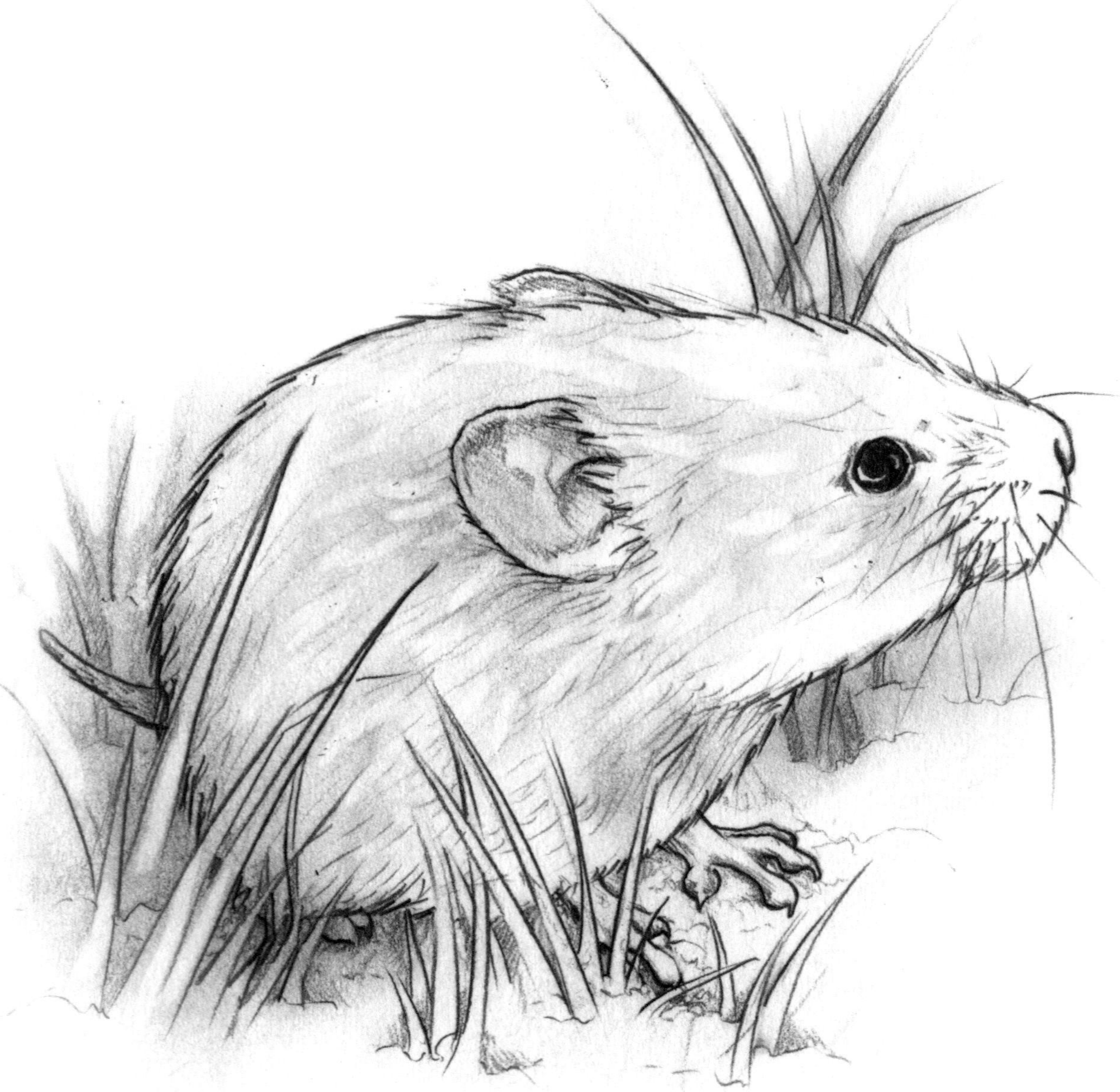

Rodents

Introduction to Squirrels, Muskrat, Lemmings, Voles, Mice, and Porcupine

Lucien Turner made the first major museum collection of mammals from the northern reaches of the Ungava Peninsula, and was the first serious scientific observer of rodents in the area. His notes and collections, sent back to the Smithsonian, documented 11 species of rodent from the region. These included the two largest of Canada's rodents, the beaver (*Castor canadensis*) and porcupine (*Erethizon dorsatum*), as well as three species of squirrels, the semi-aquatic muskrat, and five species of mouse-like rodents (voles and lemmings). Though several additional rodent species occur in Labrador (Harper 1961), Turner collected specimens of every species now known to occur in the vicinity of Fort Chimo.

The beaver, so valuable and heavily utilized in the fur trade, was rare in northern Ungava during Turner's stay, and he secured only a single set of lower jaws for the Smithsonian collection. Turner was able to learn and write more about the porcupine, and his manuscript is an important nineteenth century contribution to porcupine natural history in eastern Canada. He secured two specimens, prepared as skins with accompanying skulls, which are still in excellent condition.

In Labrador, Turner documented the woodchuck or groundhog (*Marmota monax*) based on secondhand information, and received a specimen of the Northern flying squirrel (*Glaucomys sabrinus*) from "Planters" at the Northwest River, a skin and skull still present in the Smithsonian collection. Both of these species occur widely in more southerly sections of the Ungava Peninsula, but do not reach as far north as Fort Chimo. Based on discussions with locals, Turner also suggested that a ground squirrel (*Urocitellus parryii*), an animal he was familiar with from Alaska, occurred in Labrador, and provided a species account in his manuscript (always careful, Turner noted that his "inability to secure specimens renders the determination suppositious only.") This species has never been recorded in eastern Canada, and his account most likely arose based on confusion with descriptions of the woodchuck and its habits.

The principal squirrel of northern Canadian forests is the red squirrel (*Tamiasciurus hudsonicus*). Turner collected many red squirrel specimens both in Ungava and Labrador. His enchanting accounts of his firsthand experiences and rapport with these animals, and of his efforts to understand their natural history, form a memorable part of his manuscript. In one vignette, Turner describes watching two male squirrels quarreling with one another from adjacent spruce saplings on the very cold and otherwise quiet morning of December 2, 1882. After observing their interactions, Turner shot both animals ("I fired at the suspected intruder and a second shot robbed that portion of the Bay of all the life within it"). He describes tucking the specimens into his satchel after "a long walk up and down steep hillsides and over gullies filled with fluffy snow driven by the recent winds" to seek out the spots where his squirrels had fallen. These skins of these two squirrels are still be found in the Smithsonian's scientific collections (p. 128). Examining these specimens – holding them – brings a remarkable feeling, transporting the imagination directly back in time to that cold and crisp December morning in 1882, when Turner picked these same squirrels off the soft Ungava snow and placed them in his knapsack. Since Turner's time these same specimens have been used in a large number of scientific studies, including a recent study of isotopic chemistry of fur from these squirrel skins to evaluate how much Canada Lynx rely on squirrels in their diet (Roth et al. 2007).

Across the northern continents, the dominant group of small rodents in tundra and high-latitude forests are the arvicolines – the voles, lemmings, and their relatives – influential animals in furbearer-prey fluctuations in northern forests (Elton 1942). These are mostly small species, though the largest member of this group, the Muskrat (*Ondatra zibethicus*), weighs several pounds. Turner's succinct account for the Muskrat elegantly showcases his approach to mammalogy, with reported information derived from observations in the field, specimens collected by trapping and shooting, firsthand dissection of a stomach to identify food plants, and notes on Innu and Inuit terminology and utilization.

Turner's writings demonstrate that, while a sympathetic field observer of the natural history of rodents, he was not a specialist who could easily identify voles and lemmings. In his manuscript, Turner distinguished three kinds of small rodents among the specimens he collected in Ungava. He called these "the Lemming" (corresponding to the Ungava Collared Lemming, *Dicrostonyx hudsonius*), "the Evotomys" (corresponding to the Red-Backed Vole, *Myodes gapperi*), and "the Arvicole" or Meadow Mouse (corresponding to the Meadow Vole, *Microtus pennsylvanicus*). Mammalian taxonomists, especially C. Hart Merriam, Frederick True, and Vernon Bailey, studied Turner's Ungava and Labrador collections at the Smithsonian in the years following their receipt. Bailey (1897, 1898) described Turner's specimens of *Microtus* as a new Meadow Vole (today often recognized as a subspecies, *M. pennsylvanicus labradorius*), and his specimen of *Myodes* as a new Red-Backed Vole (today often recognized as a subspecies, *M. gapperi ungava*) (Harper 1961; Hall 1981). True (1894) identified an additional rodent species, the Northern Bog Lemming (*Synaptomys borealis*), overlooked amongst Turner's collections, and described it as a form new to science, today sometimes credited as a distinct subspecies, *S. b. innuitus* (Harper 1961; Hall 1981).

Most remarkably, Turner's collections contained one additional rodent species, never before described by scientists, that he had overlooked while in the field. Merriam (1889) relied on two of Turner's specimens (among other collections) to describe a new vole genus, *Phenacomys*. Merriam originally described each of Turner's Ungava *Phenacomys* specimens as two different new species, *P. latimanus* and *P. ungava*. Today, zoologists recognize these names as belonging to a single species, *Phenacomys ungava* (Howell 1926), the type specimen of which was collected at Fort Chimo by Turner. Turner had, in effect, discovered a notable group of rodents new to science without even realizing it. This should not at all diminish appraisals of his field acumen; Merriam's initial confusion in studying Turner's specimens demonstrates how challenging studies of rodent variation can be, even to specialists!

Porcupine
Nine-month-old juvenile porcupine. Winter 2010. Photo by Charles Ver Straeten.

Woodchuck or Groundhog

Arctomys monax (Linn.) Schreber.

[1378]

Order Name used by Turner	*Rodentia*
Current Order Name	*Rodentia*
Family Name used by Turner	*Not recorded*
Current Family Name	*Sciuridae*
Scientific Name used by Turner	*Arctomys monax (Linn.) Schreber.*
Current Scientific Name	*Marmota monax* (Linnaeus)
Syllabary	*Unknown in Ungava*
Modern Nunavimmiutitut Term	*Unknown in Ungava*
Eskimo Term by Turner	*Unknown in Ungava*
Definition of Eskimo Term by Turner	*Unknown in Ungava*

2013 Species Distribution

Descriptions by Turner

Several whitemen informed me that the woodchuck (they gave it the name of Ground Hog) is not uncommon on the dry hillsides west and southwest of the head of Hamilton Inlet. It is said to be common from Mingan westward. Specimens were not procured by me. [Under the heading of "Ground Squirrel, *Spermophilus empetra empetra*"] An animal occurs commonly in the vicinity of Northwest River, at the head of Hamilton Inlet, which was so accurately described as to lead me to conclude it was this species. Inability to secure specimens renders the determination suppositions only.[9]

---

[9] It is unlikely that the ground squirrel (now *Urocitellus parryii*), for which Turner provided a separate species account, was present in the Labrador and Ungava region as Turner supposed, and this allusion also likely refers to the woodchuck.

ᓇᐹᕐᑐᓯᐅᑦ
Hudsonian Squirrel
napaartusiut

Sciurus hudsonius.

[1380]

Order Name used by Turner	*Rodentia*
Current Order Name	*Rodentia*
Family Name used by Turner	*Not Recorded*
Current Family Name	*Sciuridae*
Scientific Name used by Turner	*Sciurus hudsonius.*
Current Scientific Name	*Tamiasciurus hudsonicus* (Erxleben)
Syllabary	ᓇᐹᕐᑐᓯᐅᑦ
Modern Nunavimmiutitut Term	*Napaartusiut* (common name in southern Ungava Bay = "tree squirrel")
Eskimo Term by Turner	*na pák ta śyut*
Definition of Eskimo Term by Turner	Hudsonian squirrel (*S. hudsonius*) [p. 2683]; means tree-dwellers to Hudson Strait Innuit. Generic term for squirrel [p. 1380]
Innu Term by Turner	*A nis tsú ku tsas* [p. 708]; *a nis chu ku chash* [p. 1830]

2013 SPECIES DISTRIBUTION

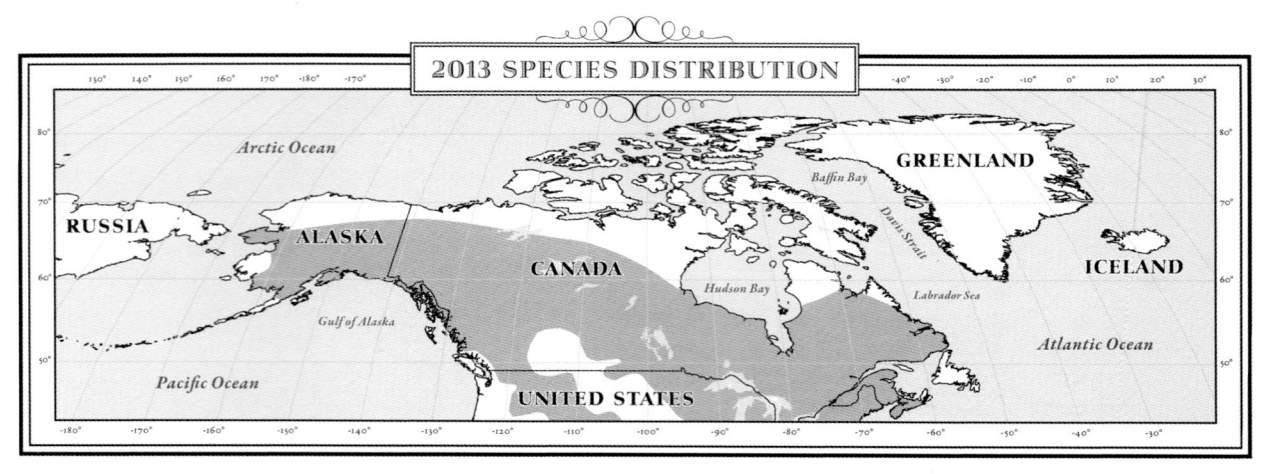

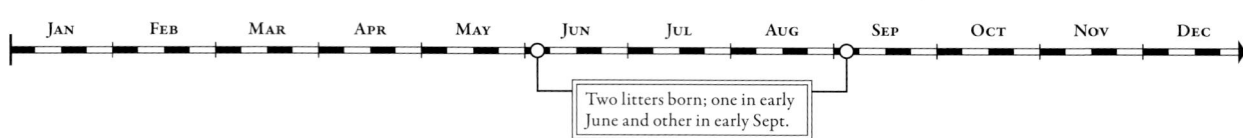

| Jan | Feb | Mar | Apr | May | Jun | Jul | Aug | Sep | Oct | Nov | Dec |

Two litters born; one in early June and other in early Sept.

Chapter Two: Mammals of Ungava and Labrador

Descriptions by Turner

The Hudson Strait Eskimo give the name of *Na pák ta syút* or tree dweller, to this animal. The Labrador Eskimo know it by the name of *Sĭk sĭk*. This name is identical with that applied by the Malimyut, of Norton Sound, Alaska, to this species of rodent. The Únalit of the southern shores of that sound, however, name it *Chĭ gĭk*. The latter names are, of course, but an imitation of its note.

The Hudsonian or Red Squirrel is plentiful in the Ungava district as far as the verge of the timber line. In the vicinity of Ft. Chimo they are quite common, as many as half a dozen may be found in a day's tramp. Toward the south they become excessively abundant and in the vicinity of Northwest River they are so plentiful that they become a pest during certain years.

The food supply either gives out in certain localities or else [1381] they are seized with a desire to wander. Entire districts will be nearly depopulated by these creatures wandering until they arrive at a land of plenty where they remain for a few years and return to the place they had left. During these visits they appear in such numbers that they commit sad havoc among the barrels and bags of biscuits and flour, for nothing is safe from them.

Red Squirrel Skulls Collected by Turner
USNM A23350 and A23356, Division of Mammals, NMNH. Photo by Angela Frost, 2011.

It was related to me that at Northwest River they were so abundant on one of these occasions as to steal nearly all the biscuit from the storehouses and that it was not a rare sight to find these squirrels scampering off with a huge biscuit in its mouth. A premium on them set the Indian boys at work and the squirrels soon ceased their depredations.

This little animal is not at all shy of man and exhibits a large amount of curiosity which is not satisfied until it follows often for a mile, from tree to tree, [1382] often but a few feet from the person, squeaking its querulous note as it springs from one tree to another or leaps across the intervening space clear of bushes.

It is affirmed by the Indians that two litters are brought forth each year, one in early June and the second in early September. A young male scarcely half grown was secured November 11th. It was fearless, approaching so closely that a shot from any gun would have ruined it. I endeavored to retreat but the creature followed and only by rapid running could I obtain any distance sufficient to warrant me in shooting. I was often mislead by the sudden reappearance of one of these animals following me that I considered it to be another and then a third one.

The note uttered by this squirrel is a peculiar trilling or fluttering sound which renders it, at times, extremely difficult to locate. Another sound uttered when the animal

"The galleries formed by these creatures from one locality to another, often a hundred yards apart, are of the most intricate pattern"

has been frightened and [1383] retreated to its burrow, sounds very much like the clicking nose made of a person to urge a horse into greater speed. The imitation of this sound if persevered in will invariably cause the squirrel to reappear.

The nest for the summer is built of twigs of spruce or larch and often placed within those peculiar growths, on conifers, known as "boquets." These growths are apparently a local disease and from the surrounding healthy, wood start innumerable fine twigs and form a peculiar bunch. The squirrel cuts out the central ones and within forms its nest. The structure is not bulky, rarely more often than a foot in diameter and nearly circular in form. The interior is lined with fine grasses and in these places the young are brought forth. Three to five young form a litter and generally two broods each year. The last brood of a year do not have young the succeeding year and the first brood has but one litter the following year.

The males are polygamous and in the [1384] rutting season are quarrelsome, frequently fighting until their cries may be heard a great distance. In their engagements they appear to use the claws with more effect than their teeth. It is not unusual to find males in late March with their heads quite bloody and their ears slit into fringes.

On the 2nd of December I was out hunting in a locality known as "Hunting Bay," not far from the post. While I was on the high bank forming the eastern wall of that locality I heard a couple of these squirrels making a great ado among the trees several hundred feet below me and at the bottom of the "Bay." The snow was very deep and so dry and powdery that my snowshoes sank deep in the fluffy mass among the intricate mazes of tangled willow and alder, nearly covered with snow; and, among the tops exposed above the general surface I floundered in my [1385] haste to secure the objects of my search. On my approach the squirrels gave a screeching chatter which led me to discover their location. I soon observed them sitting each on top of a spruce sapling of near twenty-five feet in height. The trees separated only a couple of feet which brought the terminal bunch of cones near each other. The animals sat facing each other and were industriously husking the seeds from the cones. They were seemingly not disturbed at my approach and gave me ample time to observe their actions while eating their breakfast. They appeared to be in ill humor over something for they paused in their occupation and gave each other a severe scolding, shaking their tails with the customary jerk. I expected them to jump at each other but as hunger was with them a peacemaker that cold, yes very cold morning, although the sun shone yet, as if mocking for even the [1386] frost crystals conjealed on the whiskers of those creatures and reflected myriads of brilliant scintillations. I suspected that one of the two had invaded the chosen clump of trees which the other was to search for food for his breakfast.

I watched them and as they began to show signs of sating, I fired at the suspected intruder and a second shot robbed that portion of the "Bay" of all the life within it.

A long walk up and down steep hillsides and over gullies filled with fluffy snow driven by the recent winds now awaited me. Putting the specimens in the satchel disclosed them to be males and this may have been the cause of their angry notes.

At the mouth of the Larch River (110 miles from the mouth of the Koksoak River), I found these squirrels very abundant. The timber is much heavier there than in the vicinity of Ft. Chimo and affords a better supply of food for them. [1387] Their tracks were everywhere to be seen on the surface of the snow. I saw a great number of holes leading to the burrows under the heavy sheet of snow. The galleries formed by these creatures from one locality to another, often a hundred yards apart, are of the most intricate pattern. Numerous chambers were excavated and probably formed resting places for the creatures. Bunches of leaves and twigs formed dry retreats where a few hours nap might be indulged in or a tempting spruce cone could be separated and the seeds devoured.

The food appears to consist of the seeds from the cones of the spruce and larch trees with probably a few grass seeds. Berries appear also to be eaten if the fact of finding stains from that fruit, on the lips is an indication.

After much labor I succeeded in discovering the difference between a common runway and a gallery leading to the storehouses [1388] of these creatures and was amply rewarded by the results disclosed.

It was a matter of greatest surprise to find what an amount of labor had been performed by these squirrels to lay by a store of food to serve them when the dreary days of winter sent relentless storms of snow driven by piercingly cold winds causing the hardier animals of the earth to seek shelter until the sun again shone on the changed scene. The squirrel by its industry gave little heed to that passing above the generous mantle which served as a protection for it. Here to this store, where great boughs of spruce laden with cones forming a weight many times that of the creature hoarding them, the squirrel repaired and at its leisure satisfied the craving of its hunger.

A number of these stores were inspected and in some instances the heap certainly contained more than a wheelbarrow load [1389] of food.

During the bright, calm days it is not unusual to find the squirrel sitting in the sun warming itself. On approach a fluttering note is sounded and when uttered in a place whence no sound is expected it is as startling as unexpected. When alarmed it will instantly scamper to its hole near the body of a tree and if a few minutes of time are awaited it will appear. I have also heard them clucking under the snow as though alarmed as I walked over the surface.

It was my desire to obtain the young to rear as pets but I failed to arouse sufficient enthusiasm among the Indian boys to induce them to bring them to me.

There is but little individual variation in size among the adults of either sex and during the different seasons only will noteworthy changes of pellage occur. The weight of a freshly killed specimen varies from seven to eleven [1390] ounces.

The average weight is about nine ounces. I did not taste the flesh but suspect it to be equal to that of the other species.

Numerous specimens were secured for scientific purposes.

The extreme length from tip of nose to end of hairs of tail varies from 13 to 15 inches of which the tail, hair tip included, is but slightly less than half the entire length. The height varies from 2.3 inches over the shoulder to 2.53 over the lumbar region.

The incisors are reddish yellow.

Iris	bright blue-black
Undersurface of toes	dusky
Pads	soiled white
Soles	dark gray
Hair of back	reddish and black
Hair of sides	more grayish
Hair of belly	light gray or soiled white
Tail	reddish, black and gray
Legs	deeper reddish
Whiskers	Black
Eyelids	yellowish white
Lips	Pinkish
Nose	Black
Claws	reddish-brown tipped with clear horn

Red Squirrels
Arrangement of red squirrel specimens collected by Turner at Ft. Chimo. Division of Mammals, NMNH.

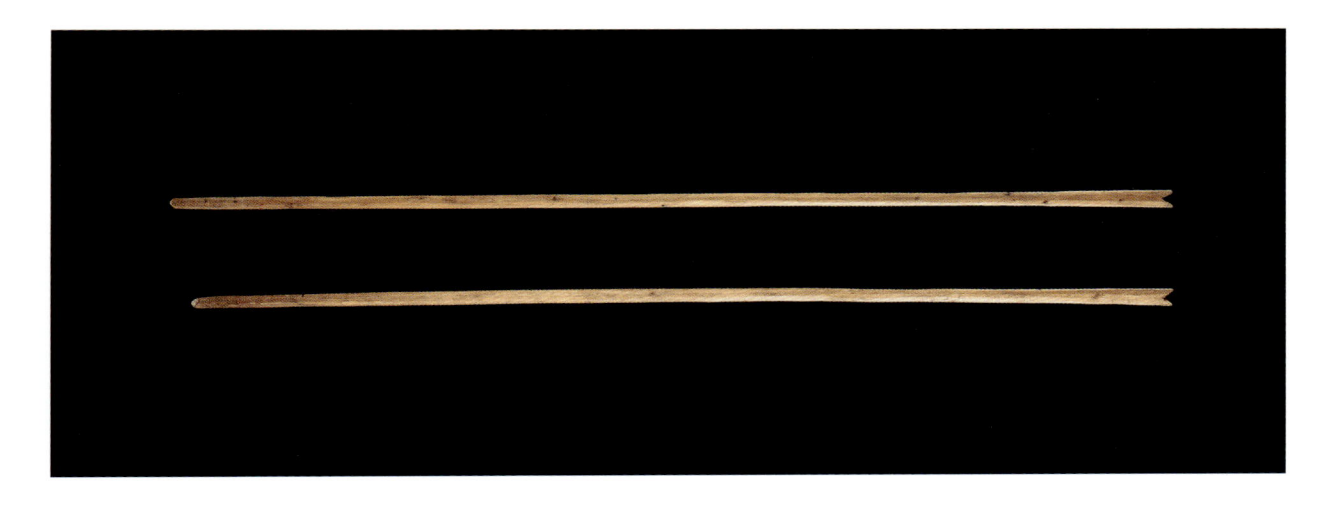

Squirrel Arrows
These arrows collected by Turner in 1884 at Ft. Chimo, were fashioned by the Inuit from driftwood to hunt squirrels and hares. USNM E90138/3137, Museum Support Center, NMNH. Photo by Donald Hurlbert, 2013.

Squirrel Ethnography
Turner

SQUIRREL OR HARE ARROW (1887A: 175-176)
Arrow containing original number 3137 is a shaft of wood without feathered vane or specialized head or point. The purpose of this arrow is for obtaining small game such as a wood hare or the little Hudsonian Squirrel from the top of the trees which do not attain a height of over fifty feet.

It will be observed that the notch which fits against the bowstring is at right angle to the lay of the vane and the edges of the iron head.

Having observed repeatedly the manner in which both the northern Indians and invariably the Innuit hold the arrow when it is to be projected by the bow it is deemed worthwhile to describe the method which insures the greatest effectiveness.

The bow is held in the left hand with the palm outward and the back of the hand to the left while the thumb is within or to the right.

The arrow is now placed so that it will lie on the knuckle of the index finger of the left hand; the notch is placed against the string and with the four fingers of the right hand, palm within and back of the hand turned out, the notched end of the arrow lies between the two middle fingers and holds it firmly against the string.

The bow is now turned so that the ends will be up and down so as to obstruct less of the line of view. The left arm is straightened and by the strength of the four fingers the bow is bent to the required curve and then the string is released. The thumb of the right hand plays no part in the movement except as a guide to place the arrow against the string as does the index finger of the left hand thrown over the arrow to steady it until ready to be released.

The moment when to release the strain is the most important in the use of this weapon, for if it be done too quick or the bow bent too much the force is diminished that much. Practice alone determines the amount of strain for each distance and penetration.

Origin of the White Hairs on the Under-Eyelid of the Hudsonian Squirrel
An Innu story, Turner (1887b: 625-626)

A reindeer once called all the quadrupeds together and announced that he would give them names according to the importance of each one before him. He arranged them in a circle and began with the large ones of the group. To the bear he gave the name *Naskwh* to the wolf, *Meh'ekan*; to the fox, *Mechéshu*; to the Wolverine he gave the name *kwekwéchu*; and, finally came to the little red squirrel that sat quivering upon a limb of a tree near by. The reindeer inquired of the tiny creature what name it would prefer. The Squirrel chose the name *Naskwh*, the name of the bear. The assembly of animals laughed heartily at the ambition of the insignificant creature. The reindeer replied that the Squirrel could not have such a name as that; but, it must be content with *A'nis ch'u ku tsas*. The squirrel began to cry, and wept so long that the hair of the lower eyelid turned white from the scalding tears and that part remains white even to this day.

Red Squirrel Skins
Two red squirrels collected by Turner at Ft. Chimo on 2 December 1882. (Left) USNM 14169 and (Right) 14168, Division of Mammals, NMNH. Photo by Angela Frost, 2011.

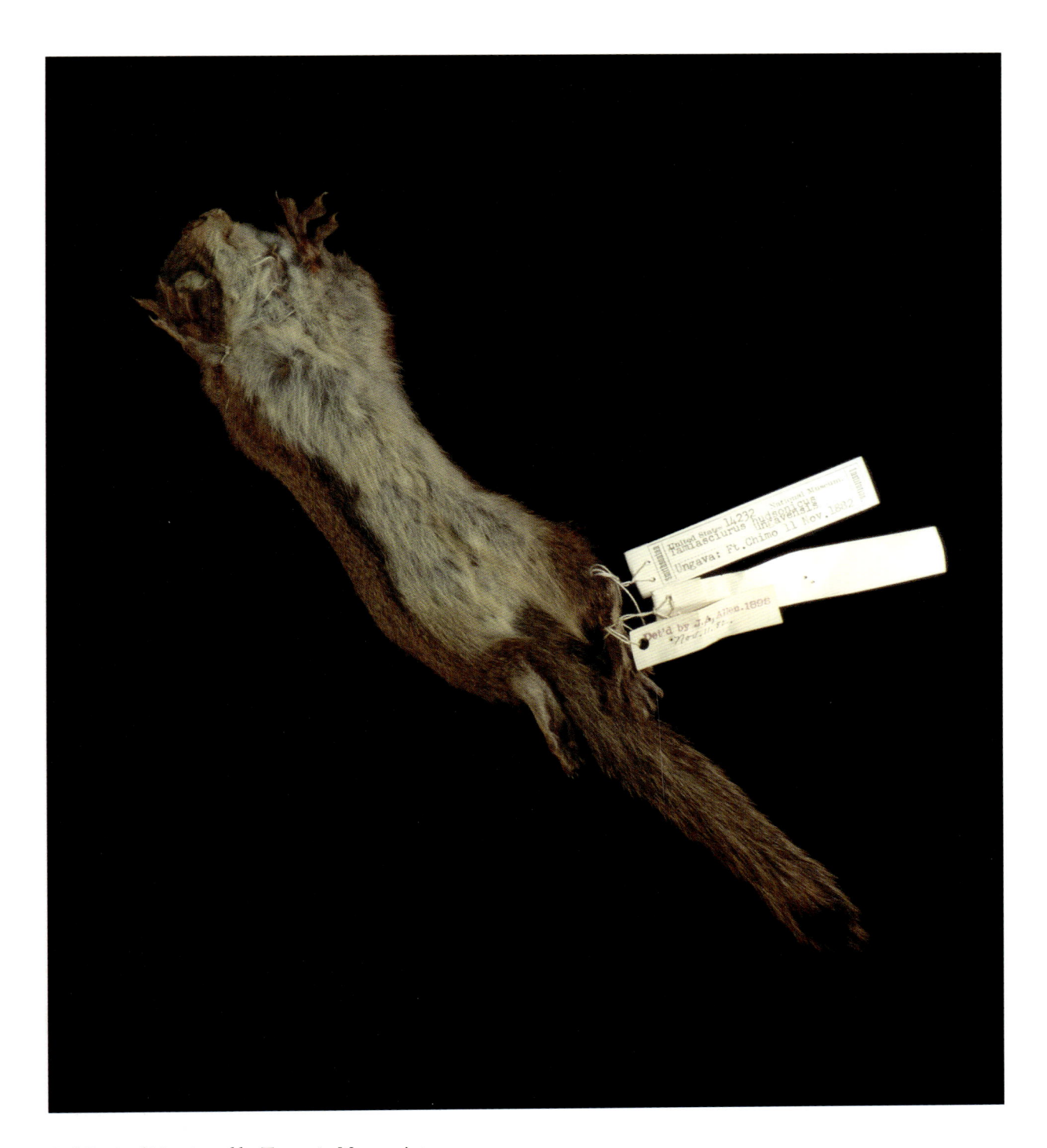

Red Squirrel Mentioned by Turner in Manuscript
Turner wrote about hunting this specimen: "A young male scarcely half grown was secured November 11th. It was fearless, approaching so closely that a shot from any gun would have ruined it. I endeavored to retreat but the creature followed and only by rapid running could I obtain any distance sufficient to warrant me in shooting." USNM 14232, Division of Mammals, NMNH. Photo by Angela Frost, 2011.

Flying Squirrel

Sciuropterus volucella

[1391]

Order Name used by Turner	*Rodentia*
Current Order Name	*Rodentia*
Family Name used by Turner	*Not recorded*
Current Family Name	*Sciuridae*
Scientific Name used by Turner	*Sciuropterus volucella*
Current Scientific Name	*Glaucomys sabrinus* (Shaw)
Syllabary	*Unknown and unnamed in Ungava*
Modern Nunavimmiutitut Term	*Unknown and unnamed in Ungava*
Eskimo Term by Turner	*Unknown and unnamed in Ungava*
Definition of Eskimo Term by Turner	*Unknown and unnamed in Ungava*

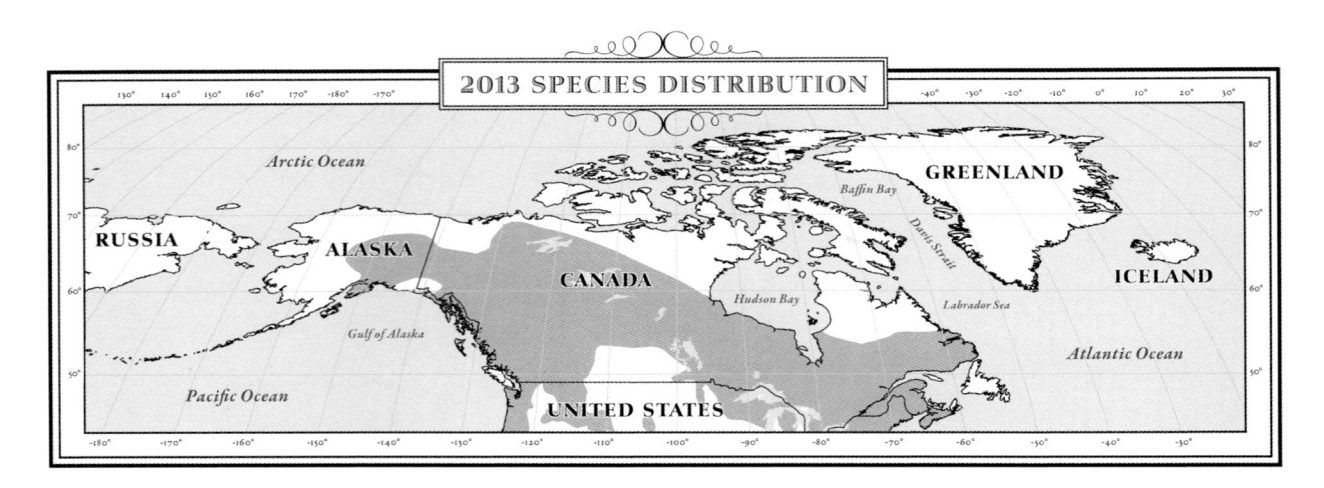

2013 SPECIES DISTRIBUTION

Chapter Two: Mammals of Ungava and Labrador

Descriptions by Turner

A single individual of the Flying Squirrel was obtained at Rigolet. It was taken near midway between Rigolet and the station of Northwest River.

I was informed by several credible persons that the Flying Squirrel is not at all rare in the woods south of the "Heighth of Land." It is not known to occur north of latitude 55 degrees.

Not having seen a living specimen, I am unable to give anything in relation to its habits.

The specimen procured, from one of the "Planters," is a female; and, the mammae are distended to such degree as to indicate that she was suckling young at the time of her capture.

The coloration is mouse-blue above and lighter below. The size is nearly twice as great as that of the Southern Flying Squirrel.

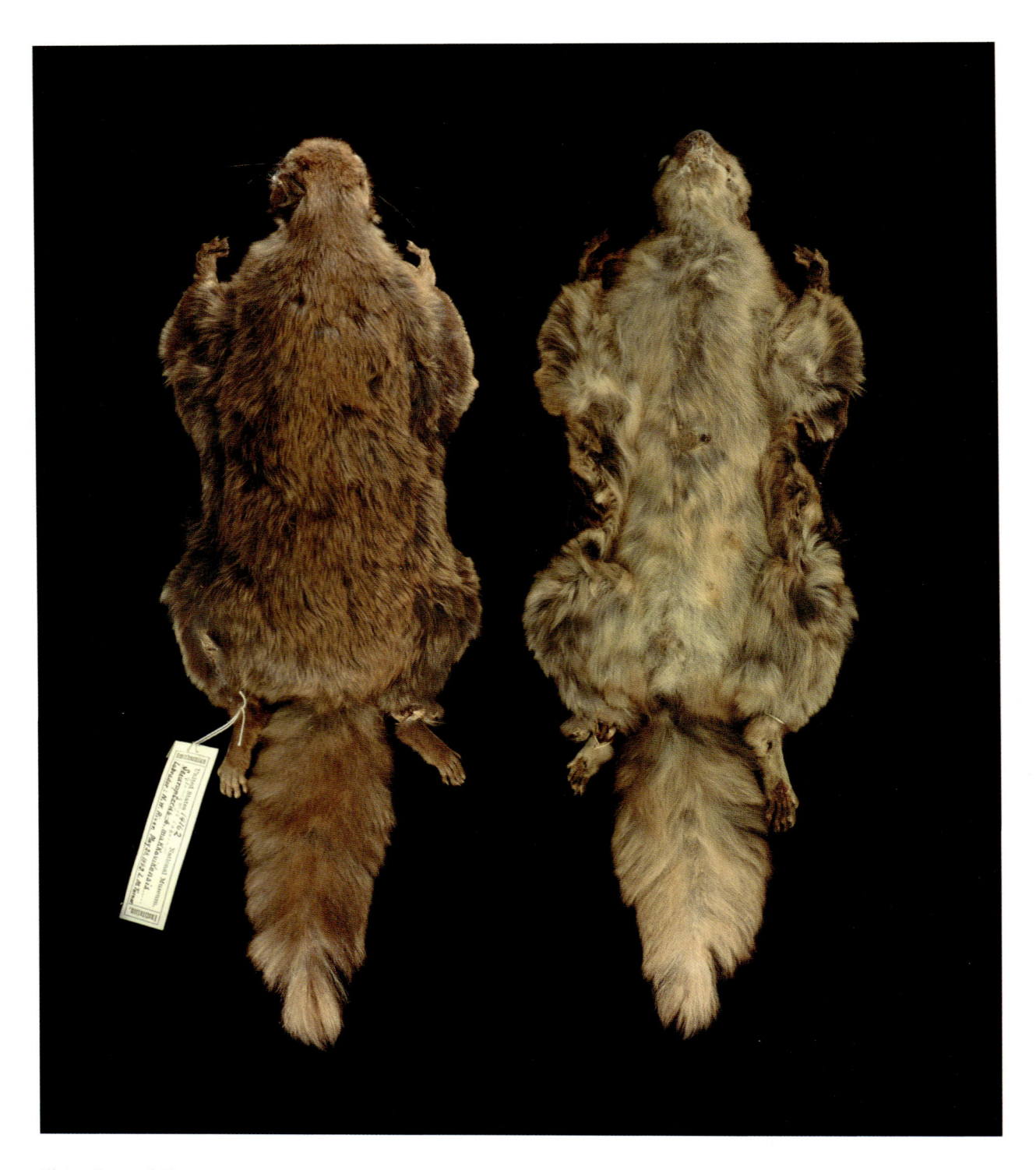

Flying Squirrel Skin
This was the only flying squirrel collected by Turner during his time in the region. USNM 14162, Division of Mammals, NMNH. Photo by Angela Frost, 2011.

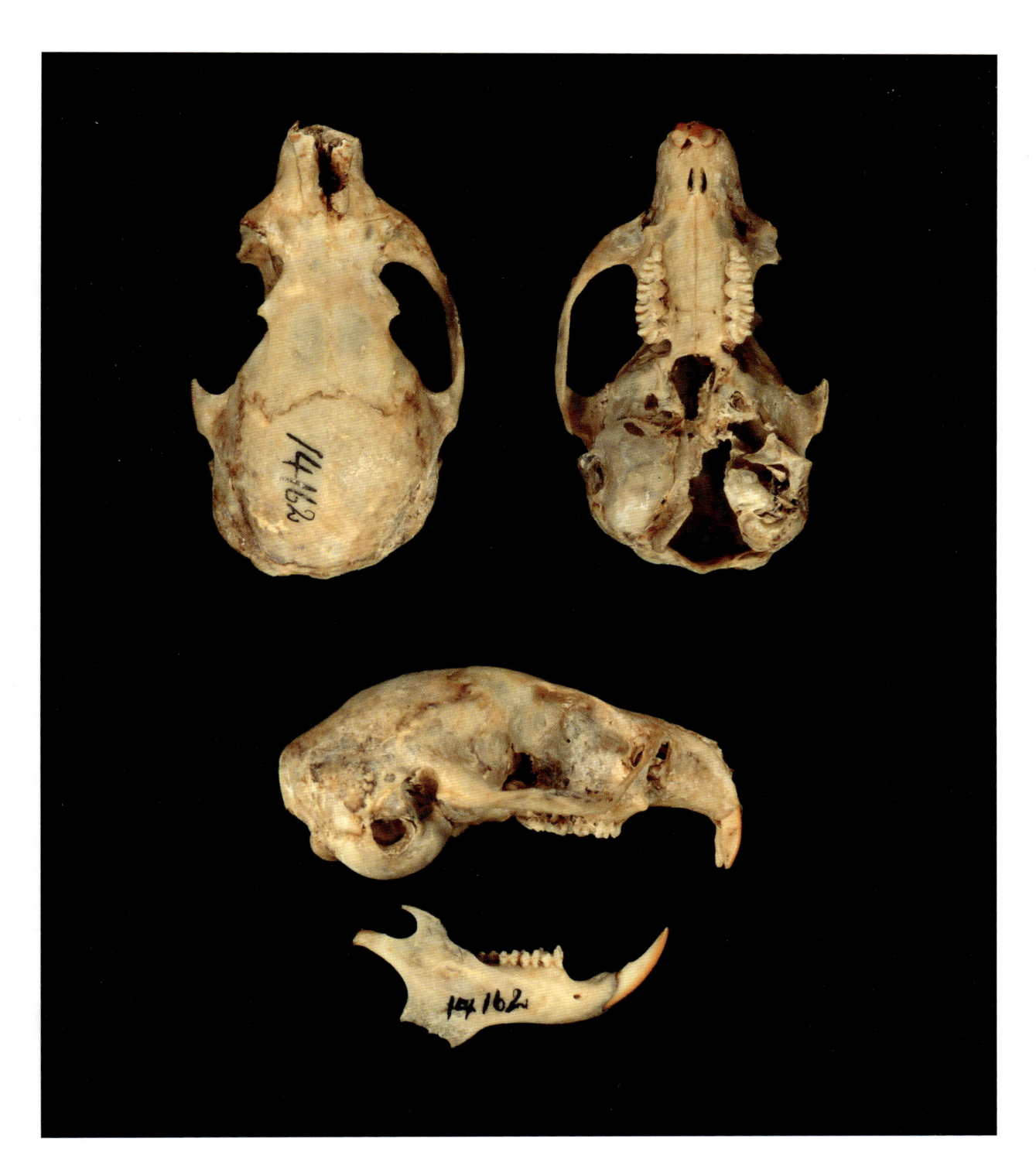

Flying Squirrel Skull
USNM 14162, Division of Mammals, NMNH. Photo by Angela Frost, 2011.

ᑭᒋᐊᖅ

Beaver

kigiaq

Castor fiber Linné.

[1392]

Order Name used by Turner	*Rodentia*
Current Order Name	*Rodentia*
Family Name used by Turner	*Castoridae*
Current Family Name	*Castoridae*
Scientific Name used by Turner	*Castor fiber Linné.*
Current Scientific Name	*Castor canadensis* Kuhl
Syllabary	ᑭᒋᐊᖅ
Modern Nunavimmiutitut Term	*kigiaq (cf. Schneider p.132 - kigiaq = beaver [Castor canadensis]*
Eskimo Term by Turner	*ki yûk*
Definition of Eskimo Term by Turner	Beaver [p. 2313]
Innu Term by Turner	*a miskwh* [p. 708]

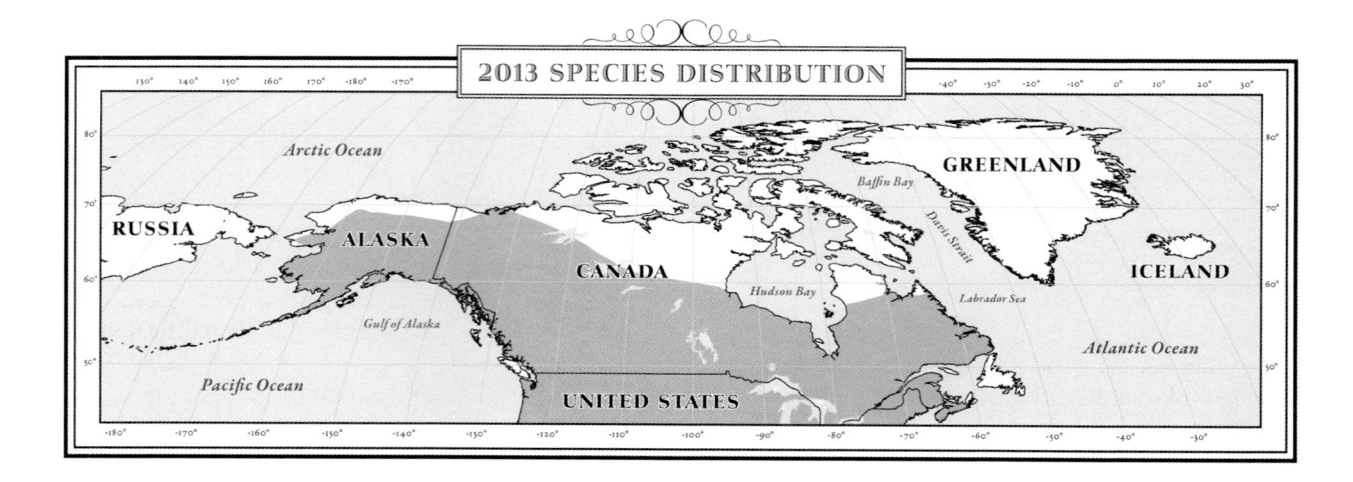

2013 SPECIES DISTRIBUTION

Chapter Two: Mammals of Ungava and Labrador

Descriptions by Turner

The beaver is very rare in the Ungava district. But few are to be found north of the "Heighth of Land." It is quite rare that as many as fifty skins find their way to Ft. Chimo.

The few that are brought there are obtained from the lakes lying between the headwaters of George's River and the Koksoak River.[10]

The quality of the fur is not best. The hair appearing scant and the fur short and sparse.

The beaver is taken in several ways. Their lodges are broken open and the animals killed within the structure. The means of egress being, of course, stopped. The water is sometimes shut off from them and as they perceive the water decreasing they emerge from their retreat to enter a kind of purse-net which is described in another connection. South of latitude 55 degrees the beaver becomes plentiful and in certain localities abundant. The flesh of the beaver is highly prized [1393] by the Indians. They also use the large incisor teeth of this animal to sharpen their knives.

The breeding habits for this region were not determined.

The Indians apply the name *A misk* to the beaver. The Eskimo give the beaver the name *Ḱi yûk* in allusion to its biting (or gnawing) habit.

---

10. In surveys conducted with Inuit hunters from Kangiqsualujjuaq in 1984, it was reported that beavers were hunted mostly in the fall upstream of a place called Helen's Falls on the George River (Makivik 1984).

Beaver Ethnography
Turner

BEAVER NETS. NASKOPIE INDIANS; UNGAVA DIST., H. B. T. (1887B: 430-432)

The scarcity of the beaver in the Ungava District renders its capture not only a prize but a source of envy from the others less fortunate; hence every means is adopted to secure the creature if its haunt be discovered. Rarely is it found alone as there are (in this locality only do I refer to) usually an adult male and female and, perhaps, their young. The Beaver is well known to be quite sociable in its natural condition for its peculiar habits tend to make manifest that several of them are required to undertake the wonderful labors performed by them.

When it has constructed its dam and erected the lodge in which it stores its food and passes the time away while at leisure, if not sporting at the surface of the water, there will be found one or two doorways each submerged or covered by the water.

The Indian watches and when he is assured that the creature is within the hut he blocks up one of the holes, if there be more than one, and places a net of peculiar construction over the other exit. The Beaver remains within so long as there is no unnecessary demonstration without but would escape if the wall of the structure was pounded upon.

The net is a square or slightly oblong piece of netting made from deerskin thongs as this material has the property of stretching and thus further disconcerts the actions of the timid animal. Around the outer meshes of the net is run a long thong the ends of which are arranged that when the beaver presses, in its escape from the lodge, against the central portions of the net the strings will draw the net corners and edges together and entrap the Beaver in it. The struggles of the creature tend to pull the tighter on the strings which are fastened to the side of the lodge or elsewhere as may be.

The strings have a ring encircling them so that they must necessarily be quite long and draw through that ring while the net is closing so that the length of the strings may be several feet when the Beaver is entrapped and only a few inches when the net is spread out. The Beaver of course is soon drowned. Those within hearing the struggle are not apt to attempt to make an attempt to escape by that opening. They crouch in silent fear upon the ledge where they sit when eating or huddle in the water compartment while the Indian breaks into the hut and thrusting his arm among the shuddering creatures seizes one by the hind leg and withdraws it, taking the precaution to keep his arm in motion, for if it stops for half a second the Beaver doubles up and seizes the arm or hand and snaps a piece out or amputates a finger with the ease of a hatchet chopping it off.

A club is at hand and a blow over the head knocks the life out of it.

The Indian and His Beaver Wife
An Innu story, Turner (1887b: 587-592)

On a bright spring morning an Indian was walking along the bank of a large lake that lay not far distant from a river. A Beaver swam toward him; and as the Indian was about to shoot her she cried out, "Do not shoot, I have something to say to you." The hunter inquired of her what she wished to tell him. The Beaver asked him, "Would you have me for a wife?" The man replied, "I cannot live in the water, or eat the bark of willows for food." The Beaver smiled, and told him, "You will not know you are in the water if you follow me; and when you are eating of my food you will not think it is willow bark." I have a nice house to live in; and the water surrounds it all the year, but never enters it." The man then added, "My brother will search for me, and will laugh at me for living with a Beaver. He will never know where I am."

The Beaver said to him "Take off your garments and place them with your weapons on the bank and follow me. Never mind your brother, for if he finds you, he will not laugh at you." The man did as he was directed and began to wade in the water. He soon began to swim and did not feel the water touching his body. The Beaver now came back to him and they swam side by side until they reached her home near the middle of the lake. The Beaver said to him, "There is my house. You will find it as good as your tent and as warm."

They entered the lodge; and, after they had passed two nights there, his brother began to search for him. He went along the bank and discovered his clothing and weapons. The brother was alarmed lest his poor brother had been drowned. He took the garments and returned to his tent and there told his wife that he feared his brother had been drowned.

The next morning the brother related to his wife that he had dreamed the lost brother was living in the middle of the lake with a Beaver. She would have to make some new clothing for him and he would take them to the lake and bring his brother home with him. The next day the clothing was prepared and ready for the lost brother. He directed her to tie them into a bundle and have them ready as he would start early in the morning. Other Indians offered to accompany him but he told them to remain; for, if you come I cannot induce my brother to return.

The next morning he started to search for the lodge of the Beaver, who had his brother as a husband. He soon found the Beaver's lodge and then began to drain the lake into the river so that he could get at the lodge in which they lived.

Two children had been born from the Beaver during this time. When the water became drained so low that the brother could wade to the lodge he entered the water and began to tear down the mud walls of the structure. He pounded on the back of the house and heard movements within, and thus knew that it was occupied.

The father told his children to go out or else they would be killed when the house fell. When they went out the father and mother heard the uncle kill his nephews by striking them with a club. The wife knew she also would be killed and asked her husband to keep the skin of her right arm; and, if he loved her they would meet again. He promised to do so, and she went out of the house. A blow on the head killed her and the husband began to cry when he know that his good wife was dead.

The brother now began to destroy the house and in a few minutes had a large hole in the top of it. The one within inquired of the other, "What are you doing? The air is cold and I am freezing." The Indian replied,

"I have brought some clothing for you so that you will not be cold." The husband asked him to throw the garments into the lodge. The Indian now saw that his brother was covered with hair like a beaver, but asked him to come home with him to his tent where he might live and forget the Beaver. The brother consented on condition that nothing should ever be said or done to make him angry. The Indian promised that nothing should ever be done to make him angry. The husband then put on the clothes his brother had brought to him and came out of the lodge.

The Indian tied the legs of the Beavers together and slung the bodies on his back and they returned to the tent.

On the way returning they found other beavers and killed a great many of them. Taking them home he threw them down and directed the women to skin them.

The husband asked his brother to save the skin of the right arm of his Beaver wife. The brother brought him the arm and the husband then gave it to an old woman to skin, telling her to dry the skin and return it to him. The old woman skinned it and in a few minutes had dried it. The man now took it and put it in the fold of his belt. The others noticed this but made no remark. The flesh of the beavers was now cooked and they feasted long upon it. They prevailed upon the husband to have some of the meat but he refused to touch it. At last he become so hungry and they asked him to have some of the meat. He replied that he would eat only the flesh of the male beaver. They gave him a portion and when he tasted of it he took a second piece, which was that of a female beaver. He tasted of it and instantly a huge river rushed from his side. The other Indians rushed out of the tent to save themselves. They looked down the river and saw the husband swimming away by the side of the Beaver wife.

Recollections on Inuit Use of Beaver
Tivi Etok (Heyes 2010)

A long time ago there were hardly any beavers along the Korac and George Rivers. Now we see them more often; they are coming back slowly. They are not coming back fast like geese, but are returning at the pace of a turtle. They are making more babies. Animals can make their own houses, even in running water. But not Inuit. If there were lots of beaver up the river they would cover the stream in their houses. The beaver is more stronger than carpenters because they can build and protect themselves. When the beaver has a baby and it cannot be helped they leave it alone. We used to eat beaver meat and its white fat. The fur was used for mitts and hats. The best way to eat beaver is to boil it or fry it. I used to watch Indians roast it and boil it. If I was very hungry I would have eaten it raw; but we usually boiled it.

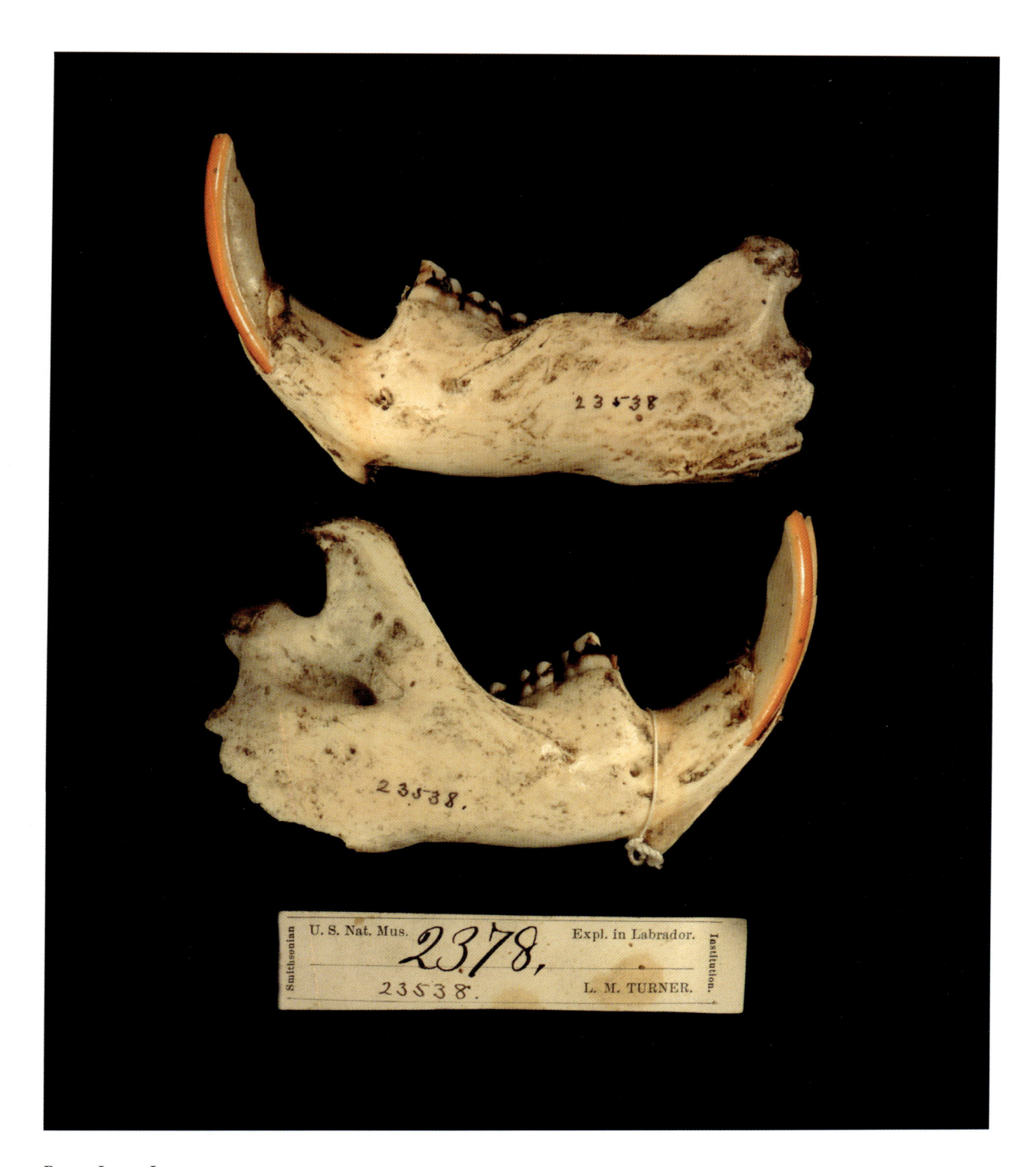

Beaver Lower Jaws
These jaws are the only beaver specimens that Turner collected. The long curved teeth are used to gnaw trees. USNM A23538, Division of Mammals, NMNH. Photo by Angela Frost, 2011.

Muskrat
kivvaluk

Fiber zibithecus, (Linné) Cuvier.

[1394]

Order Name used by Turner	*Rodentia*
Current Order Name	*Rodentia*
Family Name used by Turner	*Muridae*
Current Family Name	*Cricetidae*
Scientific Name used by Turner	*Fiber zibithecus*, (Linné) Cuvier.
Current Scientific Name	*Ondatra zibethicus* (Linnaeus)
Syllabary	ᑭᕝᕙᓗᒃ
Modern Nunavimmiutitut Term	*kivvaluk* (cf. Schneider p.131 - *kivvaluk* = muskrat [*Ondatra zibethicus*])
Eskimo Term by Turner	ki vá lu
Definition of Eskimo Term by Turner	Muskrat (*Fiber zibethicus*); a word cognate with that for beaver [p. 1396; 2311]
Innu Term by Turner	wa chesk [p. 708]; utsaskwh [p. 709]

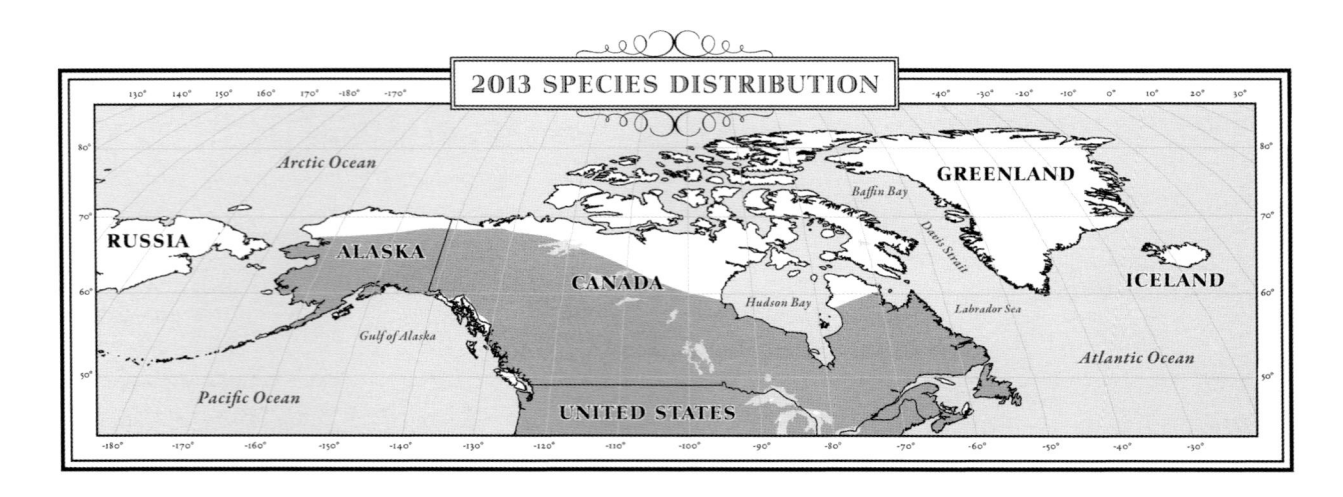

2013 SPECIES DISTRIBUTION

Chapter Two: Mammals of Ungava and Labrador

Descriptions by Turner

The muskrat is not rare in the Ungava district. It ranges as far as the timber line and beyond this I could obtain no evidence of its occurrence. A few are to be found in the immediate vicinity of Ft. Chimo.[11] They are known to breed in Whitefish Lake, some three miles east of Ft. Chimo. This expanse of water is about two and a half miles long and about seven hundred yards wide. The situation is somewhat peculiar, being surrounded on one side by high rocky ridges, but short distance from the water. The eastern side is low, bounded by trees of small growth, the northern end swampy and full of jagged rocks. Along certain portions of the lower end (north, for these two streams form an outlet), sufficient food of vegetable nature is to be found to afford subsistence for a family of these rodents. Their piles of grass and other food are often placed directly upon the ice of the lake in winter and by [1395] placing a trap near their supply of food two of the muskrats were obtained. They were adults, a male and female. Below are appended the respective measurements taken while the animals were in the flesh.

---

11. Inuit hunters from Kangiqsualujjuaq report that the best place to set traps for muskrats is on flat rocks upon which muskrats have defecated. Muskrat are known to occur upstream of Helen's Falls on the George River (Makivik 1984).

	MALE	FEMALE
Tip of nose to root of tail [12]	12.00	10.75
Caudal length	9.60	9.75
Circum mid. Of belly	10.70	8.60
Between eyes	0.87	0.90
Between nostrils	0.13	0.11
Eye to ear	0.22	0.21
Eye to tip of nose	1.30	1.10
Cleft of mouth	0.90	0.80
Color of eye	blue black	blue black
Color of claw tips	white	white
Color of claw base	pink	pink

Color of teeth	Low incisors	Pale yellow
	Upper	Reddish yellow
Lips	Silvery bluish-gray	
Nose	Black	
Body	Above	Rust brownish black
	Under parts	Slightly paler

12. Measurements in above table are in inches.

Other individuals were observed up the Koksoak River. Among the line of islands locally know as the "Juniper" Islands (so named [1396] because they have no Juniper trees growing on them) the muskrat was frequently observed. One was shot as it was industriously collecting a species of grass, *Zostera*, for food. Dissection showed its stomach to be nearly full of this substance. The animal swam back and forth selecting the tender shoots and had quite a bunch in its mouth. This plant appears, with few kinds of coarse, marsh grass, to form their only food. The breeding season was not determined. The parents obtained from Whitefish Lake had three young which deserted the locality as soon as the adults were caught in the steel trap.

The Indians give the name of *wa chĕsk* to the muskrat and also apply that name to certain ones of themselves.

The Eskimo recognize it as *ki válu*; a word cognate with that for the Beaver.

The skin of this animal is sometimes converted into a cap for the head and worn by either Indian or Eskimo.

Muskrat
A muskrat collected by Turner at the "Forks" near Ft. Chimo on 10 October 1882. This specimen is one of three muskrat skins collected by Turner in the region. USNM 14165, Division of Mammals, NMNH. Photo by Scott Heyes, 2011.

Muskrat Skin
The dorsal and ventral view of a muskrat skin collected by Turner at Ft. Chimo, 1882. USNM 14163, Division of Mammals, NMNH. Photo by Scott Heyes, 2011.

LEMMING

avinngaq

Cuniculus torquatus

[1397]

ORDER NAME USED BY TURNER	*Rodentia*
CURRENT ORDER NAME	*Rodentia*
FAMILY NAME USED BY TURNER	*Muridae*
CURRENT FAMILY NAME	*Cricetidae*
SCIENTIFIC NAME USED BY TURNER	*Cuniculus torquatus*
CURRENT SCIENTIFIC NAME	*Dicrostonyx hudsonius* (Pallas)
	Synaptomys borealis (Richardson)
SYLLABARY	ᐊᕕᙳᖅ
MODERN NUNAVIMMIUTITUT TERM	avinngaq (cf. Spalding 1998, p.15 -avinngaq = lemming)
ESKIMO TERM BY TURNER	aviṅ ûk
DEFINITION OF ESKIMO TERM BY TURNER	Lemming (Cuniculus torquatus). Innuit girls make bedblankets for their dolls from the skins of these. [p. 1401; 2207]

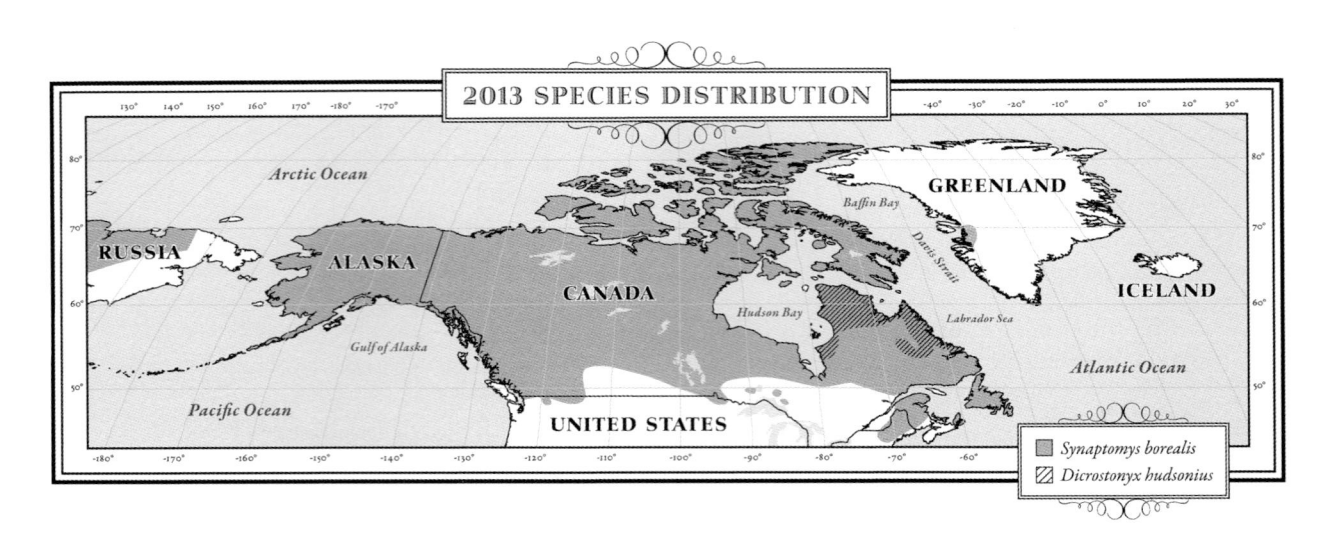

2013 SPECIES DISTRIBUTION

Legend:
- *Synaptomys borealis*
- *Dicrostonyx hudsonius*

CHAPTER TWO: MAMMALS OF UNGAVA AND LABRADOR

Descriptions by Turner

The lemming is found somewhat sparingly on the lower tracts and but rarely on the wet grounds. The tops of the dryer hills and ledges of the mountains being its favorite resorts. The animal, being migratory, is more plentiful in certain districts at times than it may be again for years, there being no regularity of its apparently erratic movement. All these creatures do not forsake a given locality for a greater or less number remain, and again they may be so numerous as to be seen anywhere.

Their dens are usually under the moss and Empetrum layers and to search for the ends of the runways is an almost interminable task as their paths cross and re-cross in such a puzzling manner as to render the search nearly fruitless.

I have on but one occasion found the nest of these creatures. Its size [1398] was so great as to astonish me.

A cavity beneath the larger of vegetation above had been excavated by the roots soil and other matter being removed

Collection Notes

Two species are included/confused in Turner's account, *Dicrostonyx hudsonius* and *Synaptomys borealis*. Turner collected the type specimen of the subspecies *Synaptomys borealis innuitus* True, 1894 (type locality is Ft. Chimo).

affording a space over a foot in diameter. The nest material was of fine and coarse grasses matted together in such manner that the roof of the structure was self supporting. When I tore away the mass of *Empetrum* roots the affair was disclosed. The depth was about five inches and little more than a foot in exterior diameter; the walls being about an inch thick. The interior was quite dry and although thoroughly inspected, I was not able to conclude whether it was a structure for the size of the male alone or for the female and young. From appearances I suspected it to be a residence or dormitory. It had been recently occupied but at that time deserted. The alarm conveyed of [1399] the trembling masses of vegetable roots was sufficient to give timely notice for the escape of the occupants.

The greater number of these creatures were obtained as they were found on the moss either sunning themselves or in a sort of sleep for of the many I have seen I have never seen one in motion.

The white people of Labrador term these creatures "Mountain Mice" and assert that they throw themselves on their back and kick and squeak on the approach of danger. I have never heard them utter a sound.

Individuals were procured from May 8th to late October. I have not seen them during the depth of winter and suspect that they do not but rarely appear above the surface after the snows have fallen. In the spring when the snow melts and fills their runways with water the creature is then most [1400] plentiful on the ground.

The number of young or the number of broods reared each year was a subject on which I could obtain no reliable information. An adult female was found drowned, May 12, 1883; her mammae were quite large and gave evidence that about that time a brood was being nursed by her [USNM 14208]. Young, scarcely larger than a common mouse, were found on the same date and determined to be at least a year old. Another individual was procured May 20 and was found to be two years old [USNM 190378]; the creatures not attaining their full size until the completion of the third year.

The Lemming in this region does not become white in winter; the pellage is lighter tipped becoming darker below, having for the dorsal and upper side color an ashy reddish-brown. Along the back is a dark narrow stripe, often obscure. The lower sides of the adult in winter are much lighter [1401] while the abdomen, throat, and sides of face

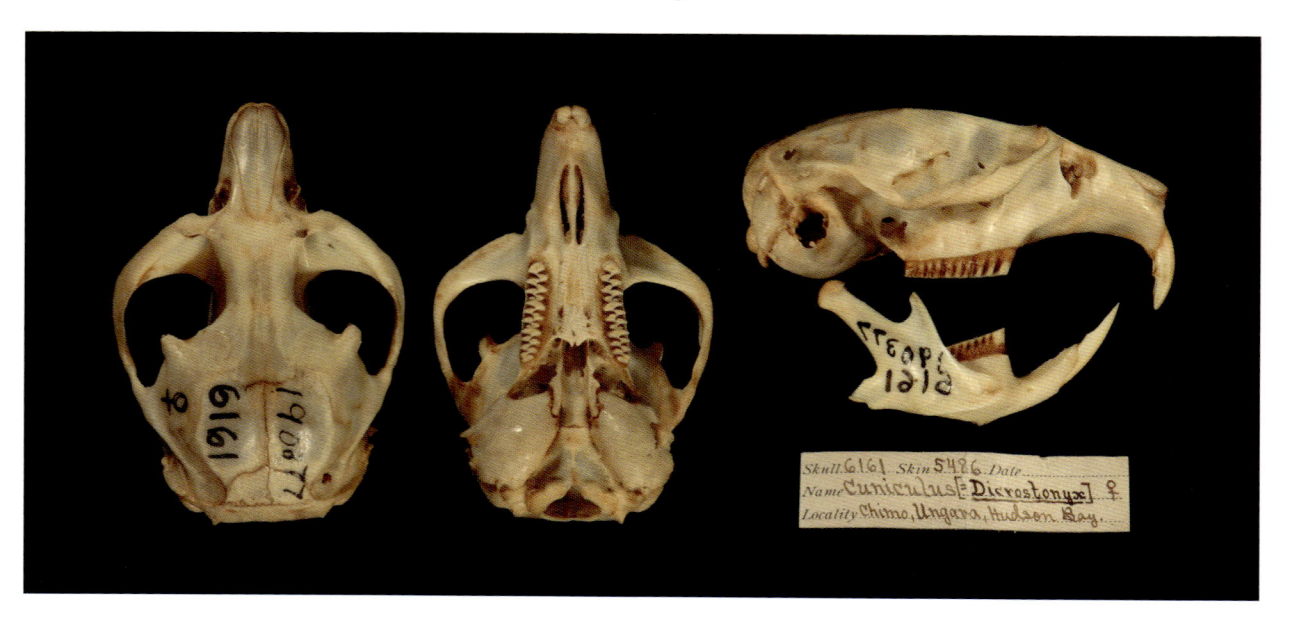

Female Lemming Skull
Collected by Turner at Ft. Chimo on 8 May 1883. The skins of lemmings and squirrels were prepared by Inuit elders as blankets for children's dolls. These blankets mimicked the bedding used for infants, which were made from soft skins such as "those of the reindeer fawn or the hare." (Turner 1887a: 180). USNM 190377, Division of Mammals, NMNH. Photo by Angela Frost.

Fluid Specimens of Lemmings
Collected and preserved by Turner at Ft. Chimo on 14 October 1882. The specimens contain the original tags. Division of Mammals Collection, NMNH. Photos by Scott Heyes.

are considerably lighter grayish. The stiffish hairs on feet, tail and about the nose are often of a glistening whitish. In the summer the animal is considerably darker, having a reddish-brown shade on the sides and back. At certain seasons the pellage is spotted with rows of rounded dots of darker or lighter color than the surrounding hairs. The skins vary so much in color of pellage that a minute description of the species would be out of place here. The Eskimo girls make bed blankets for their dolls from the skins of this lemming. The Eskimo apply the name *A víng úk* to this species.[13]

The food of the lemming was not entirely determined but as it is not found about the houses it is not known to be destructive to other than such food as grasses and roots, which it finds amongst the heathers and mosses.

13. Inuit elder, Sophie Jararuse Keelan, of Kangiqsualujjuaq reports that lemming and rabbit skins have medicinal qualities that can can be used to treat boils and cuts. The lemming or rabbit skin is lightly scrapped off and applied to the affected area. Inuit still use this method today (Pers. Comm., 2012).

ᓄᓂᕚᒃᑲᖅ
EVOTOMYS
nunivakkaq

Evotomys

[1402]

Order Name used by Turner	*Rodentia*
Current Order Name	*Rodentia*
Family Name used by Turner	*Muridae*
Current Family Name	*Cricetidae*
Scientific Name used by Turner	*Evotomys*
Current Scientific Name	*Myodes gapperi* (Vigors)
Syllabary	ᓄᓂᕚᒃᑲᖅ
Modern Nunavimmiutitut Term	*nunivakkaq* (cf. Schneider p.225 -nunivakkaq = house mouse [Mus musculus] ...)
Eskimo Term by Turner	*nú ni vû kûk*
Definition of Eskimo Term by Turner	*Not recorded*

2013 SPECIES DISTRIBUTION

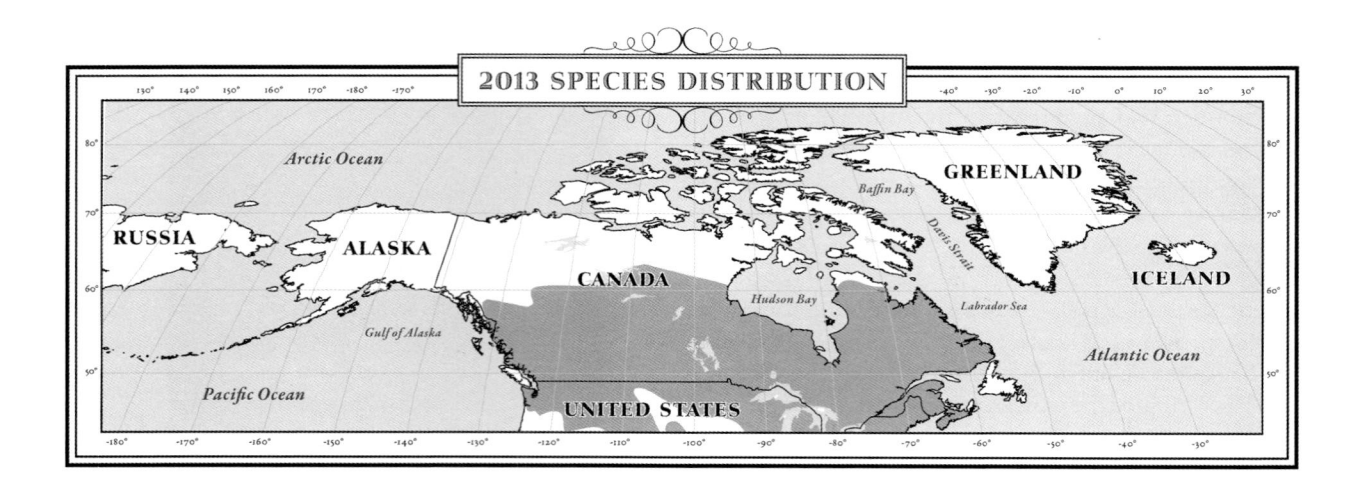

CHAPTER TWO: MAMMALS OF UNGAVA AND LABRADOR

Descriptions by Turner

This species [*Myodes gapperi,* Red Backed Vole] is not so plentiful as the Meadow Mouse. It is not known to have any different habits than the other arvicoles.

The species is not distinguished by the Indians and Eskimo; the former, however, assert that there is a kind of mouse which dies as soon as it comes upon the path of a person. To that species they apply the term *kwe kive ta puk á shu* (a name fully as long as the creature itself).

"...there is a kind of mouse which dies as soon as it comes upon the path of a person"

Collection Notes

In Turner's time, *Evotomys* was the genus name used to describe the red-backed vole, *Myodes gapperi.* The genus name *Myodes* is used today in preference to *Evotomys*. Turner collected the type specimen of *Myodes gapperi ungava* Bailey, 1897 (type locality is Ft. Chimo).

Myodes (Evotomys) **Skin and Skull, Type Specimen**
This is the type specimen of Evotomys ungava, *described by Bailey (1897) as a new species, but now classified as a subspecies of* Myodes gapperi. *This skull and skin was collected by Turner on 5 December 1883 at Ft. Chimo. USNM 186492, Type Collection, Division of Mammals, NMNH. Photo by Renee Regan.*

***Myodes* Habitat**
The spongy and matted edges of lakes and rivers in Ungava provide ideal habitat for Myodes. *Photo taken near Kangiqsualujjuaq by Scott Heyes, 2012.*

RODENTS

ᓄᓂᕙᒃᑲᖅ

ARVICOLE

nunivakkaq[14]

A. riparius

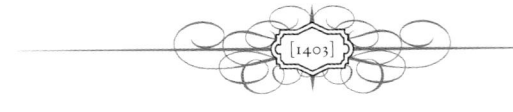

[1403]

ORDER NAME USED BY TURNER	*Rodentia*
CURRENT ORDER NAME	*Rodentia*
FAMILY NAME USED BY TURNER	*Muridae*
CURRENT FAMILY NAME	*Cricetidae*
SCIENTIFIC NAME USED BY TURNER	*Arvicola riparius*
CURRENT SCIENTIFIC NAME	*Microtus pennsylvanicus* (Ord)
	Phenacomys ungava Merriam
SYLLABARY	ᓄᓂᕙᒃᑲᖅ
MODERN NUNAVIMMIUTITUT TERM	*nunivakkaq* (cf. Schneider p.225
	-nunivakkaq = house mouse
	[*Mus musculus*] ...)
ESKIMO TERM BY TURNER	*nú ni vû kûk* (Meadow Mouse p. 1402)
DEFINITION OF ESKIMO TERM BY TURNER	*Arvicole (A. riparius)* [p. 2722]
INNU TERM BY TURNER	*a púk a shísh* (Mouse) [p. 708]

2013 SPECIES DISTRIBUTION

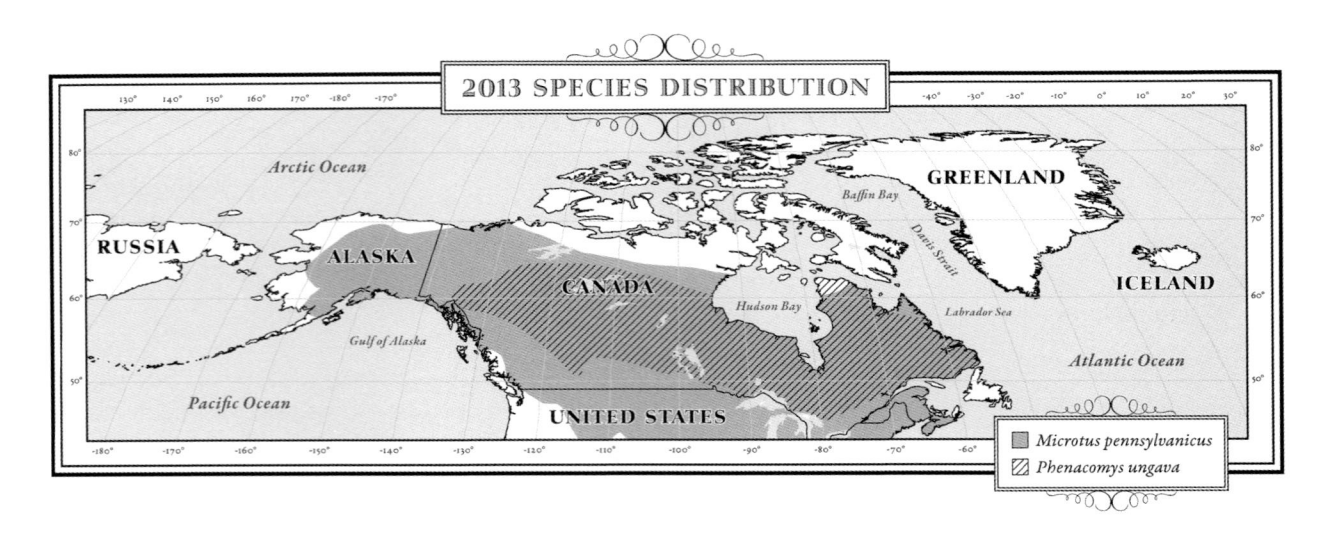

- *Microtus pennsylvanicus*
- *Phenacomys ungava*

---

14. Note that the Inuit apparently do not distinguish between *Myodes gapperi* and *Microtus pennsylvanicus*. They use the term *nunivakkaq* to describe both species.

CHAPTER TWO: MAMMALS OF UNGAVA AND LABRADOR

Descriptions by Turner

This arvicole [1403] is quite plentiful throughout the region, being by far the most abundant of all the rodents, inhabiting the wooded as well as the treeless areas. The lower grounds appear to be preferred and on no occasion did I observe its signs above an elevation of 1,500 feet. Along the ledges of the ravines and amongst the mosses and lichens the runways are everywhere visible. It infests the warehouses and often does considerable damage to the goods stored therein. The bags of seal oil collected by the Eskimo and left exposed temporarily on the rocks of the beach are often cut and the contents flow away. The general habits of the creatures could not be learned. I have on several occasions found nests, constructed by some species

Collection Notes

In Turner's time, *Arvicola* was used for many vole species, including the species now known as *Microtus pennsylvanicus*. Turner collected many specimens of *Microtus pennsylvanicus* while stationed at Ft. Chimo, including the type specimen of *Microtus pennsylvanicus labradorius* Bailey, 1898 (type locality is Ft. Chimo). He also collected the first known specimens of a newly discovered (at the time) species of vole now known as *Phenacomys ungava* Merriam, 1889 (type locality is Ft. Chimo). It is interesting that Turner did not distinguish or write about this additional species in his manuscript; he was apparently unable to distinguish it from other voles in the field.

of mouse, lying on the [1404] ground, exposed to plain view. As I never saw them occupied I suspected they were temporary winter nests built after the winter's 2 inches mantle of snow had fallen and thus accounted for their exposed position, as they were built on the ground at that time when digging in the frozen soil could have been impossible. The deserted nests were favorite resorts for bumble bees to form their few cells and deposit their eggs in.

The Eskimo name of this mouse is *nú ni ŭ kŭk*. The Indians term it *a púk a shísh*. Great numbers of this species were obtained at Ft. Chimo, several at Davis Inlet and one at Rigolet.

Phenacomys Skin and Skull, Type Specimen
This is the type specimen of Phenacomys ungava, *collected by Turner at Ft. Chimo, probably in 1884, and described as a new genus and species of vole by Merriam (1889). USNM 186488, Type Collection, Division of Mammals, NMNH. Photo by Renee Regan.*

Recollections on Inuit Use of Mice
Tivi Etok (Heyes 2010)

We used to boil them; but only when we were hungry. We even ate lemmings, too. These were very bony. The small mouse is very stupid. In old times, when the summer starts all the animals arrive. The mice used to crawl all over our legs at night in our tents during the summer time. We used to have caribou mattresses in the tent. The mouse used to make a house for itself from our caribou skin mattresses. The mice used to eat our scraps. In some years there were plagues of mice, but in other years there were few. In terms of pets, we used to sometimes carry lemmings. Mice were not as friendly; they used to scratch and they had a fast bite. They could even bite through a mitt; we could even get hurt from a bite from them. They are very strong; they even make a trail. They could even cut up the meat like Inuit using their sharp teeth. The mouse can eat anything; even left over food.

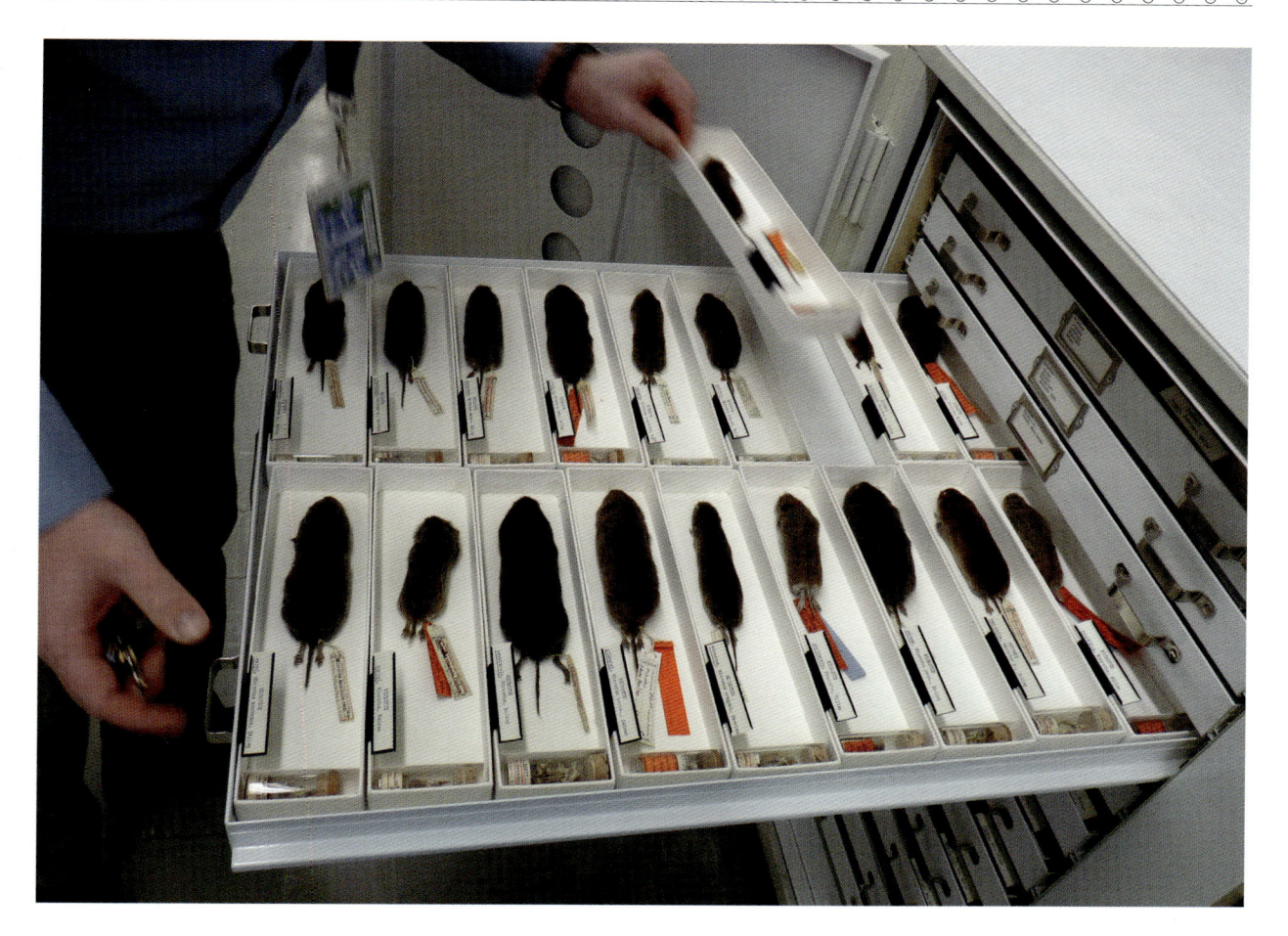

Type Specimens of Voles
Type specimens (specimens on which scientific names are originally based), of various species of voles, some collected by Turner. Type Collection, Division of Mammals, NMNH. Photo by Angela Frost, 2011.

Vole Skulls in Glass Vials
Type Collection, Division of Mammals, NMNH. Photo by Scott Heyes, 2011.

Vole Skin and Skulls
Specimens Collected by Turner at Ft. Chimo. Division of Mammals, NMNH. Photo by Angela Frost, 2011.

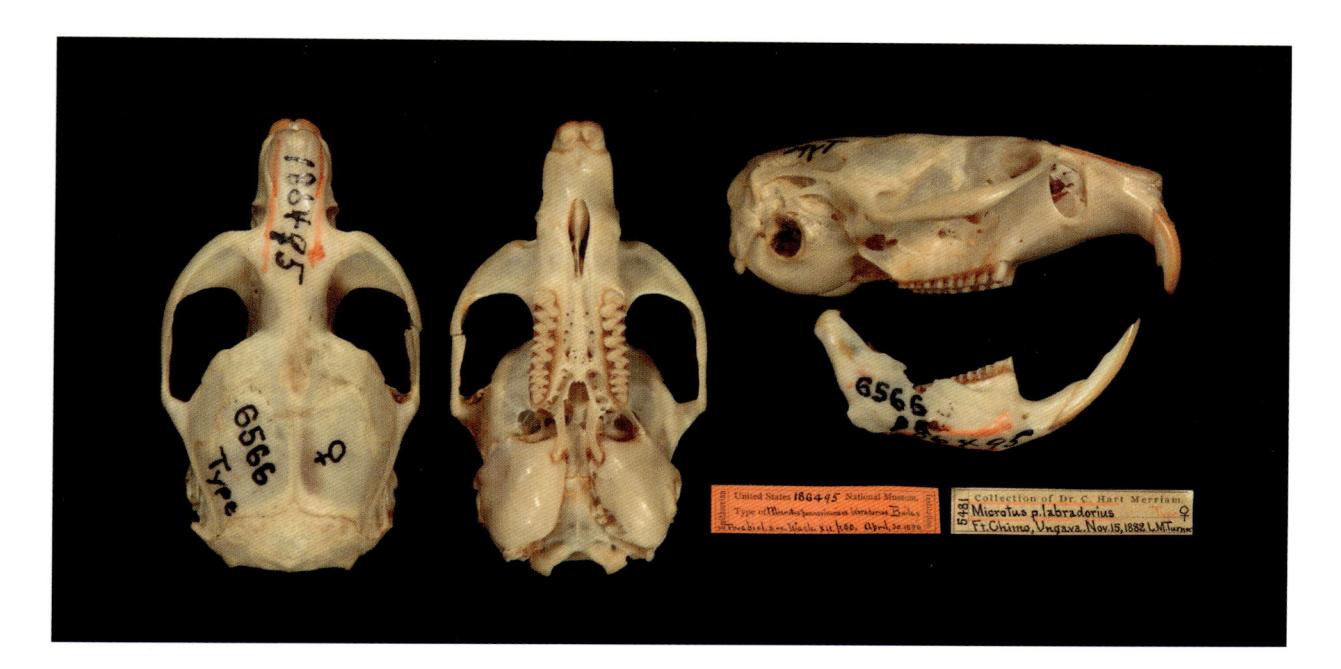

Vole Skull, Type Specimen

This skull of a female Microtus pennsylvanicus *was collected by Turner on 15 November 1882 at Ft. Chimo. This is the Type specimen of* Microtus pennsylvanicus labradorius, *which was described by Bailey (1898) as a new subspecies. Turner (1887a: 27-28) indicated that the Tahagmyut Inuit on the east side of Hudson Bay, like Inuit from Ungava, did not consider mice as a food source. He wrote that: "The flock of all creatures is eaten, excepting that of the mouse, ermine, raven and the lower forms of life from the cold waters of the Strait. A decided preference is had for the reindeer, seal, walrus and the various cetaceans occurring there." USNM 186495, Type Collection, Division of Mammals, NMNH. Photo by Angela Frost, 2011.*

ᐃᓛᖁᑦᓯᖅ

CANADIAN PORCUPINE

ilaaqutsiq

Erethizon dorsatus

[1405]

ORDER NAME USED BY TURNER	*Rodentia*
CURRENT ORDER NAME	*Rodentia*
FAMILY NAME USED BY TURNER	*Hystricidae*
CURRENT FAMILY NAME	*Erethizontidae*
SCIENTIFIC NAME USED BY TURNER	*Erethizon dorsatus*
CURRENT SCIENTIFIC NAME	*Erethizon dorsatum* (Linnaeus)
SYLLABARY	ᐃᓛᖁᑦᓯᖅ
MODERN NUNAVIMMIUTITUT TERM	*ilaaqutsiq* (cf. Schneider p.61 -ilaaqutsiq = porcupine [Erithizon dorsatum])
ESKIMO TERM BY TURNER	*i lá ku chúk; i lá ku syûk*
DEFINITION OF ESKIMO TERM BY TURNER	Canadian porcupine (*Erethizon*) [p. 2231]; Canadian porcupine [p. 1405]
INNU TERM BY TURNER	*kakwh*

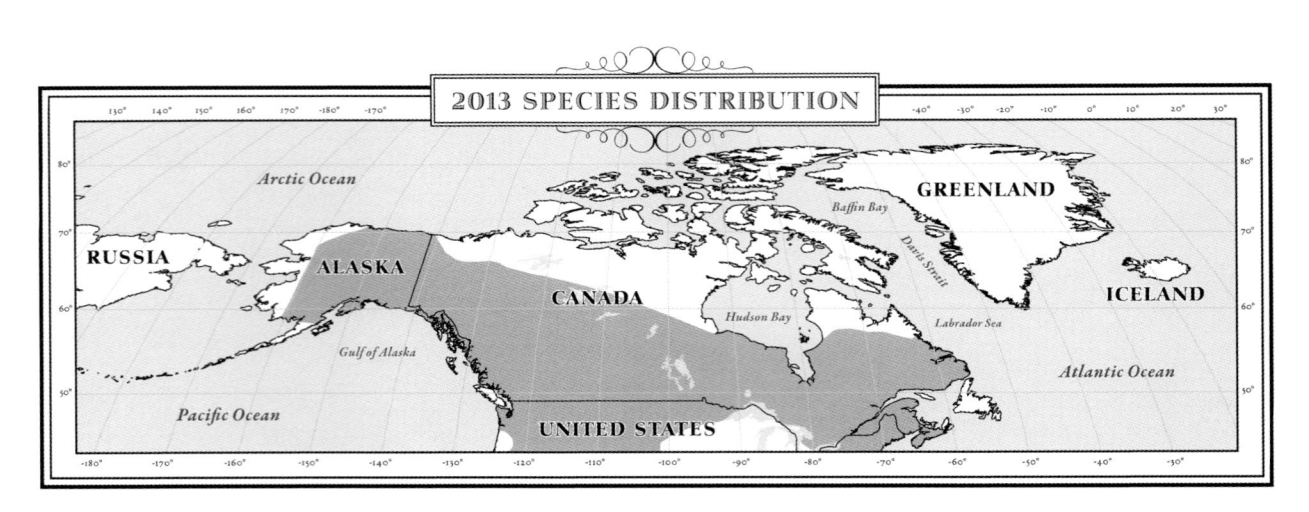

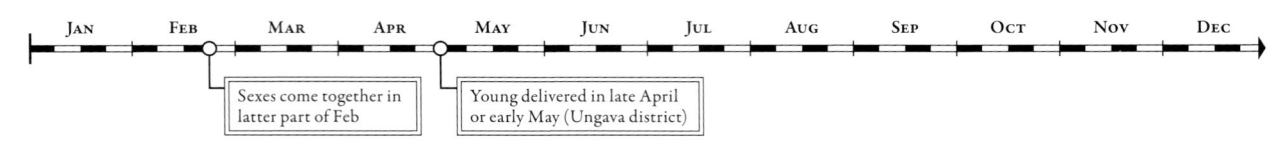

2013 SPECIES DISTRIBUTION

| Jan | Feb | Mar | Apr | May | Jun | Jul | Aug | Sep | Oct | Nov | Dec |

- Sexes come together in latter part of Feb
- Young delivered in late April or early May (Ungava district)

CHAPTER TWO: MAMMALS OF UNGAVA AND LABRADOR

Descriptions by Turner

The Canadian porcupine is known to the Eskimo by the name *i lá ku syúk*; and by the Indians is called *kaq*. In the immediate vicinity of Ft. Chimo the porcupine is very rare as the timber limit ceases near there and the tracts covered with trees are so scattered and so sparsely grown with wood that sufficient protection is not afforded to these animals. About seventy miles from the coast the porcupine begins to become common and south of the "Heighth of Land" they are very numerous.

As an illustration of their rarity near the coastline of the Hudson Strait region many Eskimo assured me they had never seen one of these animals. Along the Labrador coast it is found as far north as latitude 58° and south of 54°. On the southern [1406] slope of the "Heighth of Land" and to the Gulf shore the animal has its center of abundance for the entire region.

Their food consists of spruce, larch, pine, birch and a few other trees. The principal food is obtained from the spruce. Their presence may be known by the chips fallen about the tree; the top of which has been eaten and the tender bark of the upper growth peeled off. Of course each tree decorticated dies above that part and the tree then assumes a peculiar growth which immediately attracts attention.

The porcupine is more or less sociable in its habits, although their sociability doubtless depends much upon a common instinct to seek the better localities for food and there two or more of them are to be found. The Indians assert that where one is found another is not far off. When solitary they [1407] are considered as wanderers in search of food. Generally a male and female are found near each other. The sexes are stated to come together in the latter part of February and the young delivered in late April or early May for the Ungava district.

Two young are not common and three very rare. A single young at a birth is the rule. The newly delivered young is a queer object, having as much shape as a flat-iron. The color of the young is a deep, rather lustrous, brownish black. The hair is quite short and smooth with sparse long hairs irregularly disposed. They are quite helpless and remain with the parent until the middle of the ensuing winter when they are able to shift for themselves. The spines do not appear until after the fourth month. It is stated by Indians that the spines do not attain their full length until the animal is more than [1408] a year old. The retreats occupied not only as a protection from the severity of the weather but also as a den in which to bring forth their young is under a prostrate tree overgrown with accumulations of moss and fallen branches. Within this burrow the animal prepares a snug home of leaves and grass. Sometimes two or more nests are found in a single retreat and each of these is occupied by a single adult, the male and female each repairing to a separate nest when dwelling together. I was informed that the male remains at least a year with his chosen female and would thus appear to be monogamous.

The porcupine rarely travels a great distance and the greater part of its wanderings in search of food is made in the late fall before the snows of winter set in. The legs are quite short and the deep snows in the timbered tracts, where the snow rarely becomes encrusted, [1409] would seriously impede progress. The track made by the porcupine resembles that made by a stout stick thrust into the snow. The weight of the body causing a deep furrow plowed through the yielding snow.

Their motions are quite sluggish and even on unobstructed ground they progress no faster than a moderate run of a man. They appear to be indifferent to get out of the way or attempt to escape. They are well aware of the formidable weapons of defense which they carry on the back and tail. With their teeth they are able to amputate a finger or snip the toes from a thoughtless dog.

When irritated the porcupine immediately turns its tail toward the assailant and switches it in swiftest manner. The back is humped and with the backward movement of the animal it strives to press its body and tail against the attacking object.

[1410] A misguided and impetuous canine meets with a reception that makes a lasting impression as it wonders how much a thing shaking like a bundle of dry straw could so suddenly sting it in a hundred places. The poor dog is now covered about the lips and face with spines so loosely set in the skin of the porcupine as to become detached on the least touch. The feet of the dog pain so intolerably that a veriest howl of agony is set up and each movement of the dog only increases the torture of scores of needles penetrating the skin. The dog learns his lesson the first day and rarely attacks a porcupine the next time. The spines must be pulled out or serious results attend.

The outer point of the spine is sharper than a needle and along its point are barbs from which there is no release. It often happens that less injury results from pushing the spine through [1411] and extracting it point first.

The Indians have small dogs which they teach to scour the woods for porcupines and when the location is discovered the dog sets up a bark which brings the owner of the dog to the spot. If the animal is suddenly attacked it strives to roll into a ball with the nose directed toward the hind legs and then covered with the tail. The skin is thrown forward to elevate the spines which now become only too well apparent. An Indian kills it with a small stick, striking it on the nose which is said to be the "tender" spot of a porcupine. In the southern portion of the region the Mountaineers (Indians) use the flesh of this animal as their principal food when the reindeer are temporarily absent from the barren tracts of that country.[15] The spines and hair are removed by scorching it in hot ashes or by pouring hot water over the body to loosen the hair and spines. The body is then cut open and [1412] care must be exercised not to cut a certain part lest the flesh becomes tainted and totally unfit for food. The ani-

---

15. Ungava Inuit report that porcupine were hunted only when food was scarce. Inuit elder, Joseph Morgan, in an interview on the topic, indicated that he once shot and ate three porcupine to support his family during a period of starvation (Makivik 1984).

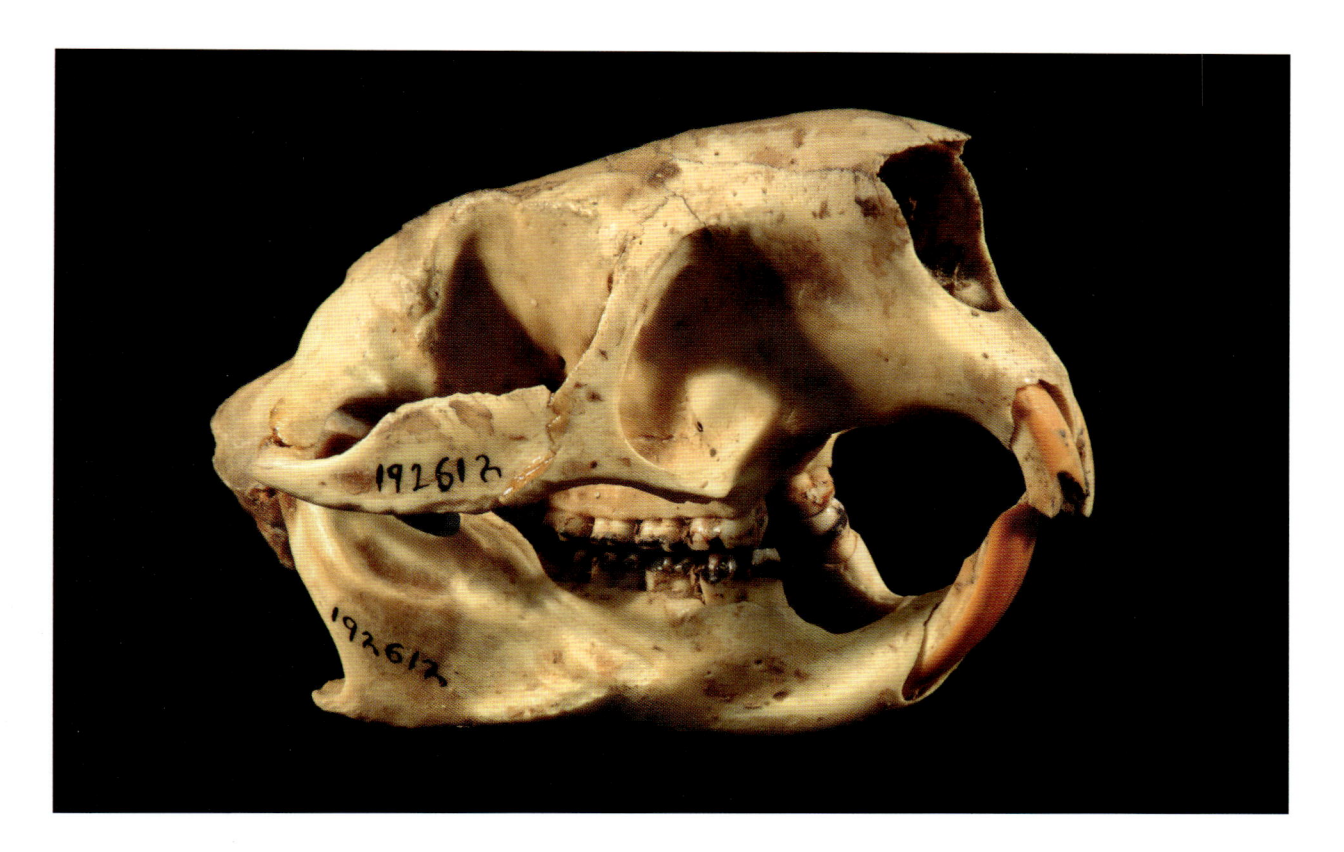

Porcupine Skull Collected by Turner at Ft. Chimo
With their sharp teeth, porcupines often ringbark trees in the boreal forests of Northern Quebec and Labrador. Many trees bear the scars of a hungry porcupine. USNM 14157/192612, Division of Mammals, NMNH. Photo by Angela Frost, 2011.

mal is usually quite fat and when well prepared forms an excellent food for either native or white person. The taste of the flesh is peculiar, resembling that of a hare and pig.

The porcupine has several natural enemies, among them the Marten and Pekan or "Fisher." The Marten finds the porcupine a difficult animal to kill as its weight is too great for such a small beast to manage. It is said to attack and wound the throat which causes the porcupine to double-up, whereupon the Marten springs away and awaits another opportunity; successively attacking it until it succumbs. The Pekan, however, being strong is able to inflict a fatal wound on the under parts at the first attack and being even more agile than its relative (the Marten), [1413] rarely fails to capture its prey. The Indians destroy far more than any other enemy to this animal and as only one or two are brought forth at a time they must be quite numerous to afford food for so many people. The distribution of this animal is very extensive; as far north as the limit of trees and on the west to the head of Norton Sound in Alaska. At the present day they are but rare as far south as Maryland, northern Virginia, Ohio, and the Great Lake states. The western animal differing, however, from the one under consideration, as a species, but not in habits generally.

The hair of the Canada porcupine is of a stiffish nature, of an inch to more than several inches in length on the body, especially the dorsal region, and short and smooth on the legs and nose. The hair of the lower portion of the body is shorter and sparser, somewhat lighter than on the back and much lighter than on the legs and [1414] face where it is nearly black. There appears to be for the Ungava animal but little difference for seasonal changes. In the summer the hairs are thinner and slightly lighter in color except on the feet which apparently do not change the color of the hair on that portion. It is said to renew the lost spines although this is a matter which I could not determine.

Recollections on Inuit Use of Porcupine
Tivi Etok (Heyes 2010)

I've never tasted them. They are all around this area (Kangiqsualujjuaq). They are scary and dangerous even when they are slow. We never used to bother them because of their dangerous quills. I have been told that when people wanted to eat them they would first need to burn off the quills over a fire.

Porcupine Brush
A cleaning brush made from porcupine quills collected by Turner at Ft. Chimo on 9 January 1884. USNM E90295, Museum Support Center, NMNH. Photo by Angela Frost, 2011.

Porcupine Skin Collected by Turner at Ft. Chimo
USNM 14158/192613, Division of Mammals, NMNH. Photo by Angela Frost, 2011.

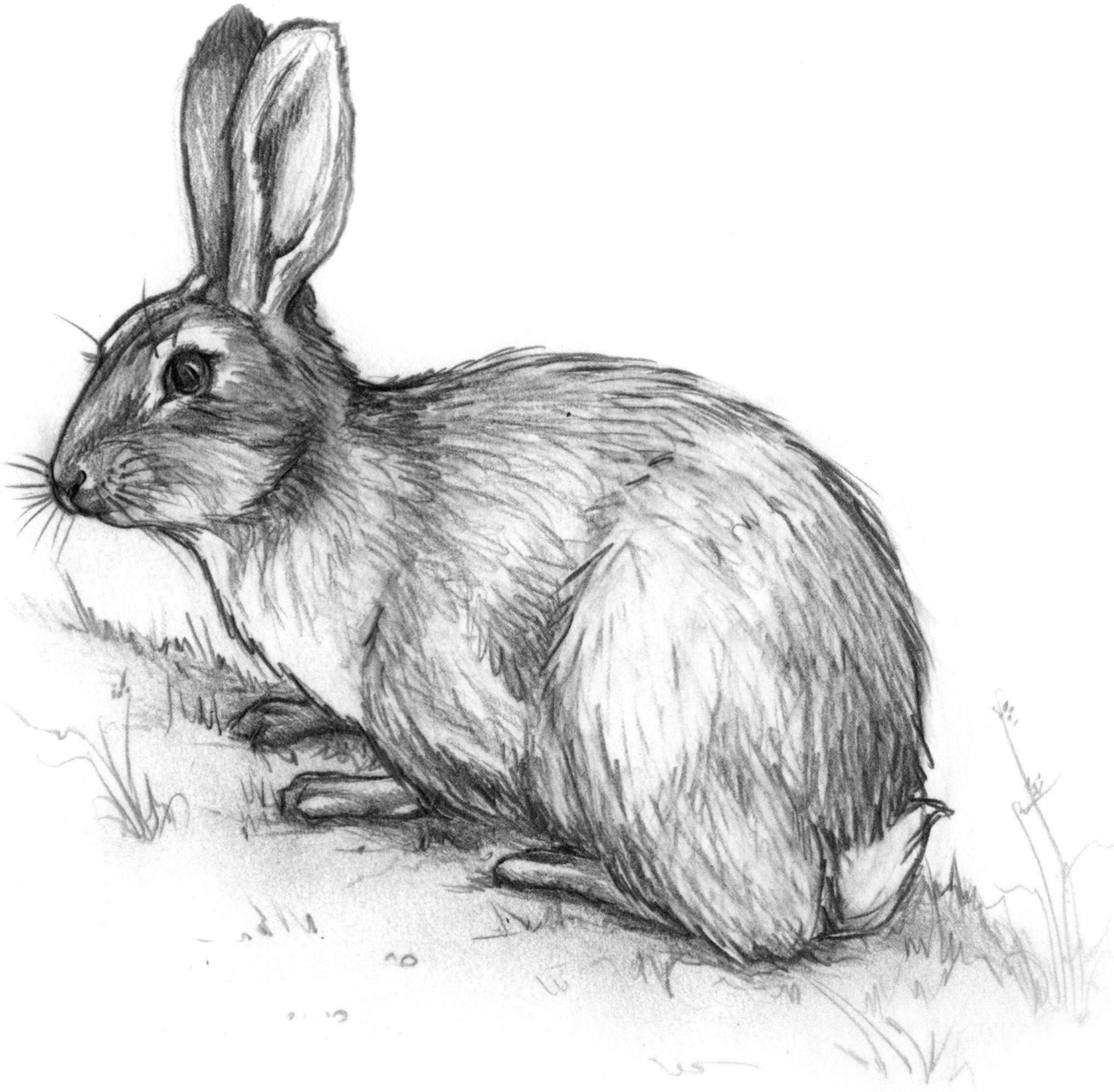

LAGOMORPHS

Introduction to Hares

The lagomorphs comprise the hares, rabbits, and pikas, a handsome group of animals distantly related to rodents (and indeed, they were included with the rodents in the scheme of Turner's original manuscript). There are two species of hares in northeastern Canada. Though both are classified in the genus *Lepus* (i.e., they are hares, rather than true rabbits): one (*Lepus arcticus,* the arctic hare) is traditionally called a hare, while the other (*Lepus americanus,* the snowshoe or varying hare) is instead often traditionally called a rabbit. Turner uses this terminology, distinguishing between the two. The "hare" is the larger of the two and lives mainly north of the treeline, while the "rabbit" is smaller, and largely a forest animal. Turner provides winsome portraits of their natural history (especially of the former species) and his experiences hunting them. Both have brown pelage in the summer, turning to white in winter, with molts in spring and autumn. The white fur of the "rabbit" is gray at the bases (seen when the fur is parted), but that of the "hare" is pure white, from base to tip. Turner's "rabbit," the snowshoe hare, is famous for its boom-bust oscillations in population density (to which similar cycles in their primary predator, the Canada lynx, are intimately linked), which have a periodicity of roughly ten years in northern Canada (Elton and Nicholson 1942, Krebs et al. 2001). Turner's stay in Ungava was at the northern limits of the range of the snowshoe hare, and at a period when it was not especially abundant.

Turner portrays the importance of hares in Inuit life and folklore, richly attested further in his collected artifacts and ethnographic accounts. The cultural importance of the hare persists in the Inuit world today. Considerable scientific importance is also attached to Turner's collection of Ungava hare specimens stored at the Smithsonian. These were used especially by Edward W. Nelson (1909), a biologist and chief of the U.S. Biological Survey, based at the Smithsonian, in his influential review of the classification of North American lagomorphs, which modernized understanding of their taxonomy. Specimens collected by Turner (USNM A23132 and USNM 14149) were also earlier used by Gerrit S. Miller (1899), a Smithsonian mammal curator, in describing a new species of hare, *Lepus labradorius,* but this scientific name is now regarded as a synonym (or subspecies, *L. a. labradorius*) of the arctic hare (Harper 1961, Hall 1981, Wilson and Reeder 2005).

ᐅᑲᓕᖅ
HARE
ukaliq
L. glacialis

{1415}

Order Name used by Turner	*Rodentia*
Current Order Name	*Lagomorpha*
Family Name used by Turner	*Leporidae*
Current Family Name	*Leporidae*
Scientific Name used by Turner	*Lepus glacialis*
Current Scientific Name	*Lepus arcticus* Ross
Syllabary	ᐅᑲᓕᖅ
Modern Nunavimmiutitut Term	*ukaliq* (cf. Schneider p.439 - *ukaliq* = arctic hare [Lepus arcticus])
Eskimo Term by Turner	*ú ka lik*
Definition of Eskimo Term by Turner	Hare (*L. glacialis*) [p. 2837; 1421]
Innu Term by Turner	*wa pús* (Hare); *wá pu shush* (Wood Hare) [p. 708]; *mi kum nash* (summer pellage) [p. 709]

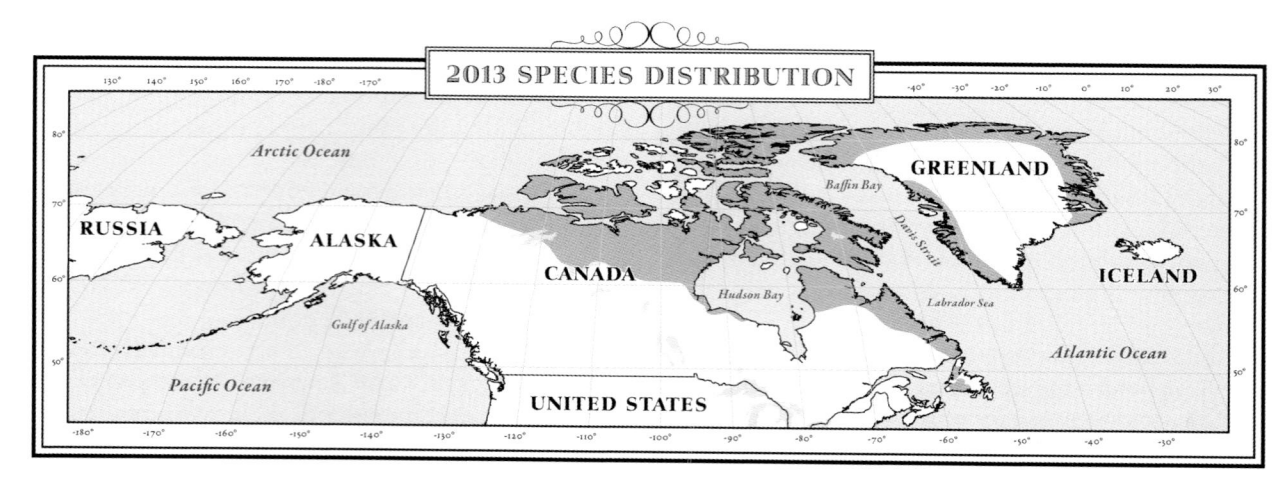

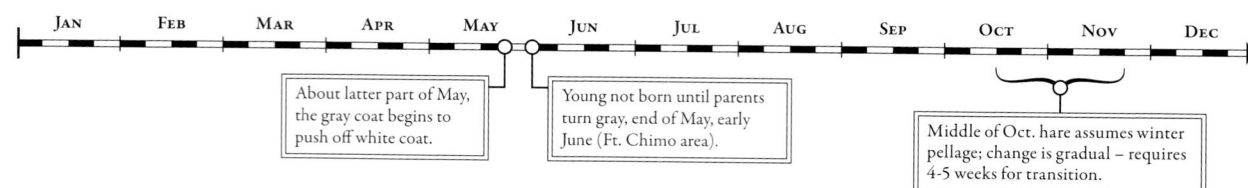

2013 SPECIES DISTRIBUTION

About latter part of May, the gray coat begins to push off white coat.

Young not born until parents turn gray, end of May, early June (Ft. Chimo area).

Middle of Oct. hare assumes winter pellage; change is gradual – requires 4-5 weeks for transition.

Chapter Two: Mammals of Ungava and Labrador

Descriptions by Turner

The Polar Hare confines itself to the Barren ground tracts of the region. They seldom occupy the timbered portions and then only temporarily when traversing the land in search of food or the opposite sex.

The barren coast and even the islands near the mainland, from which the sea freezes to the islands and thus forms a convenient means of travelling to those outer places, are the localities mostly resorted to and inhabited by this hare. My first acquaintance with this animal was at Davis' Inlet on the Labrador coast. At Rigolet their presence was known from the signs everywhere visible.

On Solomon's Island, near the entrance of that Inlet, I was hunting specimens in company with Captain Alex Gray of the S.S. *Labrador*. After a few hours spent in

Collection Notes

Turner collected the type specimen of the subspecies *Lepus arcticus labradorius* Miller, 1889 (type locality is Ft. Chimo).

searching for birds and other objects my attention was aroused by the cry "There goes a Hare." It was above me and on a level shelf of rock. I was [1416] so astonished that I forgot I had my gun in my hand. It was the first Arctic Hare I had ever seen and it appeared as large as a horse. In a few minutes I decided to obtain a second sign; and, having a knowledge of the hill she ran around I followed. As we neared the place on the return, around the circular base of the rock, she started up. A shot stopped her just as she was disappearing behind a rock. I picked her up and discovered she was the mother of five, nursing young. As she was in the beautiful gray summer pellage, July 19th, I saved her skin [USNM 14151]. The flesh was prepared by the cook, for the table and I am certain I have seldom found anything tougher.

Everywhere along the coast signs of Hares were plentiful and after arriving at Ft. Chimo, I found those mammals quite abundant. The flesh is quite palatable when properly cooked but becomes tough [1417] and dry if not properly attended to while being cooked. When roasted, after stuffing it, it becomes one of the best articles of food afforded of that country.

During the summer they feed on grasses and other plants and in the winter the buds of alder, willow, and other shrubs and trees, forming their principal diet.

They begin to feed during the late afternoon in summer; and if, in passing a pile of jagged rocks, one will approach cautiously, a hare may be seen feeding. The least alarm sends it off in a twinkling. I saw one feeding at the base of a cliff. I was too close to make a good shot with my Winchester rifle so the first ball passed over its neck and flattened, with a thud, against a wall a few feet beyond. That thud frightened the creature together with reverberation of the report that it merely squatted as though expecting the entire hill to fall. A second shot grazed its back, [1418] causing it to jump in a most astonishing manner.

That same afternoon I was ahead of the two Eskimo accompanying me, when a Hare started up too far off for my shotgun to reach. I called the man to bring my rifle but he was so slow that the Hare had got off several hundred yards. I fired a ball beyond it to turn its course. It did so and the way that hare sped up the side of a hill was not slow. I had a stand where I could view five-hundred

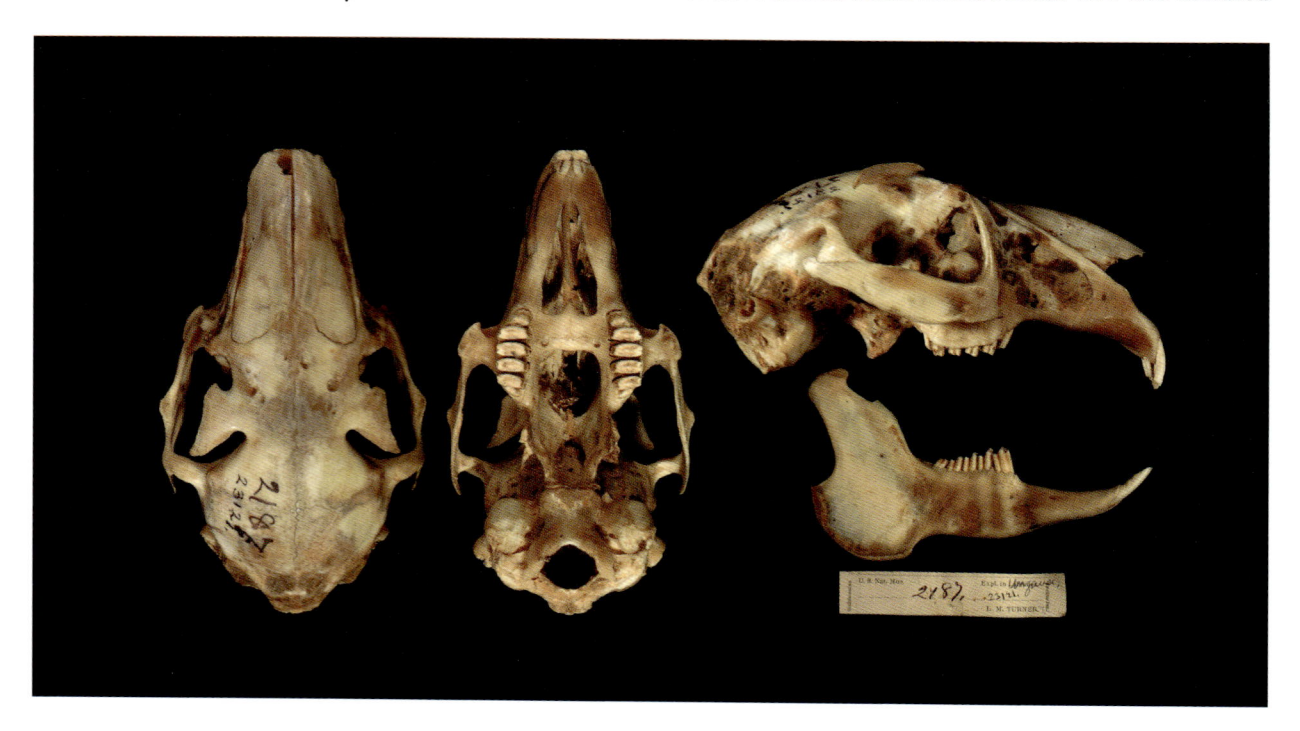

Arctic Hare Skull
Turner collected 17 specimens of Arctic hares while in the region. USNM A23121, Division of Mammals, NMNH. Photo by Angela Frost, 2011.

yards unobstructed in advance of the animal. I told the Eskimo to observe the place where the ball struck. I fired at 400 yards and struck within two feet of the Hare now flying along. The instant the ball struck the ground the Hare suddenly remembered it had business on the farther side of the hill. I never saw a mammal run so fast. It certainly leaped twenty feet at a bound. The natives with me [1419] enjoyed the fun heartily. I have heard it said that the white fox is able to catch a Hare. It may do so at times but no fox would have caught the Hare I shot at that time.

The Hare is quite cunning at times and if it discovers that it is being tracked it will nearly always follow the person trying to find it. A hunter who knows this habit will always return to where he started and near that spot will find the Hare.

About the middle of October the Hare, in the vicinity of Ft. Chimo, assumes the winter pellage of purest white.[16] The change is gradual and requires about four or five weeks to effect the transition. About the latter part of May the gray coat is beginning to push the white coat off. The winter coat falls off in large patches. When sitting on the snow the white Hare is difficult to distinguish. The shining black eyes and black tips of the long ears [1420] are the only parts other than white. The Ptarmigan has a cream tinge to its feathers during the life of the bird and this renders it easy to perceive those birds at a distance but not so with the Hare. It is exactly like the snow and unless it be in motion it is very difficult to detect.

The natives assert that the young are not born until the parent turns gray which at Ft. Chimo is the last of May or early June. Two to five young and two litters each season are said to be the breeding habits of this Hare.

The weight of the animal is from seven to eleven pounds.

Near Ft. Chimo and about halfway to the "Chapel" is an immense erratic boulder of size equal to a medium room. This rock is resting on a gentle slope at a point just where the foundation slopes suddenly for several hundred feet. Near the southeast corner [1421] of this rock, Mr. James Irvine had set a fox trap and from it took so many Hares that I gave the name of Hare Rock to that monument erected by glacial action.

The Eskimo give the name of *Ú ka lik* to the hare. The Indians call it *Wa pús*.

---

16. Inuit elder Nick Ittulak in a 1981 interview indicated that he would hunt hares for dog food near Keglo Bay in the fall and Hebron in the winter. Ittulak used the white fur of the hare as a camouflage device when seal hunting. He would lay the fur on his head to conceal his dark hair and profile when approaching a seal on the ice (Makivik 1984).

Hare Ethnography
Turner

SHAMAN USE OF HARES AND OTHER RESOURCES TO TREAT DISEASES (1887a: 68-70)

Disease is supposed to be the evil work of a spirit having its residence in the afflicted part and may be driven from one part of the body to another and be captured or forced to forsake its victim by the potency of the shaman who alone is capable of dealing with the more powerful spirits. The offices of that individual are performed under a series of incantations, contortions of his body accompanied by mumblings or singing; his eyes are blindfolded or a blanket is thrown over him and the patient or the place made dark by excluding all light. By a series of manipulations of a character as varied and mysterious as the inventive faculty will allow the patient undergoes a treatment oftentimes as rough as the suffering of the disease itself. The application of gum from the spruce tree to a wound, binding the lungs of a freshly killed hare upon the part, placing a few hairs of some creature over it to withdraw the pain are frequently resorted to.

The shaman may decree that the person slit the ear of a favorite dog or cut of the tail of that brute and dispose of it or wear the harness of the dog, put himself under the influence of some beast or bird by wearing a portion of that creature about him, to enable the affliction to pass into it and thus afford relief.

No natural remedies, except those noted, were ascertained to be applied to alleviate the distress. Relief from the stock of medicines at the disposition of the whiteman are but little sought for. A few ointments, plasters and outward applications are sometimes asked for.

THE FUR OF HARE WORN ONLY BY INUIT WOMEN (1887A: 202-203)

The material used to form a protection for the hands differs according to the season and character of the weather. Long haired mittens would retain so much water, during wet weather, as to be worse than the exposure to a temperature that may be borne without great inconvenience. So that it is only during cold dry weather that fur or hair mittens may be worn. Among the Innuit the mammals are socially divided into the classes of what may be distinguished as noble and inferior beasts.

The former are used though not exclusively by the men while the latter may be worn only by the women. No man would debase himself by wearing a particle of the fur of the hare or of the white fox; the skins of those timid creatures are reserved for women. Either sex may wear the skins of all other mammals, but, at times, under restrictions based upon superstition.

As remarked in another connection mittens of tanned sealskin as worn by the men while in their canoes or performing other labor where the air is colder than the water splashing upon their members. The women wear mittens of fur from the hare or white fox to keep their hands warm only when walking along or travelling; a wolf skin fringe may encircle the wrist to exclude the cold from the sleeve of the garment. The palm of these mittens may be of sealskin or buckskin, the latter always of Indian tanning and smoking for the Innuit does not prepare the hide of the reindeer into the condition known as buckskin.

The skins of the reindeer are also used but always with the hair on; the body of the deer furnishing skin for mittens for general purposes. If for a man driving dogs and requiring the whip to urge or control those brutes, the palm would be filled with the long hairs and prevent a sufficient grasp of the whip handle. To overcome this the palm generally consists of a piece of sealskin sewed in. Mittens made of sealskin with the hair on are quite rare, none ever having been observed by me. A fringe of white bear skin may be sewed around the wrist of any kind of mitten.

The skin of the forelegs of the reindeer is used for a special purpose as well as general wear by both sexes. The hair of those limbs of the deer are not of the same character as obtained on the remainder of the body; in the winter they are larger and coarser than in summer hence are not so well adapted for the purpose as the leg skins of the animal killed in late summer. The skin is tanned in a manner similar to that of preparing pelts for other garments, and cut into the proper shape to satisfy the Innuit hand. While these mittens are used for general purposes they are specially fitted for another use. When the snowhuts are to be constructed the mittens are put on for the hair admirably protects the skin from becoming wet by the heat of the hand melting the snow to be lifted into place or crumbled between the hands and thrust into the aperture left between the blocks of snow. Mittens of another material become cold which although may not be appreciated while the labor is going on, yet the next day will be felt. If the Innuit, with whom a White man may be travelling, has any regard for the latter he will inform him to desist from the attempt to hasten an operation that will result in unexplainable suffering on the morrow's cold air.

An inspection of the mittens will reveal the fact that the Innuit fashions the thumb-piece on a peculiar manner as that when joined to the palm piece produces such a short length of thumb covering as to be very inconvenient for a White man who spreads his thumb away from the palm while the Innuit directs its tip toward the center of the hand within. Another feature peculiar to these mittens is the shortness of the wrist. It is not unusual to discern quite a space of bare wrist exposed unless the sleeve and the wrist of the mitten each be furnished with a fringe of wolf or dog skin whose long hairs mitigate the cold.

Cup and Ball Game Used by Ungava Inuit
This game was known as ajagaq *(syllabics: ᐊᔭᒐᖅ). Length 3.5″. Turner (1887a: 694) recorded the spelling as* Ai ú gaut. *Collected by Turner at Ft. Chimo, Quebec, 1884. USNM E90227, Museum Support Center, NMNH. Photo by Angela Frost, 2011.*

CUP AND BALL GAME, NORTHERN INNUIT OF SOUTH SIDE OF WESTERN HUDSON STRAIT (1887A: 140-141)

A pointed piece of bone, either a piece of ivory shaped like a bodkin or else one of the leg bones of a hare is tied by a short thong to the skull of a hare or a piece of ivory having an irregular series of pits on which the weight is to fall and be impaled on the point of the catcher or peg.

In the case of a hare's head being used the string is attached to the opening where the nasal bones have been removed. The ivory piece is hitched so that the heavier end is next to the player while the smaller end is outward and thus renders the object more difficult to be caught on the peg.

In using these playthings the object is to catch the skull or ivory piece on the peg as many times in succession as the skill of the player may allow.

A failure gives another person an opportunity to test his skill. I have never heard that any wager is laid on the result and so far as it being a gambling game I can only state that adults or those of mature age rarely occupy themselves with an object devised by and made use of only by boys and girls; the latter more frequently than the former.

It appears to be a very simple matter to impale the skull of a hare having so many perforations, either natural or created, in the walls of the skull, but the string, by which the peg is attached, is scarcely longer than the peg the short arc of a circle described in endeavoring to throw it so as to fall on the peg may be essayed ever so many times ere success will be attained.

HARE HEAD "CATCHER", INNUIT, SOUTH OF HUDSON STRAIT (1887A: 209-210)

This object is one of the few diversions of the Innuit youths and maidens. It is, however, considered more of a woman's game than that for a boy; the latter preferring the games of sturdier character.

It is not infrequent to observe two or more girls with this affair endeavoring to catch the skull upon the bone point. The player saves the skull of a hare which has served at some repast and upon removing the flesh from it certain bones of the skull are broken and others displaced so as to create more orifices or openings in the skull plates. A short string is now secured around the jaw just back of the incisor teeth. The cord is then affixed to a piece of bone, usually the radius of the fore-leg of the hare and by sharpening the outer end it serves to transfix the skull by entering one of the numerous openings when the string-end of the bone is held similarly to the manner of grasping a pen or pencil when writing, except that the bone is held nearly upright as to be in position for the skull to fall upon it when a pendant below the hand it is gently swung from the body and falls on the point, either to slip away or to be impaled. The shortness of the string gives a motion of only few inches and if it be too rapid the pointer has no time to be thrust toward it.

It occasions great amusement if one fails repeatedly to impale the skull; while to repeat the catching a number of times is greeted with exclamations of surprise. I have never known any wager laid up on the possibilities of the issue and from the character of those (children) seen engaged with it I should conclude it to be essentially a child's game.

Arctic Hare Skin

The winter coat of the arctic hare collected by Turner at Ft. Chimo, 1882. This specimen may have been exhibited at the World's Industrial and Cotton Centennial Exposition in 1884-1885, which was held in New Orleans, Louisiana. The specimen shows signs that it was once mounted for display, and a statement in the Annual Report of the Board of Regents of the Smithsonian Institution (True 1885: 130) suggests that it was exhibited: "From British America the most important accession is the collection of skins and skeletons made by Mr. Lucien M. Turner in the vicinity of Hudson's Bay. The series included very fine skins of the Polar Hare, Lepus timidus, *one of which was sent to the New Orleans Exhibition." USNM 14150, Division of Mammals, NMNH. Photo by Angela Frost, 2011.*

THE CHILD AND THE HARE
An Inuit story, Turner (1887a: 299)

A small child was so tormented by its parents and others on account of its long large ears that it decided to hide itself from mankind. It went away and hid behind a rock and was transformed into a hare. When it perceives a person approaching it lays its ears on its back so they may not be seen. The hare has no tail because the child had none.

THE VENTURESOME HARE
An Innu story, Turner (1887b: 582-586)

A young Hare lived with his aged grandmother. They were very poor, having, at times, no fire and seldom an abundance of food to eat. She was too old to provide food and the few sticks he could pick up near the tent door made but little fire to keep them warm in the tent full of holes; and, which she was unable to mend because she had no deerskins with which to patch it.

They had been for several days without food and even longer without fire. The only thing that remained in the tent was a net which belonged to the young Hare. He endeavoured to cheer his grandmother, and so often told her such impossible things that she paid but little attention to his wild schemes which he was awaiting the day to carry out. The pangs of hunger caused him to announce to his grandmother that he would take the net and catch some fish from the lake nearby. The poor, old woman was amazed at such a statement and said: "Every day you conceive some new project, but of all, the one for a Hare to catch fish is the greatest. You are aware that the Hare never wets its fur. We cannot eat raw fish; and, there is no fire over which to cook them."

The Hare arose early the next morning and put the net in order. He then went to the lake and set it where he had often seen the fish making ripples in the water. After placing the net in position he returned to the tent and slept. On the following day he went to the net and found it so full of fish that he could not draw it ashore. Every mesh in the net was holding a fish and many more were swimming in the water it surrounded. The Hare cut a hole in the side of the net to allow some of the fish to escape; in order that he would then be able to drag the remainder on shore. He did so and when they were hauled on shore he found he had so many fish that he erected stays and hung the fish up to dry. A large number were put in a pile to be taken home. The net was now stretched out and mended. When it was dry he rolled it into a bundle and put it under his arm. He took the fish and the net to the tent and showed his grandmother what he had done. She was well pleased at the prospect of food but regretted they had no fire. The young Hare replied "Never mind; there is an Indian camp on the other side of the river and I shall go to them and get some fire."

The poor, old grandmother now thought her grandson was certainly insane that he would dare visit the midst of an Indian camp to procure fire. She said, "My child, you know the Indian is our worst enemy; far more to be feared then the owl or hawk or even the fox that crouches behind the rocks and bushes along our pathway." The Hare made no reply, but took his net under his arm and went to the river bank, which separated the Indian camp from his home. When he arrived there he gave a loud shout and a number of huge whales came puffing and plunging up the river. At the command of the Hare they arranged themselves so that by springing from the back of one to the other he was able to cross to the other side.

He now dismissed the whales, but enjoined upon them to come at an instant's call. Just at this moment a troop of Indian boys and girls came rushing from the weeds and grass that grow near the water's edge. The noise frightened the poor Hare nearly out of his skin. He suddenly thought of a plan to outwit them. He sprang into the water and wetted his fur, then rolled in the sand and dust until he was covered with dirt. He now stretched himself along the ground and pretended to be dead. As the children came noisily along one perceived the Hare and tossed a stone at him, remarking, "There is a dead Hare, let us take it home." One of his companions said, "Do you not all think that it has been drowned and the river has cast it on the shore? See how dirty it is as though dead for several days." The boy picked it up and dragged it to the tent. He laid it down near the fireplace in the center of the tent and said, "There is a Hare that we got along the river bank." An old man told one of the girls to take the skin from the Hare and prepare the flesh for food. The girl did not immediately do so as she was playing with the other children.

The Hare was so frightened that his heart thumped against his side; and, he wished that the fire-brands would break and scatter the fire so he could get a coal. He now opened half of one eye and looked about. He saw an opening only in the top of the tent, the place where the smoke goes out.

The old man again told the child to skin the Hare. The creature now thought its last moment had arrived; at the same instant, however, the fire-sticks broke and as they fell a shower of sparks sent one on the net which he still held under his arm.

The Indian girl started to pick up the Hare, whereupon the animal gave a bound through the smoke hole and ran towards the river. The Indians ran, yelling at the tops of their voices, frightening the creature so badly that he forgot to call the whales; and with a prodigious bound he leaped across the stream and quietly entered his tent; there he coolly said to his grandmother, "Here is the fire."

The Greedy Hare and the Lame Frog
An Innu story, Turner (1887b: 593-600)

One evening, as the sun was sinking low behind the hills, a Hare thought to take a walk on the snowy hillside near his home. It was not often that he wandered far, for he was of a timid disposition and fearful lest he be ensnared into the jaws of a prowling wolf or fox, which might be lurking behind the rocks that jutted, here and there, from the earth.

As he strolled along he saw a tent which belonged to some Indians. Not perceiving any sign of life without, or hearing a sound of voices within, he became emboldened and approached the structure. Peering within, through one of the holes in the tent, he saw a frog.

The Hare went to the door of the tent; and throwing the flap to one side, addressed the Frog, "Brother, what are you doing?" The frog recognized the voice of the Hare and answered, "I am playing with the ashes. My brothers are away hunting. I have to remain here because I have a sore leg, and cannot go far."

The Hare offered to carry the frog on his back. The Frog consented; and with a toss the Hare placed the Frog on his own shoulders; adding, "This is the way I shall carry you." The Hare carried the Frog to his own tent, and having placed him within went off to get his supper. While nibbling at the tender grass shoots he looked up and directly in front of him he saw a large column of smoke issuing from among the willows that grew along the creek at the foot of the hill.

Terrified beyond measure at this unexpected discovery he exclaimed, "I have forgotten my crooked-knife and must go to get it." He flew along the ground at such a rapid rate that his ears touched his shoulders. Arriving, almost breathless, at the entrance to his home, the Frog observed his agitation and inquired what ailed him that he should appear so frightened. The Hare answered that he had seen a great smoke coming from among the willows along the creek.

The Frog laughed heartily at the Hare and told him it was the smoke issuing from the beavers dwelling there; taunting the Hare for its fear, added, "Running away from beavers when they are good to eat! Although, my brothers could never catch them they frequently sent me and I always obtained as many as I wanted. They carried me to the houses of the beavers and I would easily kill them."

The Hare was amazed at the assertion of the Frog and immediately offered to carry him to the place where the beavers lived. The Frog acceded and was placed upon the back of the Hare and quickly taken to the bank of the creek.

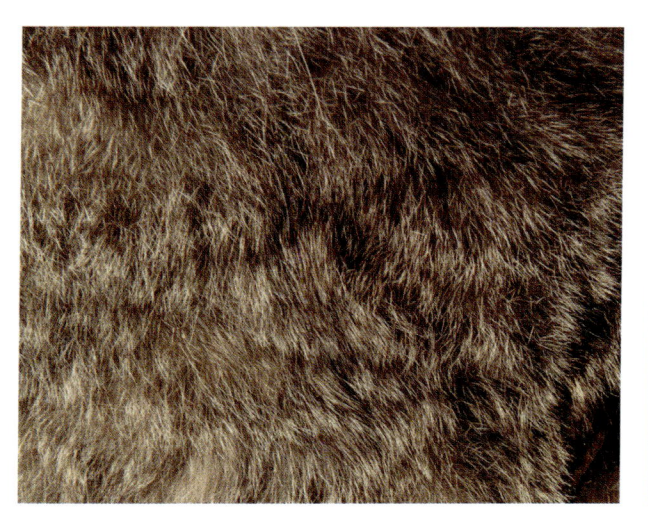

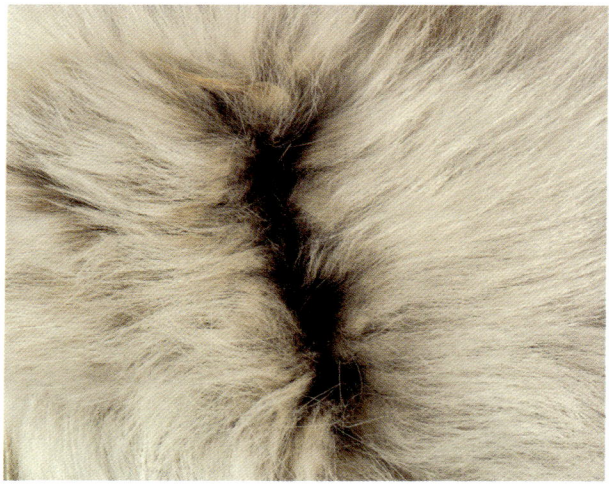

Underfur of Snowshoe Hare
Details of the summer and winter coats of a snowshoe hare collected by Turner at Ft. Chimo in 1884 and 1882, respectively. USNM 14793 (summer coat) and USNM 14153 (winter coat). NMNH Museum Support Center, Suitland, Maryland. Photo by Angela Frost, 2011.

The Frog directed the Hare to place a log across the stream; and, against it a dam was made to shut off the water from around the lodges where the beavers dwelt. Where this was done the Hare was further directed to break into the mud walls of the structure so that he, the Frog, could seize hold of the creatures within the hut. The Frog, however, loosened some of the sticks, so that the beaver should escape and thus cause the Hare to become angry. When the Hare saw the beavers escaping he took the Frog by the shoulder and rudely shoved him into the water to drown him.

The Hare was not aware the Frog could live in the water as well as upon the land, but thought he had killed the poor Frog. He felt very sorry and returned to his tent and wept bitterly at the loss of his brother. He kept calling his name until sleep overpowered him.

The Frog, in the meantime, swam under the water and killed all the beavers, dragged them on the land and then to the home of the Hare. When he arrived with the load of beavers he put them on the ground near the door and began to play with the side of the entrance. He did not wish to enter, so he called to the Hare to bring him a piece of fire as it was very cold outside. The Hare supposing that the Frog was dead thought it some enemy simulating the voice of his brother and endeavouring to entice him outside that he might be killed and eaten.

The Frog again called to him and the Hare threw out a coal. The Frog said, "Brother there is no fire on this coal. How do you expect me to cook the beavers which I have brought with me?" The Hare now recognising the voice of his brother the Frog and brought him inside of the tent and placed him near the fire as the Frog was nearly frozen.

The Frog began to moan and complain of pains through his shoulder. The Hare inquired if the beavers had bitten him. The Frog replied, "No, it was you who gave me a hard push and it has hurt my shoulder." The Hare was very sorry and said he did not intend to hurt him, but help him jump into the water to catch the beavers.

The Frog now directed the Hare to prepare the beavers, take off their skins and to cook the flesh for supper. The Hare went out to do as was told him, but he began to eat the flesh and continued to eat throughout the night. When he had eaten all the meat he went for a short walk; and, on arriving at the top of the hill he looked into the valley beyond and saw a great smoke curling from the level land that lay along it. He was much frightened and quickly ran back; his heart nearly in his mouth, and quite out of breath.

The Frog perceiving his alarm inquired what was the matter. The Hare replied that he had forgotten his crooked-knife and had run back home for it. The Frog asked if he had seen anything unusual. The Hare quietly observed he had seen a great smoke in the valley beyond the high hill. The Frog laughed and said, "Again you are frightened at smoke." The smoke issues from the home of the reindeer which dwell there. My brothers often sent me to kill deer whenever they wanted meat for food. They could never kill them and I was always sent to kill them."

The Hare was again surprised at the courage of the Frog and said he would gladly carry him to the plain where the deer lived, as he had a great desire for some meat from those animals. He now offered to carry the Frog. The Frog, however, said, "You need not carry me all the way. Make me a snow-shoe for my foot that does not pain me and I think I can catch the deer by hopping on one leg." The Hare made the snow-shoe and then threw the Frog on his back. He took him to the willows that skirted the stream flowing down the valley. Here they rested for a while.

The Frog directed the Hare to lie hidden in the snow until a signal should be given for him to come. The Frog fastened the snow-shoe to his foot and started. He hopped with such prodigious leaps as to astonish the spell-bound Hare. The Frog sprang to the side of the deer and slayed them in an instant. He then set to work to skin them. He cut the head from one of the largest and stuck it in the snow, looking toward the direction whence the Hare should come. The head presented a ghastly sight in the cold air.

The lungs of one of the deer were taken out and placed where they would quickly freeze. The Frog then called to his brother, the Hare, to come to him. The Hare bounded along the frozen crust of the snow and was but a moment arriving at the place where the head of the deer was placed. The Hare saw the object and was so terror-stricken that he remained transfixed with fear, screaming, "Brother, he sees me; Brother, he sees me." The Frog replied, "You foolish, timid creature why do you fear the head of a dead deer? Come on; I have a nice piece of fat for you." The Hare sprang to the side of the Frog and eagerly seized the fat and began to devour it. (It was the frozen lung of the deer). The Frog now prepared a shelter for the night. After they had slept a short time the Frog was aroused by the moans and piteous groans of the Hare, which had been seized with violent spasms.

The Hare vomited all the remainder of the night and suffered great pain from having eaten so much of the frozen lungs of the deer, that when it began to thaw out in his stomach the colic was the result and served as a punishment for having eaten the beavers the night before.

Arctic Hare Skulls Collected by Turner
USNM A23130, A23129, and A23121, Division of Mammals, NMNH. Photo by Angela Frost, 2011.

ᐅᑲᓕᐊᑦᓯᐊᖅ

Rabbit

ukaliatsiaq

L. americanus

[1422]

Order Name used by Turner	*Rodentia*
Current Order Name	*Lagomorpha*
Family Name used by Turner	*Leporidae*
Current Family Name	*Leporidae*
Scientific Name used by Turner	*Lepus americanus*
Current Scientific Name	*Lepus americanus* Erxleben
Syllabary	ᐅᑲᓕᐊᑦᓯᐊᖅ
Modern Nunavimmiutitut Term	*ukaliatsiaq* (cf. Schneider p.439 -*ukliatsiaq* = [...] snowshoe rabbit)
Eskimo Term by Turner	*ú ka lá chi úk*
Definition of Eskimo Term by Turner	Hare (*Lepus americanus*) [p. 2837]

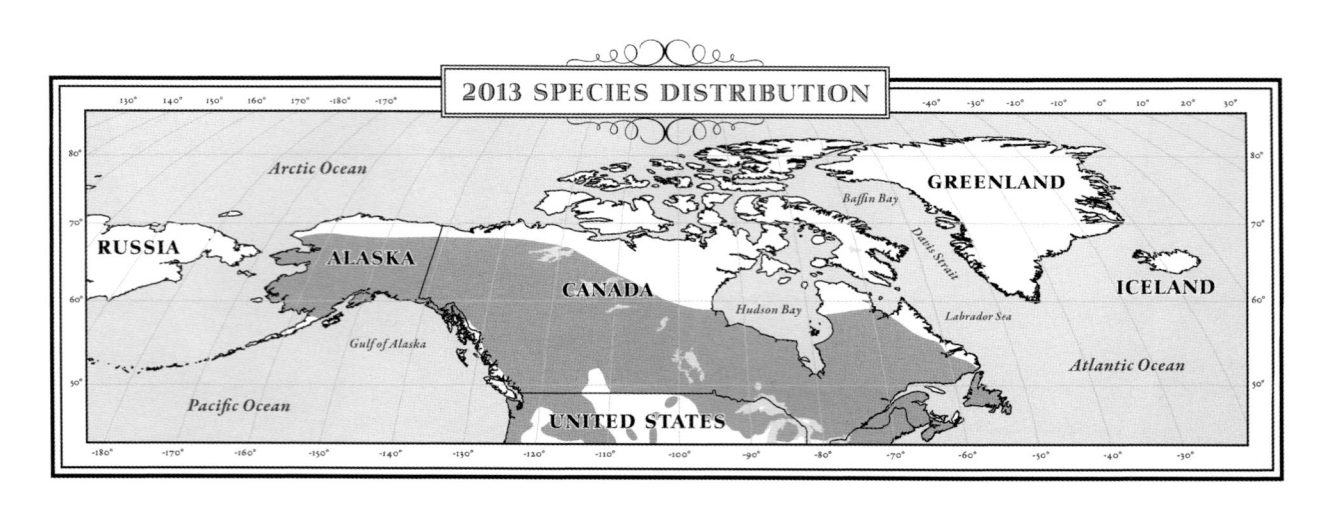

2013 SPECIES DISTRIBUTION

Chapter Two: Mammals of Ungava and Labrador

Descriptions by Turner

The Rabbit occurs only sparingly near Ft. Chimo. In fact but one specimen was secured north of the post. Some four or five miles to the southward in a locality known as Hunting Bay a few of these animals are to be found.

They rarely wander beyond the timbered tracts and not until a hundred or so miles from the coast is reached do they become at all plentiful. About the mouth of the Larch River they are extremely plentiful.

Their runways or paths, among the clumps of trees in that locality, traversed every direction and from the tortuous manner in which they recross each other one would safely conclude all the rabbits in the country had travelled those woods the night before. It is simply impossible to follow their tracks as they soon become lost among the number either just started up ready to join the fun or startled from under the drooping boughs of the trees, [1423] heavily laden with snow forming safe retreats where the cunning thing lies with ears laid back on its neck and only its shining black eye to disclose its presence among the mass of snow and white fur. As I was unfortunate in pursuit of these animals I gave them up after finding so many tracks that confused me and often caused me to wander round and round in a circle while an Indian boy would take his bow and arrow and return before me with as many as he could carry.

Steeltraps are often placed in their runways and rarely fails to secure them. A few terminal buds of spruce are scattered nearby and forms the only bait used. The Indian women snare them successfully and use them for food. The flesh acquires a terebinthine odor from the spruce and larch twigs forming its principal food at the season when sought for by the people. The animal acquires a considerable amount of [1424] fat during the winter but becomes very poor during the summer. Its food then consists of grass and leaves of various kinds.

I could not satisfactorily learn the number of broods they bring forth or the number of young at a brood. A young individual only one-third grown was brought to me in the latter part of June. I suspected the animal to be about six weeks old.

The size of the hinder limbs is great in proportion to its body and the track made by those members often equals that made by its larger relative the Polar Hare. The track made by the Rabbit is more pointed than that of the Hare and by this means alone may they be distinguished.

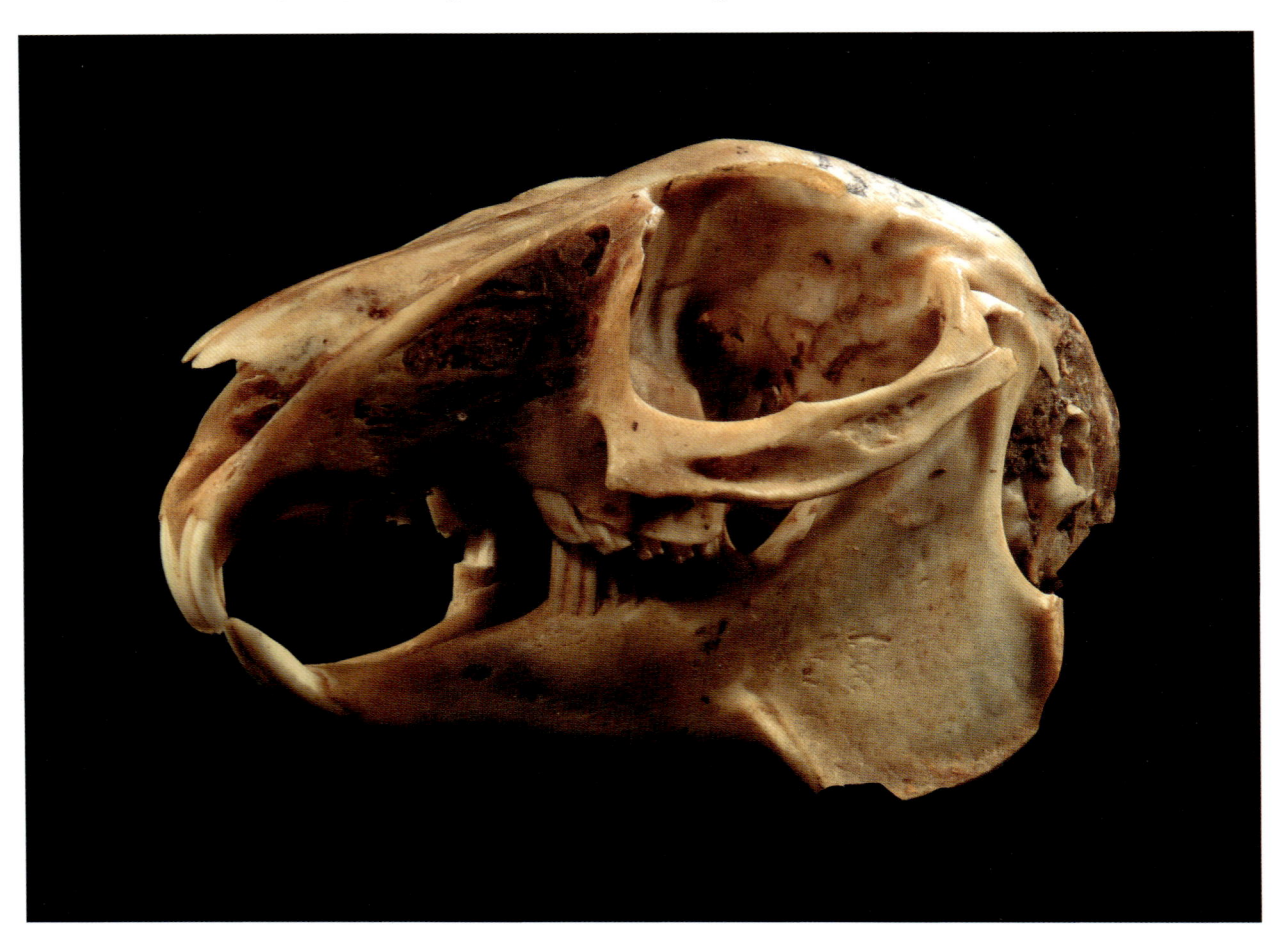

Articulated Hare ("Rabbit") Skull
The skull of Lepus americanus. *Collected by Turner in Ungava Bay. USNM A23123. NMNH Museum Support Center, Suitland, Maryland. Photo by Angela Frost, 2011.*

Shipping Manifest Showing Turner's Ungava/Labrador Collection

List of packages and contents sent by Turner to the Smithsonian Institution for the year 1884 from Ft. Chimo. The manifest indicates the amount and types of contents that were shipped to the Smithsonian, such as an Inuit kayak. This object has not been located in the collection. Source: SI Archives, Record Unit 305, Accession 15388.

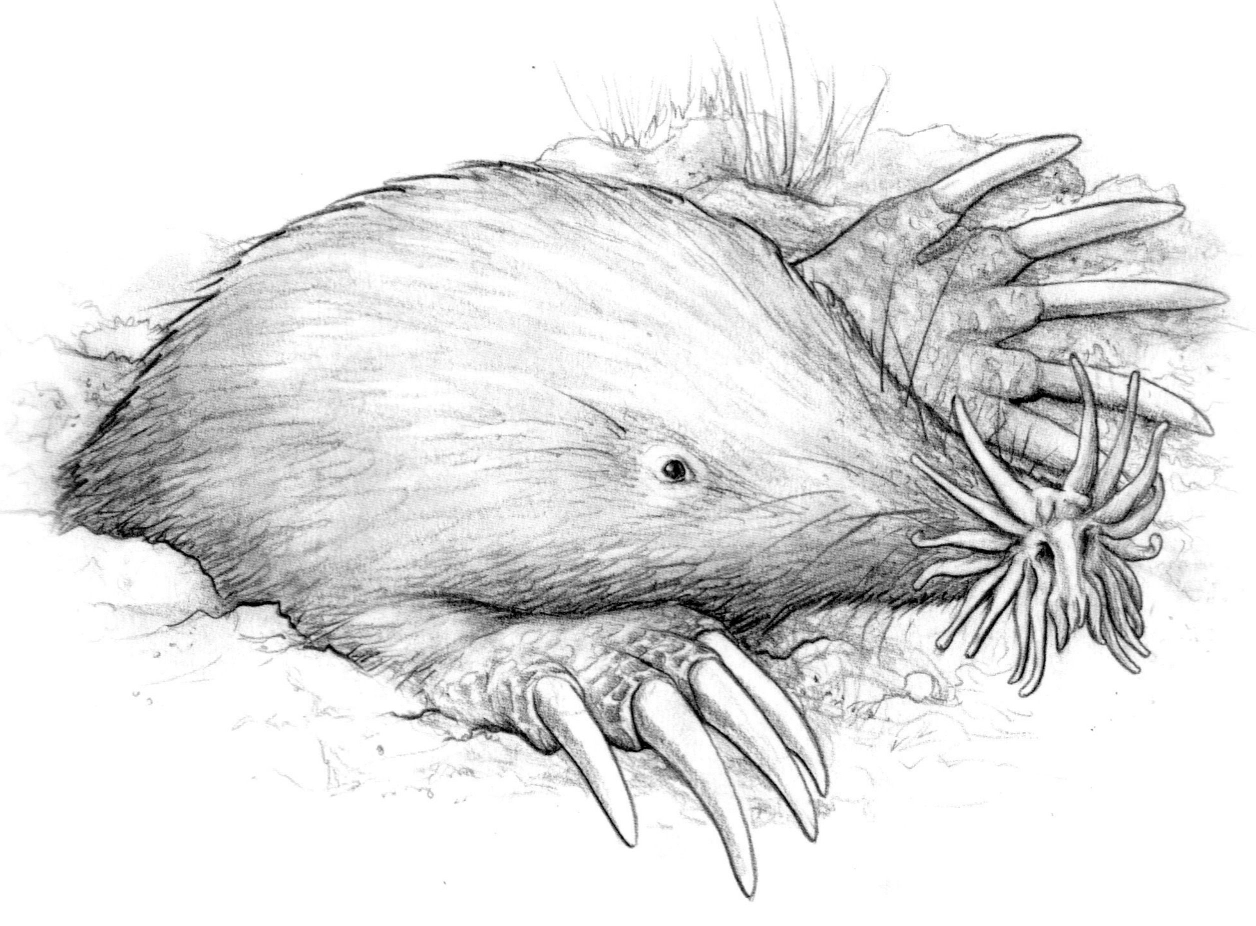

Insectivorous Mammals

Introduction to Insectivores and Bats

Turner expressed familiarity with two kinds of insect-eating mammals in Labrador and Ungava – shrews (occurring widely in Ungava), which he called *Sorex*, and moles (occurring only in the southern part of the peninsula), which he called *Scalops*. Turner's brief notes on shrews, correctly noting their lifeways, indicate his familiarity with them at Ft. Chimo. Tiny and easily overlooked ("scarcely larger than a bumblebee," says Turner), shrews usually have minimal cultural significance. He mentions a very interesting name applied to shrews by the Inuit, *ug zhúng nuk,* literally denoting a diminutive version of the bearded seal (*Erignathus barbatus*) – presumably a reference to the soft, velvety, dark gray fur shared by shrews and seals.

At Ft. Chimo, Turner collected specimens of two species of shrews, the Cinereus or masked shrew (*Sorex cinereus*) and the pygmy shrew (*Sorex hoyi*), as documented by specimens that he sent to the Smithsonian. Only one additional species of shrew occurs in eastern Canada as far north as Ft. Chimo – the American water shrew, *Sorex palustris*. During a visit to Ft. Chimo in 1947, American mammalogist David H. Johnson, a curator at the Smithsonian, collected and described the only specimens of this remarkable shrew known from the northern part of the Ungava Peninsula. Finding that his specimens from Ft. Chimo differed in size and coloration from all other water shrews, Johnson used them as the basis for a description of a new subspecies, *S. palustris turneri,* fittingly named after Turner. (Johnson (1951) wrote: "It is named in honor of Lucien McShan Turner, who in 1882 and 1883, while attached to the U.S. Signal Service, made the most important collection of mammals in the Ungava Region.") This remains the only mammal named in Turner's honor, and the name is still in use in taxonomy today (Hall 1981, Beneski and Stinson 1987; Wilson and Reeder 2005). One other shrew species, the arctic shrew (*Sorex arcticus*), inhabits much of southern Ungava (Harper 1961), and two others, the smoky shrew (*Sorex fumeus*) and the northern short-tailed shrew (*Blarina brevicauda*), reach the extreme south of the peninsula (Peterson 1966).

Turner's mention of a mole in Labrador stems from secondhand reports he received "from the vicinity of the Northwest River." Turner called the mole *Scalops,* the genus name then in use at the time for two of the common moles of eastern North America, the eastern mole

(*Scalopus aquaticus*) and the hairy-tailed mole (*Parascalops breweri*), with which Turner would have been familiar from his time living in the northeastern United States. Reports from the Northwest River could, however, only refer to the remarkable, semi-aquatic star-nosed mole (*Condylura cristata*), the most cold-adapted of the moles, and the only mole in Labrador (Turner 1961, Wilson and Ruff 1999).

Another group of insectivorous mammals found in eastern Canada are bats (classified in the mammalian order Chiroptera and, in Canada, represented by several species in the family Vespertilionidae). Turner made no notes on bats in his manuscript and thus presumably did not encounter bats (or firsthand knowledge of them) during his tenure in eastern Canada. Only one species of bat, the hoary bat (*Lasiurus cinereus*), occurs as far north as Ungava Bay, though rarely (Kays and Wilson 2009; Wilson and Ruff 1999). The hoary bat is a migratory species that would be present at such high latitudes only during the summer. Two other species, the little brown bat, *Myotis lucifugus*, and the northern long-eared bat, *M. septentrionalis*, both of which are active in the warmer months of the year and hibernate during the winter, are present further south in the Ungava Peninsula and Turner could have encountered them during his travels. Sadly, these two latter species are declining rapidly in eastern Canada from the spread of a fungal disease, White-Nose Syndrome, apparently introduced from European caves (Foley et al. 2011) – another new and unpredictable environmental change affecting wildlife in Canada's north.

Little Brown Bat
Close-up of nose with fungus. New York, October 2008. Photo courtesy Ryan von Linden/New York Department of Environmental Conservation. U.S. Fish and Wildlife Service.

Northern Long-Eared Bat
Myotis septentrionalis. *Taken December 4, 2007. Photo credit: Al Hicks/New York Department of Environmental Conservation. U.S. Fish and Wildlife Service.*

ᐅᑦᔪᓇᖅ / ᐅᔾᔪᓇᖅ
SHREW
utjunaq / ujjunaq

Sorex

[1426]

ORDER NAME USED BY TURNER	*Insectivora*
CURRENT ORDER NAME	*Soricomorpha*
FAMILY NAME USED BY TURNER	*Soricidae*
CURRENT FAMILY NAME	*Soricidae*
SCIENTIFIC NAME USED BY TURNER	*Sorex*
CURRENT SCIENTIFIC NAME	*Sorex cinereus* Kerr
	Sorex hoyi Baird
SYLLABARY	ᐅᑦᔪᓇᖅ ; ᐅᔾᔪᓇᖅ
MODERN NUNAVIMMIUTITUT TERM	*utjunaq ; ujjunaq (cf. Schneider p.468 -ujjunaq = shrew [Sorex cinereus])*
ESKIMO TERM BY TURNER	*ug zhuṅ nak; ug zhuṅ núk*
DEFINITION OF ESKIMO TERM BY TURNER	*Shrew (Sorex) [p. 2896; 1427]*

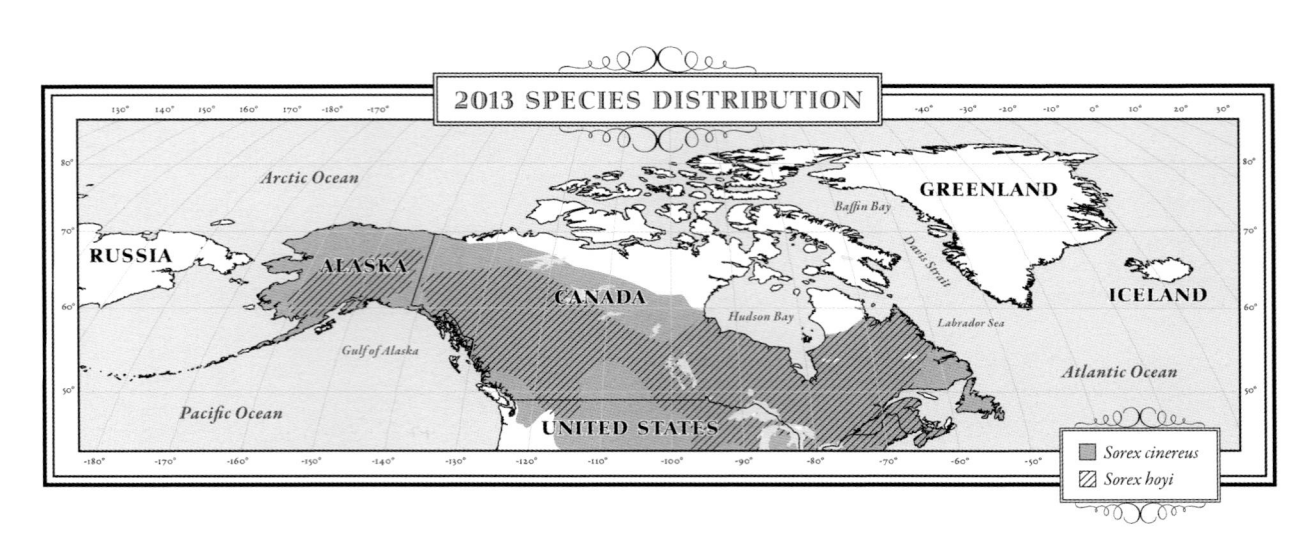

2013 SPECIES DISTRIBUTION

- ■ *Sorex cinereus*
- ▨ *Sorex hoyi*

CHAPTER TWO: MAMMALS OF UNGAVA AND LABRADOR

Descriptions by Turner

Shrews are tolerably common but as the tiny creatures, scarcely larger than a bumble bee, are so easily overlooked it is not an index of their absence because they are not seen.

The Shrews pass a good portion of their life amongst the rank patches of grasses and the denser thickets amongst the roots of which they search for worms and insects. They are often to be seen in the houses and an admirable plan to entrap them is to put a bottle on the floor. They soon enter it and as they are not timid the affair may be picked up.

The track in the freshly fallen snow is so insignificant that one day to affect an ignorance of the creatures I remarked to an Eskimo that the tracks were certainly those of a walrus. He replied "Probably a mosquito." (The tracks are certainly not so large as some of [1427] the Ft. Chimo mosquitoes.)

Where or how these Shrews breed was not discovered.

The Eskimo term them *ug zhúng nuk*, meaning minute *uǵ zuk*, or Square-flipper Seal, *Erignathus barbatus*.

Collection Notes

Five specimens of *Sorex cinereus* and one specimen of *Sorex hoyi* were collected by Turner at Ft. Chimo. These are located at the Division of Mammals, NMNH.

MOLE

Scalops

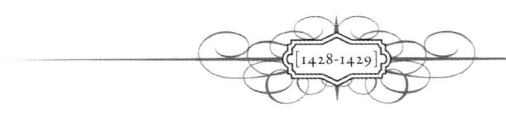

[1428-1429]

ORDER NAME USED BY TURNER	*Insectivora*
CURRENT ORDER NAME	*Soricomorpha*
FAMILY NAME USED BY TURNER	*Talpidae*
CURRENT FAMILY NAME	*Talpidae*
SCIENTIFIC NAME USED BY TURNER	*Scalops*
CURRENT SCIENTIFIC NAME	*Condylura cristata* (Linnaeus)
SYLLABARY	*Unknown and unnamed in Ungava*
MODERN NUNAVIMMIUTITUT TERM	*Unknown and unnamed in Ungava*
ESKIMO TERM BY TURNER	*Unknown and unnamed in Ungava*
DEFINITION OF ESKIMO TERM BY TURNER	*Unknown and unnamed in Ungava*

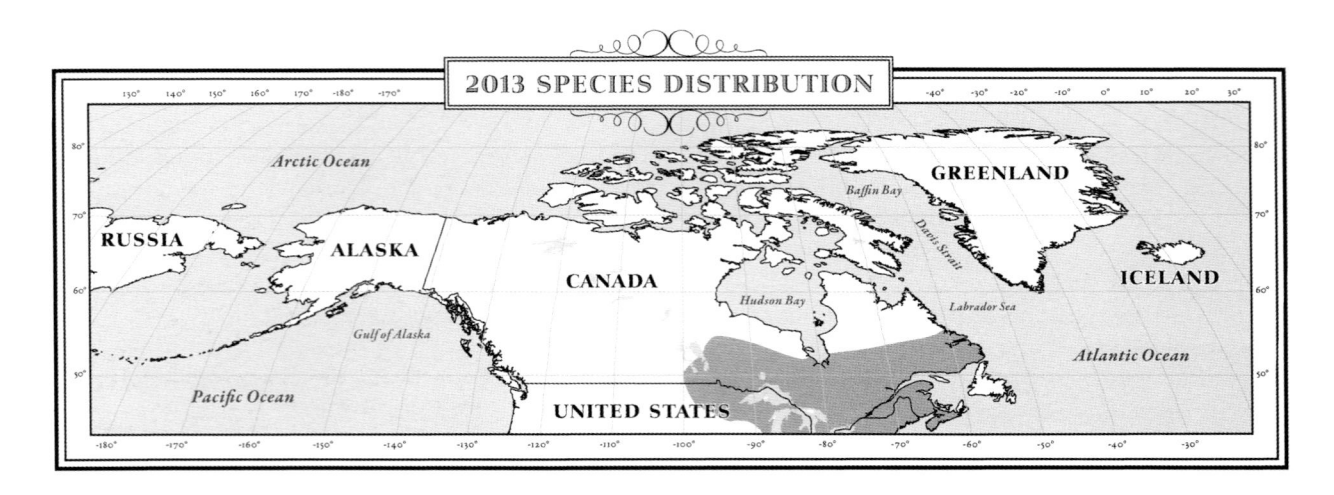

2013 SPECIES DISTRIBUTION

CHAPTER TWO: MAMMALS OF UNGAVA AND LABRADOR

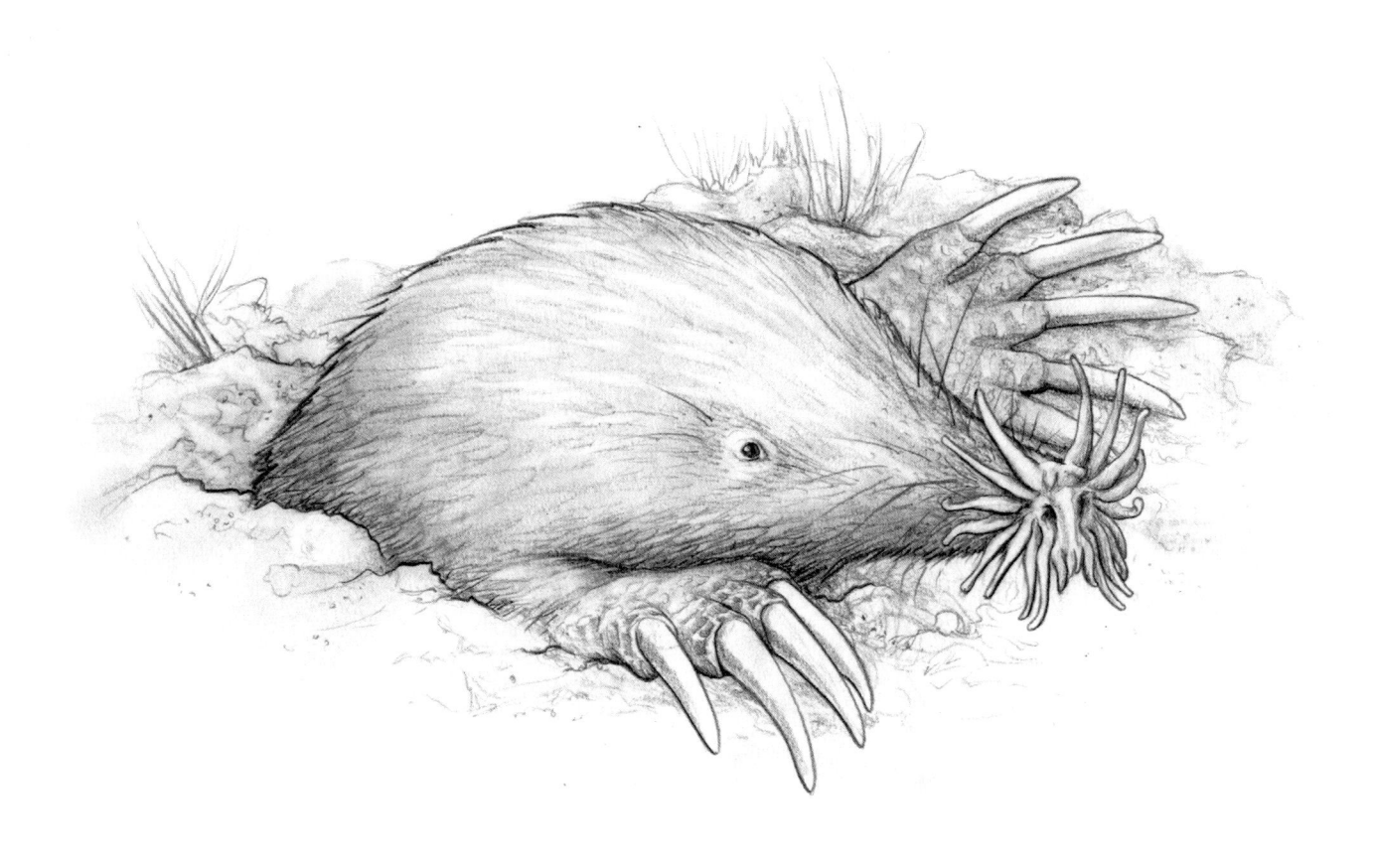

Descriptions by Turner

A species of Mole is reported to me to be not rare in the vicinity of Northwest River.

The habit of the creature making its underground passages among the dryer tracts of that locality as reported to me certainly refer[s] to the mole.

Collection Notes

Turner did not collect any mole specimens while in Labrador and Ungava, and he did not record the Inuit or Innu names for this species. Recent interviews with Inuit from Ungava on mammals indicate that this mammal is not known to them. The star-nosed mole, *Condylura cristata*, is the only mole in the Labrador region: "the northern limit of the range, as known to date, extends across the peninsula from Rigolet to Whiteman Lake and Little Whale River" (Harper 1961).

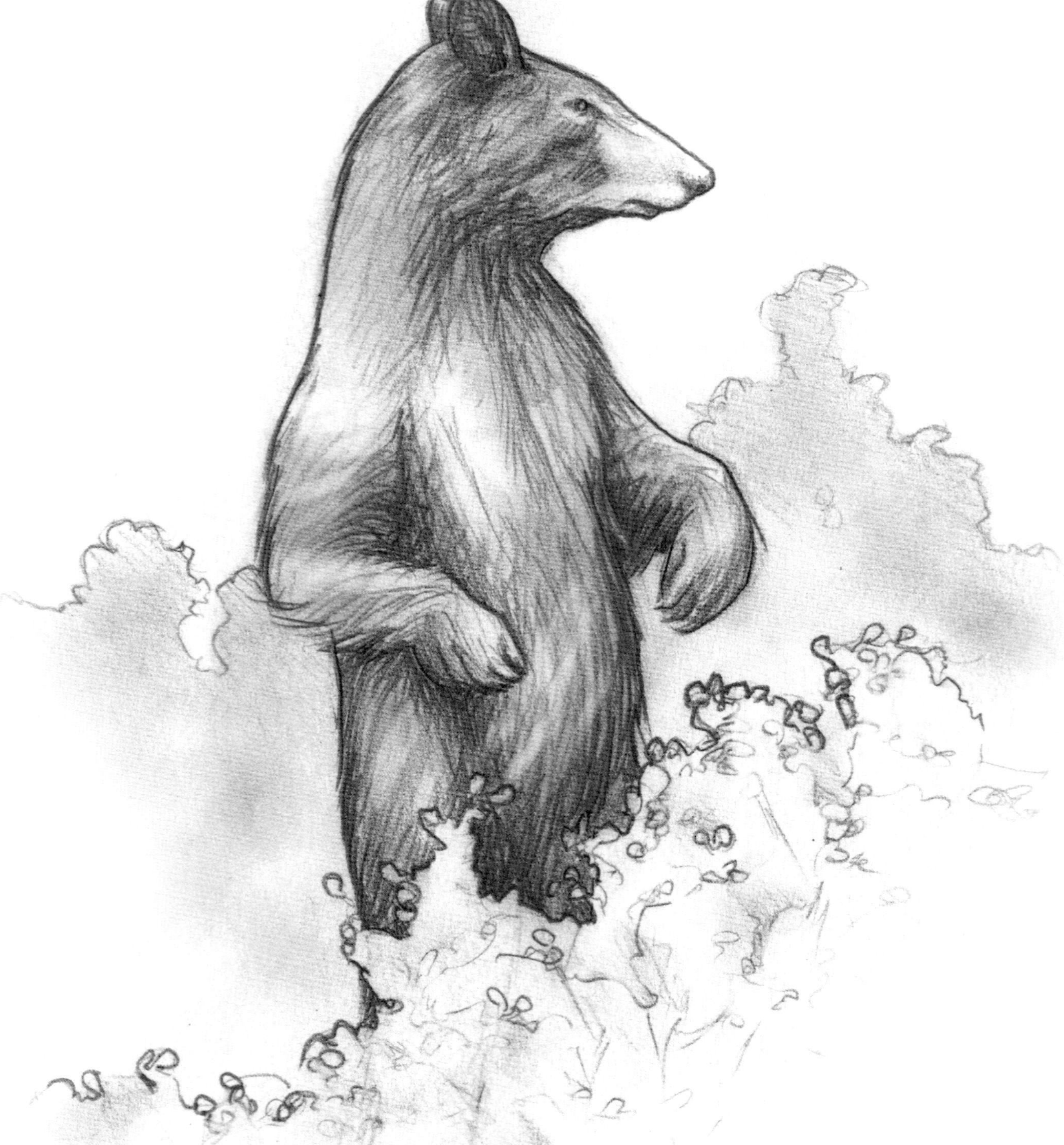

CARNIVORES

Introduction to Carnivores

Together, Turner's notes on the Carnivora ("flesh-eating mammals") form a large part of his manuscript, and this is to be expected. The carnivores of Ungava include both the terrestrial carnivores – the dog, cat, bear, weasel, and other related taxonomic families, as well as the sea-going pinnipeds – seals and walruses. The economics of Fort Chimo were founded in largest part on harvesting skins from fur-bearing carnivores such as foxes and marten. Ungava's indigenous economies always relied heavily on carnivores, whether furbearers, seals and walruses, or the hardiness of the north's one domestic animal, the sledge dog. Turner's manuscript provides a valuable window into the biology and utilization of all of these species in eastern Canada in the late nineteenth century.

Turner took particular interest in pinnipeds, especially in the Inuit vocabulary applied to them and the ways in which they were hunted and utilized. In his manuscript, he treated all species known from the vicinity of Ungava Bay and the Labrador coast – six seals, plus the walrus (*Odobenus rosmarus*). Turner referred to the harbor seal (*Phoca vitulina*) also as the "Fresh-water Seal," mentioning its occurrence in freshwater lakes in Ungava. Harbor seals utilize marine habitats throughout the northern continents, but also occur in various freshwater lakes throughout Ungava. Doutt (1942, 1954) documented the distinctive features of harbor seals isolated in the land-locked Lacs des Loups Marins (Seal Lakes) of western Ungava, now recognized as a distinct and endangered subspecies, *P. v. mellonae* (see Smith et al. 1994, 1996; Smith 1999), and Harper (1961) discussed additional freshwater lakes throughout Ungava where harbor seals occurred historically. Only one pinniped specimen collected by Turner, a striking walrus skull, remains in the Smithsonian collections.

Two species that Turner profiled in short species accounts, the raccoon (*Procyon lotor*) and the eastern striped skunk (*Mephitis mephitis*), penetrate only the southern reaches of Ungava, and do not occur in the areas Turner visited in Labrador and Ungava Bay. The fisher or pekan (*Martes pennanti*), a member of the weasel family, has a similar distribution in forests in the south of the peninsula. Although he did not provide a separate species account for the Fisher, Turner discussed it in his account of the porcupine, describing its unique adeptness at hunting that well-defended animal ("the pekan, however, being strong is able to inflect a fatal wound on the under parts at the first attack and being even more agile than its relative, the marten, rarely fails to capture its prey."). Turner found the mink (*Neovison vison*) and the American marten (*Martes americana*), skins of which were very valuable in the fur trade, to be uncommon near Fort Chimo during his stay there. He collected no mink specimens and only a single skull of the marten, and was able to report only brief notes on these species. He learned more about the natural history of the otter (*Lontra canadensis*), respectfully characterized as an "agile animal whose bite is something like the snip of a pair of shears and the clasp of a steel trap." Turner collected the skin and skull of a female otter for the Smithsonian, described in detail in his notes, but only the skull can now be found in the collection.

Two small species of weasels, the ermine or stoat (*Mustela erminea*) and the least weasel (*Mustela nivalis*), occur in the Ungava Peninsula. Turner provided arresting notes on the ermine, including firsthand observations of behavior based in part on keeping an injured female ermine in his cabin. He did not think the least weasel was present near Ft. Chimo, but a single weathered skull from his Ft. Chimo collections (likely salvaged off the ground) is from a least weasel, and represents one of the most northeasterly records of the species in North America (Harper 1961; Hall 1982).

Two fox species, the arctic fox (*Vulpes lagopus*) and the red fox (*Vulpes vulpes*) occur in the Ungava Peninsula, traditionally the most important source of skins in the Ungava fur trade (Elton 1942). Turner retained many skulls of both species for the Smithsonian collections, but secured few of the valuable skins. Skins of these two species of foxes come in a rather bewildering number of colors. The arctic fox has both a white and a "blue" (bluish-gray) morph, and these coat colors transform to brown and a darker charcoal color, respectively, in summer. Turner correctly surmised that these various seasonal and color forms were variants of a single species. He was more confused by the red, cross, and silver or black foxes. Though these too are all color variants of a single species, most commonly known as the red fox, each variety brought a different price in the fur trade (with the silver or black phase being especially valuable), and Turner gave each of them separate species accounts. He had heard claims that all three types were supposedly the same species ("dens have been discovered containing all the three"), but noted that "owing to the engrossing nature of other work I was unable to satisfy myself of the truth of these assertions."

Wolves (*Canis lupus*) are the principal large wild mammalian predator throughout the Ungava Peninsula, and their movements and population sizes are no doubt tied to increases and decreases in the number of caribou in the region (Bergerud et al. 2008), as Turner noted. The wolves of northern Quebec and Labrador are often recognized as a distinct subspecies, the Labrador wolf, *Canis lupus labradorius* (Hall 1982). This subspecies was named by Smithsonian taxonomist Edward A. Goldman (1944) based on Turner's wolf specimens collected at Ft. Chimo – one of Turner's many contributions to museum-based mammalogy.

Turner's manuscript includes important source material for understanding the biology of Inuit and Innu dogs and their importance to Inuit and Innu culture and livelihoods. His important notes on sledge dog disease pre-date the veterinary understanding that rabies plays a principal role in these epidemics (Mørk and Prestrud 2004). Turner's manuscript adds an additional data point to the study of cycles of epidemic canine disease in the Canadian Arctic that, buried until now in Turner's unpublished notes, has been out of view in previous studies (Elton 1931, 1942).

The only wild cat in the Ungava Peninsula is the Canada lynx (*Lynx canadensis*), a specialist predator of hares that also hunts other small and medium-sized animals. The lynx occurs throughout the forests of the Ungava Peninsula but is very infrequently encountered. Turner learned relatively little about its life in Ungava, where it has been considered rare from Turner's time to today (Elton 1942; Harper 1961; Bergerud et al. 2008).

Turner's observations on bears are especially notable. Three bear species were present in the Ungava Peninsula in Turner's time – the polar bear (*Ursus maritimus*) of the northern ice and coasts, the American black bear (*Ursus americanus*) of the forests, and the brown or grizzly bear (*Ursus arctos*), which Turner and his contemporaries called the "barren ground bear." Turner's notes on this last species are of great importance and interest, because the isolated far north of Ungava was the only place east of the Mississippi River that the brown bear occurred in modern times and it likely became extinct in Ungava not long after Turner's visit (though some reliable reports suggest this bear may have persisted into the late 1940s). McLean (1849) was the first naturalist to discuss the presence of brown bears in eastern Canada. Turner, familiar with the brown bear from his time in Alaska and aware of McLean's observations, compiled reliable notes on the natural history of the brown bear in Ungava and confirmed its existence by firsthand examination of skins, which he could not secure for the Smithsonian. ("I saw skins of it and was unable to discover any appreciable difference between them and those of the Barren Ground Bear from other localities.") Other important accounts of the brown bear in Ungava/Labrador were published in the late 1800s (Bell 1884, 1895; Stirling 1884; Low 1896, 1897; Bangs 1898). For most of the twentieth century, historical occurrence of the brown bear in Ungava was questioned by authorities because no specimens were available in museums (Strong 1930; Anderson 1934, 1948; Allen 1942), though the two most careful reviewers of the subject (Elton 1954; Harper 1961) found the circumstantial evidence for historical occurrence compelling. More recently, documentation of skeletal remains of *Ursus arctos* from northern Labrador (Spiess 1976; Spiess and Cox 1976; Loring and Spiess 2007) leaves no doubt that the species was present in the region in recent centuries. Loring and Spiess (2007) discussed Turner's unpublished manuscript notes on the barren ground bear for the first time, citing these as additional core documentation for the recent occurrence of this species in Ungava. Although Turner mentioned the occurrence of the brown bear in Ungava several times in passing (Turner 1885, 1888b, 2001 [1894]), the fact that his full mammalogical notes were long overlooked surely delayed zoologists' acceptance of the fact that an isolated population of the species occurred in Ungava until the late nineteenth century, a very interesting biogeographic phenomenon. Even today, with the matter now settled, most references regarding the distribution of brown bears still do not mention the species' recent occurrence and extirpation in the northern reaches of Quebec and Labrador.

Turner's extensive notes on the wolverine (using the spelling "wolverene") also provide an important window into Ungava natural history, given the animal's decline and possible extinction in the region. In Turner's time and later (Low 1896; Bangs 1912) the wolverine was considered reasonably common in eastern Canada, but it has been extremely rare there since the 1930s (see Harper 1961), a decline co-incident with a major drop in the caribou population (Bergerud et al. 2008). Despite reported sightings every year, there have been no confirmed records of the wolverine in Labrador or Quebec for several decades, and it may no longer occur in Ungava (COSEWIC 2003; Slough 2007). The wolverine, both a scavenger and a predator, is a true symbol of the northern wilderness with a fearsome reputation. With remarkable strength and tenacity, it can overcome much larger prey, including injured or exhausted large ungulates, and drive away much larger predators, including wolves and bears. In Turner's time the wolverine was important to fur trappers both as an animal whose skin was traded, and as a nuisance that would systematically plunder trap lines by eating trapped animals and wreak havoc in unattended cabins and camps. A wolverine discussed by Turner, shot at Ft. Chimo, had used its strength to escape earlier from a steel trap, leaving the claw of one toe behind ("this individual had lost a toe-claw and was doubtless the same that had visited a trap some three weeks before and in that trap a claw was found"). The skin of this specimen is still in the Smithsonian collections – still identifiable by its missing claw. As with the brown bear, the extirpation of the wolverine from Ungava, perhaps already come to pass, renders eastern Canada's mammal fauna less complete, its forests and its tundra less wild.

ᙯᓯᒋᐊᖅ

Harbor Seal or Fresh-water Seal
qasigiaq

Phoca vitulina, Linné.

[1431]

Order Name used by Turner	*Carnivora*
Current Order Name	*Carnivora*
Family Name used by Turner	*Phocidae*
Current Family Name	*Phocidae*
Scientific Name used by Turner	*Phoca vitulina*, Linné.
Current Scientific Name	*Phoca vitulina* Linnaeus
Syllabary	ᙯᓯᒋᐊᖅ
Modern Nunavimmiutitut Term	*qasigiaq* (cf. Spalding 1998, p.110 - *qasigiq* = ranger or freshwater seal)
Eskimo Term by Turner	*ka sĭg yak*
Definition of Eskimo Term by Turner	Freshwater Seal or harbor seal. *Phoca vitulina* [p. 1433, 2272]

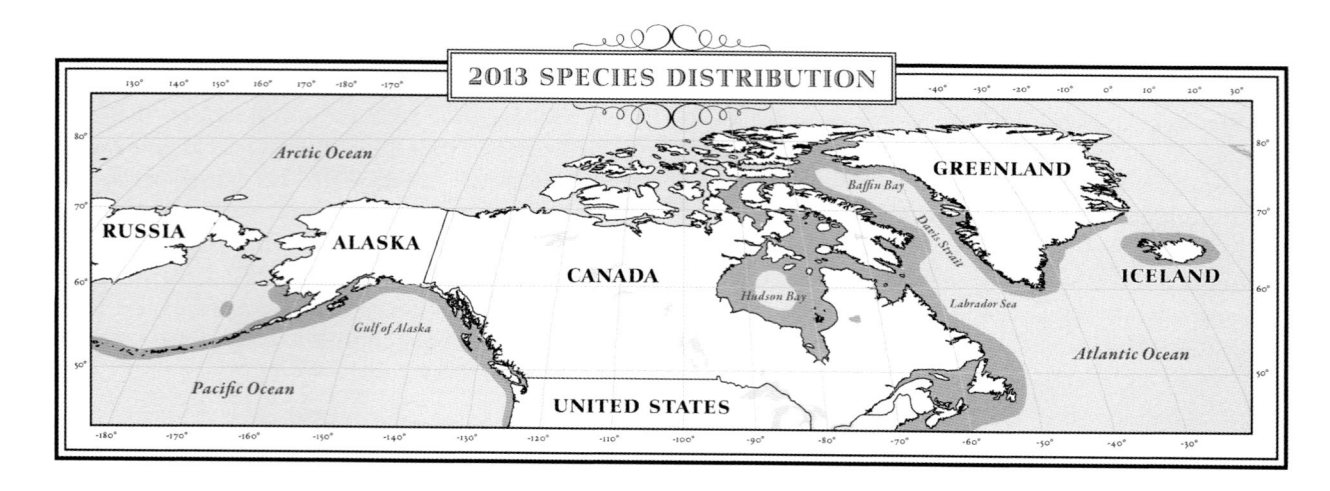

2013 SPECIES DISTRIBUTION

Chapter Two: Mammals of Ungava and Labrador

Descriptions by Turner

The Harbor Seal is quite abundant along all the coast line of the region under consideration. In certain localities it is extremely plentiful but as these resorts are rarely permanent and only temporary for two or three seasons it is unnecessary to mention them. There are, however, exceptions to this and these are the islets situated some distance from the mainland along the cost of the Strait. In Ungava Bay is one of these places where this species is known to have frequented for many years and owing to its nearly inaccessible position it is rarely disturbed, hence the seals enjoy comparable immunity from the visits of their arch destroyer, the Eskimo.

I have no information concerning their occurrence within the main body of water composing Hudson Bay but am informed by Eskimo of the eastern coast bounding that water that they are plentiful at times and [1432] scarce again at other seasons.

Collection Notes

Turner did not collect any seal specimens while in Labrador and Ungava. Owing to the striking and decorative patterns on the harbor seal, their skins continue to be used today as the preferred material for *kamiks* (sealskin boots). Populations of habor seals occur in Northern Labrador (especially near Makkovik), the George River, and in the Richmond Gulf.

The skin of this species is specially prized by the Eskimo who make various garments and innumerable other articles from its skin. The color of the hair is so variable that to record the disposition of the markings of blackish to a silvery gray with greenish tinge would be simply impossible. It is not possible to find two animals marked alike and even on different sides of the same animal the spotting will be more or less dissimilar.

The backs are considered of greatest value and command the admiration of an Eskimo when other objects would earn but a glance.

The strips of skin are often cunningly placed along parts of skin from the Harp seal and when shown in contrast present a pleasing sight.

This species of Seal frequents the fresh waters; and, in some of the larger lakes, having but short outlets leading to the salt [1433] water, they are to be found as residents and in considerable numbers.

The flesh is highly prized and is enjoyed as a special delicacy when eaten after a long period of deer meat having been alone used for food.

My opportunities for studying the seals were extremely limited and much to my regret. The Indians rarely touch the flesh or oil of any of the seals, claiming that it is too fat. The only use to which they put the oil is in preparing the gum to smear on the seams of their bark canoes. They state, and without doubt it is so, that the seal oil renders the gum less liable to break than any other fat or oil they could use for this purpose.

The Eskimos apply the name *ka sŭg yak* to this species. The younger, or extremely old ones are known by suffixes denoting age or condition.

Seal Hunting in Ungava Bay
A Kangiqsualujjuamiut hunter walks the sea edge in search of seals at the beginning of winter. Photo by Scott Heyes, 2008.

HARBOR SEAL ETHNOGRAPHY
Turner

DECORATING GARMENTS (1887A: 37)
The deerskin garments are profusely ornamented with strips of skin from the variegated black and bluish-gray haired skin of the harbor seal, *Phoca vitulina*. Strips of tin, tags of that metal, from plugs of tobacco, ivory pendants, spoon bowls, lead drops and broad bands of beads composed of many strands pendant across the breast, go to add attractiveness and represent so much value to each garment for the body of a woman.

Inuit Cap
Turner (1894: 209) noted: "[This] cap was obtained from one of the so-called "Northerners" (Inuit from Hudson Strait coast) who came to Ft. Chimo to trade. The design was copied from a white man's cap. The front and crown of the cap are made of guillemot and sea pigeon skins, and the sealskin neckpiece also is lined with these skins, so that when it is turned up the whole cap seems to be made of bird skins." USNM E90193, Museum Support Center, NMNH. Photo by Angela Frost, 2011.

HARP SEAL
qairulik

Phoca groênlandica Fabr.

[1434]

ORDER NAME USED BY TURNER	*Carnivora*
CURRENT ORDER NAME	*Carnivora*
FAMILY NAME USED BY TURNER	*Phocidae*
CURRENT FAMILY NAME	*Phocidae*
SCIENTIFIC NAME USED BY TURNER	*Phoca groênlandica Fabr.*
CURRENT SCIENTIFIC NAME	*Pagophilus groenlandicus* (Erxleben)
SYLLABARY	ᙱᐃᑐᓕᒃ
MODERN NUNAVIMMIUTITUT TERM	*qairulik* (cf. Schneider p. 278 – *qairulik* = Greenland seal [*Phoca greenlandica*]. More commonly in Canada referred to as the harp seal.
ESKIMO TERM BY TURNER	*Kai ġo lak*
DEFINITION OF ESKIMO TERM BY TURNER	*Young harp seal* [p. 2261]

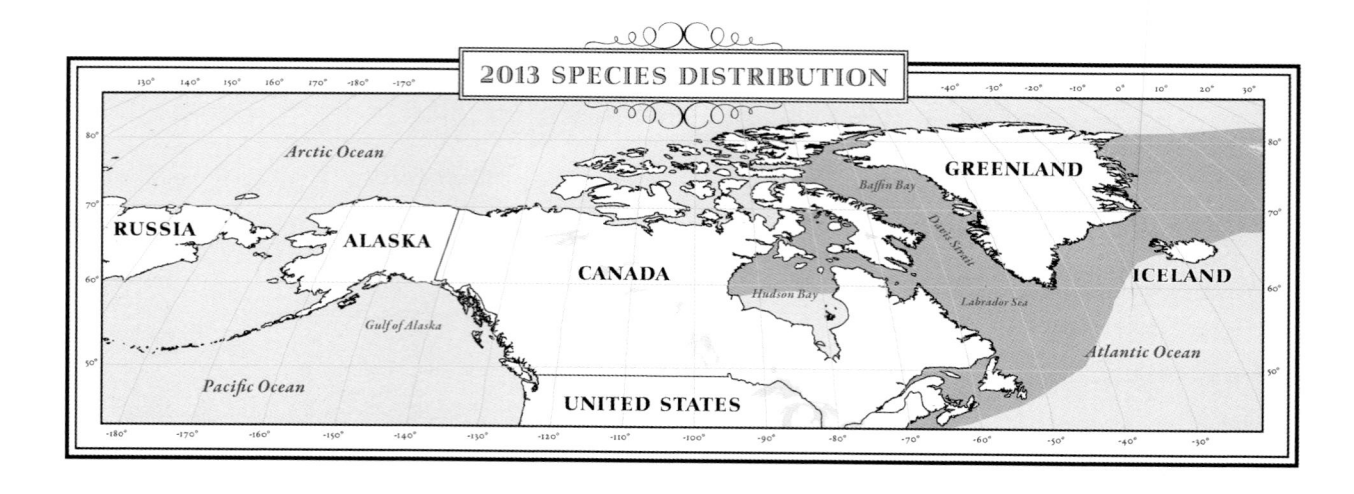

2013 SPECIES DISTRIBUTION

CHAPTER TWO: MAMMALS OF UNGAVA AND LABRADOR

Descriptions by Turner

The Harp Seal is common in all the salt water bordering the land of this region. It is rarely known to wander up the rivers beyond the brackish waters.

They appear to be rather sociable as one is rarely seen at a time, usually two or three to half a dozen near together.

This species is known to the Hudson Strait Eskimo as *Kai ó lik* and from their skins are made most of the more serviceable boots and clothing of the men and women.

The skins of the adults, however, become very stiff and require constant rubbing to render them pliable.

The flesh is not ranked high in the opinion of the Eskimo who prefer the meat of other species. Some of these Seals winter in Ungava Bay where they obtain plentiful food from the salmon-trout remaining near the tide holes, off the mouths of some of the streams, which even in the [1435] severest weather do not close. The skins are also used to form the walls of the tents for their summer dwellings.

As this species keeps well out to seaward I was unable to obtain much information in regard to it. The breeding season was not determined as the distance of Ft. Chimo from the Strait prevented me from visiting it at the proper season.

The Eskimo assert that they bring forth their young on the ice floes or else [the mother] makes an opening under the snow crust and within this chamber delivers her young. So rarely are they taken that I have not, at that season, seen a young individual in the white coat in the possession of an Eskimo.

Those obtained are usually shot from the Kaiak or lanced with the spear. The vicinity of Akpatok Island is a favorite resort for this species and many are taken there in the spring. In the numerous passes or channels [1436] rushing among the islands and rocks forming the northern extremity of Cape Chidley, this species of seal is reported to be excessively abundant in the spring months.

RINGED SEAL
natsiq

Phoca foêtida, Fab.

[1437]

ORDER NAME USED BY TURNER	*Carnivora*
CURRENT ORDER NAME	*Carnivora*
FAMILY NAME USED BY TURNER	*Phocidae*
CURRENT FAMILY NAME	*Phocidae*
SCIENTIFIC NAME USED BY TURNER	*Phoca foêtida*, Fab.
CURRENT SCIENTIFIC NAME	*Pusa hispida* (Schreber)
SYLLABARY	ᓇᑦᓯᖅ
MODERN NUNAVIMMIUTITUT TERM	*natsiq* (cf. Schneider p. 198 - *natsiq* = Ringed seal [*Phoca hispida*])
ESKIMO TERM BY TURNER	*nĭt syak; nŭ chĭk*
DEFINITION OF ESKIMO TERM BY TURNER	Ringed seal [p. 1438]; Seal (*Phoca foetidus*) [p. 2694]

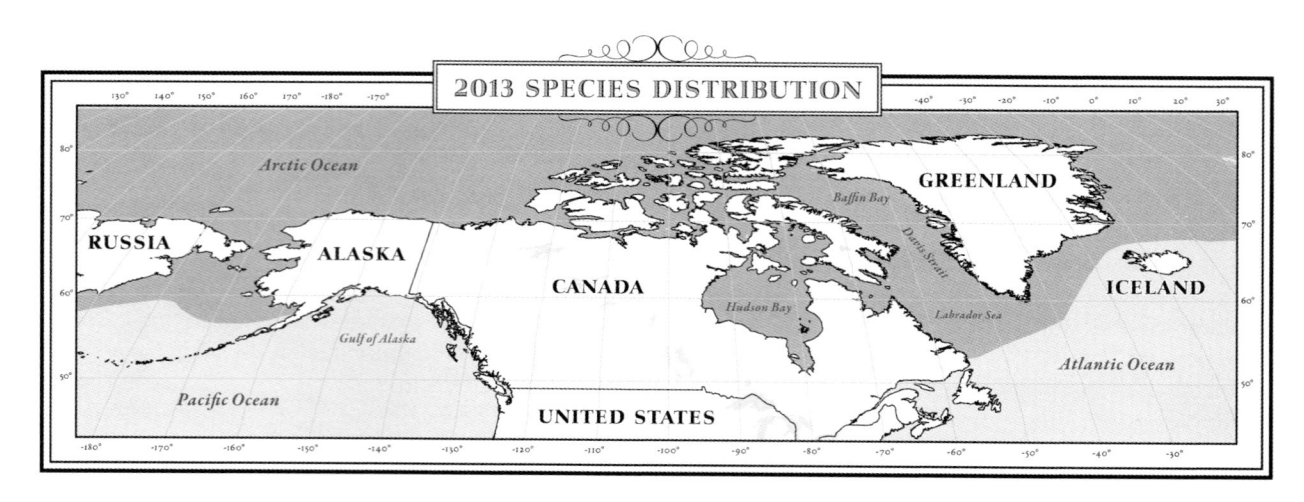

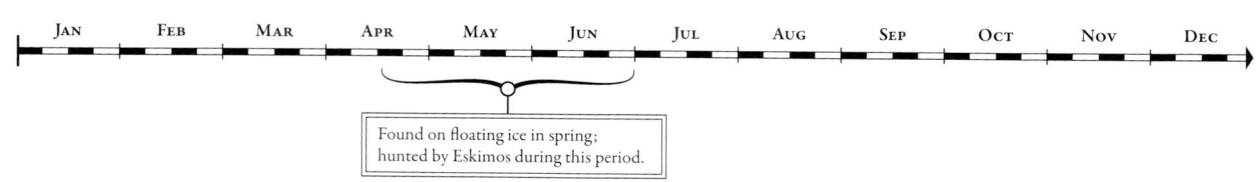

2013 SPECIES DISTRIBUTION

Found on floating ice in spring; hunted by Eskimos during this period. (Apr–Jun)

CHAPTER TWO: MAMMALS OF UNGAVA AND LABRADOR

Descriptions by Turner

The Ringed Seal is not a common inhabitant of the Strait. It is plentiful along the Labrador coast and there I had no opportunity to obtain information in regard to it.

It frequents the quiet bays and coves, seldom wandering far from land. They are eagerly sought for and are highly prized by the Eskimo for their skins and flesh. The skins are rated next to those of the Harbor seal for clothing. They are resident where found if not too much disturbed and then it is asserted they will return after their alarm has subsided. In the spring they are found on the floating ice and when seen are secured by the Eskimo.

This species seems to have little fear of the presence of man and will allow him even to stand alongside of it before making an endeavor to escape. Whether this is due to lack of the faculty of seeing and hearing or not I am unable to state. When alarmed, however, [1438] its struggles to regain the water soon bring it into that element where it disappears, swimming a great distance before coming to the surface.

I could learn nothing in regard to its breeding habits in Hudson Strait.

The Eskimos apply the name of *nit syak* to this species.

To the males under certain conditions they apply the name *ti'zhak*.

Ringed Seal Ethnography
Turner

THE SEASONAL ROUND (1887A: 102-109)

It may be early in the season and the waters yet afford but scanty sustenance to the eager crowd but when the first seal of the year is secured a feast is held as soon as the party can reach the shore and prepare it. An anxious throng gathers about the carcass and when the skin is stripped from the body the flesh is quickly devoured by the people so long held in expectancy of fresh food.

The flesh is divided according to rank and station, the choicest morsels falling to the favorite children. Ere many hours not a vestage remains except the hide so greatly coveted to repair the soles of their footwear now to undergo roughest usage over the sharp-edged gravel and sand and rocks.

Each day visibly decreases the quantity of ice and the migratory birds appear amongst the open spaces. Soon the earlier arrivals begin to lay and the eggs of all the waterfowl are sought as delicious food by the Innuit who gives no concern to the stage of development of the embryo within the shell. The islands whose barren tops scarcely rise above the swash of the waves are scanned for the nests of the gulls; the higher ones whom summit are clad with grass and weeds are searched for the nests of the eiders; while the ragged shore lines of the islands are peered among for the nests of the surf ducks, sea pigeons and puffins. The higher parts of the marshy tracts of the neighboring mainland are watched to discover the breeding places of the Canada goose or the freshwater ducks and loons. While the eggs last but little thought is given to other kinds of food for the surging sea of ice has not yet left the waters beyond and a change of wind may seal the nearer waters for days until the summer winds assert their power and slowly the mass is driven to disappear and be carried within the influence of the resistless tide currents flowing with incredible force and power that soon grinds the ice to naught. The larger seals are now beginning to appear and become plentiful; the smaller species of seals are giving birth to their offspring on the naked rocks jutting from the water. To these the Innuit journies to capture the young whose skins bring such comfort when fashioned into garments. Now must be the time when the scattered seals of the larger species begin to show here and there, to procure their pelts with which to renew the cover of the umiak and khaiak, to make soles for his boots, thongs for harness for the dog-team and a thousand other needs which only that strong skin will supply. The younger seals furnish skins for bootlegs, garments, floats, oil bags and innumerable other uses requiring lighter hides than those of the huge Square-flipper Seal *(E. barbatus)* fitted only for the heaviest purposes demanding weight and strength.

The male seals are now in the fattest condition and their oil is so greatly coveted by the Innuit not only as food but for light and fuel, lubricant or preservative. The layer of fat is stripped from the body, leaving only the muscles and frames and adheres to the pelt, which needs to be removed from the fat cut into strips and placed within the skins of the creature itself. The sealskin oil bags are prepared in a peculiar manner for the skin must be free from incisions and wear spots. The abundance of the seals indicates the number of oil bags to be required for the storage of the fat obtained beyond the actual consumption by the party. The women or the younger men, inexpert hunters or those whose poverty and lack of means prevents them from joining in the capture of the prizes, usually attend to the preparation of the bags for the reception of the fat.

The skin is stripped from the body of the seal in the following manner: the mouth of the creature is opened wide and by means of a knife the head is severed from the neck and withdrawn through that orifice. The blade is plunged within and the shoulders severed and withdrawn, the remainder of the carcass is taken out and finally the skin inverted so as to permit the bones and flesh of the flippers being removed. The adherent flesh and fat is scraped from the skin, a button shaped plug of wood or ivory is fitted in the posterior orifice and as the button is

Inuit Students Celebrate a Hunt
Kangiqsualujjuamiut school students with a ringed seal successfully hunted with the assistance of elders. Seal meat is a staple to the Inuit and the blubber provides a nutritional meal to Inuit dogs. Photo by Scott Heyes, 2008.

formed like a pulley or having a neck to it a thong must be tied securely around it to hold it in place and prevent wastage of the fluid contents. The neck of the sealskin is used as the mouth through which the fat is placed within the bag.

Piece after piece of raw fat, just stripped from the body of the seal is thrust within the bag until it is nearly full. A stout thong is now tied securely around the neck or head of the bag and the affair placed out of reach of dogs or other destructive mammals.

The sealing season is over by the first of July except such straggling individuals that may be seen at any time during the period of open water. By this time the white whales are showing their cream white sides like a piece of ice lifting with each undulation of the water. The flesh and fat of those mammals are prizes valued highly by the Innuit. The skin makes a soft cover for the khaiak or tops for his boots while the stomach affords a receptacle for the fat taken from its body. These oil bags, like those from the seal, have and represent certain values, depending on the kind of oil and the size of the receptacle. [...]

During the midsummer season there is but little to occupy the Innuit and to rid themselves of the myriads of mosquitoes which infest the camps on the mainland it is not uncommon that a community starts along the coast without special objective

Models of Traditional Hunting Implements Used to Procure Marine Mammals

Implements are displayed in a classroom at the Ulluriaq School, Kangiqsualujjuaq. School children learn about traditional hunting practices and Inuit belief systems associated with hunts. An old custom, for example, observed that: "After a seal is killed, a little fresh water is sprinkled over it before it is cut up" (Payne 1887-88: 23-230). Payne noted, however, that the Inuit did "not always carry [this custom] out, and if done in [the] presence of [non-Inuit they] would explain with a look of bashfulness that other Inuite (sic) always did so." The Inuit also believed that "seals would forsake the parts of a sea in which a human corpse lies" (PA 1873-1875, 29: 64), and that "some of the meat from a boy's first seal is saved for the midwife 'the helper' who saw him safely into the world" (Sutton 1912: 224-225). Accounts from 1861 indicate that the Inuit conducted rituals and ceremonies at the graves of their ancestors for the purpose of asking the spirit world to provide seals and reindeer during times of starvation and lack of game (PA 1861, 24: 544). In journal entries by Moravians from 1876 (PA, 30: 147), there are descriptions of Inuit architecture pertaining to seals as a construction material. With respect to Inuit who had converted to Christianity in Labrador it was observed that: "You see no more genuine Eskimo huts with windows of seal bladder. European block houses are now substituted for the former hovels (sic) (PA, 30:147)." Photo by Scott Heyes, 2010.

Kayak Under Construction (c. 1919)
An Inuit kayak frame is left to cure in the summer sun. Photo presumably by George E. Mack. Used with permission of McCord Museum, MP-1984.126.152

point obtaining such food as may be procured. The summer tent of heavy sealskins is placed in the umiak and such other articles as may contribute to their comfort are included. The stores of oil and flesh, taken in the months just gone by, are left behind if the party intend to return to that locality for the fall or winter; or it may be that the vicinity where the seals were captured is not an advantageous site for a winter camp that the stores must be removed to such place as may be determined thereafter. There are so many circumstances to be taken into consideration of earnest character that it is often a matter of deep thought as to the proper course to pursue. When all those arrangements are made the summer life of the Innuit may be one of pleasure or anxiety. In pleasurable anticipation the boat crew start from their last camping ground and journey wherever the evidence of most abundance leads them.

The remnants of the last feast are stored in the bow and stern of the boat. The rowers pull with sudden spurts of strength, merry shouts of laughter or taunts with those who struggle along the uneven shore to keep pace lest the occupants of the boat leave them far behind. Every object, appearing on the surface of the water, is scrutinized and if it possess life all is activity within the boat to give chase if it afford but a morsel of food. The head of a seal appears as a large bubble on the glassy surface of the calm sea; quiet pervades the crew and the hunter seats himself within his khaiak and stealthily paddles toward the creature straining its senses to detect the cause of the sounds that brought it to the surface. The least noise of the paddle alarms it and it slowly disappears, scarcely ruffling the water. The hunter awaits its reappearance in readiness to shoot with gun or dart a lance or harpoon into its body. The capture of a seal while on a journey of this character is usually of sufficient importance to cause the progress to become stopped for the remainder of the day.

INNUIT KHAIAKS, OR SKIN CANOES, FROM THE UNGAVA DISTRICT, H.B. TERR. MODELS (PROPERLY MINIATURES) (1887A: 239-252)

One of the chief characteristics distinguishing the Innuit from all other people of the earth is the peculiar form of the individual boat usually termed a skin canoe.

This vessel is peculiarly adapted to the special life calling of those people and without it their mode of living would be essentially changed to a manner which in itself would have none of the features now

Inuit Family Making Kayak (c. 1919)
Note the frame of the kayak under construction and the patterns of the skin parkas worn by the women. Photo by Captain George E. Mack. Used with permission of McCord Museum, MP-0000.597.527.

surrounding the Eskimo. The plan or model of the Khaiak is that which insures lightness, flexibility and strength; three important requisites in boat-building.

An exterior view, of an Ungava Khaiak from above presents the form of an elongate ovoid of twenty-two feet in length and twenty-five inches wide on the top to the rear of the manhole or at about the beginning of the posterior third of the length. From this point forward the lines gradually converge to a narrow prow or nose, decreasing behind, to the slightly upturned stern, much more rapidly. The top or deck behind the manhole is quite flat with the exception noted. The anterior third is also flat, but the posterior four feet of the middle third is elevated to form a crown, something like the upper of the foot of a shoe, in order to accommodate the limbs of the occupant.

The flare of the sides is variable from nearly perpendicular on the extreme prow to a moderate degree in the region of the manhole. Viewed from the side the greatest height of the side is twelve (12) inches at the beginning of the prow rise, while, at the widest part of the deck the side is only nine (9) inches gradually decreasing to five (5) inches at about two feet from the stern.

The bottom is quite flat and in form is nearly a counterpart of the top, but is only thirteen feet and six (13, 6) in length and nineteen (19) inches wide, or six inches less than the top over the same place, giving a side flare of only three (3) inches on each side. The prow or nose begins where the forward part of the flat bottom ends and rises gradually from a level to a height of fourteen (14) inches in a length of five feet

Sealskin Kayaks
Two sealskin Inuit kayaks on display at the Great Whale Environmental Impact Hearings in Kuujjuarapik, Nunavik, c. 1980. Photo courtesy of Peter Jacobs, Montreal.

Hunter Beside Kayak (c. 1919)
Inuk hunter with hunting equipment, harpoon and kayak. Used with permission of McCord Museum, MP-0000.597.166.

and eight inches (5,8). The length of the stern rise is three feet and one inch (3,1); thus giving a bottom length of twenty-two feet and three inches (22,3).

The greatest depth of the Khaiak is at the forward part of the manhole where it measures sixteen and a half (16.5) inches; the depth at the rear inside of the manhole is just one (1) foot, including, in both instances, the height, three (3) inches, of the hoop.

The hoop is of wood, three-fourths (3/4) of an inch thick and three (3) inches high. It has a form of a truncate ovoid, wider behind and somewhat sharp-pointed in front, measuring sixteen and a half inches wide and twenty-one and a half inches in its longer axis fore and aft. The hoop rests on the highest arched crossbeam of the deck in front and on a beam to the rear of the hole. It incloses a portion of the skin cover drawn up from within by means of stout thongs, which by a sort of herring-bone lashing inserted into eyelets in the hoop and back through loops cut in the skin, cause the hoop to sit firmly in place by means of the retractile power of the dry skin.

It will be seen that the front of the hoop is four and a half inches higher than the rear. This position gives it a decided slope backward and enables the occupant to employ his paddle in a peculiar manner. The paddle being very long, just half the length of the Khaiak, is used as follows: The hands grasp it and with a motion similar to that of a mower using a grass-scythe, that part of the mid-shaft near the hand

Kayak at Ft. Chimo (1882)
An Inuk man in a sealskin kayak on the Koksoak River at Ft. Chimo. Photo by Turner, 1882. Local # NAA INV 06531400. Source: BAE /SI GN 03218 06531400, National Anthropological Archives, Smithsonian Institution.

rests upon the rim of the hoop, and slides along and toward the side of the person as the paddle is moved from one side of the craft to the other.

This affords a fulcrum or rest for the paddle and relieves the weight of it; for the Ungava Innuit never raises the paddle from the rim when propelling the vessel, unless it be to drop it astern in order to guide the boat. Another object is gained thereby; if the paddle is lifted too high it permits the living object sought to discover the presence of the canoe. The Khaiaker, when approaching a wary object in the water, moves his paddle so that the blade uplifted will never appear above his shoulder; but when it has been sighted he dips the paddle deep and throws all his power upon the blade causing the craft to glide rapidly over the water. Another object attained by the front of the hoop being elevated is that the waves lapping over the front do not enter so readily as though the hoop were level with the deck.

In the preparation of the frame, of a Khaiak at Ungava, the first thing to be done is to procure the necessary timber, which is easily done on account of the proximity of the forests of spruce and larch. A piece with a natural curve to form the bow-piece and another to shape the stern are united to furnish the kelson. It is nearly square in form and has a top and bottom side of nearly two inches and a thickness of little more than an inch. There are two side-kelsons, each placed equidistant from the kelson and unite with it at the stem and stern; but in front the side-kelsons end where the prow of the vessel takes its rise from the bottom; at the stern they continue to the end. These two pieces are bent to the proper width and held in place by means of flat crossbars secured to the kelson and each end to the side-kelsons, the wider bars in the widest part of the boat and becoming shorter as they are placed successively forward or aft of that place. The ends of the side-kelsons are now lashed to the kelson and thus from the bottom of the frame.

The next operation is to provide two gunwales, nearly square pieces which extend the entire length of the craft. They are wider apart than the side kelsons according to the flare of the boat or at the widest part in the ratio of 19 for the bottom to 25 for the deck.

Crossbeams of flat pieces are mortised into the gunwales for and aft but at the crown they are arched to a greater degree successively to the last which forms the support for the hoop. These crossbeams are set so that the skin cover of the boat is not in contact with them and in this particular the Ungava Khaiak differs greatly from the Alaskan Khaiak in not permitting the person to step on those beams lest his weight tear the skin of the vessel. When the top frame is completed the next operation is to prepare a number of ribs, twenty-two to twenty-eight for the different sizes of those boats, from wood. They are flattish with rounded corners; their form is that of a broad U-shape, the wider ones in the broad part of the boat and the narrower ones at either end. The arms of the ribs flare more or less according to position along the bottom frame on which they rest and to which they are lashed. The upper ends of the ribs are inserted into holes in the gunwales and as arranged that they will rest between the cross-bars uniting the kelsons. When these are placed in position the work on the frame is nearly completed; the eye must guide the form in to proper shape for the Innuit employs no square level or plumb in determining the lines of a craft that has excited the wonder and amazement of all who have observed his fearlessness in mounting an angry wave with this apparently frail structure which owes its seaworthiness to the three requisites mentioned before. To prevent collapse of the skin within the rib spaces a slat, or ribbon, is laid on the outside of the ribs and secured to them by means of lashings of sinew, sealskin or whalebone (baleen). This also prevents the drying skin of the Khaiak from pressing unequally along the ribs of the boat.

The method pursued in the preparation of the skins for the rower of the Khaiak does not differ from that employed in bringing them into condition for the umiak; and may be briefly recited here by stating that

Hunting from a Kayak
Carving of a hunter approaching a beluga on his kayak by Kangiqsualujjuamiut artist, Johnny Mike Morgan, 2005. Caribou antler, leather, and soapstone. Scott Heyes Private Collection.

the skins are taken from the body of the seals, usually the Ringed Seal, or the Harp Seal for those from the adult Bearded Seal are too heavy, and placed in a pile where decomposition ensues sufficiently to cause the hair to be scraped from the surface by means of a sharp instrument of bone, ivory, or metal. The flesh side is freed from adherent muscle and fat. The skins are then soaked until they are thoroughly pliable and while wet may be stretched into almost any form.

Not every individual who may possess a Khaiak, is able to cut the skins of the mammals so as to form a cover for it. Certain persons among the Innuit attain distinction for their skill in a special kind of work; and, in each community there is a sort of master-workman who performs his work better than the others. His aid is enlisted and he superintends the cutting and fitting of the skin for the frame.

Kayak Model, Handheld Size

An Ungava Bay Inuit kayak with paddle collected by Turner at Ft. Chimo, probably in 1883. USNM E90232, Museum Support Center, NMNH. Photo courtesy of Stephen Loring.

The greatest economy must be exercised for such skins are not, at all times, to be had. Many things prevent the person from obtaining them; and among the greatest obstacle is the very fact that he has no skins with which to cover the vessel than will bear him on the water whence he procures the creatures that will furnish him with their skins for that purpose. When a sufficient number, usually four or five of goodly size, is obtained the frame is overhauled and if old the timbers are inspected and repairs made.

The skins are taken from the water, in which they have lain for several days, and placed upon the frame to be changed about until there will be the least cutting and waste. When they are satisfactorily adjusted a knife or a piece of charcoal marks the place where the cutting is to be done. The next step is to take stitches, similar to basting stitches, as that the edges of the skin may be approximated and the cover tried on the frame. If it fits snugly after making due allowance for the stretch of the skin it is taken off and is ready for the final sewing. If not fitted the parts are cut off, or, perhaps, another strip is added. The workman is usually skillful to not obviate that, unless it happens that from scarcity of good skins for the bottom he should necessarily have some pieces to be inserted which he would be careful to have come on the top or deck, for there the only water that might leak in falls from the waves and not from the presence of the weight of the craft on the water. The sole object of the skin cover is to prevent admission of the water on which it floats. To obviate this as few seams are made below the water-line as possible for the seam is the weakest part of the cover; hence, the utmost caution in sewing is necessary to prevent leakage. The fewer seams below the water-line is best and these must be placed so as to be as nearly transverse as circumstances will admit so that all the strain may come equally on each stitch. The seams of the sewing are identical with those of the umiak; a flat seam such as that used in sail making or in that of oiled garments to protect the sailor or fisherman from the water. The thread used is always of sinew, any creature furnishing a quality, but the reindeer sinew is the best.

As soon as the sewing has progressed far enough to permit the rear portions of the cover to be drawn, glove-finger fashion, over the stern of the Khaiak it is done and the top seam of the anterior part begun for if the entire cover were completed it could not be placed

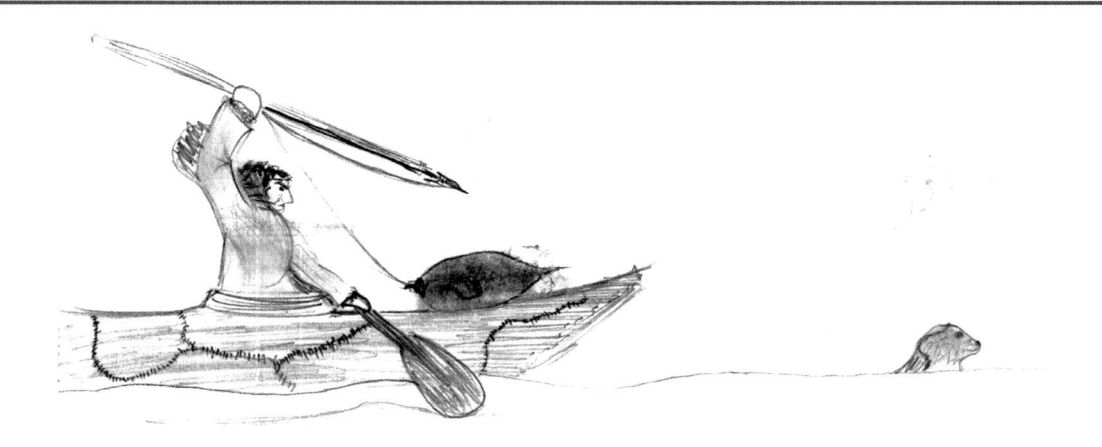

Sealskin Kayak
Illustration of a hunter harpooning a seal by Kangiqsualujjuamiu: hunter Tamasi Morgan, May 2005. Reduced from original 11″ × 17″. Lead pencil on bond paper. An avataq *(sealskin float) is attached to the rope of his harpoon so the seal will not sink when it has been struck. Benjamin Jararuse, Kangiqsualujjamiut elder, commented in 2005 that when harpooning a seal, "You should try and hit it on the lungs, because it will be less harmful. Once you hit it on the lungs it will die faster" (Heyes 2007). Scott Heyes Private Collection.*

over the frame, hence a portion must be left open to allow proper fitting. The top seam of the forward part is usually the last to be sewed, except perhaps a few stitches about the manhole skin. The two sides of the unsewed part are drawn together by means of loops of skin sewed to the inner side of the cover and through those loops a stout thong is inserted and when pulled draws the edges of the skin together and permits them to be sewed. The final sewing should be done with all possible haste for the skin on the Khaiak is rapidly drying and if not equally dried has a tendency to be askew when wet by contact with water.

The hoop is now placed in position and the skin of that part to come within it drawn up as previously stated. The Khaiak is now completed and only awaits a few days of seasoning in order that the owner may inspect it to discover any defects that will detract from its unseaworthiness for when built he is able to take his place among the best hunters and not be compelled to hunt only on land.

Before he can use the Khaiak he must furnish a means to propel it through the water. This is accomplished by means of a paddle, the constructing which differs somewhat in various localities as well as upon individual preference for a particular shape.

The paddle used by the Innuit south of the Hudson Strait is the double-bladed form; the single-blade has been so long discarded as to be forgotten. The length of the paddle is about eleven feet, having feathered blades of four feet in length and three inches wide. The middle portion, between the blades, is somewhat square with rounded corners where the hand grasps it near the blade, but flatter in the middle portion. Between the hand and the blade there is usually placed a plaited grummet of sealskin or rope to prevent the water, lifted by the blade, from flowing on the hand and wetting it; a matter of great consequence when the air is colder than the water.

In regard to the care of a Khaiak many things must be observed. It has been remarked that the seams, uniting the skins below the water line, are placed as nearly transversely as circumstances will allow. This was done because the Khaiak is subjected to straining which, in consequence of its length, exert great tension upon the entire frame and cover. Among these strains are the rise and fall of the waves, approach to land, and above all, to the handling of it when the skins are taut as the head of a bass drum, or when lax from long contact with water. When the seams are perfectly dry the sinew thread is liable to snap like cotton. When moist they tend to stretch and if inordinately strained they may not, when drying withdraw to their proper place and thus admit water when placed upon it. The pores and the interlacement of the fibres of the skin become opened or disarranged by the motion of the vessel or from the seeping of the water. These have to be diminished as much as possible and it is found that the application of oil to the exterior surface is adapted to counteract the distension of the fibres of the skin by the permeation of water through the pores.

The oil must be applied when the skin is dry and, for an obvious reason, never when wet. The animal oil leaves a kind of gum which dries and forms a pellicle upon the skin; and, until that wears off, in the course of a few days, the Khaiak will be waterproof. Frequent renewals of the oil are necessary but are often neglected; and, in the course of two years the cover becomes black so that by another half year the skin is so tender that it may be torn like a strong paper. From this remark it will be inferred that two, or at most three years is the duration of the skin cover of a Khaiak; less time is required if the owner takes no care of it for each time that a rent or break of seam occurs the affair is weakened that much.

The frame may last several years or when the first cover is worn out the owner may not have the skins to furnish a new cover and the frame is left, upon some point of land, a witness to the improvidence which is always the result of a nomadic life.

One peculiarity of the Innuit, south of the Hudson Strait, in regard to the care of their Khaiaks, differing from the western Innuit, including the Aleuts, is worth record here from the fact that they never remove the cover of their boats to scrape off the accumulations of filth and decay from the interior surface and to readjust them to the frame. In the Ungava District there is less necessity that it should be done from the fact that while the boats are used in each locality for the same general purposes as a vehicle it is not subjected to carrying such articles as would tend to increase the deposit of matter within the structure. That sand and gravel do collect within and getting between the frame and skin tend to pierce a hole in it does not appear to concern the Ungava Innuit from the fact, possibly, that the skins are thicker and deemed indestructible.

The general purposes of the Khaiak are that of a vehicle or conveyance to transport a single individual over the surface of the water to enable him to procure the necessities of life which he obtains mostly from the sea. In this vessel he seldom journeys far from the location of his temporary home and then only to procure the immediate wants. The umiak supplies the means of transportation of his worldly effects from camp-to-camp, while the Khaiak serves as a accessory to enable him to make short incursions.

The usual accompaniments of the Ungava Innuit in his Khaiak are those most essential to the procurement of various forms of life that afford him food and raiment from the water.

For the capture of the seals, white whales, walrus and an occasional white bear he must be provided with such implements, with their modifications, that will inflict a mortal wound. These implements are the spear or harpoon to be cast by means of the hand board wielded by the arm. This projectile is the only one now in use and thrown by that board. The several forms of this spear are noted in another connection.

The hand-spear or dagger with toggle-point is an accessory to the outfit and is used only when the Khaiaker is enabled to ride alongside of his quarry and give it a death stab.

The trident, usually of three iron-wire points, is used for transfixing fish (trout in the smallest stream of fresh water) or the young of the seals.

The raker is essential to reach objects which may have been pushed to such portion of the Khaiak as may not

Seal Intestines
Kangiqsualujjuamiut hunter Annie Kajuatsiak prepares the intestines of a bearded seal for consumption. In the past seal intestines were carefully dried and stitched together to make garments. Photo by Scott Heyes, 2010.

be attained by the arm. It is a flat piece of wood, having an iron hook at one end, two notches on the edge and a v-notch in the rear end. Its uses are also that of a boat-hook and assists when the Khaiaker approaches jagged ice or rocks in fending the vessel from damage.

The sealskin float to buoy a sinking object or to retard the progress of it through the water.

The reindeer spear is also an accompaniment but it has its uses at certain seasons only.

These implements just mentioned are the only ones left from the condition of the ancestors who had not the advantages accruing from the use of metals. At the present time the shotgun and rifle are a part of the outfit and the use of those weapons tend nearly to the entire rejection of the instruments employed in former times.

With a sealskin tent, an umiak to transport it, a Khaiak and a sledge and dog-team the present Innuit is a wealthy man but without either of these his life is one of a constant struggle to keep from the fate of starvation.

WATERPROOF GARMENT FROM SEAL INTESTINE, UNGAVA DIST. H.B. TERR. (1887A: 167-170)

The garment here referred to [74450/3501] is one which is rapidly going out of manufacture and use by the people of that district. Not that other protection has superseded it but that the adoption of civilized raiment in great part has caused those people to disregard the more useful of their native garments and of these the waterproof is certainly entitled to consideration, being light and sufficiently impervious to rain or the spray of waves.

The intestine of the large Square-flipper Seal (*E. barbatus*) is deemed best for the fashioning into waterproof garments; it is wider, requiring a less number of seams and nearly double the thickness of the intestines of the smaller phocids. The duct is removed from the cavity of the body and stripped of its contents. After several washings the intestine is turned inside out and thoroughly washed to free it from extraneous matter. It is now reversed and all fleshy or muscular fibres are scraped off with a dull-edged instrument. Great care being exercised to prevent tearing of the ligamentous fibres connecting the muscular walls of the intestine, these being snipped with scissors or knife whenever they become apparent. The delicate mucous membrane within requires careful removal and when done the semi translucent intestine is again washed and then distended with air. It has one end secured by a cord and upon blowing into the duct incisions may be discovered. A cord is tied around the place and the remainder is inflated and suspended to dry.

If the season of preparation is winter the exposure to the elements causes the membrane to become of a cream color and is greatly enhanced in value though not necessarily in durability; while if prepared in warm weather the heat tends to discolor the substance without impairing its quality. When dry the person slowly winds it into a roll something like a ribbon block.

When required for use the strip is split with a sharp instrument and thus affords a width double that of the flattened intestine. This is again carefully rolled into a compact fold and secured by a cord.

In the construction of a garment the principle idea is to have as few seams as possible and to secure this the strips are so arranged that the seams will extend up and down over such portions of the body of the wearer as they may fit. There are four strips which thus extend double the length of the garment; one over the shoulder down the back and in front; the second is within this while the third extends up the back and over the crown to form part of the hood while the fourth forms the back strips and the back of the hood. The remainder of the strips are only half the length of the body, the length of the neck and over the head in front or else the sleeve lengths. The principal idea being, as remarked before, to have the seams extend perpendicularly so that the water will not run through the stitching.

The edges of the pieces to be sewed together are placed evenly and then turned over, on the turned over edges of the strips is laid a narrow strip of the membrane and stitched with a sailmaker's stitch, usually termed a running stitch. The thread used being either of sinew or else a very fine strip of tanned sealskin scraped very thin but being of a black color tends to ornament the seams. In lieu of sinew a very narrow strip of the membrane itself is prepared and rolled so as to form a fine cord, somewhat resembling the twine made from paper. The thread must be coarser than the needle which perforates the membrane so as to fill the holes made by the needle.

When there is the most wear and strain the garment may be reinforced by strips of thin sealskin sewed to the garment.

When dry the waterproof is easily torn and requires careful management in adjusting it to the body but when wet it is soft and flabby adhering to the inner garments so that were it not ample in size for the wearer each movement of the arms or head would rend it beyond repair. The hood serves as a protection from the rain and wet flowing down the back of the person.

There is one serious disadvantage in the wear of these garments at sea for if the lower edge of the skirt be tied over the hole in the Khaiak the wind is certain to puff it up like an inflated bladder, seriously impeding the movement of the arms. This is especially inconvenient when the drawstring of the hood is drawn tightly over the face.

SEALSKIN MITTENS, KOKSOAK INNUIT. UNGAVA DISTRICT, HUDSON'S BAY TERRITORY (1887A: 165-166)

When the weather is too cold for the hands to be exposed and yet the water not frozen so that the Khaiak may traverse the water, the hand must be protected from the splash and rain. Mittens with hair on them would soon become a sodden mass chilling the hand beyond usefulness.

In order to protect the hand from the wet and especially the drip of the water from the paddle the Innuit wears a pair of mittens prepared from tanned sealskin. They are shaped like gants, extending well along the forearm.

The pattern of cutting out the three pieces which, when sewed together, form the complete cover, is somewhat peculiar.

The palm and underarm piece constitutes the first piece. The second piece is that for the inside of the thumb, anterior portion of the palm and part of the undercover for the fingers. The third piece embraces the outer cover for arm, back of hand over and under the tips of the fingers where the surplus skin is taken up with creases like the toe of a sealskin boot.

When worn under such circumstances as those denoted the mittens become soft and as they are nearly impervious to the continued wet they form an admirable protection to the hand.

Scaring of the Seals
An Inuit story, Turner (1887a: 280-281)

While some children were playing upon the top of a high bluff whose brow overhung the sea the elders were watching the younger ones to prevent them falling below.

The men of the village were watching the ice-covered sea in hopes that seals would appear in the rift but lately made. Directly the seals appeared but the men could not at that time get to them as the ice was too rough and broken.

The children were unaware of the presence of the seals and in their sport shouted and romped to keep themselves warm for the keen winds chilled their bare feet and legs. The villagers feared the noise made by the children would frighten the seals and cause them to disappear. The children made the more sounds as the plays were progressing and the seals took alarm and dove away. One of the men was very angry and muttered to another, "I wish the cliff would topple over and bury those noisy children for fearing the seals." In a moment the cliff overwhelmed and the children were precipitated among the falling white rocks. Here they were transformed into birds with red feet and yet dwell among the sea-washed boulders at the bases of the cliffs bordering the sea. The birds are the Guillemots or Sea Pigeons, *Cepphus grille*. The Innuit name is *Pi shu lák*.

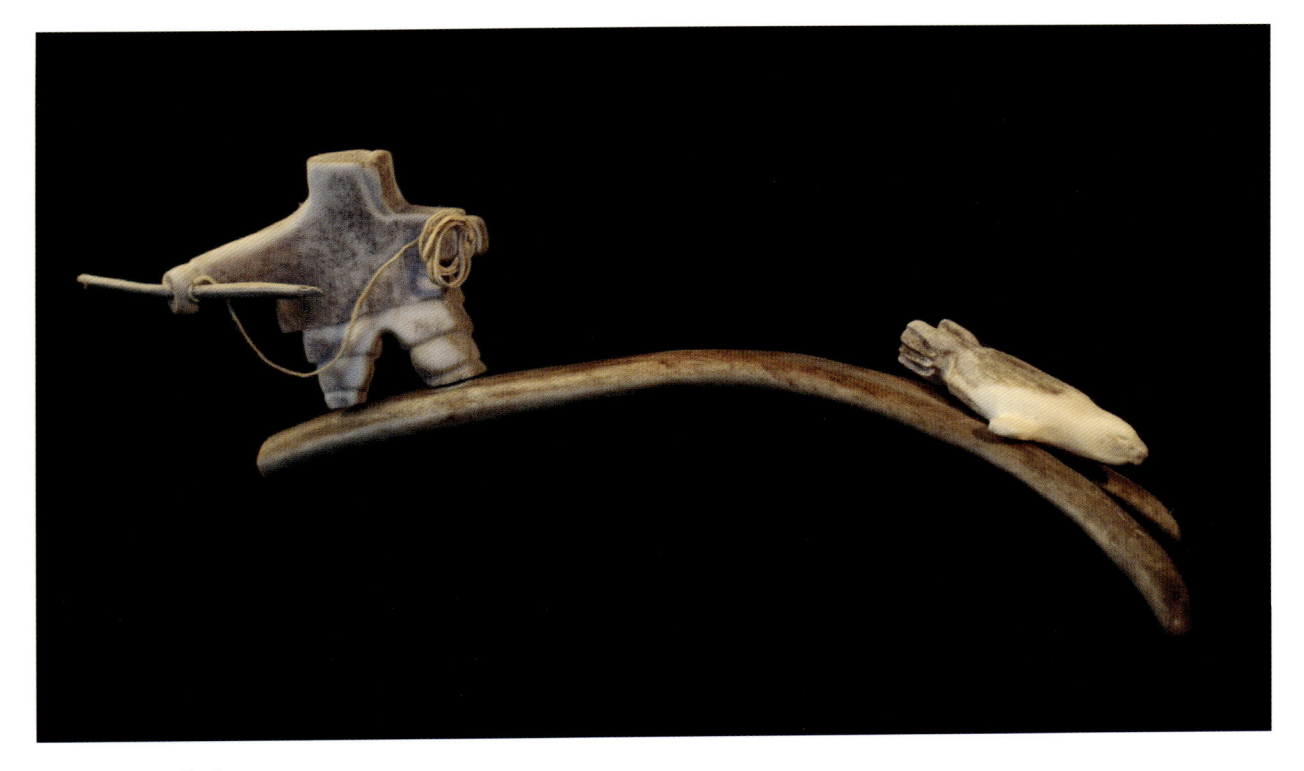

Harpooning a Seal
Carving of a hunter striking a seal by Kangiqsualujjuamiut artist Johnny Mike Morgan, 2005. Caribou antler and leather. Scott Heyes Private Collection.

Chapter Two: Mammals of Ungava and Labrador

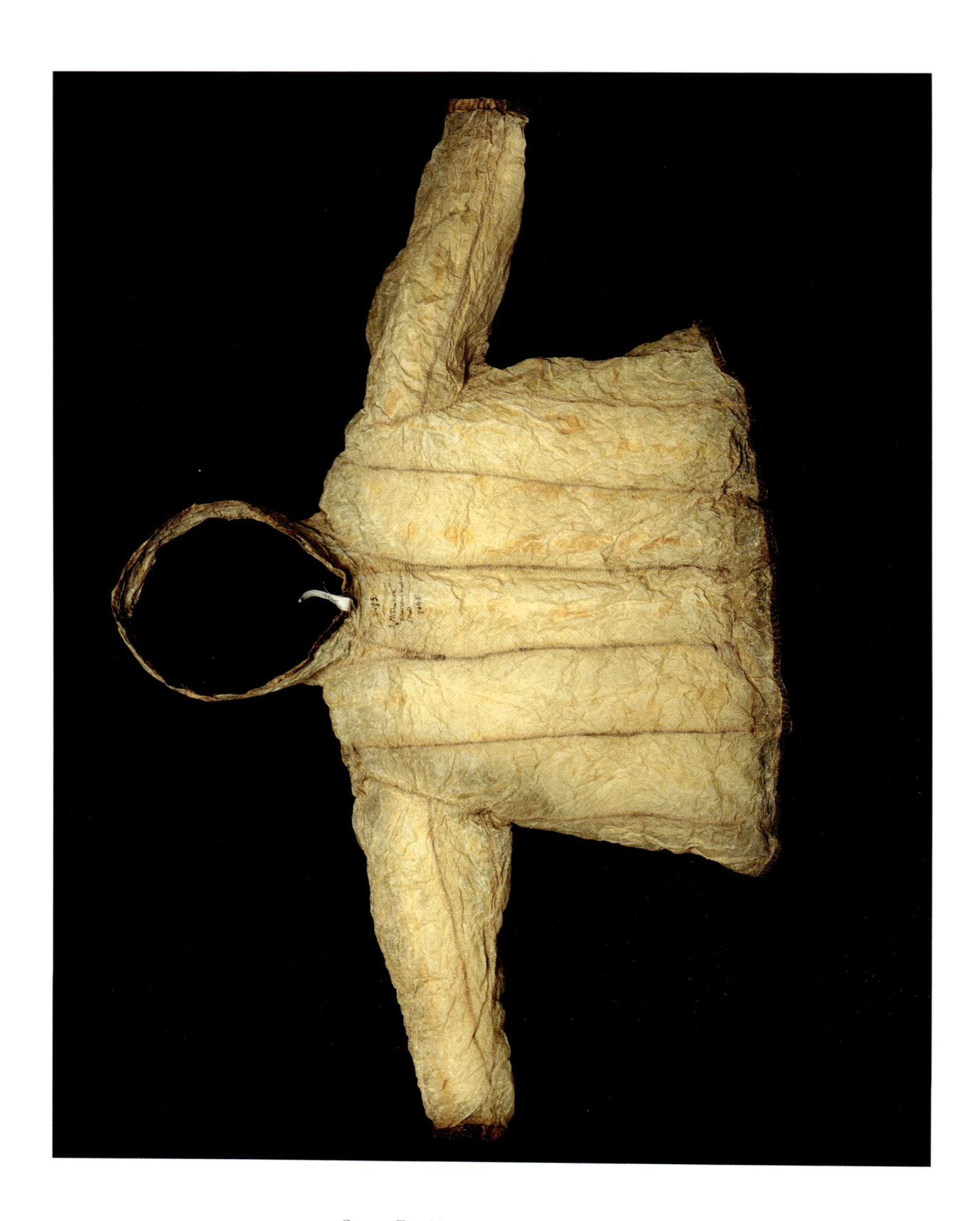

Seal Intestine Parka

Seal intestine clothing collected by Turner at Ft. Chimo, probably in 1883. Turner detailed how these garments were made by the Inuit from the intestines of bearded seals: "The duct is removed from the cavity of the body and stripped of its contents. After several washings the intestine is turned inside out and thoroughly washed to free it from extraneous matter. It is now reversed and all fleshy or muscular fibres are scraped off with a dull-edged instrument. Great care being exercised to prevent tearing of the ligamentous fibres connecting the muscular walls of the intestine, these being snipped with scissors or knife whenever they become apparent. The delicate mucous membrane within requires careful removal and when done the semi translucent intestine is again washed and then distended with air. It has one end secured by a cord and upon blowing into the duct incisions may be discovered. A cord is tied around the place and the remainder is inflated and suspended to dry." USNM E74451/3493, Museum Support Center, NMNH. Photo by Donald Hurlbert, 2013.

The Missing Seals
An Inuit story, Tivi Etok (Heyes 2007)

This story is from the Labrador coast, around Hebron. It is a story about spirits taking things from the Inuit. The spirits go hunting like Inuit but we never see them. The Inuit from Hebron have the same fish and animals as we have here in Kangiqsualujjuaq. In Hebron there are many ringed seals. They were hunted by Inuit in Hebron by setting seal-nets rather than shooting them, as we do here. The hunters were setting their seal nets to get seal meat for the winter. When the Inuit hunted with their nets they noticed that they were not getting any seals in their nets. The hunters suspected that bad spirits had been taking seals from the nets. The sea ice was starting to freeze which made it difficult for the Inuit to get close to the seals using any other method apart from the nets.

The hunters decided to sleep beside their nets one night to see why seals were not being caught in the mesh. Spirits usually take animals during the night. Hiding behind a rock, the hunters saw some skeletons paddling a kayak towards the net. The bad spirits were skeletons; they had no eyes or flesh. As the skeletons came close to the Inuit, the hunter fired their guns at the spirits. The hunters let off many rounds. But when the hunters shot at the skeletons nothing happened because the spirits were already dead. Since the spirits could not be

The Missing Seals
An illustration of a story about the "missing seals" by Kangiqsualujjuamiut elder Tivi Etok, 9 May 2005. Reduced from original, 11" × 17". Lead pencil on bond paper. Scott Heyes Private Collection.

killed the hunters decided to go home. The next morning the hunters went back to their nets. To the hunter's surprise, the hunters saw that there were lots of ringed seals in the nets. This is because the Inuit won against the bad spirits; they had been scared away by the hunters. When the hunters were taking seals out the nets there were still more seals being caught. The bad spirits had been defeated. Bad spirits are not harmful to humans but they take what we need to survive.

If we expect to hunt animals in a certain place but discover that the animals are just not there, it is most likely because the bad spirits have already been hunting there. The bad spirits are capable of hunting any animal, any time, anywhere. Bad spirits cannot be seen. But if you hide and look hard you might catch a glimpse of one. If you hear a gunshot on the Labrador coast but do not see anyone it is the bad spirits hunting. There are bad spirits that inhabit the Nachvak Fiord area. These spirits are stuck in between the living and afterlife realms. They want to go to a good place but the dark place is preventing them from entering the good realm.

First Seal Hunt
An Inuit story, Tivi Etok (Heyes 2007)

The person in the picture is me on a kayak. The drawing is about me growing up and learning how to hunt seals. It is the end of April. I practiced on young seals, because I was told that seals that had just been weaned by their mothers would not flee if I made incorrect movements. Kivakitaq is the name of the seal resting on the ice. This word means an animal that has risen up on a piece of ice just big enough for it. Older seals will move into the water straight away if they see any movement. I was shaking and sweating when I came across the seal because it was my first seal hunt. I learned how to hunt seals by following and watching my father. When the water is wavy, seals are not present on the ice. No seal will be out of the water when it is raining, but this seems strange to me because they always live in the water. You will see more seals when it is calm and areas of open water abound. When it is snowing, seals sit and bask on the ice; they are really lazy when it snows, but they really love these conditions. Seals love fresh, soft snow.

Where I grew up in Labrador the shore was very sandy and smooth. When the snow fell and the sand turned white, the seals would appear on the ice. Seals would venture onto the shore at low tide. Ice is very useful for seals and walrus. When the wind is blowing towards the land, young seals sleep on the ice.

First Seal Hunt
Illustration of Kangiqsualujjuamiut elder Tivi Etok's first seal hunt. Drawn on 11 April 2005. Reduced from original 11" × 17". Lead pencil on bond paper. Scott Heyes Private Collection.

I was told that seals have tents. They have tents near the shallow water where lots of ice is present. They feed near their homes. The seals then move in June and July to fatten up. Right now (in April) the male seals are making long, drawn-out whistling sounds. When they make the noise under water, the bubbles even make sounds. They are not making the sounds for any particular reason. After the male seal has made a sound and hops onto the ice he becomes lazy and doesn't look around. This is a good time to hunt him. When a seal bubble rises from beneath and hits the kayak that is how we realize a seal is nearby. When a seal learns that something is on the water they jump up and down and become cautious. I was told that if I want to hunt big seals I should not put my kayak over the bubbles because the seal would hear the bubbles hit the kayak and would then swim away. The bubbles move with the tides, so knowing this, I would know what side to position my kayak so that bubbles would not hit it. I could get very close to a seal if I avoided the bubbles; so close that the seal would not even see me. You had to be very diligent when hunting seals in the old days when they were the only food around.

Around this time of year I put my head down to the ice and listen as to whether seals are around. You can measure the distance of the seal from where you are depending on the strength and frequency of the sound reverberating against the ice. Animals that inhabit the water make many different sounds.

Mitilik Creatures
An Inuit story, Tivi Etok (Heyes 2007)

This story occurred around the Korac River area (Tivi's personal encounter with a Mitilik). I once saw a Mitilik; it looked like a seal. When I saw my father I told him that I went hunting and I saw a seal that looked like it had an arm. My father told me that what I saw was not a seal, it was a Mitilik. As you approach a Mitilik it can transform into any animal. If you wish for a Mitilik to be another animal just before you spear it such as a seal, then when you take it out from the ice it will turn into that animal. This Mitilik fooled me; I thought that it was a seal – even when I got close to it. After news spread around that I had seen a Mitilik people started to say that if I had got any closer it would have attacked me. This is how I found out that Mitiliks live in the Korac River. The Mitilik was as tall as me, but it appeared like a seal. The Mitiliks only live in the sea and go on the sea ice – they do not go on the land. If I was good at shooting I would have been able to say that I had killed a Mitilik. When I shot at it and missed I noticed that it dived into the water like a human would – it had the physique of a human. I have been told that Mitiliks exist so I believe that it is true. After I told people that I had seen a Mitilik, a story about Mitiliks was then told to me. Three men went hunting seals. They walked the sea-ice looking for the breathing holes. Since they were spending much time on the sea-ice, they decided to make themselves an igloo to sleep in. They camped beside the breathing hole, hoping that a seal would appear. The men were spread out near the campsite looking for breathing holes. One man arrived at the camp before the rest of the men. At night they all came together and talked about their hunting experiences of the day. One man said that when he was walking alongside the edge of the land fast ice a Mitilik came up from the sea and started attacking him. The Mitilik stabbed the man with a knife, but fortunately the hunter was quick enough to kill the Mitilik first. Just before the Mitilik died it said to the hunter, "all my relatives are going to hunt you down." The hunter was aware that the Mitilik would be vicious. The hunter told the others in the igloo that as a result of killing a Mitilik, the relatives of the Mitilik might now come and get us. The men knew that if a Mitilik was killed that the relatives of the Mitilik would appear on the sea ice and start to kill any hunters they came across. Once the men had taken in the news that a Mitilik had been killed they then left the igloo and started running towards the land even though they had not yet sighted any Mitiliks. The hunters knew that the Mitilik could seek revenge at any time. The Mitilik saw the men fleeing so they started running after them. The men were very tired of running. The Mitilik gained ground quickly and got very close to the hunters. Luckily, the hunters stepped on land just in time before the Mitilik caught up to them. The hunters were now safe, because the Mitilik do not venture onto the land. Mitilik were covered in goose down and they live only on the sea and sometimes lay on the sea ice.

Even if I did shoot the Mitilik all those years ago, I would not have eaten it because it looked like a human. There was a time when Inuit starved. I have heard that Inuit did eat other Inuit and their dogs during this period in order to survive. People may think that the stories I am telling you are not true, but I have been told these stories so it must be true. In the old days, I knew that there were creatures like Mitilik but nowadays I don't hear much about these types of creatures.

Mitilik Creatures
Illustration of the story about the Mitilik creatures that can turn into seals. The picture shows three hunters being chased by the Mitilik creatures. By Kangiqsualujjuamiut elder Tivi Etok, 10 May 2005. Reduced from original 11″ × 17″. Lead pencil on bond paper. Scott Heyes Private Collection.

Illustration of a Seal Giving Birth near Nachvak Fiord, Labrador
By Kangiqsualujjuamiut elder Sarah Pasha Annanack, 19 April 2005. Annanck described the drawing in a 2005 interview: "I was thinking of our camp in Nachvak Fiord, Labrador when I drew this picture. Close to our camp there are many cracks in the sea-ice. Seals are always popping up and down through the cracks in this location. This picture shows a seal giving birth close to a hole. The markings around the holes are blood stains which were created when the mother seal gave birth. I would have drawn this in red, but I didn't have a red pen. The seal pups pop in and out of the water very often. Seals give birth close to the hole so they can easily escape if predators approach. In springtime, the seal holes become much larger in diameter because of the melting ice. Seals keep the hole open by constantly moving in and out of the opening; if they didn't, the hole would freeze over. More than one seal uses a hole. The seal in the picture are the same type of seals except that their skins and colors are of different shades." (Heyes 2007) Reduced from original 11" × 17". Lead pencil on bond paper. Scott Heyes Private Collection.

RECOLLECTIONS ON INUIT USE OF RINGED SEAL
Tivi Etok (Heyes 2010)

The ringed seal was used for everything. The skin was used to make tents, pants, coats, mitts and kamiks. The outside layer was used for its waterproof properties. It could be used to make an *avataq*, a sealskin float. The *avataq* was made very carefully; it could not be punctured. We used to save the fat and meat of the ringed seal for winter by putting it in a bag made from sealskin. This bag had a different purpose to an *avataq*. We were not allowed to put fat inside an *avataq*. This was used as a balloon when harpooning a whale. When we shot a whale the *avataq* would follow; we could then see where the whale was heading. This was the best hunting tool that we had. It was very useful. When the whale was down for a few minutes, it would come back up and we could see where the whale was. The *avataq* was also used for harpooning a seal and other marine animals. If an *avataq* had too much fat on it, it wouldn't float. The male ringed seal is called *tiggaq*. When the female ringed seal is mating the flesh becomes very strong and smells. This smell is not as apparent in male ringed seals.

Netting and Cutting Implements

These implements collected by Turner at Ft. Chimo were made by the Inuit and Innu for purposes such as making seal nets, garments, canoe construction, snowshoes, and cuttting seal meat. Awls (top to bottom): USNM E89971; E89973; E89972. Turner reported that these awls were used for "making holes in the bark of the birch canoe where the spruce roots are used in stitching the top rail on the edge of the bark and for piercing holes in the snow-shoe frames for the insertion of the thongs used for netting" (Turner 1887b: 541). Ulu (woman's curved knifes, top to bottom): USNM E89960; E90258; E89961. Needles (top to bottom): E90166 (two pieces); E90165. Museum Support Center, NMNH. Photo by Angela Frost, 2011.

CARNIVORES

ᐅᒃᔪᒃ / ᐅᔾᔪᒃ

BEARDED SEAL OR SQUARE-FLIPPER SEAL
utjuk / ujjuk

Erignathus barbatus, Gill.

[1439]

ORDER NAME USED BY TURNER	*Carnivora*
CURRENT ORDER NAME	*Carnivora*
FAMILY NAME USED BY TURNER	*Phocidae*
CURRENT FAMILY NAME	*Phocidae*
SCIENTIFIC NAME USED BY TURNER	*Erignathus barbatus*, Gill.
CURRENT SCIENTIFIC NAME	*Erignathus barbatus* (Erxleben)
SYLLABARY	ᐅᒃᔪᒃ ; ᐅᔾᔪᒃ
MODERN NUNAVIMMIUTITUT TERM	utjuk ; ujjuk (cf. Schneider p. 468 - utjuk ; ujjuk = bearded seal [*Erignathus barbatus*]
ESKIMO TERM BY TURNER	uǵ zuk
DEFINITION OF ESKIMO TERM BY TURNER	Bearded Seal [p. 1448]; Square flipper seal (*Erignathus barbatus*) [p. 1439; 2836].
INNU TERM BY TURNER	a tikhwh [p. 708]

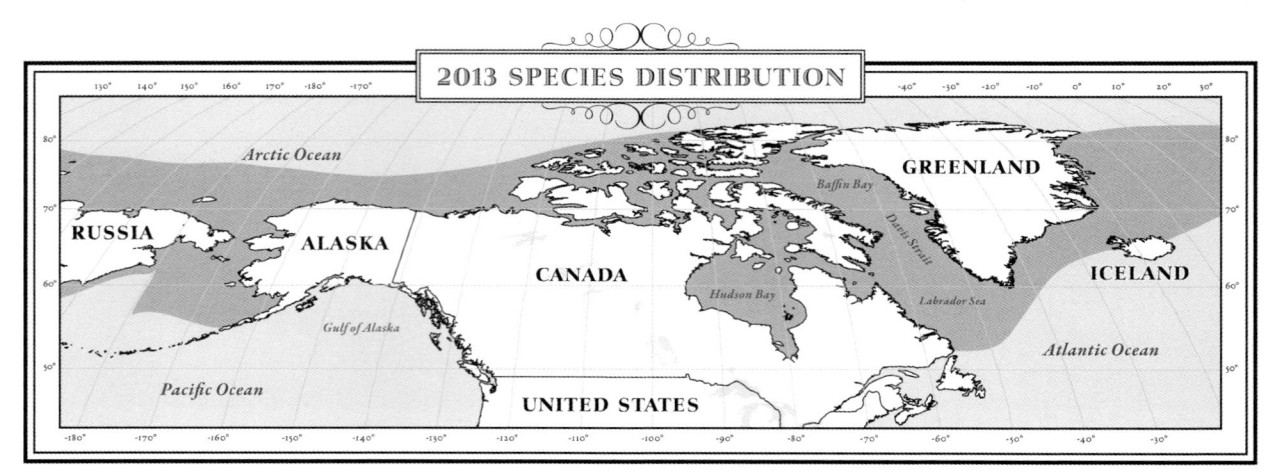

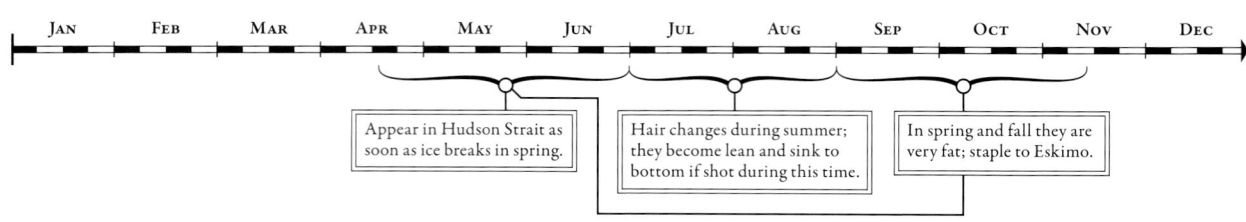

2013 SPECIES DISTRIBUTION

| Jan | Feb | Mar | Apr | May | Jun | Jul | Aug | Sep | Oct | Nov | Dec |

- Appear in Hudson Strait as soon as ice breaks in spring.
- Hair changes during summer; they become lean and sink to bottom if shot during this time.
- In spring and fall they are very fat; staple to Eskimo.

CHAPTER TWO: MAMMALS OF UNGAVA AND LABRADOR

Descriptions by Turner

The Bearded Seal is probably the most common as well as the largest species occurring in the waters of Hudson Strait. It is also plentiful along the Labrador coast.

To the Eskimo this is the most important of all the seals. Its hide forms the cover for the umiak, kaiak, and tent. Lines for harness to fasten the dogs to the sled and soles for his waterboots. The skins of other species have their uses but none are so highly valued as that of the *Uǵzuk*. The texture of the skin appears to be of different character than that of the other species and the strength of it is not to be doubted when a thong less than five-eighths of an inch wide and scarcely less in thickness will withstand the continuous strain of a team of a dozen or more lusty dogs pulling a sled, weighing with its load nearly a thousand pounds and that continued for several days at a time.

[1440] In the spring this species delights to bask in the increasing warmth of the sun slanting upon the immense fields of ice whirling here and there by the tremendous power exerted by the terrible tide forces confined within the narrow passage of water leading from Hudson Bay to the Atlantic.

The *Uǵzuk* appears to delight in being alone. The males are disagreeable and engage in terrible combat when they encroach upon the basking ground of each other. Their foreflippers are used in a manner similar to the wings of a goose. The teeth are used only to draw the opponent nearer. A wounded "aquaflipper" is not a pleasant animal to deal with if the captor be in a Kaiak.

Their strength is wonderful, enabling them to swim with surprising facility. An object may be sighted, on the still

"In the spring and fall they are very fat and contribute with their flesh and oil a goodly portion of food for the Eskimo."

water, which on nearer approach appears to be an immense bubble, as the wet hair and rounded [1441] head shining in the light give it an appearance comparable to that object. The object may disappear with scarce a ripple to disturb the calm surface. If, however, the animal becomes alarmed a sudden plunge plows the water with a splash, a kick almost turning it end over end causes it to disappear and not be seen again for several hundred yards. Their actions in the water much resemble those of the Sea Lions, *Eumetopias stelleri,* of the Pacific and is the only one of the Hair Seals which has that habit and this alone characterizing the species at a great distance. They appear in Hudson Strait as soon as the ice breaks in the spring. A few remain in the waters all the year round but keep well off the land. The Eskimo hunt them along the shore or on the inner islets the sooner freed from the mass of ice in spring. In their Kaiaks, they search the ice floes and in certain [1442] localities obtain more than they can take care of.

During the summer they change their hair and during this period they become very lean and sink the instant they are shot. In the spring and fall they are very fat and contribute with their flesh and oil a goodly portion of food for the Eskimo. The quantity of oil is extremely variable but has better keeping quality than that of the other species. The fat and oil is usually put inside of skins of this or other species of this genus. The bags are tight and loose but little of it after the skin has become well soaked as it dries and forms a stiffish gum upon the exterior which adheres so closely as to prevent farther leak. The size of the adult animal is from eight to ten feet for the male and six to eight feet for the length of the female. The color of the hair is less variable than in other species. In this it is light to dark tawny [1443] interspersed with grayish to black hairs, rarely sufficient to form spots, but giving a clouded appearance.

The head of an aged individual is peculiar, having much of the bulldog expression. The eyes are large and prominent, the muzzle heavy. The hind flippers are long and being nearly evenly edged give rise to one of the many local names by which this species is known.

I could not determine definitely the breeding habits of this species.

I have seen this species of seal as far as sixty miles up the Koksoak River. It is not rare in the river opposite Ft. Chimo.

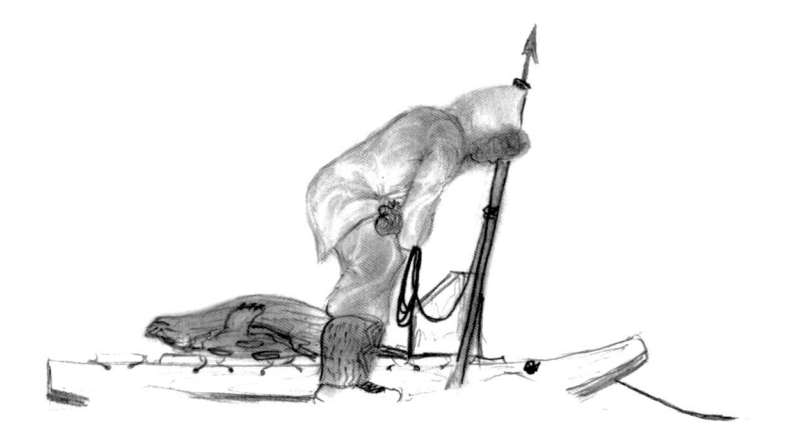

Seal Hunting
Illustration of a hunter waiting patiently beside a seal breathing hole by Kangiqsualujjuamiut hunter Tamasi Morgan, May 2005. The hunter was fortunate on this occasion to have already killed a seal near another breathing hole. Reduced from original 11" × 17". Lead pencil on bond paper. Scott Heyes Private Collection.

Bearded Seal Ethnography
Turner

AN INNUIT TENT. INNUIT SOUTH OF HUDSON STRAIT AND ALSO ALONG THE LABRADOR COAST BORDERING THE ATLANTIC.
(1887A: 218-223)

The summer dwelling of the Innuit here included varies greatly according to the means of the owner and also upon the necessity of the occasion requiring such shelter. Those Innuit of pure blood, yet uncontaminated by the white people dwelling on the Labrador coast, employ the same primitive shelters as were in vogue by their recent ancestors. Some of the more indigent of these Innuit have not the means to furnish a shelter equal to that of their more energetic neighbors and have recourse to other material to furnish the protection needed.

The ancestral type of tent is yet retained in the more primitive localities; and, as it is the one to be considered, before the modifications of that form are noted, a description, in lieu of a model, will be given, in order to arrive at a proper understanding of the typical túpik of those Innuit.

It is scarcely necessary, when treating of the Innuit, to state that the material employed to protect the occupants within is prepared from the skins of seals. The fact of these people being Innuit is sufficient to determine that point.

The skins are generally those of the Bearded Seal, *E. barbatus,* or the pelts of the large Ringed Seal, *P. foetida,* or the Harp Seal, *P. groenlandica.* The heavier skins of these seals are treated to a process of decomposition advancing far enough to loosen the hair in its follicle and permit it to be removed by the act of scraping with an edged implement of stone, bone or metal. The skins are then sewed side to side to another until a length necessary inclose an area of sufficient size to accommodate the number of persons to be sheltered within. A second piece is now made somewhat shorter to be arranged above the first or lower length. A third piece is sometimes made for a special purpose that will be better understood when the poles are arranged for the reception of the skins.

As remarked before the size of the tent depends upon the number of persons to be sheltered by it. There are, generally speaking, the head of the community,

Preparing Sealskins
Photograph by Turner in 1882 of Innu curing skins on the banks of the Koksoak River. Photoprint 8" × 5" mounted on 8" × 5". Source: SPC E Canada Naskapi NM No ACC # Cat 175484 00294100, National Anthropological Archives, Smithsonian Institution.

his father, mother, brothers and their families, sisters and their families beside many dependents, who act in the capacity of servants.

The community may number from a single family of four or five persons to as many as thirty individuals and the tent vary in length from eight or ten feet to thirty feet in length and a breadth of six to ten feet. The supports are always of wood and vary in length. They are arranged as follows, beginning at the rear; three, two, two, two or three. The back three are two shorter and one longer poles; the longer is thrust behind while the shorter are the side slopes, the next two pairs are of the same length as the anterior pair of the posterior three. The front pair, or three, are like the rear ones except they are slightly longer so as to give a gentle slope from front to rear. One or two ridge poles are also added.

In order to erect the poles and place them in position the assistance of two persons is required, one of whom binds the three rear poles with stout sealskin thongs near the upper end of the poles while the other binds the side pair, the three are put in position like a tripod with one long leg (the rear pole). The pair is now lifted and a ridge pole laid from the three to the anterior pair. The remainders of the poles are erected as the first set.

The skin is now placed as to inclose the poles; the long length at the bottom. The second shorter, width is lifted and placed around the poles so that the lower edge will overlap the lower width. The edges of the upper width do not meet at the top and a cleft is left along the ridge-pole. To prevent the snow and rain from beating in a third length of skin is thrown along those edges so as to exclude the water falling within the tent.

It will be seen that the fully equipped tent has six sides, the two ends having two each and the two long sides. If the tent be small the front ends with a pair of poles. The end is thus truncate or cut off. In either case the overlapping skin serves as a place of ingress and exit.

Sealskin Tent (1882)
Photo by Turner of a sealskin tent on a beach near Ft. Chimo. Photoprint 8" × 5" mounted on 8" × 5". SPC Arctic Eskimo Labrador NM No ACC # Cat 175484 01450200, National Anthropological Archives, Smithsonian Institution.

This structure is intended to shelter the Innuit from the melting of the snow until the snow of the next winter is of sufficient firmness to withstand the crushing power of the blocks cut out of that material to form the snow-hut for a winter habitation. The number of poles and the weight of the skins, varying from six to twelve, render the material of the tent very cumbersome and difficult of transportation. In order that it shall occupy but little space the length of skin are rolled up in a peculiar manner. The width is folded so that the outer edges will meet in the center of the piece and if the doubling thus produced is too wide a second folding of one part, two-ply, is laid upon the other and now gives four thicknesses of skin. From each end the skin is rolled so that the two bundles will meet at about the center of the length of skins. If the tent is large the weight of the piece may be as much as one hundred and fifty (150) pounds. A lusty Innuit lies on his back, having his head between the bundles and with the assistance of another rises with the burthen on his head and shoulders, taking it to the umiak, which is to carry it. The second and third pieces are served in the same way.

In order that the umiak may be properly loaded with the least strain upon the different parts the poles of the tent are placed upon the bottom frame and the skin of the tent unrolled and spread upon them thus keeping the water, which may seep between the stitching of the umiak and collecting below, from touching

Sealskin Tent (1896)
A group of Inuit at Great Whale River (Poste-de-la-Baleine), now Kuujjuarapik, 1896. Canadian Museum of Civilization, Albert P. Low, 1896.

the goods within that vessel. The internal arrangement of the occupants and properties depends also upon certain social customs of those people. The head man of the tent occupies the space under the three rear poles of the tent. Those next in importance occupy the space to the right or left and those dependent are placed near the entrance. The fire is made without the tent if the weather be warm, but if chilly the place of making it is near the center or toward the front of the tent. Here all the cooking or the work goes on. The bedding, consisting of skins, blankets or merely the bare earth is left where each person sleeps.

THE USE OF SEALS IN
UMIAK CONSTRUCTION (1887a: 88-102)

When an individual contemplates the construction of an umiak he has a task of great magnitude before him that will be readily understood as this article proceeds. The forest is searched for a tree whose natural curve will afford a piece for the bow or stern and one also nearly similar for the stern.

The shear or slope of the ends of the umiak are nearly abrupt, quite as much so as the side or rib-pieces. When such a trunk is found it is roughly hewed out and left until a convenient opportunity occurs to transport it be sledge to the winter camping place of the person. Other smaller stems having a slight curve and an elbow are sought until such number is found that will be sufficient to form the ribs of the boat. Three pieces for the bottom, or one for the keel and two for the side pieces. The rail is either a single piece or two pieces placed as herein after described.

The length of the umiak may be from fourteen to twenty-two feet in length; a breadth of four and a half feet at the widest part of the bottom and a width of five to seven feet across the widest part between the rails.

The keel is usually four inches square, or four inches thick and five inches wide, being the same size as the stern and stern posts; and may be simply continuations of those timbers joined by long laps at or near the center. This is the first piece laid; next come the bottom rails, which are square pieces of about three inches, the inner upper corner of this piece being cut out for an inch and a half deep; while the lower outer corner is rounded to prevent wearing of the rails.

The four ends of the bottom rails are now fastened, two to the stern post and two to the stem so that when spread the rails will be four to four and a half feet apart in the center. Cross pieces of varying length, from center aft or forward, are laid so that the shoulder, cut at each end of the cross pieces, will rest in the cut on the bottom side-rails and having their center on or across the top of the keel. This forms the bottom framework, and, on the strength of this portion depends the value of the vessel. The ribs are slightly curved and the lower end is cut so as to rest in the cut of the bottom siderails. The top or upper end of the ribs has a similar shoulder but on the outer edge of the rib, or the upper end of the rib may be cut out concave as there are several methods of preparing that end so as to receive the toprail or where two flat pieces are lashed together to form that rail the end of the rib is shaped to be placed between them.

Experience has demonstrated that the lashings of sealskin thongs employed to hold the various parts and joints of the frame of a vessel constructed to buffet the choppy seas, over which it must traverse, or to be subjected to the rough hauling on or from the shore at night or morning is far superior to fastenings of a rigid material such as a metal. The vessel is not furnished with the power of propulsion or the strength of frame to withstand the shock of cutting a high wave but must have as little stiffness as is consistent with firmness and safety. Holes are pierced in the various parts where necessary and a stout thong of sealskin, moist and pliable, is thrust through, over and around the parts to be brought together until they are secure from movement apart. No nails are employed even at the present day when such things are easily obtainable from the traders, excepting, perhaps, to secure the ends of the bottom side rails to the stem and stern posts. The frame is now ready to be covered with sealskin split walrus skin or the hide from a white whale. As the umiak has to withstand a great amount of rough usage the thicker skins of the large bearded seal, *E. barbatus*, are preferred.

Umiaq Model
An Ungava Bay umiaq (or sealskin boat) collected by Turner at Ft. Chimo, probably in 1883. USNM E90111, Museum Support Center, NMNH. Photo courtesy of Stephen Loring.

The difficulty of procuring these creatures causes the person, who desires to construct a umiak, a delay of a year or two before he may be able to secure six to twelve of the skins of that marine mammal, so wary in character that the most persistent efforts and exercise of caution will enable him to procure more than four or five of them in the course of a year where the seal may be so rarely seen that even a longer period may elapse; and, even then there are so many additional uses to which the skin of this most valuable seal may be put that one by one the hides are cut up for boot soles or harness for the dogs or the mice may gnaw holes in them and render them worthless. All these conditions must be taken into consideration that may necessitate the frame to lie useless for two or three years; and in the meantime deaths or other changes in the family status occur and eventually postpone the finishing until a young son has arrived at such age that he acts in his father's stead.

The skins are prepared as follows:

The flesh side is scraped until all fleshy particles and ligaments are removed; the hair is so disposed of by the process described elsewhere. The skins are not subjected to any process of tanning for use as a covering of a skin boat.

When a sufficient number of these skins have been accumulated they are placed in a pool of fresh water and allowed to macerate for several days or until they become perfectly limp. By the absorption of water the skin has become nearly double its original thickness and somewhat smaller in the area. This must be taken into consideration but while in this condition the pelts may be stretched very much so that the person who fits them for the frame must make due allowance for that property.

The edges are now trimmed so that the next skin will fit with the least waste and seam. The skins are arranged as that the heads of the skins will be directed toward the bow or else upward if placed on the sides; the flesh side of the skin being always within.

When the person, who cuts the skins for the cover, concludes that the length and width will suffice the different pieces are sewed together. The seams are such as may be termed a lap seam, that is by the edge of one skin lapping beyond the edge of the other so far that it may be turned over the edge and stitched to the other skin, the thread not appearing on the outer side. This method of sewing relieves the strain on the threads of the principal seam for when the skin becomes dry the tension is very great upon the stitches. The skin is now placed upon the frame and tested; the front and rear portions are now brought so that the edges will fit the opposite side and these are then sewed. When the sewing is completed the skin resembles in form a huge deep dish made of skin. Slits are then made several inches apart, in the edge circumference of the cover and again placed on the frame. For convenience of work the frame is now turned keel down.

Stout thongs of sealskin have been soaked in water until they are incapable of absorbing another drop of water. This thong is now passed through one of the incisions in the skin near the center of the toprail and one end of it is securely tied to the rib on the inner side and the thong drawn so as to bring the edge of the skin over and within the top rail. Passing through another slit in the cover the skin is drawn tight over the frame. Where the umiak is of large size one or two additional side rails may be secured to the outer edges of all the ribs; and as these slats are very thin they do not interfere with the outer smoothness of the cover as their purpose is then merely to

Umiaq Construction (1960)
The making of an umiaq using a timber frame and bearded seal skins at Ivujivik, 1960. Image credits, top to bottom: Umiaq construction. Lashing inner rail at stem, Ivujivik, Canadian Museum of Civilization, Eugene Y Arima, 1960, J16443; Women inspecting skins draped on umiaq framework, Ivujivik, Canadian Museum of Civilization, Eugene Y Arima, 1960, J16434; Wiviro inspecting skins placed temporarily on umiaq framework, Ivujivik, Canadian Museum of Civilization, Eugene Y Arima, 1960, J16435; umiak construction near R.C. Mission. Lacing on the skin cover to the inner rail, Ivujivik, Canadian Museum of Civilization, Eugene Y Arima, 1960, J16451.

prevent the lateral movement of the ribs increased in length by the greater size of the vessel. When the umiak is furnished with these rails the thongs which stretch the cover are usually affixed to them. Thwarts, consisting of three or more boards, are placed within the frame and as they are cut somewhat longer than the width of the boat they tend to prevent collapse of the sides and also serve as seats for the occupants.

At the stern a wider seat is shaped to fit the curve of that portion of the inner part, while a similar but narrower one serves the purpose for the bow. The rudder is shaped like that for our common boats and is moved by a tiller of such length as may suit the convenience of the steersman, who may, at times, prefer an oar to steer with.

The means of propelling a craft of this kind is by sail or by oars. The sail is simply a square sail and must needily have an aft wind to be most effective for the more nearly abeam the wind may be the greater drift or leeway is made. A head wind usually prevents travel as the umiak cannot be made to tack.

When the wind is not favorable or too light to effect progress the oars are called into use. There are generally two pairs of oars. The form of the oars partakes more nearly the nature of a sweep. The first pair of oars is placed opposite the thwart next in front of the mast. The rowlock into which the oar rests is a piece of wood, having a shallow, long notch cut in it, resting on the skin cover of the boat where it turns over the top rail at that place. The ends of the rowlock block are securely fastened with thongs to the toprail. For each oar two stout loops of sealskin extend along the toprail and through the eyelet formed by the overlapping ends of these loops the hand end of the oar is thrust far enough that the rowers may grasp it. The loop ends serve also to retain the oar if it be suddenly or thoughtlessly let go. The size of the umiak seldom permits more than two pairs of oars. Two or more persons row at an oar and as many may be at the oars as there are occupants of the boat. The shortness of the umiak and the lack of skill in rowing renders the steering very difficult, hence one who has the best knowledge of a boat is generally selected for that duty.

The care of an umiak calls for the watchfulness of all concerned in the journey. Before it is placed in the water it must be examined to ascertain whether the drying process which causes the skins to shrink and thus draw the threads in the stitches to their utmost tension and where imperfectly stitched may be torn or started and there admit a stream of water. A minute rent in the skin caused by the point of the knife when the pelt was being removed from the body of the seal; a place jagged on the sharp-edged rocks over which the hole was dragged may all have escaped detection until the drying revealed them. A patch here or there for the structure is so heavy as to require several persons to handle it that when ready to be launched or taken from the water it must be pushed and dragged over the oars laid on the beach. Sand and gravel adhere to the soft material of the cover and cut the bottom skins as the vessel is dragged over the oars or poles on which it is laid.

If the vessel is to remain out of water for several days and the skin becomes too dry, a coat of oil should be applied to fill up the open pores of the skin and again the seams should be gone over with tallow to prevent leakage. When the umiak has been in the water for several consecutive days the skin cover absorbs so much water that, especially in rainy weather, it does not dry out during the time of halting for the night hence it must be drawn out and allowed to dry until it may be oiled and that substance permeate the fibres of the skin which while wet hung in limp condition about the frame.

A greater or less quantity of water is certain to seep through and into the umiak; and, in order to prevent the cargo from becoming wet a number of poles or sticks are laid on the cross slats of the bottom within. The arrangement of the poles on the slats tends to distribute the weight of the superincumbent load, which at times, is excessive.

The umiak now being furnished with a sail and oars is ready for use on the opening of the water in spring for the Innuit may have gone up someone of the numerous streams for the winter so as to be convenient to the tracts frequented by the furbearing mammals and the reindeer; also, that fuel may be easily obtained for the coast is desolate and barren of aught but ice and snow in that season; a dreary, stormy region, affording so little sustenance that even the raven, whose maw is ever empty, rarely sails in that vicinity; and when it does it is known only by the hoarse croak expressive of disgust at finding not a morsel along its stretch.

When the bright days of April begin to thaw the snow and by the middle of May the little streams trickle over the frozen ground, slowly wearing away the ice firmly bound to the shore of the larger water courses the Innuit considers the locality where the opening of the waters will afford the most food for himself and those dependent upon his exertions.

The man consults with those around him and decides where he shall go. The belongings must be transported to the coast before all the snow has melted or the water has accumulated in deep pools on the slowly melting ice of sea and stream.

The umiak is taken from the stage where it was placed to protect it from mischievous dogs, hungry foxes or nibbling mice, the latter most destructive to skin covered vessels, and examined for seaworthiness. All repairs being attended to the boat is placed upon the sledge and taken by the dogs to the edge of the ice soon to be freed from its grasp on the land. The household effects are next brought to the place and here the people remain until the sea is opened to permit the umiak to travel along the shore. The sledge may be placed within the umiak in order to be used again where it is easier to make a portage than to risk passing a point of land extending so far into the sea as to be not yet free from the heavy ice surging back and forth along its outer limits, where the tides rush with irresistible energy and suddenly bring eddies that bear blocks of ice that threaten destruction to all on its pathway.

When all is in readiness to journey by means of the umiak, the load is placed within it as follows: the tent poles are laid lengthwise to support the goods to be carried. The skins of an old umiak or tent are placed on the poles and the heavier articles laid upon that. The lighter articles are reserved for the top and the heavier below to sustain the craft in ballast.

The crew, consisting of the lustier members of the group, seat themselves in readiness to row at the oars, while the aged and infirm stow themselves among the luggage and the remainder travel along the land where there are not streams to be crossed. The knowledge of the coast determines them as to what course to pursue to reach a certain designated place where they may exchange places with those fatigued at the oars or be prepared to assist at the landing of the vessel. As remarked before, the entire belongings accompany such a journey and the dogs that dragged the sledge to the coast must also be taken care of. Those brutes are now made to serve a good purpose beside prevented from straying after every bird that shows itself on the earth or in the air. The dogs are hitched to a long line and this to the umiak, and, under the care of the women and larger boys and girls, they drag the boat along as rapidly as though propelled by oars.

The Little People
An Inuit story, Johnny George Annanack (Heyes 2007)

The story is about elves (Inugagulliit). The Inugagulliit are trying to pull a seal that they had hunted but they are unable to move it. Two people were staying in the igloo. The Inugagulliit pulled and pulled, but the seal would not move. This is because the man who lived in the igloo spat on the seal at night and as a result the seal became stuck on the ice. The Inugagulliit tried very hard to pull the seal, they even encouraged each other by saying "again" and "again", but the seal did not move so they just left it there. The man living in the igloo did not want the Inugagulliit to have the seal because he was very hungry; he wanted to have the seal all for himself. The Inugagulliit are just like the Inuit people; they have elders and they hunt just like Inuit.

The Little People
Illustration of a story about the Little People (Inugagulliit) and a seal by Kangiqsualujjuamiut elder Johnny George Annanack, 2 May 2005. Reduced from original, 11″ × 17″. Lead pencil on bond paper. Scott Heyes Private Collection.

Gray Seal

Halichoêrus grypus Nils.

[1444]

Order Name used by Turner	*Carnivora*
Current Order Name	*Carnivora*
Family Name used by Turner	*Phocidae*
Current Family Name	*Phocidae*
Scientific Name used by Turner	*Halichoêrus grypus Nils.*
Current Scientific Name	*Halichoerus grypus* (Fabricius)
Syllabary	*Unknown and unnamed in Ungava*
Modern Nunavimmiutitut Term	*Unknown and unnamed in Ungava*
Eskimo Term by Turner	*Unknown and unnamed in Ungava*
Definition of Eskimo Term by Turner	*Unknown and unnamed in Ungava*

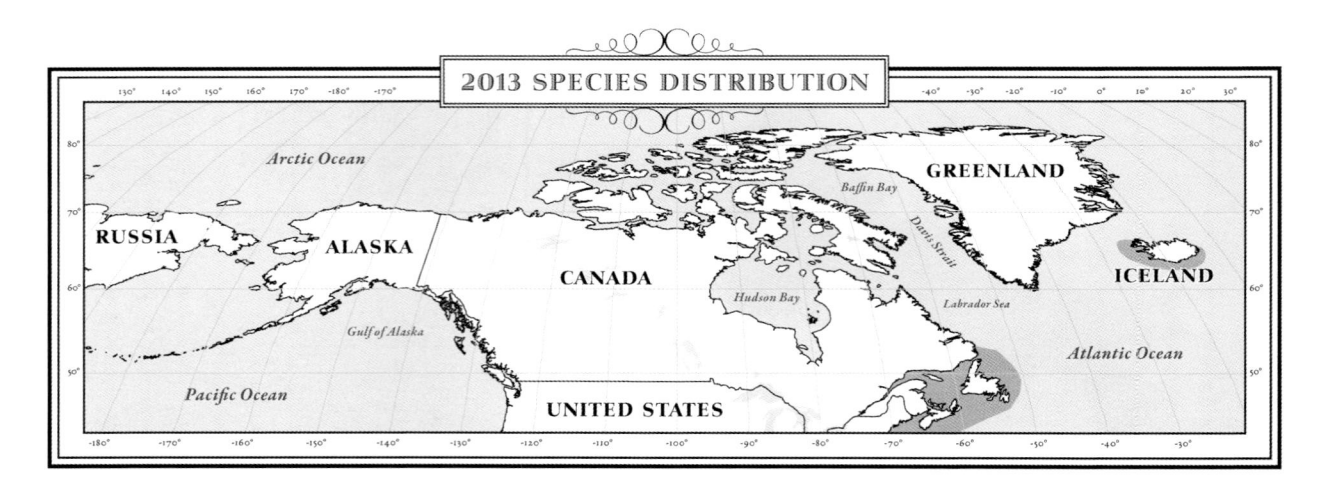

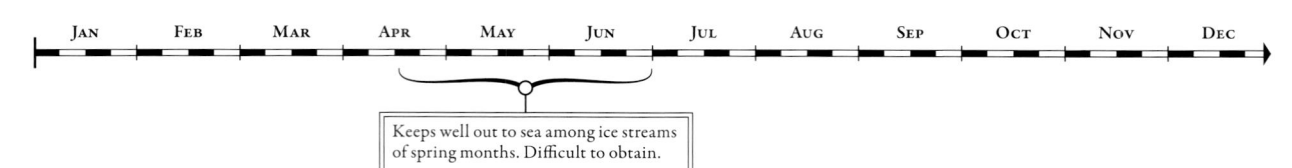

2013 SPECIES DISTRIBUTION

Keeps well out to sea among ice streams of spring months. Difficult to obtain.

Chapter Two: Mammals of Ungava and Labrador

Descriptions by Turner

The Gray Seal is, from the best information I could obtain, quite rare along the Labrador coast north of Hamilton Inlet. About the southeastern portion it is at times rather common but keeps well out to sea among the ice streams of the spring months and renders it difficult to obtain.

The Eskimo of Hudson Strait assert that a species of seal quite different from others, well known, occurs in those waters. It is quite rare and so much so that they have no name for it. Whether it is a straggler of the Gray Seal or not I am unable to state.

ᓇᑦᓯᕚᒃ

HOODED SEAL
natsivak

Cystophora cristata, Nils.

[1445]

ORDER NAME USED BY TURNER	*Carnivora*
CURRENT ORDER NAME	*Carnivora*
FAMILY NAME USED BY TURNER	*Phocidae*
CURRENT FAMILY NAME	*Phocidae*
SCIENTIFIC NAME USED BY TURNER	*Cystophora cristata, Nils.*
CURRENT SCIENTIFIC NAME	*Cystophora cristata* (Erxleben)
SYLLABARY	ᓇᑦᓯᕚᒃ
MODERN NUNAVIMMIUTITUT TERM	*natsivak (cf. Schneider p. 198 - natsivak = big or hooded seal [Cystophora cristata] rare in Inuit country)*
ESKIMO TERM BY TURNER	*nû chi vûk; na chi vak*
DEFINITION OF ESKIMO TERM BY TURNER	*Hooded seal (Cystophora) [Cystophora cristata] [p. 2694; 1445]*

2013 SPECIES DISTRIBUTION

CHAPTER TWO: MAMMALS OF UNGAVA AND LABRADOR

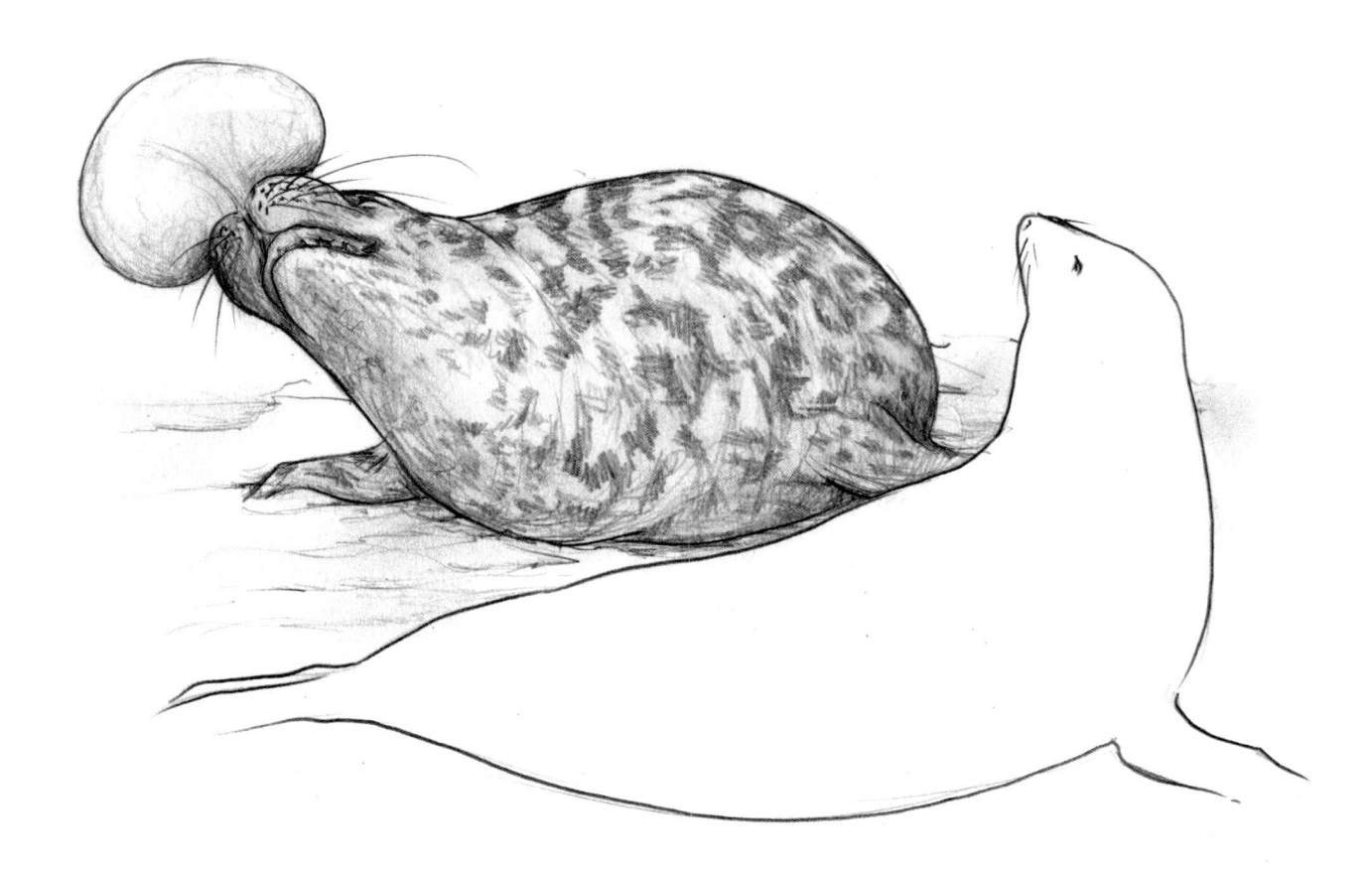

Descriptions by Turner

The Hooded Seal is plentiful on the southern and eastern shores of Labrador but apparently rarely enters Hudson Strait. Repeated inquiry convinced me that it is only a casual visitor in the waters of Ungava Bay and one of the most successful hunters assured me that he had killed but two in all his life; although he had seen several others. The Hudson Strait Eskimo appear to dread this species as they assert it to be very vicious when wounded.

Eskimo from west of Akpatok Island affirm that it is not rare in the vicinity of Cape Wegge to Cape Wolstenholme. I could not obtain satisfactory evidence of its breeding in Hudson Strait.

The Eskimo of Hudson Strait recognize this species under the name *Na chi vak*.

Recollections on Inuit Use of Hooded Seal
Tivi Etok (Heyes 2010)

These seals blow a big bubble. Many live along the Labrador Coast around Killiniq.

Seals – Terms and Definitions

Syllabary	Modern Nunavimmiutitut Term and Definition	Term by Turner	Definition recorded by Turner
ᐊᔪᑦᑕᐅᑎ ; ᐊᔪᒃᑕᐅᑎ	**ajuttauti**; *ajuktauti* (cf. Schneider p.9 - *ajuttauti* ; *ajuktauti* = hockey stick) Note: in Turner's time the term would have applied to the "thonged" stick used in traditional Inuit football to whip the ball. In Igloolik the whip is called *ajukttaq*.	*ai uk ta'ut*	A number of sealskin thongs, forming loops, fastened to a handle, used to [urge?] a football [p. 2162]
ᐊᓪᓗ ; ᐊᒡᓗ	**allu**; *aglu* (cf. Schneider p.20 - *allu, aglu* = seal breathing hole in land-fast sea ice)	*a'g lu*	Seal den (in snow on the ice of the sea) [p. 2136]
ᐊᖏᔪᐸᓗᒃ	**angijupaluk** (?) literally "almost big"	*an'g i uk pa luk*	Nearly adult seal *Erignathus barbatus* [bearded seal] [p. 2174]
ᐊᐅᒃᓵᖅ (?)	**auksaaq** (?) (cf. Peck p.46 – *auksak* = "a seal which does not dive [go down]"). Derives from *auktuq* = he bleeds.	*aú ksak*	A seal which has been struck and is unable to sink [p. 2139]
ᐊᕙᑕᖅ	**avataq** (cf. Spalding 1998, p.15 - *avataq* = inflated sealskin bag used as a buoy in [marine mammal hunting])	*a'va tûk*	Sealskin float [p. 2206]
ᐊᕙᑕᕐᔪᒃ	**Avatarjuk** (cf. MacDonald p. 89 -*Avatattiaq* = a grouping of the stars in the constellation Cassiopeia, in North Baffin Island; from *avataq*, cf. Spalding 1998, p.15 - *avataq* = inflated sealskin bag used as a buoy in [marine mammal hunting])	*a'va ta'g zuk*	Constellation of stars. Means inflated sealskin used as a float [p. 2204]
ᐊᕗᔪᖅ	**avujuq** (cf. Spalding 1998, p. 15 - *avujuq* = the bull seal calls to his mate; also p.45 - *avunek* = "everything which is brought forth by animals before the time [presumably in the sense of "premature"]"; also MacDonald p. 196 - *avunniit* [deriving from *avujuq*] refers to the "moon month" [March/April] when seal pups are born prematurely.)	*a vú yok*	A young "ringed" seal. *Phoca foetida* [p. 2204]
ᐊᕗᓐᓃᑦ	**Avunniit** (cf. MacDonald p.196 – *Avunniit* = refers to a moon-month corresponding to late March/early April when ring seal pups are aborted, or born prematurely. The month in which seal pups are usually born is known as *Nattian*.)	*a vúnit*	Period of the "ringed" seals (*Phoca foetida*) giving birth to their young [p. 2204]
ᑲᒥᒃ ; ᑲᒦᒃ	**kamik**, *kamiik* [pl.] (cf. Schneider p.119 -*kamik* = footware [all boots but not boot liners ... or shoes]	*kû mǐk*	Sealskin boot. Is the Labrador word [p. 2264]
ᒥᓚᒃᑳᖅ	**milakkaaq** (cf. Schneider p.169 – *milakkaaq* = animal that has round or oval markings on its fur)	*mi lak*	Freckle, spot on the skin (hair) of a seal [p. 2234]

Chapter Two: Mammals of Ungava and Labrador

Seals – Terms and Definitions

Syllabary	Modern Nunavimmiutitut Term and Definition	Term by Turner	Definition recorded by Turner
ᒥᓯᕋᖅ	**misiraq** (cf. Schneider p.175 - *misiraq* = oil from a sea mammal... used as a condiment or dip when eating raw [or frozen] meat). Note: the oil is usually fermented.	*mi sýû ghak*	Seal oil [p. 2236; 701]
ᓇᑦᓯᖓᔪᖅ	**Natsingajuq**: *natsiq* (seal) + *-ngajuq* ("half, part of" as in *qallunaangajuq*, "part *Qallunaaq*")	*nŭ chĭng ai ok*	Half of a Seal's body [p. 2694]
ᓇᑦᓯᖅ	**natsiq** (cf. Schneider p. 198 - *natsiq* = ringed seal [*Phoca hispida*])	*nit syak* or *nŭ chik*	Ringed seal [p. 1438]; Seal (*Phoca foetidus*) [p. 2694]
ᓇᑦᓯᑕᖅ	**natsitaq**: might have been the former pronunciation (or the dual form *natsitak*), since in more conservative dialects, the plural of some contemporary words ending in – *iaq* is *(t)tat*: cf. *timmiaq* (sing.), *timmittak* (dual.), *timittat* (pl.) **natsiaq** (cf. Schneider p.198 - *natsiaq* = young seal less than a year old). Usually refers specifically to a baby "jar" seal [*Phoca hispida*], a "white coat."	*nŭ chitak*	Nearly adult ringed seal (*Phoca foetida*) [p. 2694]
ᓇᑦᓯᕙᒐᖅ	**natsivagaq** (-*gaq* = "smaller, younger") *natsivagak*	*nŭ chi va gak*	Nearly adult Hooded Seal [p. 2694]
ᓇᑦᓯᕙᒃ	**natsivak** (cf. Schneider p. 198 - *natsivak* = big or hooded seal [*Cystophera cristata*] rare in Inuit country)	*nû chí vûk; na chí vak*	Hooded seal (*Cystophora*) [*Cystophora cristata*] [p. 2694]; Hooded seal [p. 1445]
ᓄᓚᑕᖅ	**nulataq** (cf. Schneider p.220 - *nulataq* = seal's breathing hole in the ice... convex on the outside). Usually a "frost dome" formed on the snow or ice above a seal's breathing hole.	*nu lá tûk*	Seal den in the ice before the snow falls [p. 2720]
ᓄᓂᖅ	**nuniq** (cf. Spalding 1998, p.76 - *nuniq* = female seal; cf. Inuktitut Living Dictionary – *nuniq* = "female seal without pup")	*nú nik*	A female seal. I am led to conclude that this word is applied to seals which have borne young. The explanation may, however, be questionable. [p. 2713]
ᐸᕐᖑᐊᑐᖅ	**parnguatuq**; *parnguaqtuq* (cf. Schneider p. 241 - *parnguaqtuq* = (animal) toward which one is crawling)	*paung ók tok; paung ú li ák tok*	Creeps, crawls on the belly for a seal. This is the Labrador form of expression [p. 2732]; Creep on belly for a seal [p. 2732].
ᖃᐃᕈᓕᒐᑦᓯᐊᖅ (?)	**qairuligatsiaq** (?) (cf. Schneider p.278, *qairulik* = harp seal; also p. 198 – *natsiaq* = young ["jar"] seal, less than a year old)	*Kai go lĭg a' chi ak*	Nearly adult harp seal [p. 2261]

Carnivores

Seals – Terms and Definitions

Syllabary	Modern Nunavimmiutitut Term and Definition	Term by Turner	Definition recorded by Turner
ᖃᐅᕈᓕᒃ	*qairulik* (cf. Schneider p. 278 – *qairulik* = Greenland seal [*Phoca greenlandica*]; more commonly in Canada referred to as the harp seal)	*Kai ġo lak*	Young harp seal [p. 2261]
ᖃᓕᕈᐊᖅ	*qaliruaq* (cf. Schneider [1970]: p. 262 "sealskin boot generally without fur")	*ka liǵ u ak*	Sealskin boot [p. 2264; 698]
ᖃᓯᒋᐊᖅ	*qasigiaq* (cf. Spalding 1998, p. 110 - *qasigiq* = ranger or freshwater seal)	*ka siǵ yak*	Seal. *Phoca vitulina* [p. 2272]. Freshwater or harbor seal [p. 1433]
ᖃᓯᒋᐊᑦᓯᐊᖅ	*qasigiatsiaq* (cf. Spalding 1998, p.110 - *qasigiq* = ranger or freshwater seal)	*ka siǵ i a' chi*	Young harbor seal. *Phoca vitulina* [p. 2271]
ᖁᐃᒃ	*quik* (cf. Schneider p. 317 *quik* = femur, thigh bone)	*I kwaṅg wûk*	Seal femur? [p. 2229]
ᓯᓓᖅ (?)	*silaaq* (?) (cf. Peck p.213 - *sillâk; sillaek* = "a seal which is quite white; a reindeer which is quite white in summer whilst the others are brown." According to Johnny Sam Annanack, Kangiqsualujjuaq [Pers. Comm.], the term *silaaq* applies to all abnormally white animals. Also cf. Schneider p.354 - *silarniq* = tarnished in the sun; and cf. Spalding 1998, p.134 - *silarittuq* = he is sunburned)	*s'i lak*	An entirely white seal (not understood to be an albinism or an alternative condition). A reindeer which in summer retains the lighter winter plumage is so designated. [p. 2777]
ᓯᕐᖁᖅ ; ᓯᖅᑯᖅ	*sirquq*; *siqquq* (cf. Schneider p.364 - *siqquq* = hind flipper of seals)	*siḱ kok*	Hind flippers (of seal) [p. 2778]
ᓱᐳᒥᔪᖅ	*supumijuq* (cf. Spalding 1998, p. 144 – *supuartuq* = the sea mammal [especially whales] puffs or exhales air)	*Ch'u pu ṁi ok*	Puffs (it) like a seal or whale [p. 2212]
ᑕᓕᕈᖅ	*taliruq* (cf. Schneider p.390 - *taliruq* = front flippers or a seal, bearded seal, whale)	*ta ĺi ghok*	Fore flippers (of seal) [p. 2796]
ᑎᒡᒐᖅ	*tiggaq* (cf. Schneider p.403 - *tiggaq* = bull seal whose meat stinks during rut)	*ti' gak*	A male seal. I think this word is applied only to adult males. [p. 2801]
ᑎᒡᒐᖅ	*tiggaq* (cf. Schneider p.403 - *tiggaq* = bull seal whose meat stinks during rut)	*ti'zhak*	Male ringed seal, under certain conditions [p. 1438]
ᑎᕆᓪᓗ	*tirillu* (cf. Schneider p.411 - *tirillu* = young bearded seal in spring time which will be a *pualualik* in winter and a *tirillu* in summer and autumn)	*tû ğiǵ luk*	A young seal (*Phoca barbata*) [*Erignathus barbatus*] [p. 2801]
ᑎᕆᓪᓗ	*tirillu* (cf. Schneider p.411 - *tirillu* = young bearded seal)	*tĭ ǵi luk*	Youngest bearded seal (*E. Barbatus*) [*Erignathus barbatus*] [p. 2799]

Chapter Two: Mammals of Ungava and Labrador

Seals – Terms and Definitions

Syllabary	Modern Nunavimmiutitut Term and Definition	Term by Turner	Definition recorded by Turner
ᐅᔾᔪᐊᔪᒃ	*ujjuajuk* in present-day Nunavimmiutitut (& maybe in Turner's writing, which might hint at an [incipient?] use of the law of double consonants in the 1880s); (cf. Schneider p. 468 - *utjuk* / *ujjuk* = bearded seal [*Erignathus barbatus*])	*úg zhu á zhuk*	A young stage of bearded seal several weeks old [p. 2836]
ᐅᒥᒃ	*umik* (cf. Schneider p.447 - *umik* = beard)	*u'mik*	Seal whiskers [p. 2844]
ᐅᖅᓱᖅ	*uqsuq* (cf. Schneider p.465- *uqsuq* = oil and animal fat, especially marine mammals)	*ók suk*	Seal fat [p. 702; 2728]
ᐅᑦᔪᒃ ; ᐅᔾᔪᒃ	*utjuk*; *ujjuk* (cf. Schneider p. 468 - *utjuk* ; *ujjuk* = bearded seal [*Erignathus barbatus*]	*uġ zuk*	Bearded Seal [p. 1448]; Square flipper seal *(Erignathus barbatus)* [p. 1439; 2836]

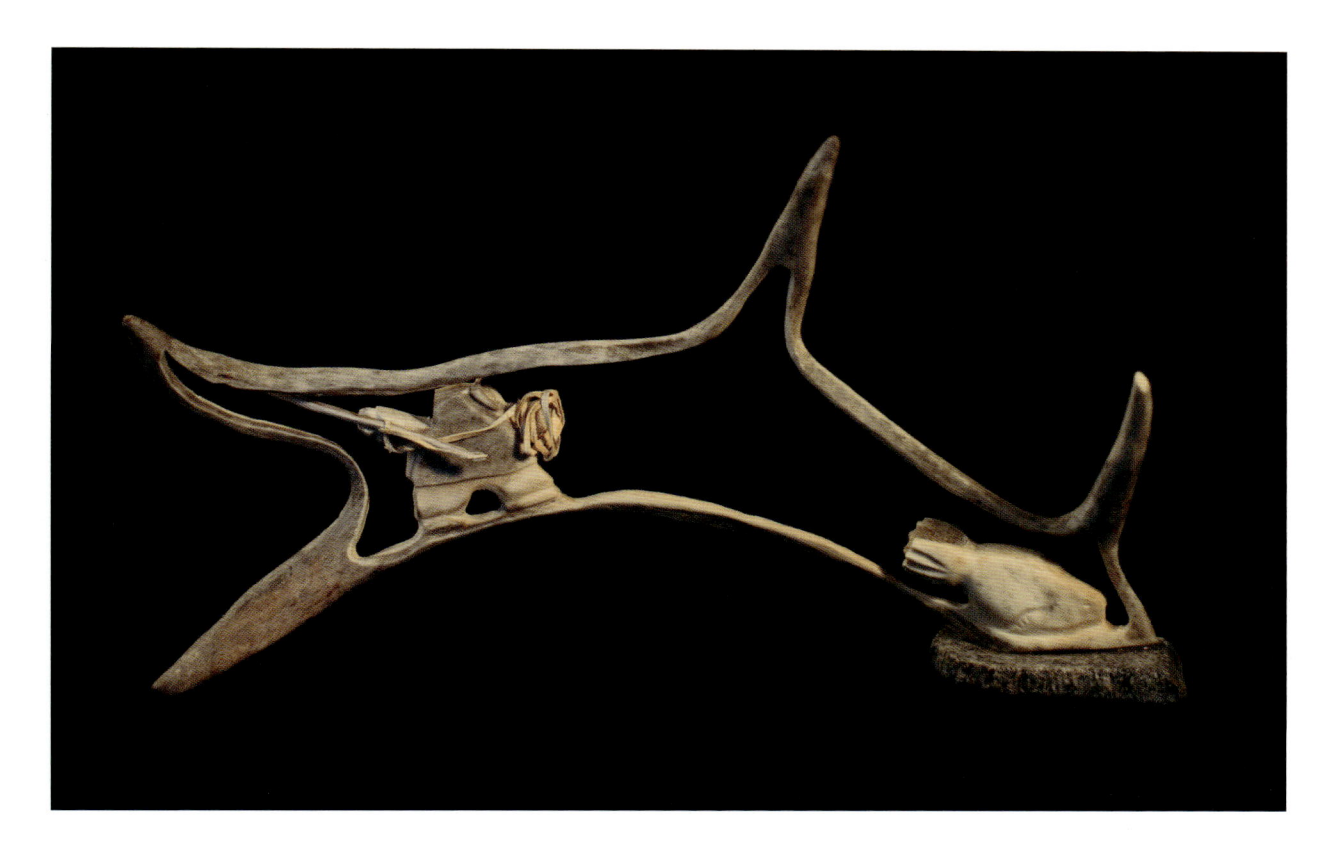

Inuit Carving Depicting Seal Hunting
By Kangiqsualujjuamiut artist Johnny Mike Morgan, 2005. Caribou antler, leather, and granite. Scott Heyes Private Collection.

Carnivores

⊲∆ᑲᖅ
WALRUSES
aiviq

Odobaenus rosmarus

[1446]

ORDER NAME USED BY TURNER	*Carnivora*
CURRENT ORDER NAME	*Carnivora*
FAMILY NAME USED BY TURNER	*Odobenidae*
CURRENT FAMILY NAME	*Odobenidae*
SCIENTIFIC NAME USED BY TURNER	*Odobaenus rosmarus*
CURRENT SCIENTIFIC NAME	*Odobenus rosmarus* (Linnaeus)
SYLLABARY	⊲∆ᑲᖅ
MODERN NUNAVIMMIUTITUT TERM	*aiviq* (cf. Schneider p.8 - *aiviq* = walrus [*Odobenus rosmarus*]
ESKIMO TERM BY TURNER	*aí vŭk*
DEFINITION OF ESKIMO TERM BY TURNER	*Walrus* [p. 2162]

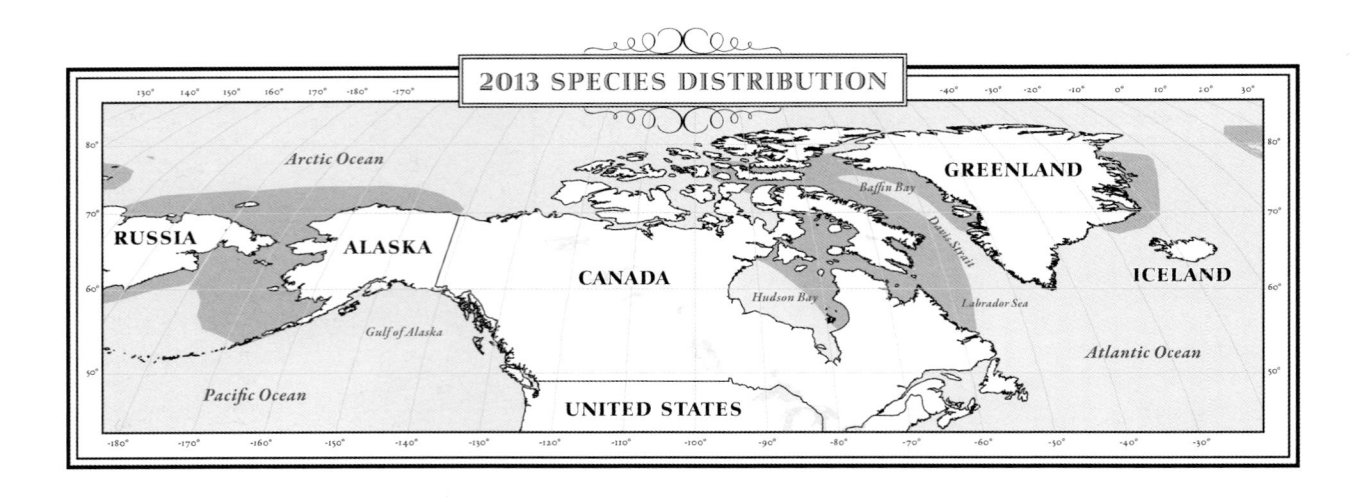

2013 SPECIES DISTRIBUTION

CHAPTER TWO: MAMMALS OF UNGAVA AND LABRADOR

Descriptions by Turner

In former years the Atlantic Walrus was known to have a much more southern extension of its distribution than it has at the present day. Evidences of its occurrence south of the Gulf of St. Lawrence are not rare among the bone heaps of the coast. At the present time an individual is not found south of Labrador and but few even pass below Nain on that coast. The great Fiord now known as Hamilton Inlet is still named *Ívûktok* or place of walrus. Along the northern portions of the coast of Labrador the walrus is yet to be found and at times in abundance. Within Hudson Strait[17] they are not uncommon and on Akpatok Island they are reported to haul up in considerable numbers. The peculiar position of that island causes a tremendous swash of contending tide currents which in certain portions of Ungava Bay are too strong even for this monster. However, under the lee of Akpatok [1447] quiet areas of water are found in these numbers of the Atlantic walrus are found. They haul up on certain portions and enjoy immunity from the attacks

---

17. Inuit elder, Noah Angnatuk, in a 1981 interview, explained that walrus were common around the Killiniq region, especially along the Labrador Coast. He recalled a story from Killiniq about an Inuk hunter who was killed by a walrus while kayaking. It is said that the walrus was a spirit animal and that it deliberately advanced towards the hunter. Angnatuk reported that an RCMP officer witnessed the incident but only reported it as an accident (Makivik 1984).

of the Eskimo who through superstitious reasons abandoned that island many years ago.

Near the northern point of land of the western part of the Ungava district is a locality known to the inhabitants of the district as *Ívûktok*. Here on some small reefs and islets, among which the tide currents increasingly whirl, the walrus makes use of his winter hauling grounds. The locality is reported to be so difficult of approach that only at certain times may the Eskimo of that vicinity repair thither to procure the animals. The place rarely freezes and when this does occur the walrus betake themselves to another locality and await the opening of the water.

It is only rarely that an individual is to be found near the mouth of the Koksoak River.[18] A single male [1448] was procured in the early spring (May/June) 1883 and the head is now in the National Museum [USNM A22014].

---
18. Inuit elder Johnny George Annanack reported in a 1981 interview that the last time a walrus was seen in the George River was 1959 (Makivik 1984).

Whenever the Eskimo finds the walrus and is able to secure it the prize is greatly valued. The flesh though tough, and to a whiteman scarcely eatable; they find, however, but little trouble in masticating the flesh and using the oil for all purposes. The skin is very serviceable, affording thongs yielding in strength alone to the *ug´zuk* or Bearded Seal. From the skin are made lines, covers for their vessels and for the tents. The skin is so thick and unwieldy that it is often split and thus affords double the quantity.

Where the animal is migratory the Eskimo perform certain ceremonies such as giving it a drink of water and other practices which were but imperfectly understood by me for lack of time to investigate them. It does not do to be in too great haste with such things lest suspicion of ulterior motives rests upon the inquirer.

[1449] I was unable to discover the character of the food upon which the walrus, in Hudson Strait, subsist, it would have been specially interesting as my own investigations led me to conclude that the lower forms of marine life were, in that water, very scarce.

Harpoon Tip
Detail of a harpoon tip, probably made from walrus ivory. Ivory is stronger than plastic and other manufactured materials, which become brittle in the cold. Walrus and narwhal ivory are still used today by Inuit mushers as hinges, toggles, and tie-off points due to their ruggedness and strength in cold conditions. Collected by Turner at Ft. Chimo, 9 January 1884. USNM E90220, Museum Support Center, NMNH. Photo by Angela Frost, 2011.

Walrus Ethnography
Turner

SEALSKIN BUOY OR FLOAT. AVATAQ. INNUIT OF SOUTH SIDE OF HUDSON STRAIT (1887a: 121-122)

This object resembles, when inflated, a sealskin of oil and is prepared in nearly the same manner as the receptacle for oil. The difference being that the plug or button employed to close the anal orifice usually has a mouthpiece or else a perforation in its center through which to inflate it when required. The aperture then being closed by a peg. The neck of the skin is left on it and a thong tied around it. To this neck is attached another thong to affix it to whatever may be desired.

If a swimming walrus or other marine mammal of huge size be struck it will be lost unless the blow be immediately fatal which, on account of the conditions of the chase, is so seldom as to be unimportant to other than the fortunate hunter. The hunter well knows what game he may expect to encounter and goes prepared for it. The float is attached to the harpoon line and thrown overboard as soon as it can be done. Care must be taken lest the coil of line foul and overturn the frail Khaiak. The resistance of the inflated skin impedes the stricken mammal and when its last breath expires and the body sinks the buoy remains to indicate the position.

When the umiak is being drawn along the shore by a team of dogs the tracking line must necessarily be so long that the boat may swing out from the shore to float with its burden. Persons are delegated to free the line from the jutting stones or huge rocks and when the water is free from such obstructions the buoy attached to the central portion of the line prevents the rapidity of current or other motion from causing the line to sink so deep that a great resistance would be required otherwise to overcome it. The buoy then forms the most valuable accessory to the equipment of travel.

The small boys must, in their sports, imitate the actions of the men and to please them the father or elder brother constructs toys to resemble the large object known as *avatak*.

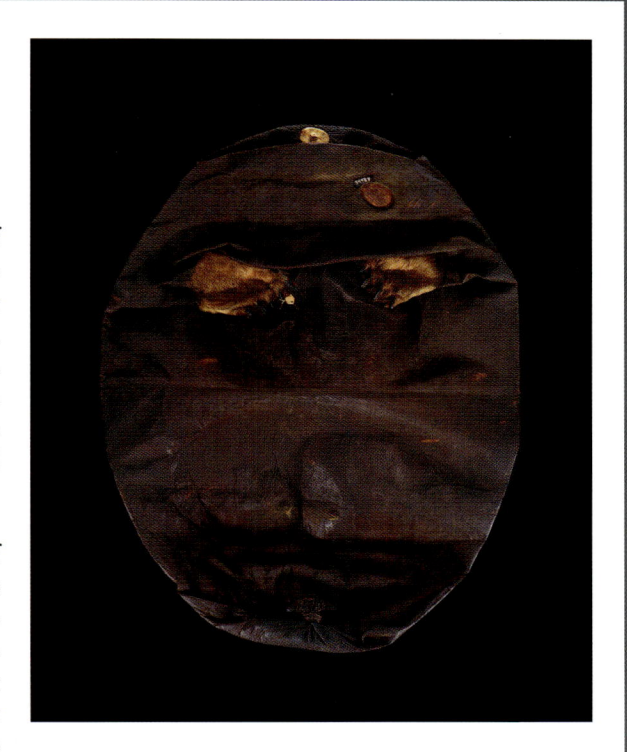

Sealskin Float Used for Walrus Hunting
Floats made from the inflated skin of seals were tied to harpoons by Inuit hunters. The floats make it easier to retrieve walrus and seals when hunted, which tended to sink when harpooned. This float was collected by Turner at Ft. Chimo, probably in 1883. USNM E74489, Museum Support Center, NMNH. Photo by Donald Hurlbert, 2013.

IVORY ORNAMENTS FOR EDGE OF GARMENTS. NORTHERN INNUIT. S. AND W. SIDE OF HUDSON STRAIT. (1887a: 124)

The objects here represented [USNM E90234/3266] are carved pieces of walrus ivory cut into pyriform or conical shape, more or less tapering to truncate small end, through which is made a small hole for the insertion of a stout thread of sinew. The pieces are arranged along the edge of the skirt of the garment of either man or woman while strands of them are placed along the back, across the breast and on the front flap of the woman's dress. They are so placed that the free end of each piece may strike against the next and produce a rattling sound.

Ivory Ornaments

Walrus-ivory ornaments collected by Turner at Ft. Chimo, probably in 1883. Turner wrote that "[The] carved pieces of walrus ivory [are] cut into pyriform or conical shape...through which is made a small hole for the insertion of a stout thread of sinew. The pieces are arranged along the edge of the skirt of the garment of either man or woman while strands of them are placed along the back, across the breast and on the front flap of the woman's dress." USNM E90234/3266, Museum Support Center, NMNH. Photo by Donald Hurlbert, 2013.

At the present day these objects have been discarded and drops made of pewter are substituted, as is shown on the woman's garment 90237/3005, by those people less remote from the trading stations. The Innuit along the south side of western Hudson Strait yet have these appendages to their garments.

Deer Skin Woman's Outer Dress, Ornamented

A woman's parka made from caribou skin, which was collected by Turner at Ft. Chimo, 1883-1884. The parka is adorned with seed beads, pewter spoons, and lead weights (presumably sourced from fishing sinkers). Turner noted that this form of ornamentation had replaced the use of walrus-ivory ornamentation on male and female parkas. USNM E90237/3005, Museum Support Center, NMNH. Photo by Donald Hurlbert, 2013.

The number of these objects required to complete the lines indicated amounts to several hundred, adding no inconsiderable weight to the garment.

The Walrus and Grass

An Inuit story, Turner (1887a: 400 [correctly 300])

Ages ago all the land was covered with water and as it receded from the land the seaweeds stood out and became trees, bushes shrubs and the various trailing bushes of the land. The grass had its origin from an act of a walrus. The exact manner was not understood as the person was interrupted while relating the manner and she was never able to recite the affair again!

Origin of the Walrus and White Bear
An Inuit story, Turner (1887a: 282)

A long time ago some people desired to change the place of their habitation and went to another locality. Among these people was a poor woman who had no relatives and was a charge upon the people and dwelt with them only upon their charity. The people put all their affairs into a boat and then set out not knowing where they went. They seized the woman and cast her overboard. She struggled to regain the boat and when she grasped the edge of the vessel with her hands they cut off her fingers, which fell into the water and were immediately changed into seals, walrus and white bears. The woman, in her agony... her determination to have revenge for the cruelty perpetuated upon her. The thumb became a walrus, one of the fingers became a white bear. The white bear dwells on the land and pursues his prey in water and when he perceives a person the feeling of revenge is again... upon the person whom the bear believes to have been one of those who mutilated the unfortunate woman whose fingers were the origin of the bear.

Story of the Orphan Boy
An Inuit story, Turner (1887a: 282-291)

A poor boy who had neither father or mother was very ill-treated by the people with whom he lived. He was kept in the entry way of the hut like a dog; and was allowed to eat only the refuse cast aside by the other people. Scraps of walrus hide were his principal food and it was so tough he could scarcely tear it with his teeth and as he had no knife with which to cut it he constantly wished for a knife. A little girl, one of the household, would at times, when unknown to the others, take a knife to him and also gave him food of a better quality. Her kind attentions pleased the boy very much. Being treated in such manner as was done by the others caused the boy to long for means of escaping such a life. Help seemed distant except what came from the girl smaller than himself. He endeavoured to... some plan but his many thoughts came to nothing; no help for him.

One night when all the sky was clear he was looking at the moon and thought he could discern the face of a man in it and he begged the man to come to his relief. The moon appeared to grow bigger and in a little while a huge man stood alongside of the now frightened boy. The man seized the boy and beat him unmercifully so that his screams were heard by the household within. But as they recognised the voice of Ka-uj-iyuk it did not concern them how much such a good for nothing was beaten. The more the boy was beaten the larger he became and so strong that he could take a large boulder in the palm of his hand and hold it as easily as common man does a bullet for his gun.

The moon-man then told the boy that he was now strong enough to do as he liked. The boy shrank to the size he formerly was but retained his prodigious strength. The two friends parted and the boy went into the house, a place he had seldom entered, the people demanded to know what he meant by coming in. He seized a large piece of wood and broke it as though it was a splinter. The terrified people saw how immense his strength was and they attempted to run out. As they ran past him he seized them by the head or heels and lashed the life out of them. Some promised all their belongings if he would spare their lives. The boy continued until all but the little girls were killed. The little girl alone was saved because she had done him no harm.

The story of the orphan boy is one from Labrador and the Innuit assert that the occurrence took place at a locality known as Okhak, now a missionary station. They are able to show the rock which presents the appearance of dried blood and brains adhering to it.

Ceremonies Pertaining to Mammals
An Inuit story, Turner (1887a: 65-66)

A ceremony attends the catch of all the more important first-obtained mammals from the sea. A walrus must be given a drink of freshwater when it has expired from wounds inflicted by man. Charms of various kinds are worn suspended from the garments or about the neck or carried in the hunting bays to ward off evil influences. Other charms are worn as representations of departed relations. These charms are, in most instances, the resemblance of the totem to which the person, now weakening to its representation, formerly belonged. The wolf, bear, fox, reindeer, gully, raven, loon, and ptarmigan are the living forms of those objects now recognised. Many of the outward signs of these totems have disappeared and linger only in associations so inextricably confused by marriage and death that they are not accountable for by the people now wearing them.

The Incident at Akpotak Island
An Inuit story, Turner (1887a: 33-36)

The primitive superstitions, rites and beliefs of these people [Tahagmyut] have undergone no change and only few opportunities presented themselves to ascertain facts from them. Fatalities work potently upon their minds and as an illustration relate the following obtained direct from them. No inducement will cause them to accept anything procured from Akpatok Island. In the year 1859 (at that time, and from 1843 to 1866, the trading station of Ft. Chimo had been abandoned for lack of profit at that place) a vessel was wrecked near Akpatok Island. The commander with others in one boat and the first officer in another boat left the wreck. A dispute about direction to pursue to reach the nearest white settlement caused the boats to be separated. The second boat finally arrived at Nackvak on the Labrador Coast and that part of the crew was saved. The captain's boat landed on Akpatok Island, which was then inhabited by Inuit. Here the crew remained and received kindly attentions until such attentions were misconstrued into the appearance of a familiarity which led to some of the crew giving greatest offence by being free with the wives of the Innuit who had rescued them. Matters went from bad to worse and the sailors became emboldened until the Innuit could no longer restrain their anger and resolved to massacre the whitemen. One of the men was so badly frozen as to be unable to walk. He escaped death at that time and one account says he wandered away and perished while another relation states he was bound with thongs and placed in the cold where he perished. Some of the Innuit hesitated to participate in the massacre but were taunted by their companions. They consented reluctantly and stated a visitation of some kind would afflict them for killing whitemen. I have heard of several of the participants of that murder. They always appear uneasy when they visited Ft. Chimo and were anxious to depart. Several years after the commission of that deed, the island was visited by a protracted storm of rain which fell on the frozen ground and was converted into ice several inches thick. This prevented the reindeer and other mammals, which abounded then, from procuring food and they perished. The Innuit were prevented from leaving their tents and all perished from starvation. Their friends repaired to that locality for hunting walrus and discovered the misfortune which had befallen their people and concluded it to be a visitation in consequence of the massacre perpetuated the preceding fall. Since that time no Innuit will dwell on Akpatok Island or accept anything which has been brought from it.

Adult Male Walrus Skull

This was the only walrus skull collected by Turner. The customs Inuit observed in relation to the hunting of walrus and other animals were described by Payne (1887): "Here, as elsewhere, the Eskimo take two days rest after killing a walrus, and become very indignant if asked to do work during this time. During the walrus season they will not put needle into deer skin and, although often pressed, nothing would induce them to do so. Nor will they sew anything when one of the family is ill." USNM A22014, Division of Mammals, NMNH. Photo by Angela Frost, 2011.

The Walrus with Small Tusks
An Inuit story, Tivi Etok (Heyes 2007)

There were so many walruses at a place where three families lived. There were also plenty of ringed seals, bearded seals, and lots of caribou. The caribou would cross the island nearby and annoy the hunters. The hunters couldn't sleep during the night because the caribou would pass through the island so often. One of the hunters wanted to catch walrus for the winter. Walruses came in all sizes, but this man wanted big walruses. When the caribou finally stopped crossing, he started walrus hunting on his *qajaq* since it wasn't that far from where they were camped in a place called Ulliq. He started heading for the walruses when a small walrus came up to the *qajaq* and said "Harpoon me", and it kept telling the hunter the same thing over and over. The man took a good look at the small walrus and said "I don't want to harpoon you, you have such small tusks". So the small walrus headed back to its group of walrus still speaking in Inuktitut saying "He doesn't want us, he doesn't want us". The man was not supposed to decline a walrus that was offering himself. The little walrus told all the other walruses that were resting on the beach that the man didn't want the walrus, and then they all started to roll down to the water and headed for the deeper waters, making the water wavy. Heading for the Ikulliaq, they were all talking "he doesn't want us, he doesn't want us". The caribou were leaving too, so the man ended up with nothing to kill. So that's what happened to the man. He should've killed the little walrus when it offered itself to him. Since the man did not take the offer of the little walrus, all the animals, seals, walruses, whales, caribou all disappeared, leaving the hunter empty handed. They went through a hard time afterwards. It happens to everyone sometimes, we don't always catch animals, so the people that lived in those days all died of starvation. I think there were 3 or 4 houses that they had tried to build to make a little village, but the people all died of famine. The houses that were there are now almost covered over, and they are not allowed to be dug up because of the history. That's how the story goes about this man who refused the walrus that was offering himself.

Walrus Hunting
An Inuit story, Benjamin Jararuse (Heyes 2007)

I used to follow my father and/or relatives when they went hunting, not realizing that I was learning their skills. What ever it was, if I saw it, I was learning, but now I'm getting old and I cannot do this anymore.

I also know about walrus hunting because my uncle used to hunt walruses a lot. He was Lodi's grandfather, my father's older brother. He has worked a lot in his life. He used to harpoon walruses on a *qajaq*, or when they were resting on ice patches. There are always ice patches and we will always have them. I know how much they used to hunt walruses, on the ice edge too, they would wait patiently for a walrus to come up for air, my father and his brother have worked very hard and hunted walruses.

The younger brother would wait with the harpoon in his hand while the older brother would be ready to help. When the walrus would come, he would be ready to harpoon it. Once the walrus was hit, the other end of the harpoon got stuck to the ice. The ice was too thick and couldn't break. The rope was connected to the walrus and the one stuck to the ice, but the ice cracked and the walrus started dragging the ice with the men on it. It went towards the other brother, and then it stopped. They waited to see what happened next. They had no rifles back then, only harpoons and traditional hunting gear, they didn't even think of rifles. Then the walrus finally died, and was pulled out of the water, by poking a harpoon onto the ice and using skin ropes to pull it out. They made it look so easy to do. That's how we used to catch our game. I have done it myself along with my cousins. Today, they use rifles. If you wanted to kill the walrus with one shot, you should aim for the head, and when it is hit, it goes down under water. When it is hit, you wouldn't want to be near the beast, and the second time it goes down and struggles, it attacks the ice with its tusks and the third time, it goes on the ice, and when it does, then the person can finally shoot to kill. That's how they used to hunt walruses.

WALRUS SKULL AND THE CARIBOU
An Inuit story, Elijah Sam Annanack (Makivik 1984)

A hunter near Alluviaq (Abloviak Fiord) was trying to sleep but was unable to because the caribou were travelling near his tent. The hunter put a walrus head on the caribou track and said to the caribou: "Stay away from this place." Soon after, the caribou began to disappear from the area. The caribou made their way south to the Whale River in one line. Caribou never appeared again in the Alluviaq area. There was an old woman who used to say that the caribou would return one day. There is evidence that caribou used to be here. You can still see their old migratory tracks between the berry bushes in the summertime. The Annanack ancestor used to say, "The caribou will come back when their food grows." The old people used to say that animals come and go from an area and I think this is true. There used to be many walrus around here; for a long time there have been very few in the area. But they are starting to come back now around Killiniq. It's very possible that they would come back to this same area where they used to be. People, a long time ago, used to say that the walrus would come back and you can feel this is true as they begin to come near Killiniq.

Walrus Carving
Soapstone carving of a walrus by Quaqtamiut elder, David Okpik, 2000. Okpik has hunted many walrus and polar bears around Akpatok Island. Scott Heyes Private Collection.

Walruses on an Ice-floe
A pod of walruses basking on ice in the Labrador Sea near Killiniq. In Turner's time, walruses were sometimes referred to as "river horses." Photo courtesy of Daniel Annanack, 2012.

Walrus Tusks Collected by Turner

Three juvenile walrus tusks. In Turner's manuscript (1887a: 120), he describes these walrus tusks as being "trophies, and also to serve as material from which to fashion small articles". In his description of the tusks, he states: "Various portions of the teeth of the marine and terrestrial mammals are used to form many articles which, in the absence of metals, wood and stone have either too great brittleness or else soft and easily worn to withstand the pressure or shock to which the object subjected. The great size of the walrus tusk enables the Innuit to fashion innumerable objects from it while for smaller affairs the teeth of other mammals are equally suitable. In the absence of metal with which to shape the ivory sharp edged stones cut into shape and rough grained stones rub it to the required thickness." USNM E90229, Museum Support Center, NMNH. Photo by Angela Frost, 2011.

ᐊᑕᕐᖅ / ᐊᒃᓴᖅ

Black Bear
atsaq / aksaq

Ursus americanus, Pall.

[1450]

Order Name used by Turner	*Carnivora*
Current Order Name	*Carnivora*
Family Name used by Turner	*Ursidae*
Current Family Name	*Ursidae*
Scientific Name used by Turner	*Ursus americanus, Pall.*
Current Scientific Name	*Ursus americanus* Pallas
Syllabary	ᐊᑕᕐᖅ; ᐊᒃᓴᖅ
Modern Nunavimmiutitut Term	*atsaq / aksaq (cf. Schneider p.47 - atsaq / aksaq = [...] in Ungava, the black bear, Ursus americanus)*
Eskimo Term by Turner	*ák hla*
Definition of Eskimo Term by Turner	*Black bear [p. 2163]*
Innu Term by Turner	*Naskwh [p. 708]; Maskwh [p. 1451]*

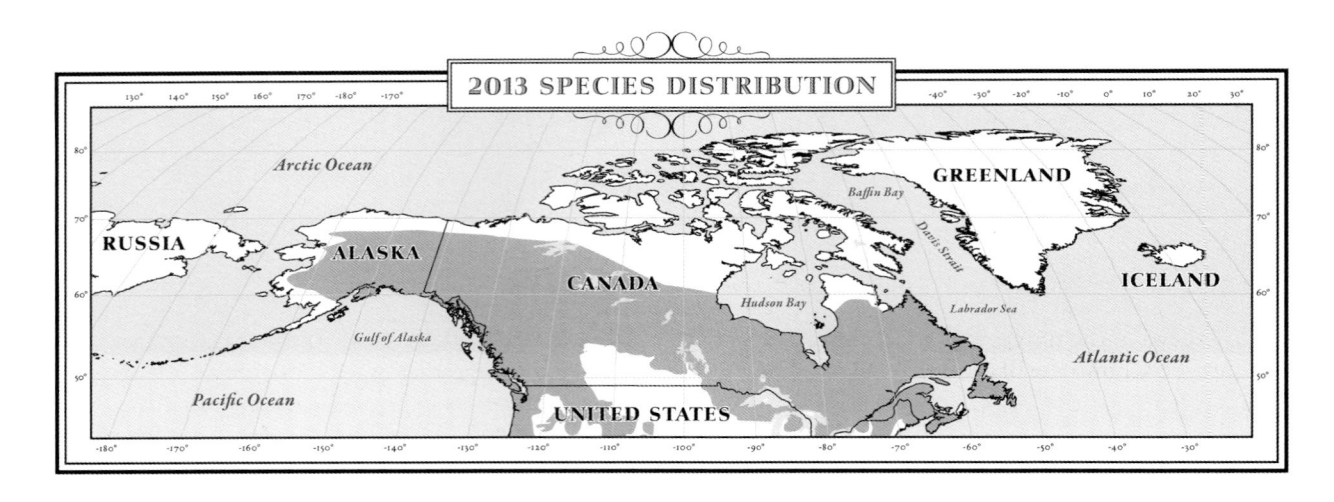

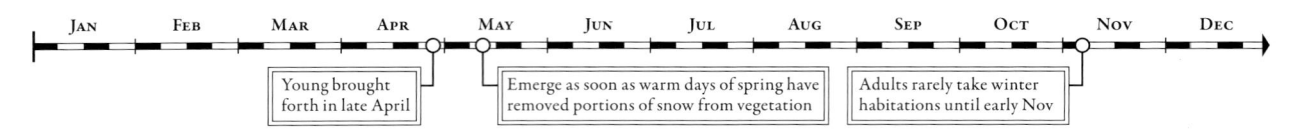

2013 SPECIES DISTRIBUTION

- Young brought forth in late April
- Emerge as soon as warm days of spring have removed portions of snow from vegetation
- Adults rarely take winter habitations until early Nov

Chapter Two: Mammals of Ungava and Labrador

Descriptions by Turner

North of latitude 58° the Black Bear is but rarely found in the region herewith included. It never wanders beyond the timber line and as this limit – it is very near Ft. Chimo the Black Bear is seldom known to occur even within fifty miles south of that locality.

I saw but one individual and that was nearly sixty miles above Ft. Chimo and fully eighty-five miles from the mouth of the Koksoak River. This creature was on another rock ridge separated from the one I was on by a deep valley, which was crossed and where was found no traces of the animal which but little while before was plainly seen sitting on its haunches endeavoring to locate the sound of voices which the still air so distinctly bore toward it. It was doubtless watching the movements of a huge buck reindeer which was near the hill top. I secured the deer and [1451] saw nothing further of the bear. Another party off some distance to the westward saw it, in the course of an hour after, making hasty tracks across the country.

The Indians occasionally obtain them toward the headwaters of the Koksoak River. The animal is said to be common in the region drained by the Great Whale River and north of Lake Mistissing. South of the "Heighth of Land" it becomes quite plentiful being frequently seen ranging to the shores of the Atlantic south of Davis Inlet to the Gulf.

The East Main Indians preserve the lower lip of the bears killed by them, as a trophy. These portions of skin are tanned, painted and otherwise ornamented. The Eskimo apply the name of *Ák hla* to the Black Bear. The Indians give them the name *Maskwh*.

Bear Canine Teeth
Perforated beer teeth, possibly used as a charm, collected by Turner at Ft. Chimo. These were once on display at the Peabody Museum at Harvard University in 1888. USNM E90231, Museum Support Center, NMNH. Photo by Angela Frost, 2011.

The Fate of the Bear Who Deceived Her Children
An Innu story, Turner (1887b: 612-614)

A bear had two children whom she constantly deceived about the approach of spring. They were so anxious for spring to come that they might go outside of their narrow den and gambol in the bright sunshine, smell the perfume of the flowers and taste the tender grass and herbs which their mother had so often described to them but always with the charge that they must not appear outside until she gave them liberty to do so; for the severe cold would make their noses and ears tingle with pain. The mother went out each day; and, on her return the cubs first asked, "Has the spring come?" The mother answered, "No, my children, it is yet cold outside of the home which shelters you from the storms of the winter which is yet upon us. Soon, however, it will be bright and joyous spring. The green grass and leaves will deck the ground with beauty, and their fragrance fill the air; the berries will ripen and then food will be plentiful and the earth dry and warm. Await my bidding and the spring will come."

The little cubs were so impatient when the mother always repeated the same story of the springtime yet so distant. They were dutiful children and had no thought of disobeying their mother. One day, however, they waited long for the return of their mother. When she came back she lay down and fell asleep. Her mouth opened and the cub saw, between her teeth, the remains of the green leaves she had eaten. They knew their mother had deceived them. They resolved to go out and see for themselves. As soon as they appeared beyond the entrance to the den they saw the wonderous beauties of the forests and heard the little birds sing so gaily among the trees and bushes that they knew spring had come. They wandered far away, they were so charmed by their new surroundings that they forgot that the sun had nearly set. The mother awakened and finding her children away, ran out to call them. The tone of her voice was not assuring so they hid in the bushes. The mother was now alarmed and rushed along, exclaiming, "my children the spring has come; the spring has come." She continued her cry until nightfall. The elder cub said to the younger, "I hope the devil of the woods will kill her for telling us a falsehood that deprived us of so many days of pleasure."

In a moment a wild scream startled the little cubs nearly from their skins. The mother cried out, "My sons the devil is killing me; the devil is killing me." The cubs heard the blows falling heavily and fast upon her head. They said to each other that it served her right.

The next morning they went their way and dwelt by themselves.

Barren Ground Bear

Ursus richardsoni, Aud. & Bach.

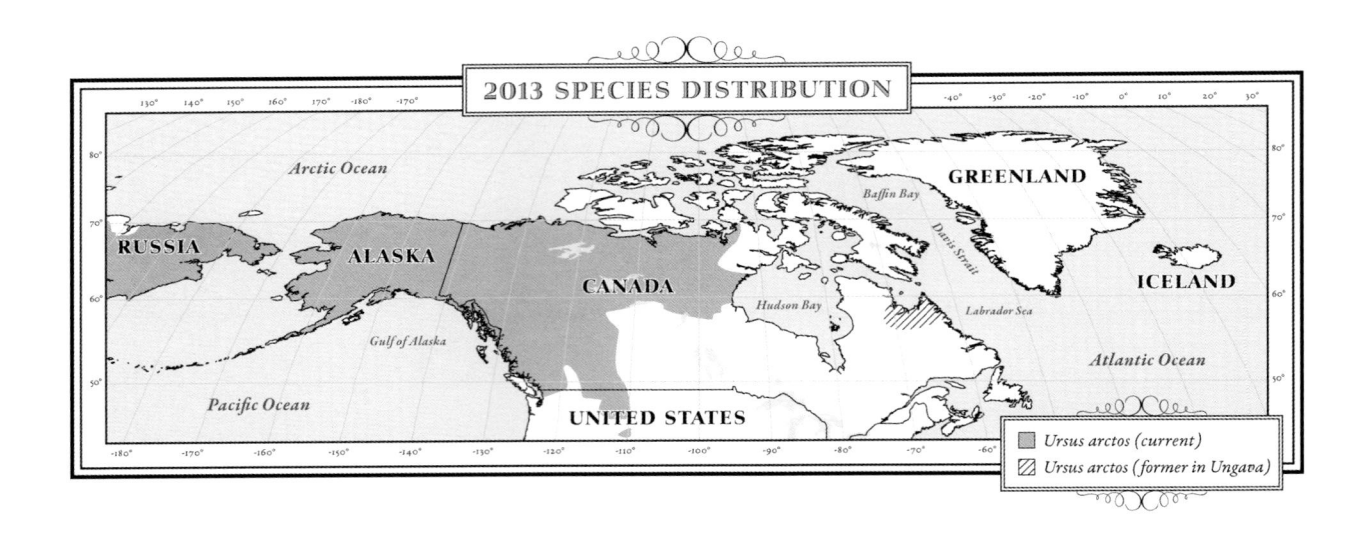

[1452]

Order Name used by Turner	*Carnivora*
Current Order Name	*Carnivora*
Family Name used by Turner	*Ursidae*
Current Family Name	*Ursidae*
Scientific Name used by Turner	*Ursus richardsoni, Aud. & Bach.*
Current Scientific Name	*Ursus arctos* Linnaeus
Syllabary	*Unknown and unnamed in Ungava*
Modern Nunavimmiutitut Term	*Unknown and unnamed in Ungava*
Eskimo Term by Turner	*Unknown and unnamed in Ungava*
Definition of Eskimo Term by Turner	*Unknown and unnamed in Ungava*

2013 Species Distribution

Ursus arctos (current)
Ursus arctos (former in Ungava)

Chapter Two: Mammals of Ungava and Labrador

Descriptions by Turner

A species of bear supposed to be the Barren Ground Bear[19] is well known to inhabit the sparsely timbered tracts along George's River from within thirty miles of its mouth to the headwaters. This animal is not plentiful, although common enough and too common to suit some of the natives who have a wholesome dread of it. It may be somewhat strange but it is nevertheless a certainty that it is not an inhabitant of the Koksoak valley south of latitude 56 degrees, but confines itself in the more northern portion of its range to the area between the coast range of hills along the Labrador coast and the George's River

19. Also known as Grizzly or Brown Bear

Collection Notes

Turner did not collect any barren ground bear or grizzly bear specimens while in Ungava and Labrador. The species is considered locally extinct.

valley, ascending that region to the headwaters and thence striking across the country to the westward north of the "Heighth of Land". South of 55 degrees it is not known to occur that I have any trustworthy information of. The Indians [1453] affirm that only within recent years has this animal taken a freak to extend its range to the westward of the headwaters of George's River.

(On pages 108 and 109, second volume, of Twenty-Five Years in the Hudson's Bay Territory written by John M'Lean is found enumerated the varieties of animals of the larger kind inhabiting the Ungava district. This list includes "Black, brown, grizzly, and polar bears." Page 109 of that volume contains a short discussion on the grizzly bear.[20])

The coloration of the Brown or Barren Ground Bear is so variable as at times to be dirty-yellowish brown to a dark grizzly.

I was informed that this animal is extremely savage, rushing upon its foe with a ferocity characterized by no other species of Bear. Its eyesight is limited from the position of the eye hence the animal has its vision directed only immediately in front of it. At the distance of thirty yards [1454] it is, while feeding, incapable of seeing the approach of a hunter who takes advantage of this defect and approaches only while it is engaged in feeding upon berries and other vegetable products. When the animal raises its head the person remains motionless and easily escapes detection. If, however, the animal observes anything moving on the horizon it immediately goes to that direction and if the track is discovered it relentlessly pursues the person. Only under most favorable circumstances will an Indian attack it. Eskimo seldom traverse the area occupied by this huge beast.

A single young (rarely two) is brought forth in late April, not attaining adult size for three or four years. The adults rarely take to their winter habitations until early November and emerge as soon as the warm days of spring have removed portions of the snow from the tracts overgrown with [1455] bushes which afford a precarious living for the animal which when it came from its den was apparently as fat as when it lay down for its five months of semi-lethargic condition. In the course of three weeks the animal appears, in spring, as a huge mass of skin and bones, tottering over the uneven surface, scanning every foot of ground for a morsel of food to satiate its ravenous appetite.

The flesh of everything is consumed as food by this species.

Unfortunately, I was unable to procure a specimen of this animal although a large reward was offered for an individual. I saw two skins of it and was unable to discover any appreciable difference between them and those of the Barren Ground Bear from other localities.

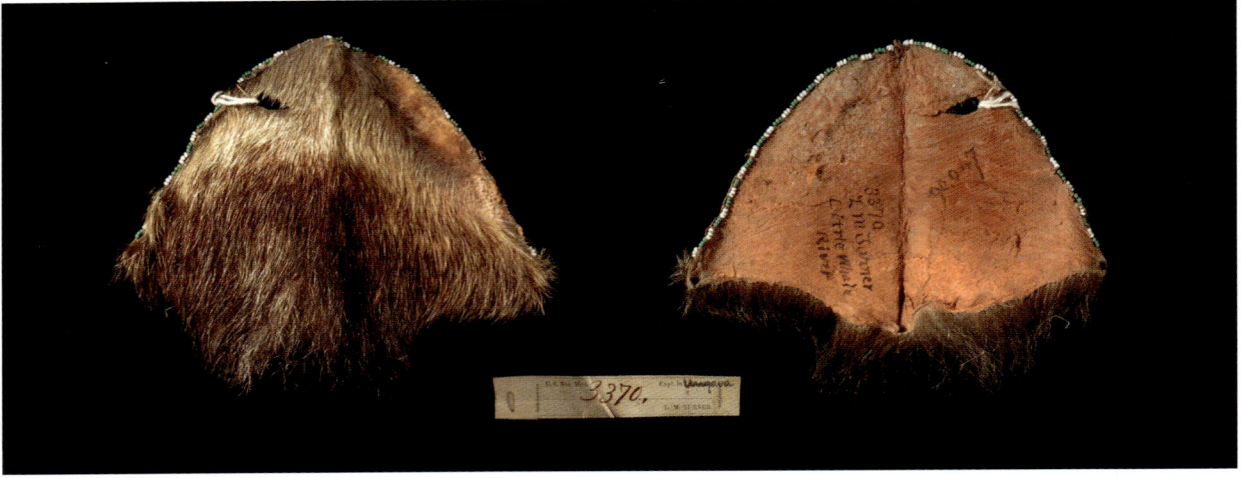

---

20. This is an original note by Turner.

Bear Ornament
*A bear lip ornament used by the Innu, possibly as a charm, collected by Turner at Ft. Chimo. It is unclear whether these are made from the skin of a black bear (*Ursus americanus*) or a brown bear (*Ursus arctos*). USNM E90047, Museum Support Center, NMNH. Photo by Angela Frost, 2011.*

"I was informed that this animal is extremely savage, rushing, upon its foe with a ferocity characterized by no other species of bear"

Bear Ornament with Beads
*A beaded bear lip ornament used by the Innu, possibly as a charm, collected by Turner at Ft. Chimo. Turner recorded that the ornament originated from the Little Whale River. He apparently believed that the ornament was made from the skin of a brown bear (*Ursus arctos*), writing: "These mementos are procured with great difficulty from the hunter who has risked his life in the struggles attending the capture of these beasts, for the barren-ground bear of that region is not a timid creature like the Black Bear; and unless the hunter is well prepared for the animal he would do well to let it alone" (Turner 1894: 275). USNM E90037, Museum Support Center, NMNH. Photo by Angela Frost, 2011.*

CARNIVORES

ᓇᓄᖅ

Polar Bear

nanuq

Thalassarctos maritimus, Gray.

[1456]

Order Name used by Turner	*Carnivora*
Current Order Name	*Carnivora*
Family Name used by Turner	*Ursidae*
Current Family Name	*Ursidae*
Scientific Name used by Turner	*Thalassarctos maritimus, Gray.*
Current Scientific Name	*Ursus maritimus* Phipps
Syllabary	ᓇᓄᖅ
Modern Nunavimmiutitut Term	*nanuq (cf. Schneider p.192 - nanuq = polar bear [Ursus maritimus])*
Eskimo Term by Turner	*nań ok*
Definition of Eskimo Term by Turner	*Polar Bear (Thalassarctos maritimus, Gray.) [p. 2683]*

2013 Species Distribution

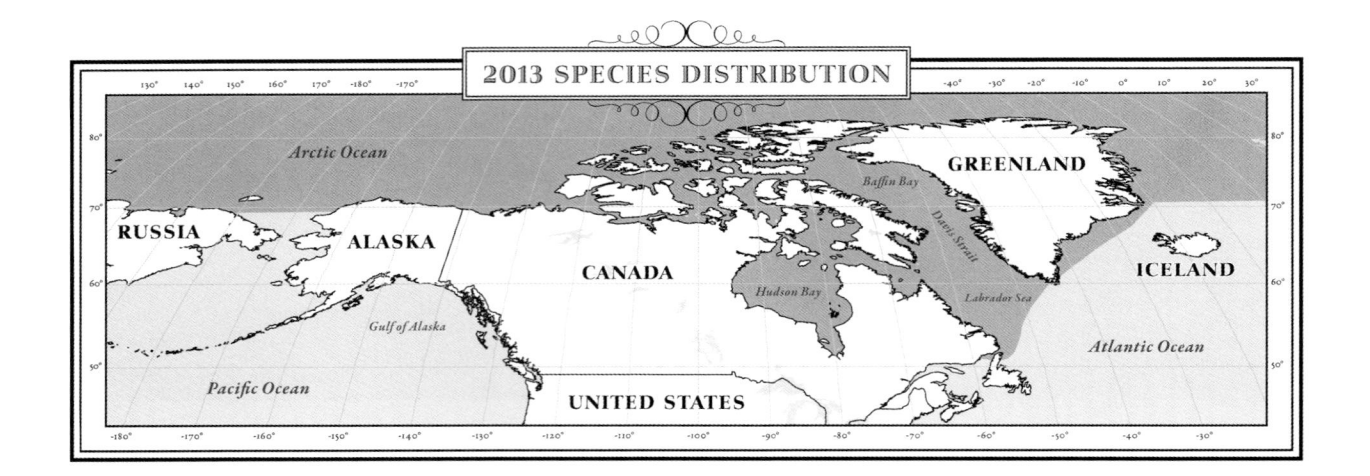

Chapter Two: Mammals of Ungava and Labrador

Descriptions by Turner

The southern range of the white or Polar bear on the Atlantic coast is as far as Newfoundland to which it is carried by the icebergs from the north. It is not rare as far south as Nain on the Labrador coast and becomes common in the Cape Chidley vicinity. It rarely wanders to the southern portion of Ungava Bay, several individuals have, however, been seen as far up the Koksoak River as Ft. Chimo and in the month of February 1883 the tracks of one was seen near the "Rapids" some thirty-five miles south of Ft. Chimo.

Occasional individuals thus become apparently lost and make extensive wanderings ere they regain the seashore. During these wanderings the animal is often so reduced for want of food that its condition is easily known from its tracks in the snow. It frequently lies down and again resumes its wavering course [1457] always seeking the sea away from which they appear to be uneasy and restless.

Great numbers are reported to be upon Akpatok Island since its desertion of the Eskimo formerly dwelling there. The favorable situation of the northeastern extremity of this island affords access to it from the masses of ice swept back and forth by the almost irresistible tide currents of that strip of water.

"The Eskimo snow houses are sometimes attacked by this bear which throws itself upon the structure crushing it down and devouring the inmates."

Fighting with a Polar Bear
An illustration of a story about a hunter fending off a polar bear by Kangiqsualujjuamiut hunter George Don Annanack, 12 May 2005. In a 2005 interview (Heyes 2007) Annanack explained that this drawing depicts an Inuk hunter who was suddenly awoken by a polar bear that tried to knock down the igloo while he was sleeping in the early hours of the morning. The hunter attempted to kill the bear with his harpoon. Reduced from original, 11" × 17". Lead pencil on bond paper. Scott Heyes Private Collection.

The Eskimo use the skin of this animal for a variety of purposes and from the peculiar adaptation of the hair of the beast to shed water the people take advantage of that fact and encase their footwear with portions of the skin and patch the knees of their pants with it to protect them from the water and snow; for when on bended knees for hours at a time scarcely daring to breathe lest the noise frighten the seal momentarily expected to make its appearance and receive the thrust amply pays the Innuit for the hardship and exposure he has undergone.

[1458] The Polar Bear is not savage and when two persons engage in pursuit of it the animal is certain to fall to their lot. The manner is to conceal oneself while the other goes out to "drive" it. The bear either takes flight past the hidden person or else follows the other who leads it near the concealed person who delivers a ball where it will do most good. A boy is often concealed partly while the elder person exposes himself. A boy who had in a single day killed two of these animals and for the first time shot at a Polar Bear related to me that his heart beat as loud as an Indian drum.

The Eskimo caution one not to pursue a direct course in avoiding the attack of this animal but await until the animal makes a sudden lunge and then spring to one side, [1459] which allows the impetuosity of the animal to throw it heel over head and that but few repetitions of this maneuver will disgust the creature and cause it to give up the chase.

Along Hudson Strait the Polar Bear is known to bring forth its young but it probably does not do so much south of the shores of the Strait. A skull was obtained on Gull or Beacon Island on the south west side of the mouth of George's River [See USNM A23207]. The habits of this animal are so well known that I need not repeat them here.

The Eskimo snow houses are sometimes attacked by this bear which throws itself upon the structure crushing it down and devouring the inmates. A woman related that she and two children were in one of these snow houses when a bear suddenly appeared at the doorway. She had presence of mind enough to offer it a [1460] piece of seal meat which she was cutting. The animal took it from her hand and walked off without further molestation to the people. It was an amusing sight to see the woman withdraw her hand as she imitated the motion of feeding the bear. She remarked that she was glad her hand is her own or in other words that the bear did not take it along with the meat.

The Eskimo term this bear as *nań nok*. Many island and other parts of land have this name as either a prefix or part of the name under which the location is known.

Chapter Two: Mammals of Ungava and Labrador

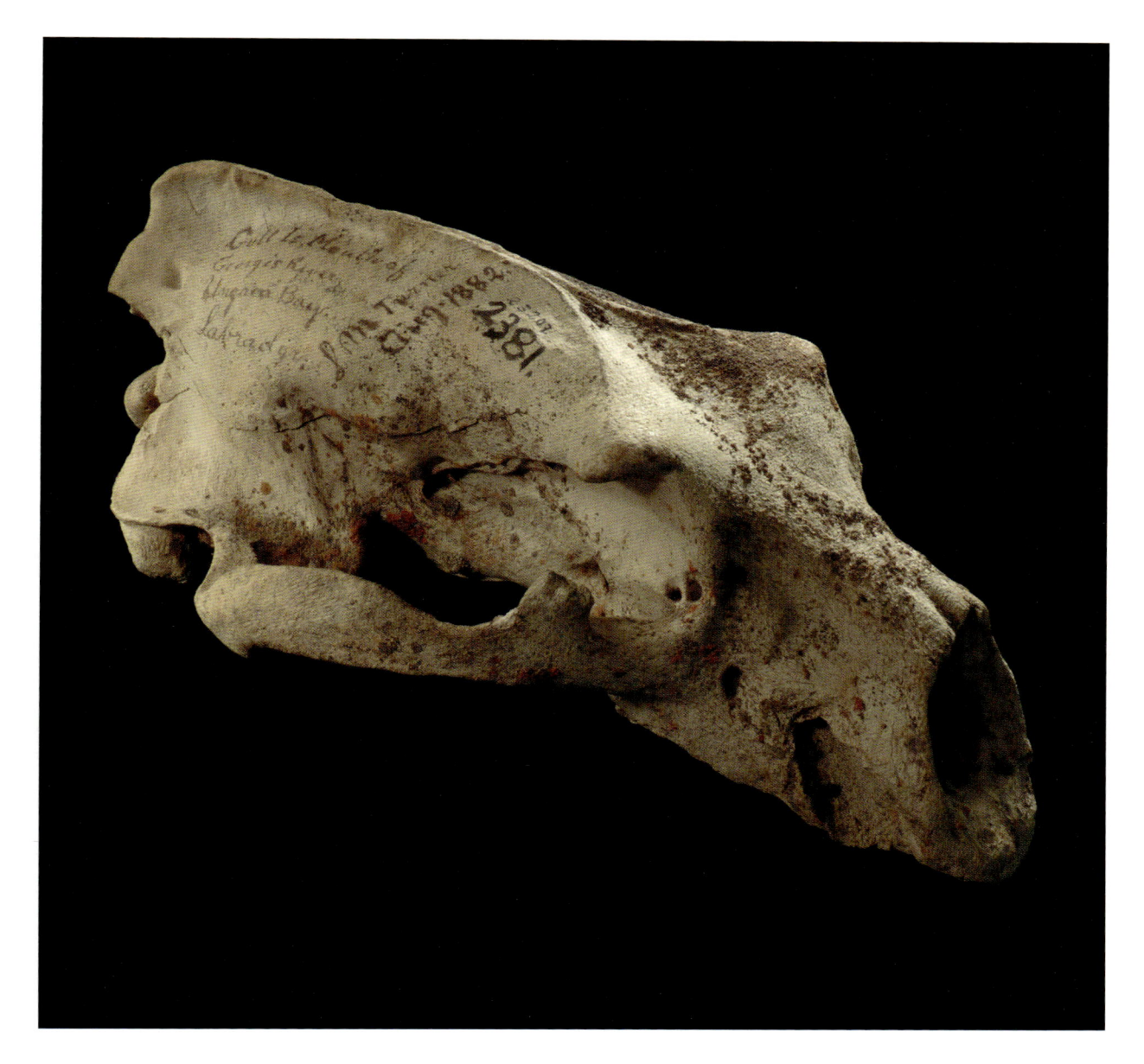

Polar Bear Skull

This is the only polar bear skull collected by Turner. It is a "pick-up" skull, still covered in lichens, that Turner must have found on the ground. It was collected on Gull (or Beacon) Island, near the mouth of George River, in August 1882. Polar bears are still commonly seen on this island. USNM A23207, Division of Mammals, NMNH. Photo by Angela Frost, 2011.

Turner (1887a: 64-65) wrote about animals and the spirit world, including polar bears: "Each mammal is supposed to have an individual spirit and also a class spirit. The spirit of a monstrous white bear is the guardian of the reindeer. It dwells in a huge cave in the vicinity of Cape Chidley. I may add that the reindeer repair to those hills to bring forth their young and may have given origin to the belief. This reindeer spirit may be propitiated by the shaman, who decrees this or that among the people lest he be offended and withhold the material forms of those creatures which give food and raiment to these people." Turner collected charms and ornaments used by Inuit shamans from Ungava Bay, including a "white bear skin belt," called Kak shúngaut *(Turner 1887a: 232).*

The Giant Polar Bear
An Inuit story, Paul Jararuse (Heyes 2005)

There is an old saying, actually my father's ancestors went up this crossing, they actually seen a beast in the water, out in Labrador coast somewhere. When he was on his kayak he ran into the Labrador coast, because there are high mountains. When he was kayaking he ran into something white when it was supposed to be deep and dark. They used to call this Nanukulluk. This means it is a big creature, a big polar bear, we call Nanuk a polar bear, Nanukulluk is a big creature shaped like a polar bear. He ran into this white thing in the Atlantic sea off the Labrador coast. He noticed that it was a huge polar bear under water. When he saw it he got so scared that he started paddling out as fast as he could before it surfaced. That is true, because he really witnessed it. A huge, huge one. Yes, it can be bigger than that. I have seen maybe 16-17 footer, when I killed that bear. But I have killed a 14 footer before. There is an old saying too, every animal has a huge, let's say there is a shark, there is a huge shark out there, there is a huge polar bear out there. I believe in them too sometimes because I have seen this huge animal not too long ago. I know there is a huge, huge polar bear out there. There used to be a lot of polar bear dens up there near Killiniq in the small bay. Big dens. They claw out their dens, like this and they make their den. It can be so high. I have been in one but there was no polar bear inside of it. We tried to go to a polar bear den with one guy; we thought there was a polar bear inside because we know there is a story that the men used to go inside the dens when the polar bears were inside. They used to go in, go behind the polar bear. There is an old saying 'they can never get across it' because they don't want to get their den blocked; not much room to part. What they used to do is they would go behind the polar bear, go inside them, they used to push it when someone was waiting to harpoon it when it came over. In the old days they used mostly spear and harpoon or a knife with a wooden handle and struck it inside them. To the heart.

Polar Bear with an Inuk Wife
An Inuit story, Tivi Etok (Heyes 2010)

There was a big polar bear that had an Inuk wife. When he retuned home from hunting he turned into an Inuk. When he went back out hunting, he became a polar bear. Nobody knew this or saw him change into a polar bear. One day a group of hunters on a dog team visited the region in pursuit of seals. When the hunters saw the tracks of the polar bear they crept up to the bear and hunted it. But none of them knew that he had a wife. These were not like real animals. I didn't like these animals. God made animals and God made Inuk. But I don't believe that these animals turned into Inuit. This was like the work of the devil. It was trickery. The wife of the polar bear was waiting for her husband for a long time; but he had been caught. She didn't know that her husband was killed. He never came back. Some lemmings, polar bear, some fox, black bear, and wolverines were like tricksters that could become Inuk. These were not like real animals.

Rough Ice and Nanuklurk, the Big Polar Bear
An Inuit story, Noah Angnatuk (Makivik 1986)

When the bay is frozen on the Labrador coast the ice begins to make waves and breaks up even when there is no wind. I believe that there is a creature in the ocean along the Labrador Coast that causes the ice to break up. This creature is a huge polar bear called Nanuklurk. When Nanuklurk moves his huge body around, it generates very large waves, even when the weather conditions are nice.

Carnivores

271

The Man Who Wanted to Be Eaten
An Inuit story, Tivi Etok (Heyes 2007)

This story came from my father. It originates from Tasijuaq (Leaf River). It's about my father's friend. The older brother of my father's friend was killed by a polar bear while hunting during a time of starvation. The other brother was angry that his brother was killed. The brother came across a large polar bear and he started chasing him. The man wanted to be killed by a polar bear (since his older brother was killed by one). The polar bear ran away from the man, up a mountain. The man caught up to the polar bear and tried to put his hand inside the bear's mouth. But, the bear did not want to eat the man, so the bear covered his mouth with his paws. The bear had already eaten one member of the family; he didn't want to eat the other brother. Yet the man still tried to put his elbow in the bear's mouth, but the bear refrained from eating him. Since he didn't get eaten, the man decided to go home. The next morning the man went over the hill to see whether he could see a seal on the ice. He tried to look for a seal with the scope on his rifle, but saw a bear instead. The man then started running back home; the bear frightened him. When he almost reached the tent, he saw a baby playing. The man started playing with the child. The man could not remember why he had been running after he had been playing with the child for a while. All day long he could not remember what he had been fleeing from. At bedtime the man remembered what he had been running from; but by that time the polar bear had moved on. My father really enjoyed being around this guy, he was a very kind and friendly man.

The Man Who Wanted to Be Eaten
An illustration of a story about a man who wanted to be eaten by a polar bear by Kangiqsualujjuamiut elder Tivi Etok, 28 April 2005. Reduced from original, 11" × 17". Lead pencil on bond paper. Scott Heyes Private Collection.

The Spirit Wind
A carving made from caribou antler by Kangiqsualujjuamiut artist Daniel Annanack, 2005. The carving tells a story about the spirit wind blowing the animals into existence including the polar bear. Scott Heyes Private Collection.

CARNIVORES

Recollections on Inuit Use of Polar Bears
Tivi Etok (Heyes 2010)

I have seen many of them. Some polar bears could cover themselves in mud. These sorts of polar bears are very dangerous. Polar bears are not always pure white. We would always hunt the ones that were covered in mud. We wouldn't hesitate to hunt these; they were very dangerous. Some polar bears can pretend to be ice. The seal goes towards the bear, thinking that it is ice. The polar bear is very strong; even in its mind. The polar bear skin could be made into many things: mitts, pants, mattresses. When we laid on polar bear mattresses in our tent we were not cold in the winter.

When a polar bear is covered with mud we would shoot it, but the bullet would richochet. We would try and shoot them in the section of fur that was not covered in mud. Polar bears could build houses for themselves in the winter. In Inuit way, when the polar bear is out from the house the Inuit used to go and see the polar bear den/cave. When the polar bear is inside its den they are not scary. When an Inuk crawls into the den, like a baby, very slowly, the polar bear will come out of its den. A hunter would be waiting at the entrance of the den with a harpoon. My father broke and entered a den once by going inside and killing a bear.

Some polar bears are very dangerous; some aren't. There are different kinds of bears. They are like humans, with different personalities. Some are dangerous; some are friendly. Some are good. Some don't attack. When you go to a polar bear den you can tell what type of polar bear it is. If a rock is thrown inside the den the dangerous and scary ones will come out very fast. If we put a packsack inside the scary ones would attack it. This is how we would test how scary the bear was. The ones that would not touch it were frightened. These were not scary ones.

The scary ones would have their mouths wide open and would run towards Inuit. In Inuit way, you can also tell the difference between scary and timid polar bears by the way that its jaw is facing as it comes towards you. If the jaw is facing to the right then this means the bear is dangerous. This type of bear will attack you and throw you. The polar bear cannot turn to the left with its jaw; so any bear that confronts an Inuk with its jaw open to the left will not be dangerous. My son, Adamie, was attacked by a polar bear a few years ago; but I have not been attacked before. In Inuit way we used to watch the bottom jaw; this was the best way to see whether a polar bear was dangerous or not.

Tivi Etok (Makivik 1984)

Polar bears are the best divers, they can be underwater just like a seal. When a polar bear hunts a seal, it carries it inland on its back. I have found seal bones high up in the mountains, carried all the way to these highlands by polar bears.

Thomas Etok (Makivik 1984)

I have killed approximately 20 of them (as of 1981). Once I killed a bear with only one bullet. When I ran out of bullets, I buried my gun and cached it. Whenever I killed a polar bear, I tied a rope to its mouth and put the bear into the river where the water was deep. Then I pulled it back up and it would be black with leeches and other things from the water. The women would then clean the skin and the dogs would bring it to Nackvak Fiord. I have killed 5 bears at Alluviaq Fiord. A few years ago, I killed one bear near Payne Bay.

Polar Bear Hunting

An Inuk hunter from Kangiqsualujjuaq takes aim at a polar bear near the sea ice on the eastern shores of Ungava Bay. The hunter missed. As part of an initiation to being recognized as a learned hunter, young men are expected to hunt a polar bear. Turner (1887a: 231) described that certain types of arrows were used by Ungava Inuit for killing large game such as polar bears. On the use of the bow and arrows he remarked: "[This] means of procuring game has nearly passed out of existence, except where by some misfortune the individual is reduced to such circumstances as to re-employ that weapon in providing his immediate wants. The gun has superseded the implements of former days and now the bow and arrow is to be seen in the hands of small boys only. The character of the game required a special arrow and in some instances a particular form of bow for the white bear, wolf and reindeer could not be killed with the same arrow that would knock the life from a ptarmigan. The larger the game was the rule. The arrows under consideration are but eighteen inches in their entire length. The shaft feathers are placed with great care; everything indicating great strength. The points were of stone, bone (antler) or ivory as the case might be." Photo by Scott Heyes, 2004.

Raccoons

Procyon lotor (Linné) Storr.

[1461]

Order Name used by Turner	*Carnivora*
Current Order Name	*Carnivora*
Family Name used by Turner	*Procyonidae*
Current Family Name	*Procyonidae*
Scientific Name used by Turner	*Procyon lotor (Linné) Storr.*
Current Scientific Name	*Procyon lotor* (Linnaeus)
Syllabary	*Unknown and unnamed in Ungava*
Modern Nunavimmiutitut Term	*Unknown and unnamed in Ungava*
Eskimo Term by Turner	*Unknown and unnamed in Ungava*
Definition of Eskimo Term by Turner	*Unknown and unnamed in Ungava*

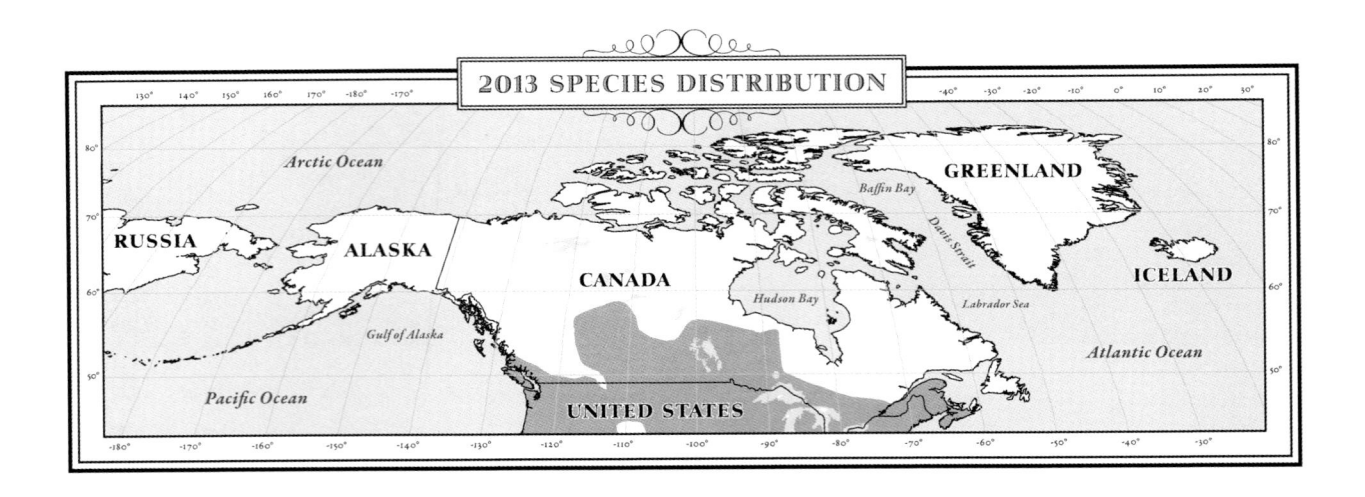

2013 Species Distribution

Chapter Two: Mammals of Ungava and Labrador

Descriptions by Turner

Several traders informed me of the occurrence of the raccoon in the area lying between Mingan and Northwest River. There can be scarcely reasonable doubt of its occurrence in the southwestern portion of the region under discussion, although specimens were not procured to substantiate the occurrence.

ᐸᒥᐅᓕᒃ / ᐸᒥᐅᖅᑐᖅ

Otter
pamiulik / pamiuqtuuq

Lutra canadensis, (Tur.) F. Cuv.

[1462]

Order Name used by Turner	Carnivora
Current Order Name	Carnivora
Family Name used by Turner	Mustelidae
Current Family Name	Mustelidae
Scientific Name used by Turner	*Lutra canadensis*, (Tur.) F. Cuv.
Current Scientific Name	*Lontra canadensis* (Schreber)
Syllabary	ᐸᒥᐅᓕᒃ
Modern Nunavimmiutitut Term	*pamiulik* (cf. Schneider p.238 – *pamiulik* = [...] something that has a tail); *pamiuqtuuq* (c.f. Schneider p. 238 = American Otter)
Eskimo Term by Turner	*pûm yú lik*
Definition of Eskimo Term by Turner	Otter [p. 1467]
Innu Term by Turner	*Ńtsuk* [p. 708]

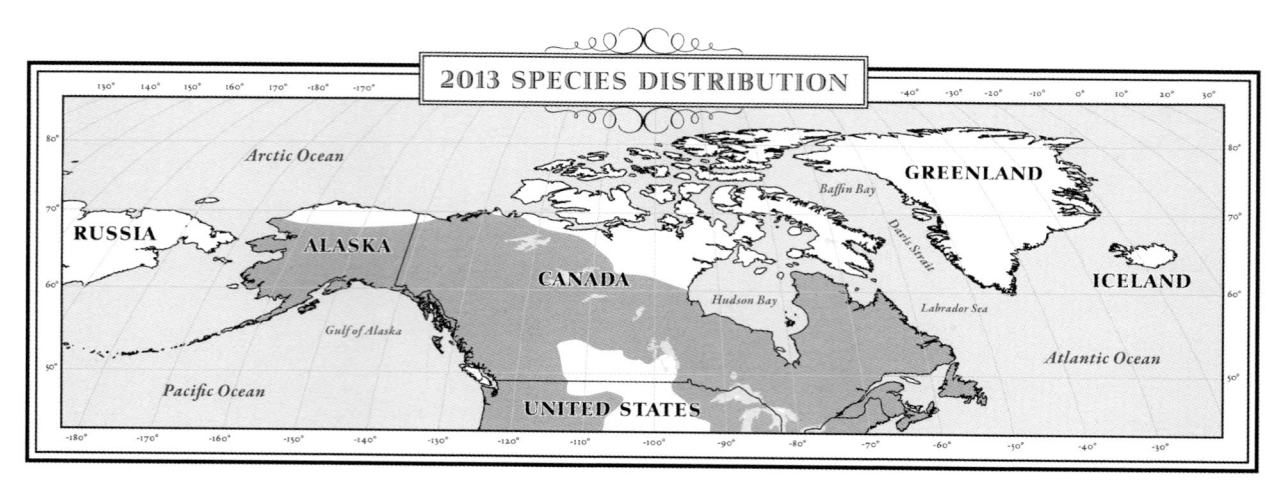

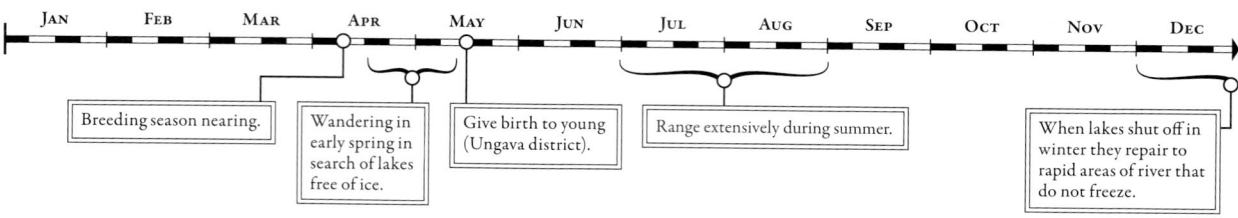

2013 Species Distribution

- Breeding season nearing.
- Wandering in early spring in search of lakes free of ice.
- Give birth to young (Ungava district).
- Range extensively during summer.
- When lakes shut off in winter they repair to rapid areas of river that do not freeze.

Chapter Two: Mammals of Ungava and Labrador

Descriptions by Turner

The North American Otter ranges throughout the wooded portions of the region. I am not aware that it frequents the barren grounds unless it may do so for only a short distance. Certain open areas devoid of trees are at times occupied by one or more of these animals but their occupation of that tract is only temporary. During the summer they range quite extensively and in the early spring are wandering in search of lakes whose favorable situation cause them to be the sooner free from ice. As the winter season successively shuts in the open lakes the otter repairs to the more rapid portions of the rivers which do not, until late in the winter, freeze over. Around these open holes, caused by the rapid water below, the tracks of the otter may be seen where it emerged to eat a fish in the open air or gambol in sport on the [1463] frozen snow. During these bright days the otter is often seen by the natives who immediately give chase, endeavoring to stop its retreat by getting between the hole in the ice and the animal. When attacked the otter gives a sturdy fight. Facing a dog without fear and only a good dog will successfully cope with this agile animal whose bite is something like the snip of a pair of shears and the clasp of a steel trap.

The Indians watch for these animals as they swim on the surface of the water and shoot them with large shot which they purchase for that purpose. But few are taken in steel traps as the animal is too wary to enter it unless by purest accident.

The sense of sight is good but not so acute as the sense of smell. The hunter must exercise the greatest caution lest the animal get scent of him. The hearing is also very acute and only the [1464] muffled cocking of the gun will insure the continuance of the animal on the water.

The motions of the animal are very quick and when pursued it leaps over great spaces. Near the northeast end of Whitefish Lake I observed where an otter had leaped in the snow a distance of fourteen feet and had to pass over a top of a slight ridge which caused the animal to rise not less than four feet in the air to avoid touching the snow. Other bounds were, on the level ground fully as great. What its object, unless sport, could have been I was unable to even conjecture as there were no signs of anything to alarm it and nothing in the way of food.

Less than one hundred skins of this animal find their way to the post of Ft. Chimo. The quality of the skins is excellent and will compare favorably with those of almost any other locality.

[1465] A single specimen was procured by me [USNM 14156]. It was about to complete its second year of age. The following measurements [see table] were taken in the flesh. No. 1616. Female. April 7, 1883. Near mouth of North or Larch River.

Tail regularly tapering and two-fifths as deep as wide. Claws pale bluish, diaphanous above and whitish on basal portion of lower part of arch of claws, forming a triangular, whitish mark on all the claws. Nose pad dull brownish-black. Eyes blue-black. Color above lustrous black, few white, small spots on neck and crown. Under the neck and jaws whitish brown; belly black with silvery reflection. Legs black. Sole pads chocolate brown. Corneous growths on the pads of hind feet were whitish.

As the breeding season was nearing I carefully searched for embryos. Nothing of the kind was found. The Indians informed me that she [1467] would not breed until another year older; and would then have two or three young and never over five. The young are quite helpless at birth and grow but slowly. They do not separate from the mother until the middle of the succeeding winter. The time of giving birth to their young is reported to be in early May for the Ungava district.

I learned nothing about its distribution in the southern portions of the region under discussion. The Eskimo give this animal the name *pum yú lik* in allusion to its tail.

	INCHES
Extreme length, nose to tip of tail	43.50
Nose to root of tail	27.50
Tail	16.0
Tail tip hairs	0.37
Nose to base of skull	4.70
Nose to anterior corner of eye	1.50
Nose to posterior corner of eye	2.00
Length of eye slit	0.50
Eye to ear	1.80
Heighth of ear	0.41
Width of base of ear	0.60
Between ears (anterior base)	3.10
Circumference head over eyes	7.00
Circumference over ear	10.00
Circumference base of skull	10.00
Circumference neck (anterior portion)	10.00
Circumference neck (posterior portion)	11.70
Circumference body (back of shoulders)	12.70
Circumference middle of body	17.50
Circumference over navel	15.00
Circumference over kidneys	14.00
Heighth over shoulders	7.00
Heighth over kidneys	9.00
Heighth over hips	8.10
Circumference radius	4.50
Circumference tibia	3.50
Length of forearm	3.00
Length of femur	4.70
Circumference of tail (one inch from body)	7.50

Recollections on Inuit Use of Otter
Tivi Etok (Heyes 2010)

Otters are good hunters of fish. They even catch ugly fish. They can eat anything. They hunt in salt water near the beach and also in streams and lakes. We used to trap them and sometimes we would wait for them to come up through a breathing hole, just like we would when hunting a seal. The meat is good to eat.

Female Otter Skull
This is the only otter specimen collected by Turner in the region. It was collected at Ft. Chimo in April 1883. Turner observed: "The otter is one of the most wary mammals requiring all the cunning and craft of the Indian [Innu] to entrap him by even the most approved methods. The senses of smell and sight are acute and scarcely less so that of hearing. The taint of the hunter's hand may be long detected after the object touched after exposure to the air" (Turner 1887b: 497). Inuit hunters from Kangiqsualujjuaq report that otters are typically hunted in the fall, especially along the Tunulic and Korac Rivers, and wherever fish abound (Makivik 1984). USNM 14156, Division of Mammals, NMNH. Photo by Angela Frost, 2011.

ᓂᐅᓪᓗᒃ

Skunk

niulluk

Mephitis mephitica

[1468]

Order Name used by Turner	*Carnivora*
Current Order Name	*Carnivora*
Family Name used by Turner	*Mustelidae*
Current Family Name	*Mephitidae*
Scientific Name used by Turner	*Mephitis mephitica*
Current Scientific Name	*Mephitis mephitis* (Schreber)
Syllabary	ᓂᐅᓪᓗᒃ (?)
Modern Nunavimmiutitut Term	*niulluk* (lit. [has] a bad leg)
Eskimo Term by Turner	*ni ú dlúk*
Definition of Eskimo Term by Turner	Skunk [p. 2689]. Turner noted that the Inuit word could not be fully determined and that the word was not in common use. He said the word was applied to animal with a white belly and from the description of which a skunk, *M. mephitica* was inferred.

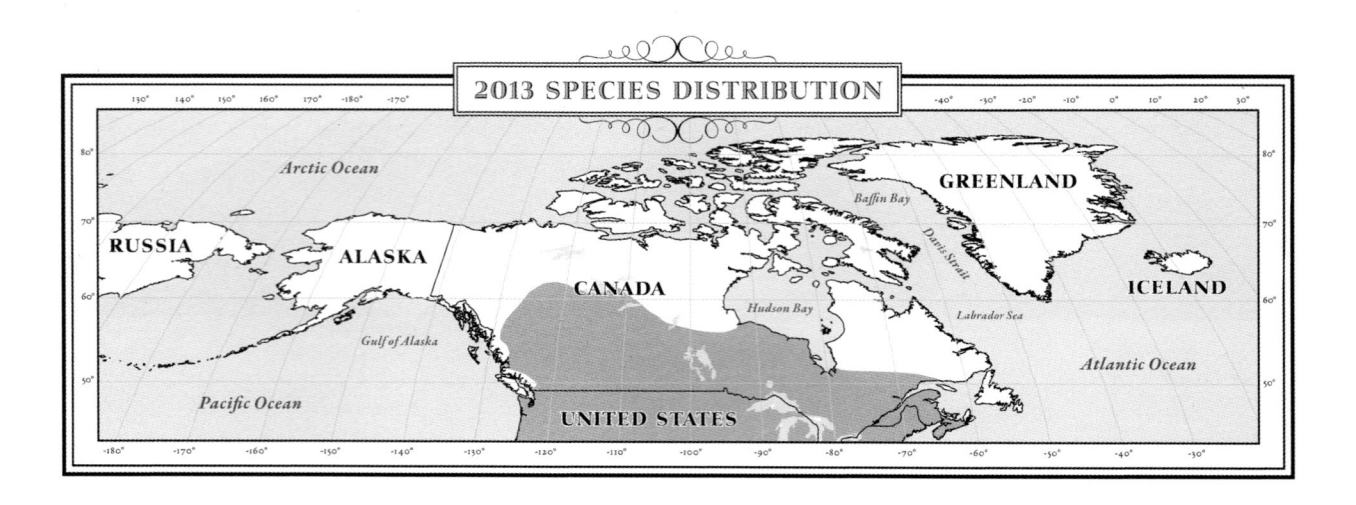

2013 Species Distribution

Chapter Two: Mammals of Ungava and Labrador

DESCRIPTIONS BY TURNER

The eastern range of the skunk extends within the region herewith included. It is reported as rare only along the Gulf Coast and somewhat more common northwest of Mingan as I was informed by a gentleman who had resided several years at that place.

ᕃᑫᔨᑫ

WOLVERENE

qavvik

Gulo luscus

[1469]

ORDER NAME USED BY TURNER	*Carnivora*
CURRENT ORDER NAME	*Carnivora*
FAMILY NAME USED BY TURNER	*Mustelidae*
CURRENT FAMILY NAME	*Mustelidae*
SCIENTIFIC NAME USED BY TURNER	*Gulo luscus*
CURRENT SCIENTIFIC NAME	*Gulo gulo* (Linnaeus)
SYLLABARY	ᕃᑫᔨᑫ; ᕃᑫᔨᑫᑕᓚᑫ
MODERN NUNAVIMMIUTITUT TERM	*qavvik* (cf. Schneider p.295 - *qavvik* = carcajou wolverine). Also *qavvikallak* (Tivi Etok, Pers. Comm., 2010)
ESKIMO TERM BY TURNER	*kû́ġ vik; kû́ vik; kû́ vĭk*
DEFINITION OF ESKIMO TERM BY TURNER	Wolverene [p. 2319; 1469]
INNU TERM BY TURNER	*kwe kúe chu* [p. 1832]

2013 SPECIES DISTRIBUTION

| Jan | Feb | Mar | Apr | May | Jun | Jul | Aug | Sep | Oct | Nov | Dec |

- Pairing season in Ungava; latter half of March.
- Young born in Ungava in late May.
- Nurse from May till beginning of Sept.

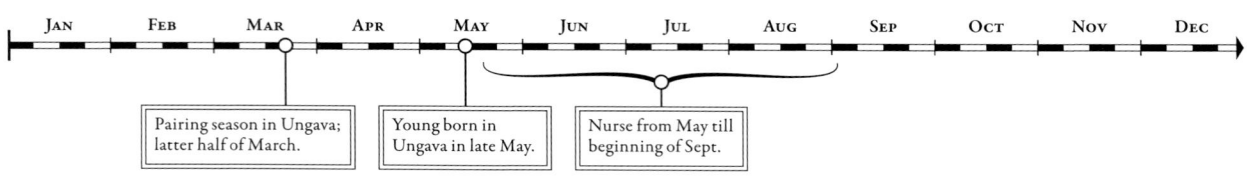

CHAPTER TWO: MAMMALS OF UNGAVA AND LABRADOR

Descriptions by Turner

The Eskimo of Hudson Strait apply the name *kûġ vik* to the Wolverene.[21] The Indian name is *kwe kive chu*. The Indian word is, without doubt, the origin of the name "Quickhatch," Carcajou, Carkajou, Karkajou of the Hudson Bay people and Canadians.

The distribution of this animal is very extensive, embracing all the region north of latitude 39° unless exceptional individuals may, in the alpine regions of the Rocky Mountains, descend farther south. In the northern tier of states it is still common and in portions of British America very abundant. The western range is far as the extension of land. I was informed that when the schooner "Eustace" in 1870 attempted to navigate "False" or Isanotaky Pass, separating the mainland of the peninsula of Alaska from [1470] the Island of Unimak (long. 163° W.) that the captain (E. E. Smith) and several others amused themselves firing at a Wolverene on one of the ledges of the overhanging cliffs forming one of the western points of that peninsula. At the station of Morzhova near the extremity of that peninsula I am informed that half a score of skins are obtained each year from the villagers who set traps for foxes and

---

21. The modern spelling is "wolverine."

"The female is extremely ferocious in defense of her young and only gives up the struggle at her last breath."

take an occasional Wolverene. This locality is important from the fact that it is situated nearly 400 miles west of the limit of trees, proving that the Wolverene is not necessarily an inhabitant of the arboreal tracts of the north. The diversified character of the wide region frequented by these brutes shows their ability to adapt themselves to the surrounding circumstances. The Wolverene is of a roving disposition, rarely travelling far [1471] at a time but by persevering accomplishes a great distance. If the supply of food is plentiful the creature remains until it has exhausted it and sets out for another tramp. They travel, under these circumstances, by day as well as by night. The short limbs do not carry it swiftly, scarcely faster than a man's rapid run yet by its untiring nature gets over a good distance in the course of the day.

Any kind of flesh is food for this animal. It is rarely satisfied by filling its stomach but practices the most disgusting defilement of all food which it is unable to consume or drag away. To its craftiness and cunning is added an almost supernatural bodily strength. The depredations committed by this brute are simply incredible. Wonderful instances of its strength displayed in destroying structures supposed to be strong enough to [1472] withstand the ravages of the black bear are torn down by this creature less than one fourth the size of Bruin. Like the bear it uses the forelegs to clasp or embrace an object it desires to seize or remove. Huge logs laid upon caches of meat or other stores have been removed by this animal and the stores either maliciously destroyed beyond use or so strongly impregnated with the stench of its ordure as to require days of exposure to remove the smell. It pursues a straight path when travelling if the character of the country will permit it. Otherwise the path is zigzag and each gully, hilltop, crevice and cranny is searched for a morsel of food. Its eyesight appears to be limited; hence, the animal frequently rises upon its haunches to view the surroundings. It is quite stubborn in its nature and rarely gives the path to any creature. It is rarely known [1473] to attack man unless he happens to be directly in its way or else when so pressed by hunger that it loses its natural instinct of self-preservation. There are few animals of the region infested by this creature that will attack it. An instance of a Wolf's respect for the Wolverene is related on another page.

When taken in a trap the Wolverene offers a most sturdy resistance and is so cunning that it appears harmless until the hunter unwarily approaches within seizing distance whereupon the animal springs upon the person and holds with the tenacity and strength of a bulldog. If the trap be not of the strongest character it will break under the great strength of the animal, which the instant it is free from the restraint of the chain, begins to travel and continues as long as it has a grain of strength. Many miles are covered [1474] by the time the hunter has visited his trap. An inspection of the locality will disclose the intent of the creature and if it has struggled so much as to nearly exhaust itself, it will be found not far off, recruiting its strength to leave that part of the county. Large rocks and stones are often piled upon the pole run through the ring of the trap-chain and snow or soil put within the crevices to freeze the stones together into a solid mass. If one of the forefeet of the animal is free it would surprise a novice in the trapping business what has been accomplished. (Relating actions of the Wolverene are very much like "Fish stories," accepted only in part.)

At Ft. Chimo during the winter 1882 and 83 the Wolverene was rather plentiful in that vicinity and quite a number were taken in traps. Ptarmigan *(Lagopus)* were specially abundant; and, doubtless, due to their presence the Wolverene [1475] was in hand. Others broke away and to show what one of the creatures may do I relate the following for fact. A trap had caught a Wolverene which was able to break away with the stick attached to the chain. A native informed me that he had seen where the animal, with the dragging trap, had crossed the river over thirty-five miles below where the trap was set. The animal had not gone on the river ice or at an intermediate point so far as could be ascertained but must have made a somewhat circuitous route through the wooded tracts; over hill and rugged ground to be at that place in less than thirty-five hours after its capture as frequent (daily) visits were made to the place where the trap was set and in which it was caught.

The natives say that the position of the eyes give the Wolverene but a short range of vision and that it constantly [1476] sways its head to each side in order to see even what is directly in front of it. The sense of smell is not acute; in fact, it appears

Wolverine Skull
Collected by Turner at Ft. Chimo, c. 1883. USNM 141911, Division of Mammals, NMNH. Photo by Angela Frost, 2011.

to stumble upon its food. A whiteman, Mr. John Miller, at Ft. Chimo went to visit his traps and found, after he had gone too far to retrace his steps, that he had but one gun cap. He saw a Wolverene making directly for him and when within few yards of him shot it. It appeared to be unconscious of his presence until the shot struck it. The marks of the teeth upon his gunstock were deep enough to prove the capability of this animal to inflict a serious wound.[22] The animal was an immature female and is now in the collection of the National Museum [USNM 14781]. A series of careful measurements are appended. This individual had lost a toe-claw and was doubtless the same that had visited a trap some three weeks before and in that trap a claw was found.

[1477] When the presence of a trap is suspected the Wolverene exhibits the greatest cunning in its devices to prevent the trap from springing. They are known to overturn the trap, dig the snow away and insert the paw under the jaw of the trap and press the tongue of the trap to spring it.

One of the officers of the Hudson Bay Company related to me that he was a witness to the manners of a Wolverene to entice a reindeer to come within springing distance. A large herd of the deer were feeding upon an open area. At one side was a *Gulo* practicing most astonishing acrobatic performances, such as rolling, tumbling, creeping and jumping toward one of the deer separated from the remainder of the herd. When the deer gazed at the strange freaks of the Wolverene the latter would move away but as soon as the deer resumed feeding it slowly made its way nearer it intended victim, doubtless awaiting a time when its ruse [1478] would insure success. The persistence of the Wolverene in tracking a wounded deer is well known. To the trapper it is a source of dread for it will not only destroy an animal already caught but evince a satanic disposition to tear the trap to pieces.

The Indians have an innate hatred for this animal and practice the most frightful cruelties upon it. Any conceivable torture that their minds can invent are perpetrated upon the animal. It is reproached with all manner of vile names. Its ancestors and relations come in for a good share of Indian cursing. The female is

---

22. Although encounters with wolverine are now considered rare in Ungava, Inuit hunters from Kangiqsualujjuaq assert that the wolverine is stronger than a polar bear. Wolverine are known to lift large rocks with their tails in search of food (Makivik 1984).

Wolverine Skin

Collected by Turner at the "rapids" near Ft. Chimo c. 1883. Turner wrote: "This individual had lost a toe-claw and was doubtless the same that had visited a trap some three weeks before and in that trap a claw was found." USNM 14781, Division of Mammals, NMNH. Photo by Angela Frost, 2011.

Chapter Two: Mammals of Ungava and Labrador

extremely ferocious in defense of her young and only gives up the struggle at her last breath. Should she be out when her young are discovered and return to find her home invaded she fights to the last moment in their defense. As might be expected from an animal having such a wide distribution the period of bringing [1479] forth their young varies as much as two months. The pairing season in the Ungava district is said to be in the latter half of March and that the sexes are attracted by the peculiar odor emitted by this mustelline animal. The female is said to rub her posterior on a tree stem or rock and that the male finding this evidence of her desire follows her trail.

The young are born, in the Ungava district in late May, giving a period of gestation of about nine weeks. Two to five young are brought forth, three being the usual number. They are quite puny and helpless for three weeks but grow rapidly after that time and soon become as playful as kittens and when old enough to emerge from the den they may be seen sporting near the entrance with their mother ever on the alert for approaching danger. The young are said to be of a soiled cream color during the earlier [1480] period of nursing and gradually assume a brownish tinge as the hair becomes longer. They nurse until the beginning of September and then wander with their mother in search of food. They soon separate and shift for themselves. They do not attain their full size until nearly three years old. After shedding the nursing coat of hair they take various patterns of color distinguishing them from any other animal. The extreme variability of color is not confined to either sex and until a certain age is attained the color of the darker parts deepens but becomes lighter with extreme age.

The origin of the peculiar pattern of coloration is illustrated by one of the Indian stories which teaches that the Frog is not a careful tailor. (See p. 293)

The weight of the animal varies from sixteen to forty pounds, the male being about one fourth heavier than the female but as much depends upon the [1481] supply of food for these voracious creatures it is rare to find one in a fat condition. The following notes and measurements were taken from a female about two-thirds grown. No. 999:

	INCHES
Extreme length (nose to root of tail)	32.37
Caudal vertebrae	7.25
Caudal hairs	5.50
Nose to eye	2.50
Length of nostril	0.62
Width of nostril	0.43
Length of orbit	0.68
Width of orbit	0.30
Color of eye	blue-black
Heighth of ear	1.40
Width of ear	1.40
Eye to ear	2.50
Cleft of mouth	2.30
Width of cleft	2.30
Width of middle of mouth	1.80
Between ears	3.92
Ear to angle of jaw	2.60
Ear to posterior corner of mouth	2.95
Heighth of head at angle of jaw	2.86
Heighth of head over eye	2.50
Nose to occiput	8.00
Width through body at shoulders	5.00
[1482] Width body through hips	4.50
Heighth at shoulders (from tip of toes)	15.25
Heighth at hips (from tip of toes)	16.00
Heighth at root of tail (from tip of toes)	13.00
Circumference at base of fore-toes	5.00
Circumference at base of hind toes	6.00

This individual was supposed to be about twenty months old and had not borne young. I observed that the lower tip of the vulva was turned up and inserted within the orifice. The glands with the anus were somewhat flattish and of less than an inch in diameter, being somewhat almond shape. The amount of secretion within the gland was considerable.

Origin of the People on the Earth
An Innu story, Turner (1887b: 608-609)

A Wolverene was walking along the bank of a river and saw, swimming in the water, a Muskrat. He asked of the latter, "Who are you? Are you a man or a woman?" The Muskrat answered, "I am a woman." The Wolverene than asked her if she would become his wife. To this, she replied, that she dwelt in the water. The Wolverene said, "You can live on the land as well as in the water." The Muskrat then consented to come and live with the Wolverene. She went out on the land and erected a tent for their home. Soon after this, a child was born. The husband said it would be a white man. In the course of time a second child was born and the father said it would be an Indian. The third was an Eskimo and the fourth an Iroquois and the fifth child was a Negro. The first child was born naturally. The Indian from the mouth of its mother. The Eskimo from behind. The Iroquois from its mother's nose and the Negro from her ears. Each of these was the first person of its kind and the father of its people. They grew to manhood and their mother called them to her. She told them, "My sons the tent is now too small to contain all of us. You must separate and each go to a different place. There you will find wives; and, whenever you need anything which you have not, go to your elder brother, the white man, and he will give it to you."

The Wolverene and the Bear
An Innu story, Turner (1887b: 615-620)

A Wolverene was taking a stroll, one evening, seeking food. As he came to the brow of the hill, on which he was walking, he looked into the valley below and perceived a sleek and fat Bear feeding upon the berries, which grew in such profusion as fairly to cover the ground with their bead-like forms. The Wolverene had been several days without food, for all that season all the creatures of the fields and forests were on the alert and were difficult to capture by the slow-motioned creature which obtained its prey by stratagem more than by fleetness of foot. Knowing that the Bear never sees what may be but little distance off while it is feeding, its paws drawing the stems towards its mouth and its eyes directed to the ground, was the opportunity for the Wolverene to steal near its side and suddenly address Bruin. The Wolverene crept up softly and inquired, "Is that you, sister?" The Bear answered, "I did not know I had a brother," but said it in a low tone so that the Wolverene would not hear it and scampered away at the sound of the voice.

The Wolverene screamed, "Come here, sister. Our father has sent me to seek for you, for you know you were lost when you were but a small thing scarcely able to run. You had wandered off while you were picking berries, several summers gone by; and, your parents have long mourned for you as dead." The Wolverene had selected a spot on the hill to where the berries grew even more plentifully than in the valley where he accosted the Bear. The Bear stopped when it hear these words; and, being easily misled believed the Wolverene had spoken the truth, although the Bear could not remember but two seasons of berries, but supposed the Wolverene knew best.

She returned to the Wolverene and was asked if she had seen the patch of berries as large and ripe which grow upon the hilltop. The Bear replied she could not see that far. The Wolverene invited the Bear to go with him to the spot and they would feast together. When they arrived at the hilltop, the Bear saw such ripe and luscious berries that she exclaimed," How could you see so far? Your eyes must be very keen to see berries so far off from where we stood below."

The Wolverene replied that his eyes were very keen-sighted, adding that their father and mother had made them so by putting bandages soaked with cranberry juice upon them; and, although it pained very much at first it soon passed off and the power to see even the smallest object at a great distance was far more to my advantage than the few hours of pain caused by the application to render them sharp-sighted.

The Bear remarked that she wished her eyes were as good as those of the Wolverene. The latter gladly offered to make them as good as those belonging to himself; and said, "If you will gather a quantity of cranberries (*Oxycoccus*) I shall prepare a sweathouse for it is necessary that you perspire freely so that the juices of the berries may go directly to work and produce the desired effect. I must, however, inform you that the pain will be excessive on the first application, but the benefit to be derived soon causes one to forget it."

The Bear expressed her willingness to undergo any pain in order that her eyesight might be improved; and gladly set about the task of gathering the red berries, which grew abundantly on the brow of the hill.

The Wolverene withdrew a short distance and began to arrange a number of stones in a circle. In the center a pile of stones was erected and near it a large stone was placed as a pillow for the head of the Bear. By the side of this lay a large sharp-edged stone. A quantity of spruce boughs was placed as a shelter to keep the air from chilling the body of the Bear while undergoing the operation.

A fire was then started and when the stones were heated the Wolverene called to the Bear to come as everything was in readiness. The Bear returned with her arms loaded with twigs bearing greatest quantity of the bright, red berries, the color of her own blood. The Wolverene directed the Bear to place the berries in a cloth and bruise them until the skins of the berries were broken. The bright, red juice streamed down the paws of the Bear as she pounded them and when she held them up to show the Wolverene she said "How much it looks like blood. It makes me shudder." The juice was expressed and collected. Cloths were dipped into it and the Wolverene directed the Bear to enter the sudatory and lie with her head upon the stone selected to serve as a pillow. She did so while the Wolverene arranged the cloths as bandages. He now entered and told her to open her eyes wide and enjoined upon her to remain quiet. A few drops of the acid juice were placed in each eye. In a moment a great quantity was dashed in each eye and under the lids. The Bear groaned with pain; but the Wolverene consoled her with the remark that it would be but a short time before the pain would pass away. The pain was now so intense that the Bear shrieked. The Wolverene said that his parents told him the greater the pain the better would be his eyesight and for that reason he endured the hurt without complaint.

The Bear willing the operation should continue; and, when made completely blind by the berry-juice the Wolverene seized the sharp stone lying near and with a with a single blow broke her skull and feasted on her body.

The Wolverene Who Wished to Fly
An Innu story, Turner (1887b: 621-624)

A Wolverene, walking through the forest, looked upward and far above the treetops he observed a flock of Brant [geese], *Branta nigricans*, flying swiftly through the air, muttering and cackling as they hurriedly journeyed to the distant north far beyond the dwelling-place of the Wolverene whom the Indians persecuted so mercilessly that he feared to traverse even the loneliest wood during the day.

As he saw the birds appearing so light-hearted and happy he bent his head and pondered how he, too, might have wings and be able to fly like the geese and ducks.

At length he devised a scheme which he thought would enable him to fly like a bird; so calling all the birds together he announced to them that he was their brother and that he should be dressed in feathers like them and fly to the most distant parts he might desire.

When the birds were summoned and collected he said, "Do you not know that I am your brother? Come to me and I will dress you up in feathers and put wings on myself. We will all be dressed in feathers." He stuck wings to his sides and said to the leader of a flock of brant that stood wondering by, "Now, let us fly." The old brant cautioned him not to look below when he heard a noise as they crossed the points of land along the course of the stream down whose length they winged their way. "Take a turn when we turn, but do not look below." When they came to the first turn the Wolverene remembered the injunction of the leader of the flock; but soon became so expert in the use of his newly-acquired members that he was forgetful of his safety. When they came to the next point of land he heard a deafening shout and lowering his head to discover the course perceived a lot of Indian tents on the ground below.

He became so confused that he tumbled to the ground "like a bundle of rays," falling so violently that life was knocked from him by the force with which he fell.

All the Indians ran up and exclaimed, "A brant has fallen." One of the old women seized the fallen creature and began to pluck its feathers. She then disembowelled it and while doing so detected such an intolerable stench that she exclaimed, "This brant is not fit to eat; it is already stinking."

She gave the carcass to one of the children to throw away. In a few minutes another old woman came in and inquired of the first what she had done with the brant which had just fallen down. The other informed her that it was so stinking that it was not fit to eat. "How could it be stinking if it has but just fallen." The other replied, "If you do not believe it go to the thicket and believe it for yourself." She went and found nothing but the body of the Wolverene.

The Wolverene and the Rock
An Innu story, Turner (1887b: 601-607)

At the close of a pleasant, summer afternoon, a Wolverene was strolling along a hillside where grew clumps of evergreens; and, at the bottom of the slope, birches and poplars fringed the stream that coursed idly through the valley below.

Absorbed with his thoughts, how to obtain his morrow's dinner, he came to a large Rock, which sat on the surface of the ground.

Walking directly up to it he accosted it with the remark, "Was that you who was walking just now?" The Rock intimated in unmistakable words that the Wolverene had stated falsely. The Wolverene replied, "You need not speak in that manner for I have seen you walking."

The Wolverene started off and dared the Rock to follow him. The animal went back, and with a smart blow of its paw slapped the face of the Rock and taunted it; bantering it to run a race with him, or see if it could catch him.

The Rock replied, "Did I not tell you I can neither run nor walk?" "But," added he "I can roll." The Wolverene laughed and said, "That is just what I want you to do." The Wolverene trotted off and looked back. The Rock gave a surge and slowly moved from its bed in the soil. Slowly it rolled over, toppled and turned. A plunge and it moved down the hillside toward the beast, which, with head turned back, was laughing at the awkward motions of the ill-shapen mass tumbling behind him.

The Rock kept along close to the heels of the brute, each momentarily increasing their speed; the one to escape and the other rolling and bounding down the slope.

The Wolverene now found that the Rock could roll faster than he could run. It began to leap over logs and stones, which seemed to make the Rock jump the faster. The beast fearing each moment to be his last, as the Rock went plunging to the valley below. A large log lay at the foot of the hill and over it the Wolverene jumped without touching it. The Rock halted for a moment; and, with a bound sprang high in the air, alighting on the tail and hinder limbs of the terrified beast it was pursuing. It came to a standstill. The Wolverene screamed with pain and exclaimed, "Go away: Get off of me. You are breaking my bones." The Rock remained firm and replied, "You tormented me; struck me and defied me to run after you. I shall not stir until someone removes me." The Wolverene retorted, "If you do not get off of me I shall call my brothers, the Wolves and Foxes, to come, and to push you off." With loud shrieks the Wolverene shouted to his brothers to come. The wolves hearing their brother's cry of distress stealthily crept toward the place, taking care to secrete their bodies among the bushes until they came near. The Foxes slipped among the grasses and weeds until they saw the wolves standing near a large Rock.

They saw their brother lying under it and inquired how he came to be in such a predicament. The Wolverene said he had been walking along the hillside and heard the Rock threaten to kill him. He then stopped and asked if he had ever offended any of the Rock's relations; and if he had forgotten the many favors he had personally done for the Rock. The Rock jumped from its bed and sprang after him; and that while springing over a log he had stumbled and the Wolverene pounced upon him while he was bandaging his toe. The wolves and foxes shook their heads and said it could not be true. The rocks of the fields were the friends of all the beasts and they doubted his word so much they concluded it served him right for some meanness he had done.

The Wolverene cried bitterly, but they gave him no help. The pain increased so much the Wolverene cried to the Rock, "If you do not get off of me I shall call my brothers, the Thunder and the Lightning. They can take you off from me if the wolves and the foxes cannot do so. These animals tried to push the Rock away, but it was so heavy they gave up the attempt.

The Wolverene now called the thunder and lightning to come to his aid. A huge black cloud appeared in the southwest sky; the air was still and hot; while the crushed brute lay panting under the load on his limbs. In a moment the trees bent their heads and the grass and bushes laid low on the ground as a gust of wind swept by. The sky became black and the angry muttering of the thunder caused the wolves and foxes to slink back to the shelter of a friendly rock until the storm would pass by. The vivid lightning flashed as it sped on the wings of the wind, darting here and there its fiery tongue, blasting a tree or shattering a rock that obstructed its pathway. It paused a moment when nearing its brother, the Wolverene, and perceiving his misfortune, rushed back to gather force, and with a dash it struck upon the Rock and shattered it into a thousand pieces; while an appalling burst of thunder announced the release of the Wolverene.

The flying pieces of rock tore the skin of the Wolverene completely from its owner's back; and, as the Wolverene gathered up the shreds of his garment he shouted to the retreating lightning, "You might have struck the Rock easier. You have caused my coat to be torn to pieces." The thunder laughed and the lightning flashed as they sped on their journey down the plain.

The naked Wolverene collected all the pieces of his coat and said, "Well, now, I must go to my sister the Frog, who dwells in the swamp at the head of the plain and have her sew my coat. He went there and found his sister sitting on the bank of the pool which was her home.

The Frog was amazed at the pitiable plight of her brother and gladly consented to mend his garment. The Wolverene laid down to take a nap while she awakened him with the good news that the work was finished. He took the coat, looked at it, saw that the stripes had been put on the wrong way. He was so angry that he slapped her ears and sent her headlong into the pond. The Wolverene put the coat on his back and started for the home of his sister, the Mouse, who dwelt on the hillside. Arriving at her home he found the tiny creature at the door ready to welcome her big brother. She laughed at the grotesque pattern of his coat and inquired, "Who sewed your coat"? The Wolverene replied that his silly sister, the frog, had put the pieces together in that manner; and, that he now wanted her, the Mouse, to sew it as it should be done. The mouse set to work and, in a short time, had the coat sewed so that the brown pieces were on the side and the black was on the back of the coat.

The Wolverene was so pleased with his sister's work that he said, "You have sewed it very well. You will live in the green grass in the summer and in a grass house in the winter." He put on his coat and walked away determined never to speak again to a Rock.

A Wolverene Teaches Birds to Dance
An Innu story, Turner (1887b: 627-631)

A Wolverene was trotting along the seashore and espied a number of ducks, geese and loons swimming in the water not distant from the land. Being very hungry he endeavoured to lay a plan by which he could entice them within his reach. At length he thought how fond such birds are of good things so he called to them, "Come here brothers; come here, I have found a pretty bee's nest and I shall give it to you if you will come on the land and dance."

The fowl eagerly swam to the land, but declared their inability to dance. The Wolverene told them he would teach them; saying "Now, shut your eyes and do not open them until we are all dancing." He continued, "I shall sing" and began to sing, *"A' ho' ho' ho' ho' m'u, A' ho' ho' ho' ho' m'u, A' ho' ho' ho' m'u hû'm, hû'm, hû'm."* The loon heard the word *hûm* so often that he partly opened one eye and saw the Wolverene snipping off the head of a duck, at the farther end of the row, every time the Wolverene said *hûm*. The headless ducks lay on their backs kicking while the Wolverene was nearly choking in his greediness but said he was laughing at the way the ducks danced.

The loon took alarm and fled to the water, screaming, "Our brother is killing us." The Wolverene sprang after the loon but it reached the water and disappeared, arising to the surface some distance off and there began to scream *A' ho' ho' ho'."* The Wolverene shouted "Hold your tongue, you red-eyed fowl." He then went back to where the ducks had been killed and began to pluck the feathers from their bodies. He prepared them for the kettle and soon had them on to boil over a bright fire.

He was so weary, from the labors of the day, that he felt very sleepy; but, hearing a noise he looked up and saw a jay hovering near. The Wolverene took a brand from the fire and cast it at the bird which only laughed and teetered up and down upon the limb where it was now sitting. The Wolverene exclaimed, "You will be telling on me you long-tongued bird." The jay then flew away to where a party of Indians were encamped and informed them, "Our brothers, the Wolverene, has killed a number of ducks and has them cooking in a kettle. If you are hungry, come with me and I will show you where the beast is; I think he is now sleeping." The Indians replied they were very hungry and would gladly accompany him. When they arrived at the place they perceived the Wolverene sleeping not far from the side of the fire. The Indians carefully removed the kettle and took from it all the meat which they quickly picked from the bones and put the latter back into the pot. When they had consumed the meat they went away, leaving the kettle on the fire.

One of the Indians proposed they should secrete themselves and observe the actions of the Wolverene when he should awaken. In a little while the Wolverene aroused himself; and, being very hungry said, "Now, I shall have my dinner." He removed the kettle from the fire and poured the broth into a pan; finding nothing but clean bones at the bottom, he remarked to himself, "I have been sleeping a long time that the flesh should boil from the bones; it shows, however, that the birds are tender and that I have made a nice pot of broth." He tasted it and found it to be very weak. The jay set up a chatter and informed the Wolverene that the Indians had eaten the meat while he had slept. The Wolverene was very angry and upbraided the bird saying, "Why did you tell the Indians; you stupid bird, I was keeping a nice piece of fat for you. You will not now get it for your impudence."

A Wolverene Goes Begging among Wolves
An Innu story, Turner (1887a: 675-692)

A Wolverene and his family dwelt, in their den in a cleft of a rock, where they were secure from the storms and the attacks of their enemies. The father was an idle character, always planning some mischief from which his cunning aided him to escape punishment.

The summer had passed and winter was now here. He had provided no food, preferring to roam about the hilltops. The wife, scarcely daring to leave her nearly helpless young, could wander but a short distance from her home to procure food either for herself of the little ones; they were poor, indeed.

When winter came the wife begged her husband to cease his idling and procure some food; for she and the children were starving. He promised to go to his brothers, the wolves, and ask their assistance.

The next morning he started away, telling his wife that he would not be gone more than four days. He travelled all the day until toward evening he saw his brothers, the wolves, running upon the ice on the river. He saw four old wolves trotting slowly along, while, at a great distance ahead there were others hurrying with all their speed. These latter he knew were hunters and would soon have some food. The Wolverene ran toward the old wolves and one of them remarked to the others, "There comes our brother, the Wolverene." The Wolverene coming up to them said, "Brothers, I am starving and my family have nothing to eat." The wolves answered that they had nothing to eat now, but the wolves in advance were on the trail of some deer and they would soon have an abundance.

The Wolverene asked them where they would camp for the night. They replied they would continue up the river until they came to a mark which would indicate the camping place for the night.

After a while they saw a mark on the bank of the river and they knew the place. They went there to await the return of the hunters whom they knew were not distant. They, in the meantime, set to work and gathered boughs from the spruce shrubs to lie upon.

In a few minutes the hunters returned and arranged the poles over which the tent skins were to be stretched. The old wolves remarked they would soon have the tent arranged to sleep in. The Wolverene began to wonder whence they would get the tent-skins to cover the poles, and the fire to keep them warm. An old wolf said, "Our brother wonders whence we will get the tenting and fire." The Wolverene replied, "I did not think that, I only thought my brothers will soon have a nice tent up and then I shall be so comfortable." The wolves sent him off to collect some dry brush twigs and when he returned he saw the tent already up. He stood outside, holding the bush in his arms. One of the wolves now told him to bring the brush within the tent. He did so, and gave it to the leader of the gang of hunters.

The leader gathered up the brush and arranged it to kindle quickly. The Wolverene wondered how it was possible for her to make a fire. An old wolf said, "Our brother wondered how you will make a fire." The Wolverene carefully watched every motion of the wolf and after she had arranged the sticks she went out to get some snow to melt for water to drink. She brought in the water and placed it near the pile of brush and in a moment sprang over the kindling; instantly it started into a blaze and was soon a bright, sparkling fire, shooting here and there among the twigs. She placed the kettle of snow on the fire and soon had some water to drink.

The Wolverene had not seen the wolves bring any deer meat home with them and began to wonder what they would cook for supper. One of the wolves said to the leader, "Our brother wonders what we will have for supper." The leader went out of the tent and in a few minutes brought in the brisket of a deer.

After the Wolverene saw the meat he said, "I did not wonder what you would have for supper; but wished I had some nice meat for supper." The flesh was cut up and placed in the kettle to boil. When it was cooked they took it out and served the Wolverene with the choicest portions of flesh and large pieces of fat.

He was asked to eat all of it, but there was so much of it that he could not consume it all. He was about to place the remainder on one of the poles of the tent when one of the wolves observed his intention and said,

"If you put the meat on the pole it will turn into bark." The Wolverene now laid the piece of meat on the brush, or floor, and when he thought none were observing him he thrust it between the pole and the tent.

The Wolverene was fatigued and soon went to sleep. One of the wolves took away the piece of meat from between the tent and the pole and inserted a piece of bark instead. In the night the Wolverene awakened and felt for the piece of meat he had placed on the tent pole, but he found a piece of bark instead of it. One of the wolves saw him do this and said, "Did I not tell you the meat would turn into bark if you put it there?" The Wolverene went to sleep again and when he awakened the wolves had another kettle of meat already boiled.

The wolves told him to arise and eat his breakfast in a hurry as they had seen fresh deer tracks and that they must pursue them for in that kind of weather the deer travels rapidly; and it might be a long time before they camped again.

He ate his breakfast and thought he would watch how they disposed of the different things they had in the tent and what they would do with the tenting itself. The wolves were smiling while they were talking among themselves and as they did so the Wolverene wondered what put them in such humor. The Woverene stepped outside of the tent and the moment his back was turned the kettle and tent vanished. He failed to discover where they had secreted it as he heard no noise.

They now started for the ice on the river. The hunter gave a yelp and a bound, then ran quickly along the trail of the deer while the four old ones and the Wolverene trotted along leisurely behind. They finally came to where the deer had taken to the land and the progress was necessarily slow. The old wolves came up and said they must follow the tracks of the hunters until they came to the mark which should indicate the site of the camp for the night. They soon came to the bones of a freshly killed deer. The Wolverene said to himself, "Well if that is the way they do I shall have to go without my supper." They soon came up to the place where the mark was set and one of the wolves said to the Wolverene, "Brother, do you see the mark which is where we are to camp for the night?" They went up and in a few minutes the hunters returned. They put up the tent poles and sent the Wolverene for some dry twigs with which to make a fire. When he returned he saw the tent was up and the wolves were engaged in stripping fat from the huge piles of meat before them, placed in the middle of the tent.

The brush was put down and the leader jumped over it and a bright fire was soon started. The leader now said to the wolves, "Let us go out and see what our brother will do when left alone with the meat." They went out and asked him if there were any holes on the tent that might let in the cold air. The instant they were out the Wolverene began devouring the choicest portions of the fat. His mouth was so full of the fat when they asked him if there was any holes in the tent he could only mumble out the word "yes." They asked him again, but this time he could not answer as he was nearly choking in his haste to swallow the food.

The wolves determined to go inside and when the Wolverene heard them he sprang to the other side of the tent and appeared to be examining the holes that would let the wind inside the tent. The wolves paid no attention to the fact their brother had eaten the best of the meat and fat. They put a kettle of meat on to boil and soon had their supper ready. Again they gave the Wolverene the choicest portions and pressed him to eat more and more. He ate so much that after he went to sleep he was seized with violent pains and vomited for a long time, becoming so weak that a high fever came on and was followed by a chill. He now asked one of the wolves for a blanket to cover himself. One of the wolves laid her bushy tail over his body. The Wolverene was very angry and told her to take her foul-smelling tail from off his body. The wolf then gave him a dressed deerskin for a blanket.

The next morning the Wolverene announced that he must return to his family, which he now supposed were starved to death. One of the old wolves directed another of the younger members of the party to prepare a sled and to load it with meat for the Wolverene to take to his family. The wolf prepared the sled and made it so long that the Wolverene could not see the end of it. When he was ready to depart the Wolverene asked the leader of the hunters to give him some fire. The wolf inquired how many nights he would be on the journey to his home. He announced that it would take four nights to reach his home.

The wolf now directed him to lie down; and when he did so the leader jumped over his body four times, and strictly enjoined upon him not to look back upon the load of meat he was to take home.

The Wolverene replied that his nose would be pointed toward his home and that he would follow that part of his head; that he did not intend to stop for rest until he had seen his family. When he started he found the sled to be quite light and he tripped merrily along. He had not been gone many hours before he concluded he would stop and see whether the wolf jumping over his body would empower him to make a fire. He gathered a lot of dry brush and carefully arranged it. He now jumped over it and a bright blaze greeted his gaze. He now saw how easy it would be for him to make a fire. He resumed his journey. At night he came to a place where he thought he would camp. He gathered brush and soon had a fire by jumping over the bundle of twigs; cooked his supper and laid down to sleep. The next morning he started off, but feeling very tired he threw away his flint and steel and said they were too heavy to carry as I shall not need them anymore for I know as well as the wolf how to make a fire.

At evening he camped; made a fire; cooked his supper and went to bed. Up to this time he had travelled but slowly and was yet far from home. He arose early and started off but feeling very tired he concluded to camp at noon. He did so and gathered some brush. He piled it nicely and said he would have a big fire this time. He jumped over it and no fire started. Again and again he jumped, each time looking back to see the blaze creep up quickly among the twigs. No fire sprang up. He jumped again until he was worn out and then abused himself for being so foolish as to throw away his flint and steel that now lay many miles behind him.

He had no other course than to return for his fire-making tool. He upbraided his folly and became so inflamed with passion that he was quite beside himself. He began the return to the place where he had left his flint and steel. He travelled all that day, the succeeding night and day before he reached the former place where he had thrown away the implements.

When he found them he hastened back for he was anxious to ascertain the safety of the sled and meat. On his arrival he camped for he was so nearly exhausted that farther journeying was not possible for him. The next morning he looked again at the sled and saw how long it was; he essayed to move it. All his wondrous strength was to no purpose; he could not budge it.

He was hungry and determined to eat some of the store on the sled and thus lighten the burden. He looked and thought the sled did not appear as long as but an hour ago. He consoled himself with the reflection that his eyes must have deceived him because he was so weary.

He took some meat and fat, devoured it and again tried to move the sled, but with no success. He was compelled to erect a stage and placed the meat, fat and other effects upon it, for he knew that his family were in greatest need of food and that he must succor them soon lest they die before he returned from the many difficulties that beset his progress. After the load was ready for placing on the stage he selected some choice pieces to take on his back, for he resolved to set out for his den with all possible haste. He started off and in a short time was out of sight.

One of the old wolves directed the younger ones to go and destroy all the meat, fat, sled and stage which the Wolverene had left behind him. They went and ate all the food, tore the sled to pieces, levelled the staging and left scarcely a trace of the possessions of the Wolverene. When the Wolverene arrived at home he told his famishing wife and children that he had obtained great quantities of food, so much so that he was compelled to leave the remainder on a stage lest the thieving wolves should destroy it. The wife begged him to give her some fat and meat as she was so weak and could scarcely stand on her feet, while the little ones were much reduced from wanting food. The all ate so much they were taken quite ill and it was two days before they recovered.

The next day the Wolverene determined to return for the supplies he had left behind him. When he returned he found the wolves had destroyed everything, leaving a few, dry bones, scraps of meat and crumbs of fat that seemed to deride the miserable creature seeking them amongst the snow. At last he exclaimed, "The wolves have ruined me; the wolves have ruined me." He gathered the bits of food lying here and there, put them in a heap and found it would make but a scanty meal. He now started home; and, when he arrived there he told his unfortunate family of his losses, stating that the wolves had stolen all the food he placed on the stage, upbraiding his relations for being such mean beasts that he declared he would never again be seen in their company or claim relationships with them.

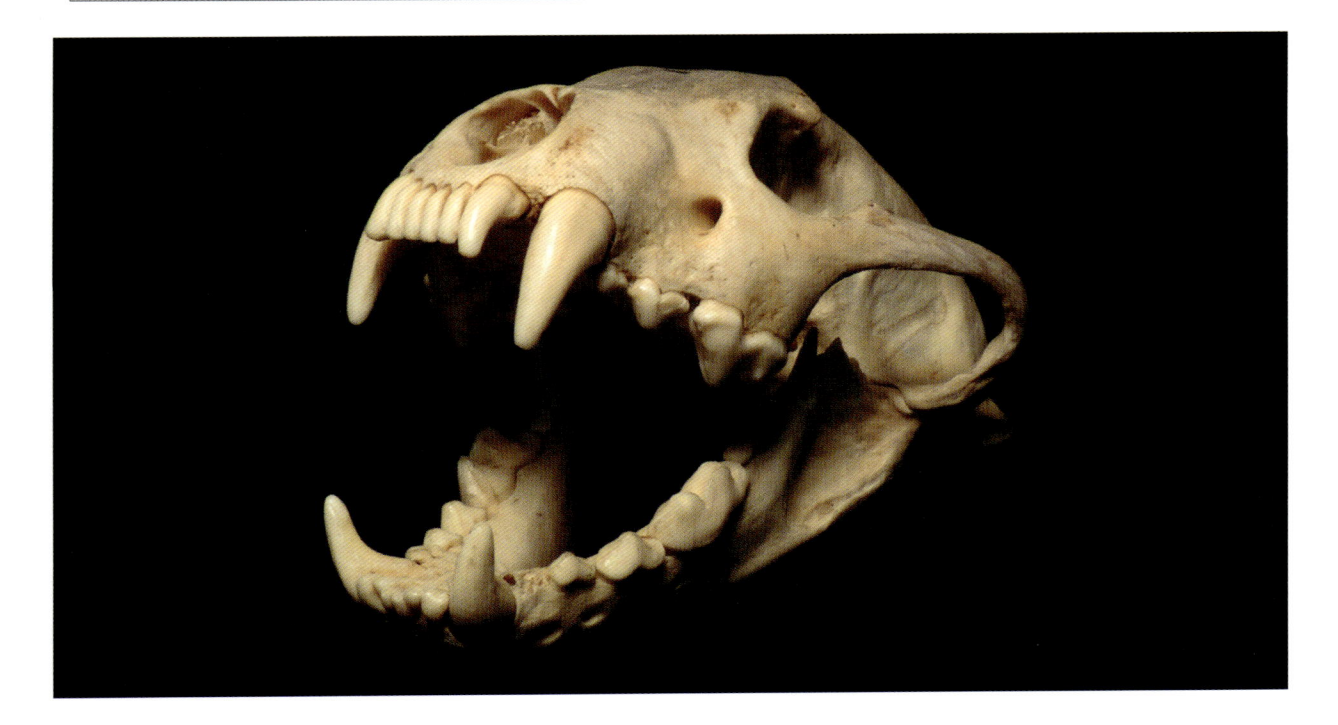

Wolverine Skull
Collected by Turner at Ft. Chimo, c. 1883. Turner wrote: "When taken in a trap the Wolverene offers a most sturdy resistance and is so cunning that it appears harmless until the hunter unwarily approaches within seizing distance whereupon the animal springs upon the person...with the...strength of a bulldog." USNM 141911, Division of Mammals, NMNH. Photo by Angela Frost, 2011.

ᐃᖃᓗᒃᓯᔾᐅᑎ
Mink
iqalugarsiuti

Putorius vison, (Schreber) Gapper.

[1483]

Order Name used by Turner	*Carnivora*
Current Order Name	*Carnivora*
Family Name used by Turner	*Mustelidae*
Current Family Name	*Mustelidae*
Scientific Name used by Turner	*Putorius vison, (Schreber) Gapper.*
Current Scientific Name	*Neovison vison* (Schreber)
Syllabary	ᐃᖃᓗᒃᓯᔾᐅᑎ; ᑯᓯᐅᑎ; ᐆᒐᕐᓯᐅᑦ
Modern Nunavimmiutitut Term	*Iqalugarsiuti; kuutsiuti; uugarsiut* (cf. Schneider pp. 93, 154, 469 respectively – all equated to mink [*Mustela vison*] in various parts of the Ungava Peninsula)
Eskimo Term by Turner	*Not Recorded*
Definition of Eskimo Term by Turner	*Not Recorded*

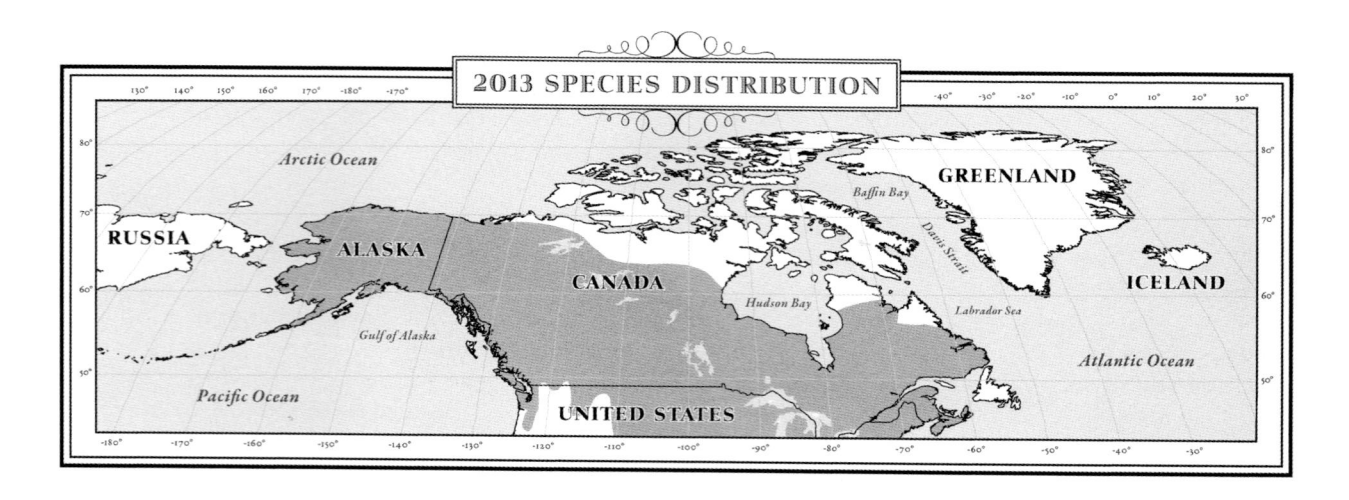

2013 SPECIES DISTRIBUTION

CHAPTER TWO: MAMMALS OF UNGAVA AND LABRADOR

Descriptions by Turner

The mink is not at all plentiful in the Ungava district rarely approaching so near the coast as the immediate vicinity of Ft. Chimo. Toward the "Height of Land" the animal becomes more plentiful and is common toward the Gulf shores. It is not abundant until the southwestern portion of the region is attained. Not more than fifty skins of the mink are annually obtained at Ft. Chimo, and these mostly obtained from the northern slopes of the "Height of Land." I could obtain no satisfactory evidence of its occurrence on the barren ground tracts skirting Hudson Strait and the northeast portion of Hudson Bay proper. The quality of the skins is poor for the extreme northern individuals while those in the southern portions are correspondingly better. The mink is in this region essentially an animal of the wooded tracts along the streams and lakes.

[1484] This is rather strange when we consider the fact that the best mink skins in the world are those obtained from the open tundra in the vicinity of Pastolik; north, but few miles, of the Yukon Delta in Alaska. The character of the streams and lakes in the Ungava district, at least, preclude the possibility of them being frequented by mink or any other animal whose principal subsistence is fish. There are no mines of fish (*Dallia pectoralis*, Bean) here in Ungava such as are to be found in the Alaskan Tundras. I was not able to obtain definite information regarding the mink in this region. I never saw a live one during the two years I was there. The Indians whom I questioned in regard to it appeared to be unacquainted with its habits to such a degree as to afford other information than is generally known or imagined.

Collection Notes:

As he noted, Turner did not secure any specimens of the mink while in Ungava. He was correct in supposing that, unlike in Alaska, the species apparently does not occur north of the treeline in the Ungava Peninsula (Harper 1961). It is thus absent from the north-eastern shores of Hudson Bay, a fact about which he wondered.

ᑎᕆᐊᖅ
Stoat or Ermine
tiriaq

Putorius erminea
Putorius vulgaris

[1485]

Order Name used by Turner	*Carnivora*
Current Order Name	*Carnivora*
Family Name used by Turner	*Mustelidae*
Current Family Name	*Mustelidae*
Scientific Name used by Turner	*Putorius erminea* and *Putorius vulgaris*
Current Scientific Name	*Mustela erminea* Linnaeus
	Mustela nivalis Linnaeus
Syllabary	ᑎᕆᐊᖅ
Modern Nunavimmiutitut Term	*tiriaq (cf. Schneider p.411 - tiriaq = great weasel or ermine [Mustela erminea])*
Eskimo Term by Turner	*tŭ ghi ak; tŭ ghi ak*
Definition of Eskimo Term by Turner	*Stoat or Ermine [p. 2319; 1493]*

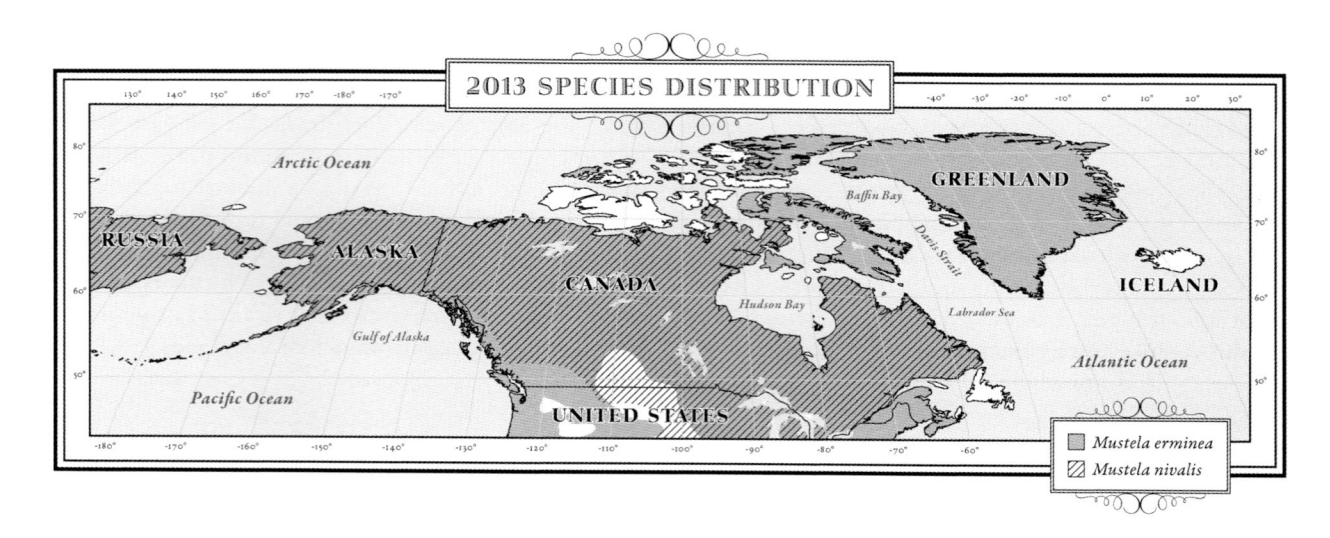

2013 Species Distribution

- ▨ *Mustela erminea*
- ▨ *Mustela nivalis*

Chapter Two: Mammals of Ungava and Labrador

302

Descriptions by Turner

The Stoat or Ermine is quite an abundant resident of the Ungava district. It appears to frequent all portions as its tracks are quite plentiful along the barren coast line of Hudson Strait. At the edge of the timber line it becomes more numerous and is there apparently as abundant as anywhere within the region. I could obtain but this species and am led to believe that *Putorius vulgaris* does not range within the vicinity of Ft. Chimo.[23] Individuals of the Ermine were obtained from numerous portions of the district and show no variation worthy of special remark, from typical specimens.

23. Turner is writing about the least weasel, *Mustela nivalis*, represented in his collections by a single skull. This was likely picked off the ground at Ft. Chimo.

The natives assert that the number of mice infesting a tract of country determines the number of Ermines. It is well known that certain species of mice are imperfectly migratory and in the Ungava district they were not nearly so plentiful as they were known to be but a few years [1486] previous to 1882.

Along the coast of the western portion of Ungava Bay the Eskimo obtain great numbers of seals and the oil taken from them is placed in bags made of the skins of the seals. As there is but little driftwood in that direction, owing to the currents of the large rivers flowing into that bay being deflected toward the eastern end of the

Hudson Strait and thus carry and drift to the eastward, the Eskimo are compelled to place the bags containing the oil upon the rocks and are thus exposed to the ravages of mice and ermines which gnaw holes in the oil bags and allow the contents to flow among the stones of the beach.

In certain years the presence of these animals renders it necessary that the Eskimo should forego the capture of seals as the mice destroy the bags as fast as they are exposed. The labor of several weeks has been known to stream away in a [1487] single night. In May 1883 an Ermine was known to have its home under one of the dwellings of the trading post of Ft. Chimo. The servant girl procured a small trap from me and protected its jaws with a piece of cloth to prevent the animal from being bruised if it should be caught. In the course of a few hours the creature was brought to me still clasped in the trap. Being called away at that moment I left it until I should return. I released the animal within my house and permitted it to have free range throughout the rooms. It evinced but little fierce disposition beyond a few spits and attempts to spring at me when I released it. In less than a quarter of an hour it came near me and began to cleanse itself from the filth voided at the instant of its capture. After an hour it began to show all signs of having forgotten its pain and terror caused by the trap. I occupied myself in skinning some ducks and threw [1488] the skins upon the floor until I had finished a number. A couple of dead mice were also lying on the floor and a pile of deerskins also lay nearby. The flesh from the body of the ducks was devoured eagerly. After it had satisfied its hunger the creature began to ramble about the place searching every nook and corner. I observed that the crack under the door caused it most solicitude. After awhile the animal was satisfied that no means of escape were at hand and now began to climb my clothing. The trap had broken the bones of one foreleg and thus prevented it from ascending an object. It soon sought amusement and took the bird skins, *Histrionicus* and *Clangula*, under the pile of deerskins. The weight of the bird skins could not have been less than five times that of the Ermine yet it dragged them quite easily to the place of concealment. The mice were then taken to the same spot [1489] and in the course of time the fragments of flesh, from the birds I was skinning, were secreted. I then carefully lifted aside one of the deerskins and found the little creature arranging them to suit itself.

"She now began to catch the large green flies, which had but lately made their appearance and eagerly swallowed them. I observed that each time she seized any object that she made a smacking sound with her lips."

By noon the Ermine had become gentle and would scamper to me when I beat a "tatoo" with my fingers on the floor. At two o'clock I found the animal snugly lying where the edge of the blankets on my bed folded back from the pillow. When aroused from its sleep it stretched its snake-like neck and with its greenish, glittering eyes which appeared to fairly revolve in their sockets gave the creature a formidable appearance but instead it slowly sank back and resumed its nap. At three o'clock the Ermine came into the larger room where I was at work. She evidently was distressed as she evinced a restless disposition. I took hold of her and found that she had a litter of small young ones somewhere. I placed her on a table near a window and as soon as she saw the outside she appeared to recognize her proximity to her young, increasing her anxiety to regain the outside. She now began to catch the large green flies, which had but lately made their appearance and eagerly swallowed them. I observed that each time she seized any object that she made a smacking sound with her lips. She appeared specially fond of the flies and certainly was not hungry as an abundance of flesh and mice lay within her taking. At five o'clock a number of Indian women came in to offer some articles to me and as they always kept the door open lest I should play some practical joke upon them the Ermine escaped between their feet.

The next morning, about 6 a.m., the limping creature was observed carrying a young one, scarcely larger than a thumb, from the building under which she was [1491] caught to another fully one hundred yards distant. During the day I saw her and attempted to call her by tapping on a board. She immediately came to my hand but some dogs, attracted by the clucking sound I made with my lips, frightened her away and was not seen after that time. In the spring of 1884 (May 12th) I was informed that an Ermine had a nest in the powder magazine. I went there and saw both the male and female. After removing a number of barrels I came upon the nest and saw one of the creatures darting in and out of the nest. I threw a hammer at her and killed the animal.

The material of which the nest was composed resembled hatshelled flax. It was grass blades, stems, and a few sticks. The interior was of finest grass broken into softest condition. The affair was shaped much like a water bottle or decanter and placed on the poles which kept the powder barrels [1492] from contact with the earth. Near the entrance lay the bodies of four mice and one shrew. Within were two shrews and one mouse, the latter partially devoured and the skin of another mouse indicated that a repast on its flesh had not long since been indulged in. A removal of the poles disclosed no less than six young Ermines, nearly twice as large as a bumble bee, lying on the damp ground where the mother had dropped them in her haste to remove them from the nest. The tiny creatures appeared dead from the cold ground but in the course of a few minutes the warmth of my hand infused activity and the squirming objects were taken into the house. I there discovered that I had disturbed the mother in the act of delivering her young as here were eight teats apparently flowing. I then inspected her and found two young still in-utero [USNM 14866]. The male was not captured [1493] he deserted the building as soon as I entered. I have every reason to believe that he was very attentive to his spouse during her approaching confinement and that he had brought to her the mice destined for her food while unable to leave her tender young. Both the adults were, at that date, in the summer pellage. I was unable to determine the exact time of the change from the white to brown condition.

It has been asserted by several writers on the subject of change of color assumed by these animals that the fur or hair is not changed but simply the color. I have had abundant opportunity to observe that the hair or fur of each is always dropped and that the brown or white is the result of a new growth appearing at the proper season. An examination of the individual hairs always showing the one color, and not partly white and brown. The Eskimo term the Ermine *Tŭghi ak*.

Recollections on Inuit Use of Stoat
Tivi Etok (Heyes 2010)

This is quite a small animal. They are white, with black tails. These are dangerous animals, too. There is one, a boss, that walks around you in winter. After they walk around you they used to call the other minks to come so they would attack the people who lived there. They would whistle to their family to come. They used to attack a person. Sometimes many tiriaq would come around a person. All the animals are strong. They would attack with their teeth. I used to trap tiriaq and sell a lot of them to the Hudson Bay Company. We would get 50 cents for one. Sometimes we would get 10 cents if the fur wasn't good. When I was young I used to see a lot. In the summer the fur goes brown. Once the snow comes, they turn white. They camouflage themselves depending on the seasons.

Chapter Two: Mammals of Ungava and Labrador

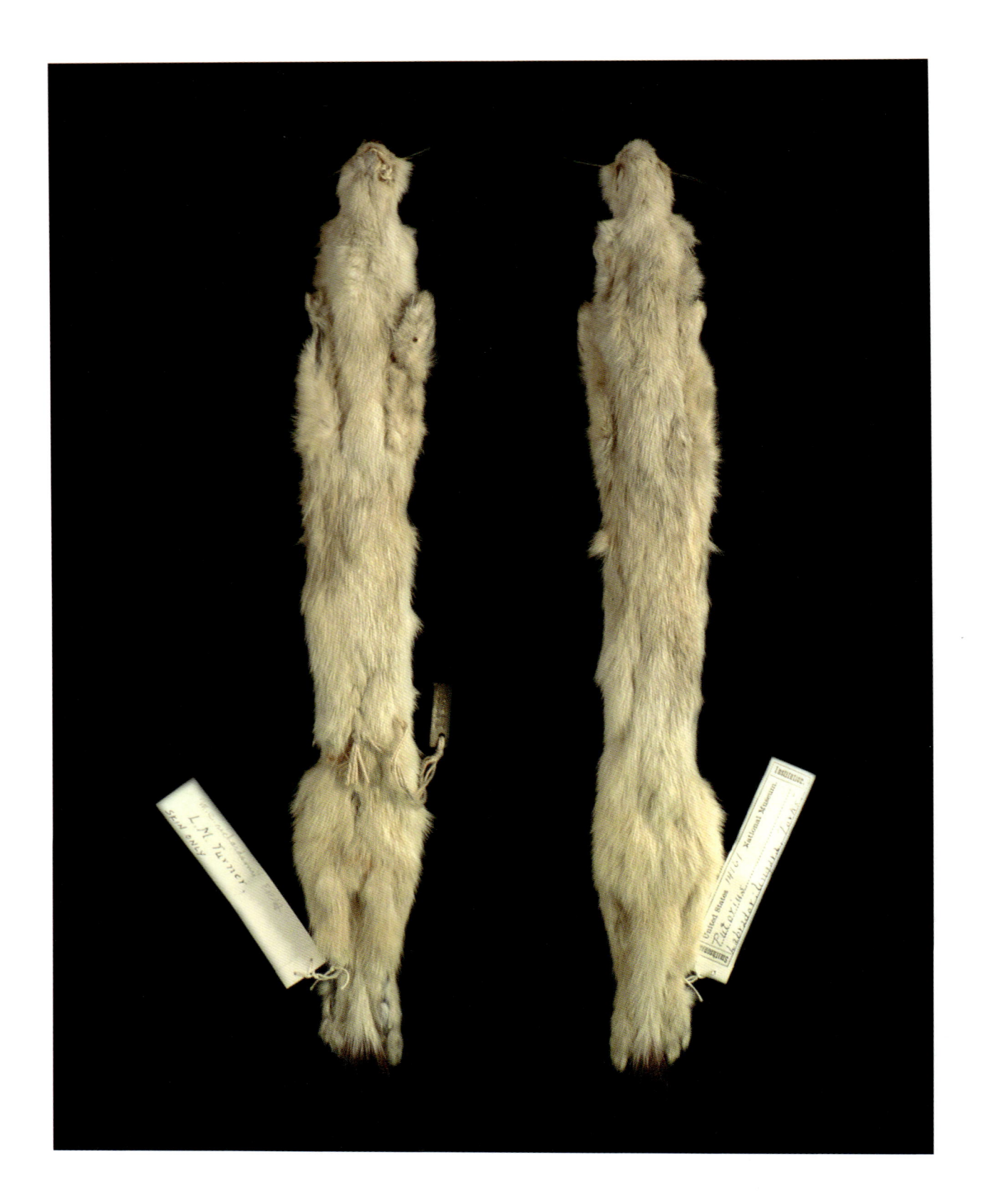

Least Weasel Skull
The skull is covered in dirt and missing teeth. Turner must have picked it up off the ground, perhaps at Ft. Chimo. It is the only least weasel (Mustela nivalis) specimen Turner collected. USNM A23135, Division of Mammals, NMNH. Photo by Angela Frost, 2011.

Winter Coat of a Stoat or Ermine (opposite)
Collected by Turner at the "Forks" near Ft. Chimo, c. 1883. Turner attempted to keep an ermine (Mustela erminea) as a pet. He wrote; "I released the animal within my house and permitted it to have free range throughout the rooms. It evinced but little fierce disposition beyond a few spits and attempts to spring at me when I released it." Turner (1887a: 27) wrote about the animals that form the staple diet for the Inuit, and made special mention of ermines: "The flock of all creatures is eaten [by the Tahagmyut], excepting that of the mouse, ermine, raven and the lower forms of life from the cold waters of the Strait." USNM 14161, Division of Mammals, NMNH. Photo by Angela Frost, 2011.

ᖃᕝᕕᐊ(ᕐ)ᔪᒃ
Marten
qavvia(r)juk

Mustela americana Turton

[1494]

Order Name used by Turner	*Carnivora*
Current Order Name	*Carnivora*
Family Name used by Turner	*Mustelidae*
Current Family Name	*Mustelidae*
Scientific Name used by Turner	*Mustela americana* Turton
Current Scientific Name	*Martes americana* (Turton)
Syllabary	ᖃᕝᕕᐊ(ᕐ)ᔪᒃ ; ᖃᕝᕕᐊ(ᕐ)ᔪᒃ (?)
Modern Nunavimmiutitut Term	*qavvia(r)juk / qavvia(r)suk(?)* ("small wolverine") *kavatsuk* (Note: possible name for marten based on *qavvik* = wolverine). *kimmiquarqutuuq* (cf. Schneider p.141 *-kimmiquarqutuuq* = [...] a penant marten [*Martes pennanti*, the fisher]?)
Eskimo Term by Turner	*ka ía chŭk; ka′f shik*
Definition of Eskimo Term by Turner	Marten [p. 2279; 1495]
Innu Term by Turner	*wa pĕs tañ*

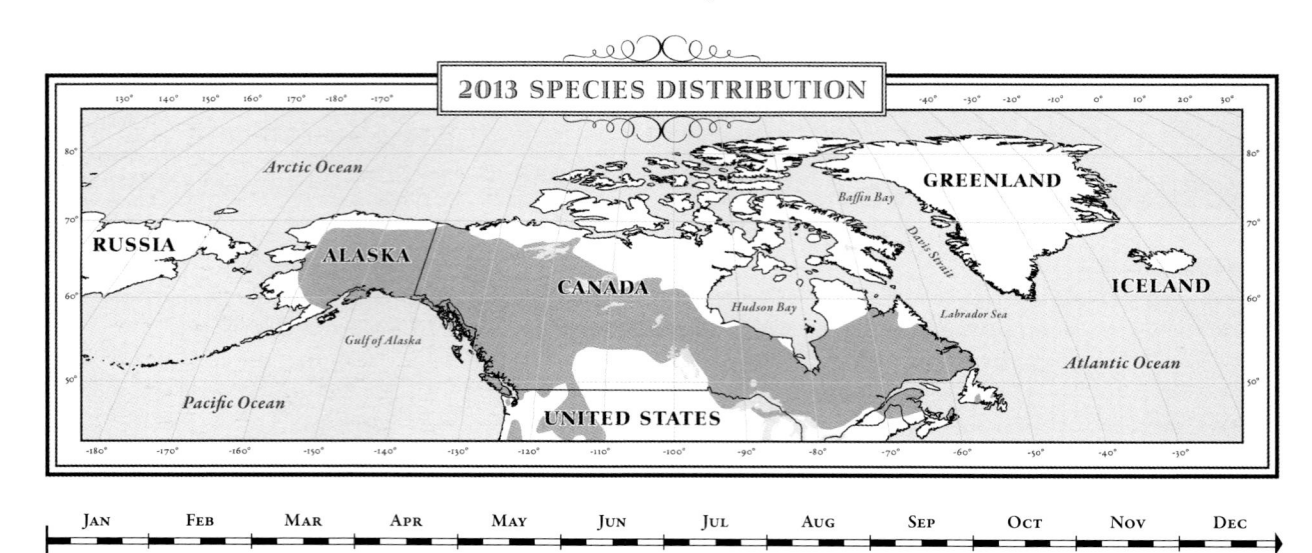

2013 Species Distribution

Jan · Feb · Mar · Apr · May · Jun · Jul · Aug · Sep · Oct · Nov · Dec

Season for taking furs begins about 1st Nov. and ends with the first rains of the spring.

Chapter Two: Mammals of Ungava and Labrador

Descriptions by Turner

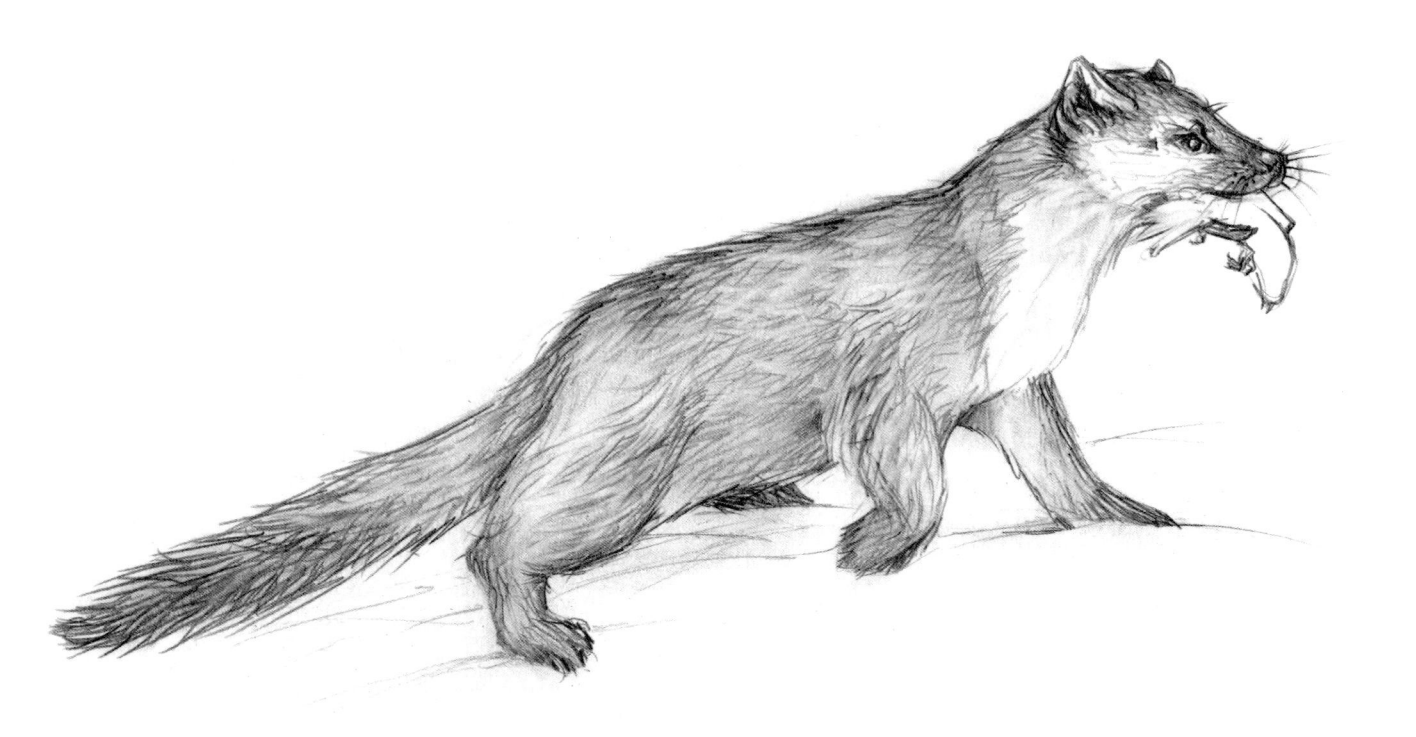

The Marten is only rare in the immediate vicinity of Ft. Chimo. Along the banks of the Koksoak River it does not appear plentiful until the headwaters of that river are reached. The region where it is said to abound is toward the headwaters of George's River where several hundred are annually procured by the Indians who organize parties specially for the purpose of hunting Marten. South of the "Heighth of Land" this animal becomes more plentiful and attains a fine quality among the denser woodlands of the southern region. The catch varies greatly, probably not so much on account of scarcity of the animals as much as the disposition and opportunity of the people. The number ranging from seven to nearly fifteen hundred skins. The season for taking the fur begins about the first of November [1495] and ends with the first rains of the spring. If a rain occurs during the good season it is said to "spot" the skins and render them of much inferior value.

The food of the Marten consists of small rodents; rabbits forming a considerable portion and birds of any kind. The breeding habits were not satisfactorily determined. In the southern portion of the region the marten is very abundant and forms one of the principle furs obtained. The value of the "North Shore" Marten is well known to be much greater than that of the South shore.

The Indians term this animal *wa pĕs taṅ*. The Eskimo call it *káf shĭk*.

Marten Ethnography
Turner

MARTEN TAIL AND BONE CATCHER NASKOPIE (NAYNAYNOTS) INDIANS; UNGAVA DIST., H. B. T. (1887B: 395-396)

The plaything here referred to is formed from five phalanged bones from the reindeer. They are hollowed to quite a thin shell and have a conical form. The apex is cut off and through the orifice thus left passes a stout, short thong to the farther end of which is attached the tail of a marten. The nearer end of the thong has affixed to it a flat bone (antler) awl shaped.

The bone point is held in the fingers, between the first and second and steadied by the thumb. The point directed from the person while the remainder dangles below the hand.

The thing is so short that the point of bone can scarcely be inserted in the larger orifice of the hollow bone nearest to it. The object is to give the string of bones a dexterous half-swing so that the point may be inserted into one or the other of the hollow cones. The marten tail serves to retard the half circle described by the affair when in motion.

Great practice alone renders one able to insert the point into the bone. I have never known a wager to be laid upon this amusement, principally for children.

SETS OF TRIGGERS USED FOR "DEADFALLS." NASKOPIE INDIANS, UNGAVA DIST., H. B. TERR. (1887B: 493-494)

In the absence of the common steel-trap the Indian must have recourse to his own ingenuity to capture the various furbearing creatures which occur within the region traversed by him.

The commoner form of trap is the one usually designated as a deadfall; two logs one below and the other above with several stakes driven on each side so as to insure such approximation of the falling log upon the lower that the weight will either crush out the life of the beast or else imprison it that escape is impossible.

For certain mammals such as the ermine, mink and marten but little regard is had to the accuracy of construction or attempt to conceal the object of the design for each of those mammals appears to be endowed with little wariness of entering a deadfall or trap. The means to accomplish the fall of the log is affected by a set of triggers of which there are numerous kinds but generally speaking they are modifications of the figure 4 form [i.e. the shape of the deadfall trap is made up of parts that resemble the shape of the number 4]. The upright piece is termed the standard, the slanting piece the flyer and the horizontal piece the tongue. The upper end of the flyer overlaps the end of the standard and thus serves to sustain the end of the superincumbent log. The lower end of the flyer is placed in a notch near the outer end of the tongue-piece. The tongue is supported by the weight of the log bearing on the flyer and this again on the tongue which is notched again so as to rest against the square corner of the standard. A piece of bait, flesh for mammals, is stuck on the inner end of the tongue and inserted between the logs so as to be opposite the stakes without and which prevent the bait being seized unless the mammal enters from the front. A slight touch disengages the tongue at the standard and this releases the mechanism, causing the upper log to fall upon the victim.

For other mammals such as Wolverene, foxes, and otters, the trap must be effectively concealed and the greatest care exercised lest the odor of the person remains to be detected by the keen sense of smell.

Unfortunately, portions of these triggers have been lost or misplaced and are, consequently, not now fitted to put in position.

The Origin of the White Spot on the Throat of the Marten
An Innu story, Turner (1887b: 610-611)

An Indian and his wife lived happily together. They had no children. A Marten dwelt in a forest nearby, and fell in love with the man's wife. He would watch when the man went out hunting and would then slip into the tent, sit by her side and endeavour to persuade her to dwell with him. One day the hunter returned unexpectedly and found the Marten sitting by his wife. The creature ran out of the tent. The Indian inquired of his wife what the Marten wanted. She answered that he was striving to induce her to leave her home and go to live in the forest and to become his wife.

The next time the Indian went out to hunt some food he told his wife to put a kettle of water on the fire to boil. The hunter went off some distance and secreted himself where he could perceive the Marten enter the tent. In a few minutes he saw the Marten steal into the tent. The man now crept up stealthily and heard the animal talking to his wife. The man rushed into the tent and exclaimed, "Marten, what are you doing here?" He seized the kettle of water and dashed it upon the creature's heart. The Marten began to scratch the burned place and ran outside. The fur on its throat was scalded off and when it grew again it was white as snow; and, from that time to the present the Marten has a white spot on its throat.

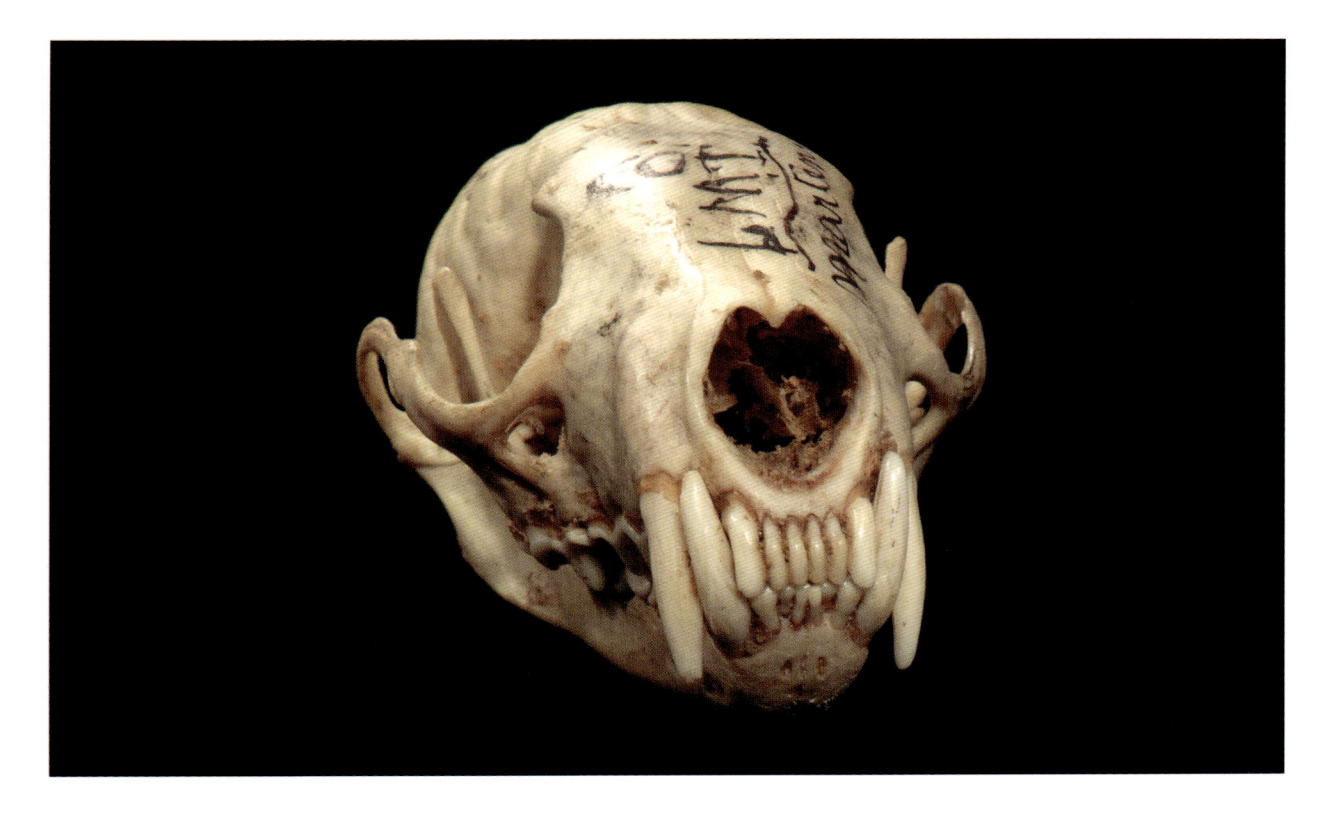

Male Marten Skull
Collected by Turner at Ft. Chimo. Turner wrote: "The food of the Marten consists of small rodents; rabbits forming a considerable portion and birds of any kind." USNM A23215, Division of Mammals, NMNH. Photo by Angela Frost, 2011.

ᑲᔪᕐᑐᖅ / ᑲᔪᖅᑐᖅ
RED FOX
kajurtuq / kajuqtuq

Vulpes fulvus fulvus, Desm.

[1496]

ORDER NAME USED BY TURNER	*Carnivora*
CURRENT ORDER NAME	*Carnivora*
FAMILY NAME USED BY TURNER	*Canidae*
CURRENT FAMILY NAME	*Canidae*
SCIENTIFIC NAME USED BY TURNER	*Vulpes fulvus fulvus, Desm.*
CURRENT SCIENTIFIC NAME	*Vulpes vulpes* (Linnaeus)
SYLLABARY	ᑲᔪᕐᑐᖅ ; ᑲᔪᖅᑐᖅ
MODERN NUNAVIMMIUTITUT TERM	*kajurtuq; kajuqtuq (cf. Schneider p.115 - kajurtuq; kajuqtuq = red fox [Vulpes vulpes]*
ESKIMO TERM BY TURNER	*kai ók tok*
DEFINITION OF ESKIMO TERM BY TURNER	*Red fox [p. 2262]*
INNU TERM BY TURNER	*wist wa we che shu [p. 1841]; wis wa we che shu [p. 1498]*

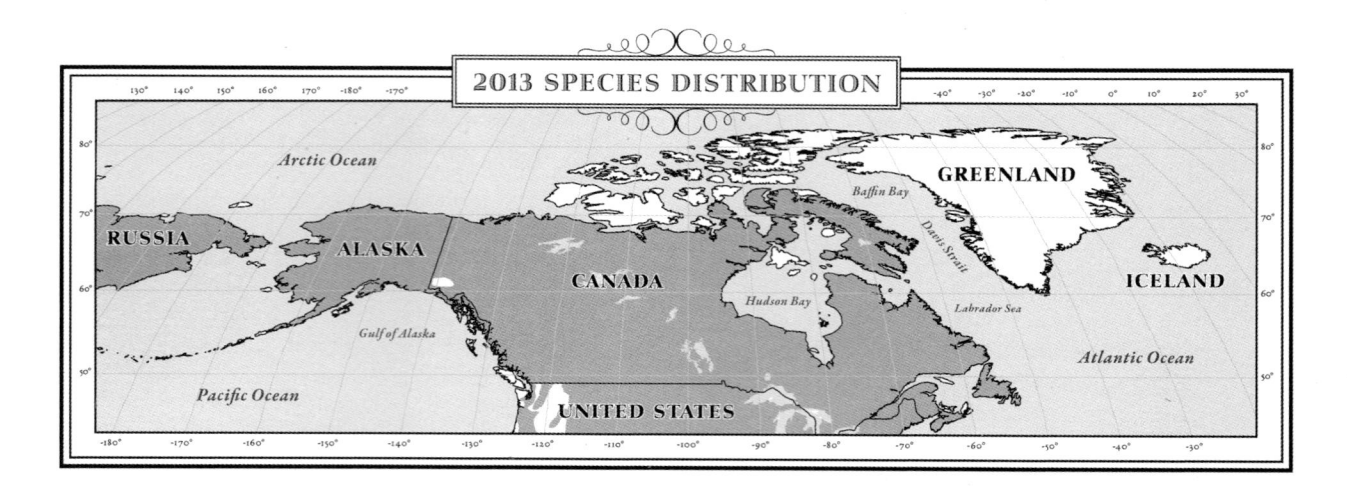

2013 SPECIES DISTRIBUTION

CHAPTER TWO: MAMMALS OF UNGAVA AND LABRADOR

Descriptions by Turner

The Red Fox[24] is plentiful throughout the region but appears to be rather scarce on the treeless areas where its place is taken by the White and the Blue Foxes. Along the streams and in the patches of willows and alders this fox prefers to make its home. In search of food it wanders along the banks, seeking mice, rabbits, hares and birds. The carcass of a deer, fallen beneath the wound inflicted by Indian or Eskimo, is a source of continual feast until the snows of winter cover it with a crust through which the fox is not able to dig.

The Red Fox in this region is not of such excellent quality as the ones obtained in Alaska. It does not also appear so large. The fur is duller and dingy, rarely is the "red" by which color it receives its name so fiery. The Red Fox is so wary that it requires much skill on the part of the trapper to successfully [1497] entrap him. If suspicion lurks in its mind that a trap is near it will sit and consider the best means of securing the coveted bait. By gradually approaching the trap the snow is carefully removed and the location of the trap disclosed. The animal carefully avoids the space large enough to be embraced by the jaws of the trap; and the remainder of the snow removed in

---

24. The red/black/silver/cross fox are color morphs of the one species, *Vulpes vulpes*, and the white and stone and blue foxes are the same species, morphs of *Vulpes lagopus*.

search of the small fragments of meat with which the place was strewed.

The value of a Red Fox is one half to twice as much as the skin of a White Fox; although, the idiosyncrasies of the fur trade value the skin of the White Fox at nearly double that of the Red in the civilized market. The unit of value in purchasing furs at Ft. Chimo is the skin of a White Fox and on this computation all other furs are purchased or bartered for. The condition of fur and, of course, the season when captured, gives added or lessened value to the fur.[25]

[1498] The general habits of the foxes are so nearly alike that separate descriptions for each species are unnecessary. The seasons for bringing forth the young is about the same in all species for the Ungava district.

The Eskimo name of the Red Fox is *kai ók tok* meaning reddish or tawny. The Indians call it wis *wá we ché shu*.

---
25. When Hudson's Bay Company Trading Posts operated in the Ungava District, Inuit hunters would travel to these posts to trade fox skins for tea, bullets, flour and other dry goods. Foxes were trapped in winter when their furs were thick. Fox hunting in spring was not carried out when they were shedding their fur (Makivik 1984).

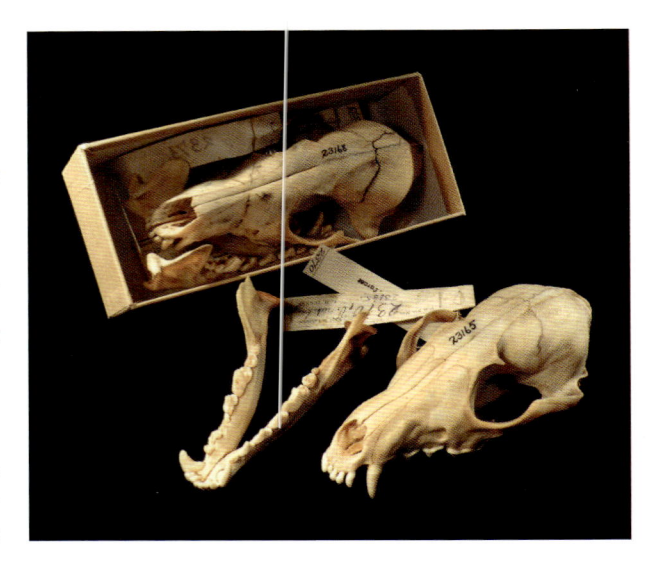

Red Fox Skulls in Box
Turner wrote of the red fox: "Along the streams and in the patches of willows and alders this fox prefers to make its home. In search of food it wanders along the banks, seeking mice, rabbits, hares and birds." Collected by Turner, probably at Ft. Chimo in 1882. USNM A23165 and A23168, Division of Mammals, NMNH. Photo by Angela Frost, 2011.

The Man and the Fox

An Inuit story, Turner (1887a: 297-298)

A man who dwelt by himself discovered that when he went from the house that it was visited by something. He determined to watch who should come to his hut while he was absent and arrange everything in nicest order as only a woman could do. His boots were always stretched and dried and so soft when he put them on in the morning. His fire was well prepared and this caused him to wonder who it could be when he was unable to find the traces of a woman about the hut.

One day he went out as though for a hunt. He made a tour and returned by another direction. As he neared his house he perceived a fox enter the door. The man approached and when entering he perceived a woman dressed in clothing of her kind and a fox skin hanging to one of the projecting sticks of the lodge. The man inquired if she was the one who had been doing those things about his house. She admitted that it was she and then consented to be his wife.

Sometime after that the husband detected a musk odor pervading the air and inquired what caused it. The woman replied it was she who emitted the odor and remarked that if he was going to find fault with her she would leave. She dashed off her dress and resumed the skin of a fox, slipped out of the door and was never again disposed to visit the hut of a man.

Fox Who Became an Inuk
An Inuit story, Tivi Etok (2010)

There was a hunter who had a kayak and no wife. He was a big man. He was hunting seal by day and went back to his tent at night. When he went back to the tent it was so neat. The bed and the stove was very neat. The man was all alone. Nobody was around. Everytime the man came back from hunting the tent was very neat. One day he pretended to go camping with his kayak. Nobody was around. He hiked to an area where he could see his tent. He hid and watched his tent. He saw a fox going to his tent. The hunter ran to the tent. But when he opened up the tent there was a lady. Suddenly he knew he had a wife. Another camper came by their tent some time later. The visiting camper indicated that he could smell a fox. It was the smell of the fox who became a lady. The visitor asked "where does this smell come from; I can smell fox." When he smelled the fox, the lady who was once a fox washed herself thoroughly and left the tent. As she ran out she became a fox again. In the old days some animals became Inuk and some Inuk became animals. But these were not like real animals.

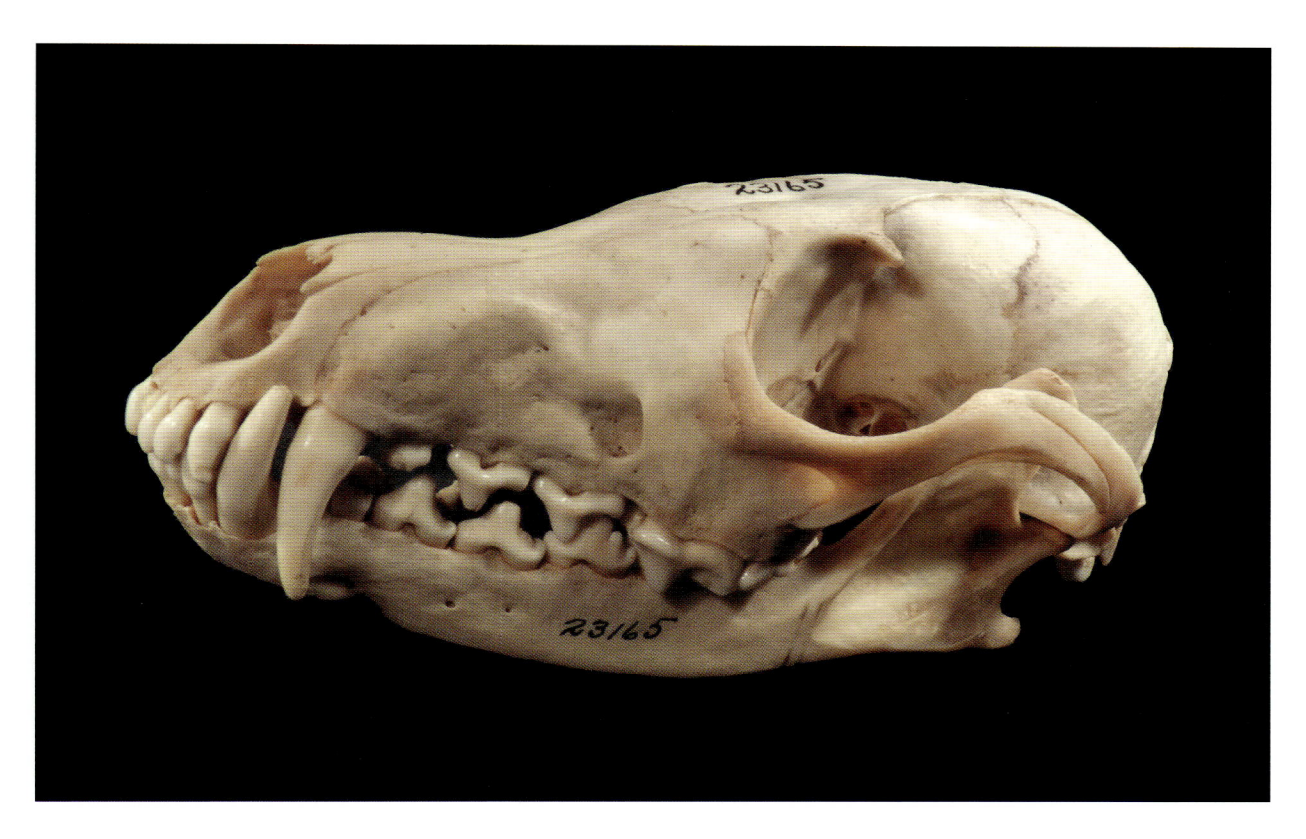

Red Fox Skull
Turner was uncertain whether the different color forms of the red fox (which he called the red, black or silver, and the cross fox) were the same species. This specimen was collected by Turner, probably at Ft. Chimo in 1882. USNM A23165, Division of Mammals, NMNH. Photo by Angela Frost, 2011.

Child in the Fox Ear
An Inuit story, Paul Jararuse (Heyes 2007)

There were two midgets that ran into a mother and daughter. Their man was out hunting. When they had igloos they used to have more than one part, sleeping area here, another part here, another part here. These two midgets they arrived and they had a little child which was kept in a fox ear. It was so small that the amautik fit into a fox's ear, like a little Inuk. When they got in, they were hiding a piece of walrus meat, but they never noticed this snow house, they didn't go to the furthest igloo but I believe they were on the third one but they had a chance to go further more. It was getting late when they arrived but this mother and her daughter were there but they just kept quiet because these two midgets they came in holding a piece of walrus meat. When they were there they did not even find out that there was a mother and daughter there. This lady she was a shaman too and they were hungry, her husband was out hunting, she was a shaman so when these midgets, when they were about to, she knew they were going to leave the next day so she spat to the floor to make the walrus meat stuck. She was in one small house and another one here and these two midgets were here. When this lady spat, this frozen walrus meat got stuck so these two midgets were not able to get it out because this Shaman woman she spat to make it stick so they couldn't take it anymore, they were hungry. So they just left the meat.

Recollections on Inuit Use of Foxes
Tivi Etok (Heyes 2010)

The red fox is bigger than other foxes. They make dens in the hard sand. When the flowers have come out in summer the foxes are inside their dens making babies. Most foxes are by the beach where there is sand. Sometimes we used to catch the babies as kids. We used to keep them for a couple of months as a pet like a dog. We would feed them. When the fur was a lot better we would kill them and sell it to the Hudson's Bay Company. The fox is very strong. They can jump very high; like Kung-Fu! They are very fast. To catch them we used to put meat on a rope so that the fox would be attracted to it. We would build a stone trap around the entrance. Once the fox grabbed the rope we would pull the rope through the trap. As the fox entered the trap we would close of the entrance with a rock so that it would be contained in the trap. This is the only way that we would capture foxes before steel traps.

In the times before I was born, the silver fox was very useful. But when the Hudson's Bay Company came along this fox was keenly targeted as its fur was sold to be used for clothing. The meat on this was good, too. The silver fox is smaller than a regular fox. It cannot breed with other foxes. They are different than a regular fox. The brown or white fox can have different colored babies. The colors of black, white or brown might result. When the silver fox is gray, however, it would have gray babies.

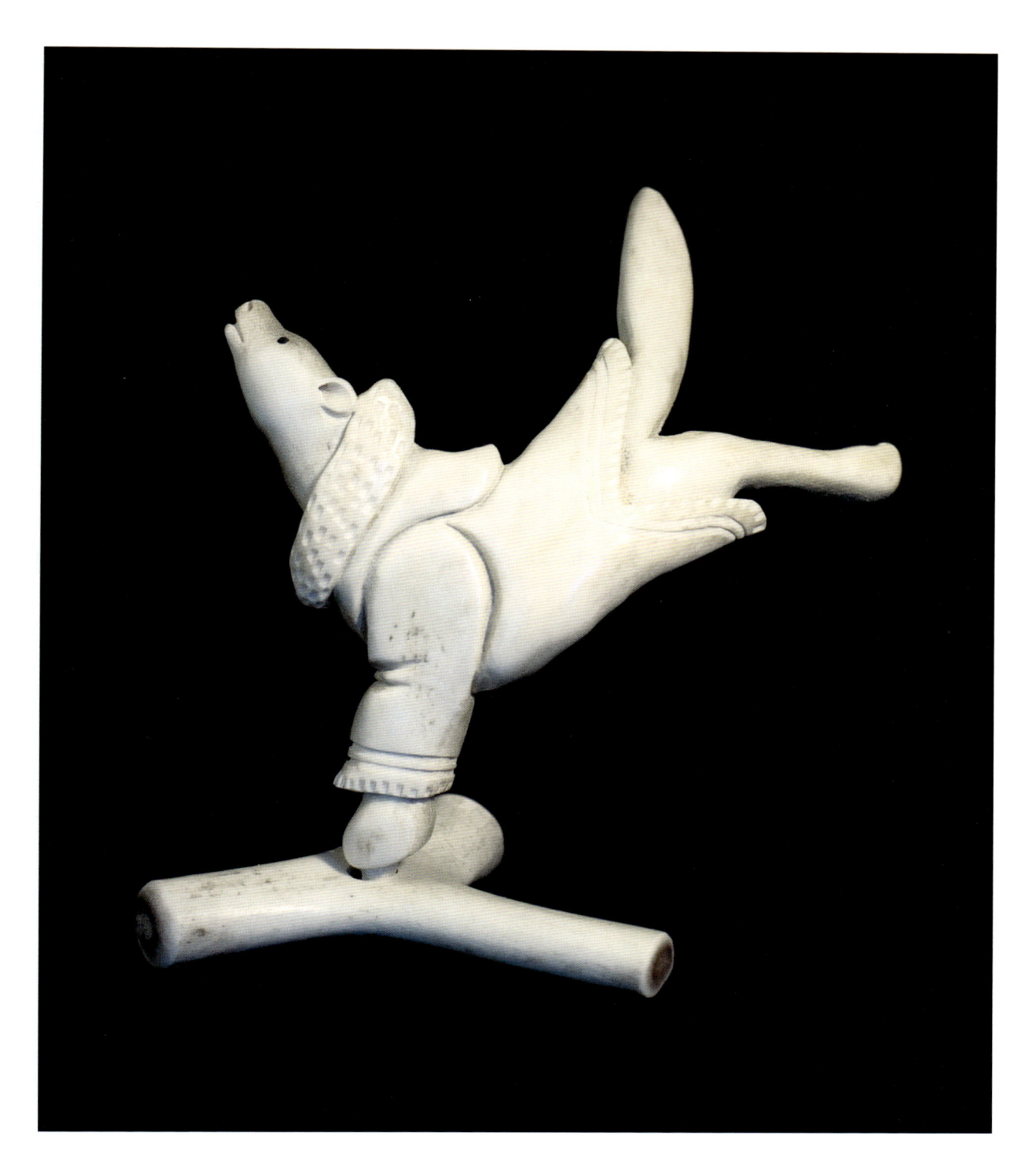

Fox Story
Carving about a story of a woman turning into a fox by Kangiqsualujjuamiut artist Daniel Annanack, 2005. Made from caribou antler. Scott Heyes Private Collection.

CARNIVORES

ᖃᕐᓂᑕᖅ

BLACK OR SILVER FOX
qirnitaq

Vulpes fulvus argentatus, Aud. & Bach.

[1499]

ORDER NAME USED BY TURNER	*Carnivora*
CURRENT ORDER NAME	*Carnivora*
FAMILY NAME USED BY TURNER	*Canidae*
CURRENT FAMILY NAME	*Canidae*
SCIENTIFIC NAME USED BY TURNER	*Vulpes fulvus argentatus,* Aud. & Bach.
CURRENT SCIENTIFIC NAME	*Vulpes vulpes* (Linnaeus)
SYLLABARY	ᖃᕐᓂᑕᖅ
MODERN NUNAVIMMIUTITUT TERM	*qirnitaq* (cf. Schneider p.310 - *qirnitaq* = that is black); *qiarngatuq* (also cf. Schneider p. 295 – silver fox)
ESKIMO TERM BY TURNER	*king ñik tok; kûng ñik tok*
DEFINITION OF ESKIMO TERM BY TURNER	Black fox (*Vulpes argentatus*) [p. 2305]; Black or silver fox (means black) [p. 1501]
INNU TERM BY TURNER	*kwese w ache shu* [p. 1832]; *Ka tsa wa che shu* [p. 1501]

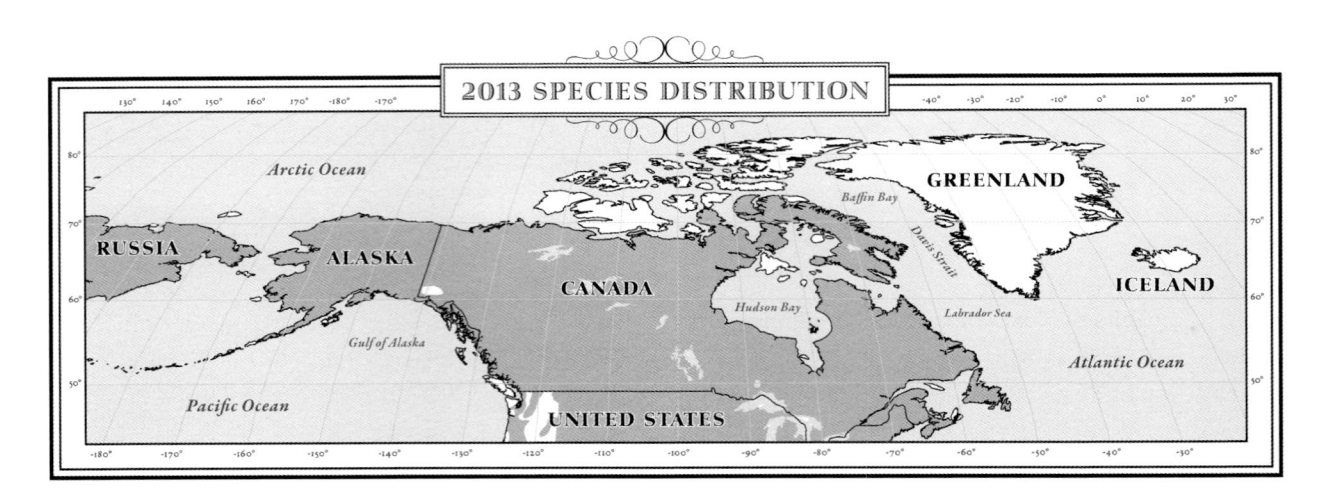

2013 SPECIES DISTRIBUTION

CHAPTER TWO: MAMMALS OF UNGAVA AND LABRADOR

Descriptions by Turner

The Black or Silver fox[26] also occurs throughout the region. It is not plentiful anywhere; although, certain tracts appear to contain more of these animals than do others apparently as favorable. The number of skins annually obtained at Ft. Chimo rarely exceeds forty and eighty was obtained but once in the history of that trading station. The greater number are procured by the "Northerners" (Eskimo dwelling along western portion of Hudson Strait). The Indians in the interior procure a score or less and a few only are taken in the immediate vicinity of Ft. Chimo.

[In general habits the foxes of that region differ so little and that depending only on particular tracts of country that their traits are not specifically remarked upon.] It is generally supposed by trappers and hunters that the White and Blue foxes are the same [1500] animal changed by local influences until they have become differentiated in pellage only. I am assured that the young of both kinds have been found in a single den and that the mother is often the opposite of the majority of her offspring. It is also affirmed that the Black and Silver are the same and these are but forms of the Cross Fox. Dens have been discovered containing all the three. The Silver or Black Fox is considerably smaller than the Cross Fox, as a cursory examination of the bones will show.

Owing to the engrossing nature of other work I was unable to satisfy myself of the truth of these assertions. I suspect that the selection of the same sand or gravel bank wherein the galleries communicate has often led to confusion when occupied by the animals known as so many species, for they do associate during the [1501] season of rearing their young. The Eskimo apply the term *tĭ gh'un yak* to all kinds of foxes and is equivalent to the English word Fox. To distinguish the species they designate them by specific names. The name applied to the Silver or Black Fox is *kŭng ńik tok*, and means Black. The Indians call it *ka tsá wa ché shu*.

---

26. The red/black/silver/cross fox are color morphs of the one species, *Vulpes vulpes,* and the white and stone and blue foxes are the same species, morphs of *Vulpes lagopus.*

ᐊᑯᓐᓇᑐᖅ

Cross Fox

akunnatuq

Vulpes decussatus

[1502]

Order Name used by Turner	*Carnivora*
Current Order Name	*Carnivora*
Family Name used by Turner	*Canidae*
Current Family Name	*Canidae*
Scientific Name used by Turner	*Vulpes decussatus*
Current Scientific Name	*Vulpes vulpes* (Linnaeus)
Syllabary	ᐊᑯᓐᓇᑐᖅ
Modern Nunavimmiutitut Term	akunnatuq (cf. Schneider p.15 - akunnatuq = lit. "between two"... cross-bred fox [...] thus named because they have a black cross on their beige fur).
Eskimo Term by Turner	á ku nak tok; a ku nák tok
Definition of Eskimo Term by Turner	Cross fox. A word having the particular meaning of, neither the one nor the other, but partaking of both; hybrid [p. 1502]; Cross fox. *Vulpes decussatus*. The word means that which is neither one nor the other, but of both [p. 2166]

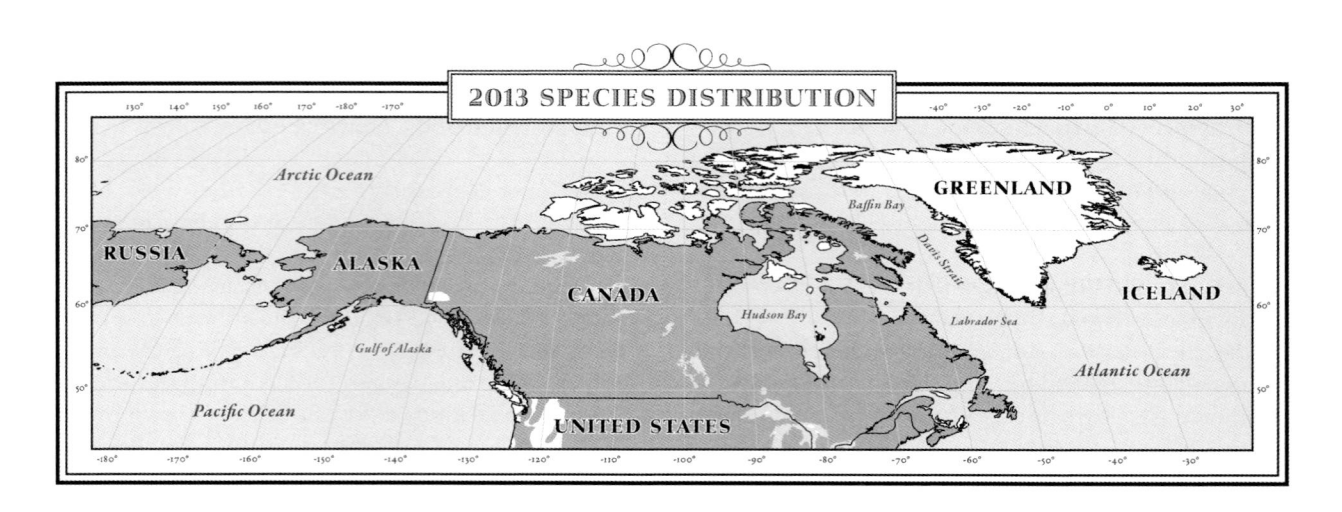

2013 SPECIES DISTRIBUTION

Chapter Two: Mammals of Ungava and Labrador

Descriptions by Turner

The Cross Fox[27] is distributed generally throughout the region, apparently more plentiful in the timbered lands, although not of a better quality than the number obtained on the barren grounds. In the vicinity of Ft. Chimo a good number of skins, most of them of superior quality, are annually taken. South of the "Heighth of Land" this species occurs plentifully. In general habits it differs but little from its congeners and in no particular [way] worthy of remark. The value of the skin of this animal is two or three times that of a Red Fox and four or six times that of a White Fox.

Some of the skins are of excellent quality; and, so far as fur is concerned, surpass some of the better grades of Silver or Black Fox skins. The Eskimo name of the Cross Fox is *Á ku nak tok* a word having the particular meaning of, neither the one nor the other but partaking of both; the word hybrid is a shorter definition.

---

27. The red/black/silver/cross fox are color morphs of the one species, *Vulpes vulpes*, and the white and stone and blue foxes are the same species, morphs of *Vulpes lagopus*.

ᖃᑯᖅᑕᔪᖅ
White, Arctic, Blue, or Stone Fox
qakuqtajuaq
Vulpes lagopus

[1503]

Order Name used by Turner	*Carnivora*
Current Order Name	*Carnivora*
Family Name used by Turner	*Canidae*
Current Family Name	*Canidae*
Scientific Name used by Turner	*Vulpes lagopus* (Linn.) Gray
Current Scientific Name	*Vulpes lagopus* (Linnaeus)
Syllabary	ᖃᑯᖅᑕᔪᖅ (?)
Modern Nunavimmiutitut Term	*qakuqtajuq (?) from qakurtaq ; qakuqtaq = white. (cf. Peck p.87 - kakkortak = "something white; a white fox"). Note in Schneider p.411 - tiriganniaq = fox in general, and specifically white fox [Vulpes lagopus]).*
Eskimo Term by Turner	*ká kok tá zhuk*
Definition of Eskimo Term by Turner	White fox [p. 2264]
Innu Term by Turner	*wa pa cha shish* [p. 1841]

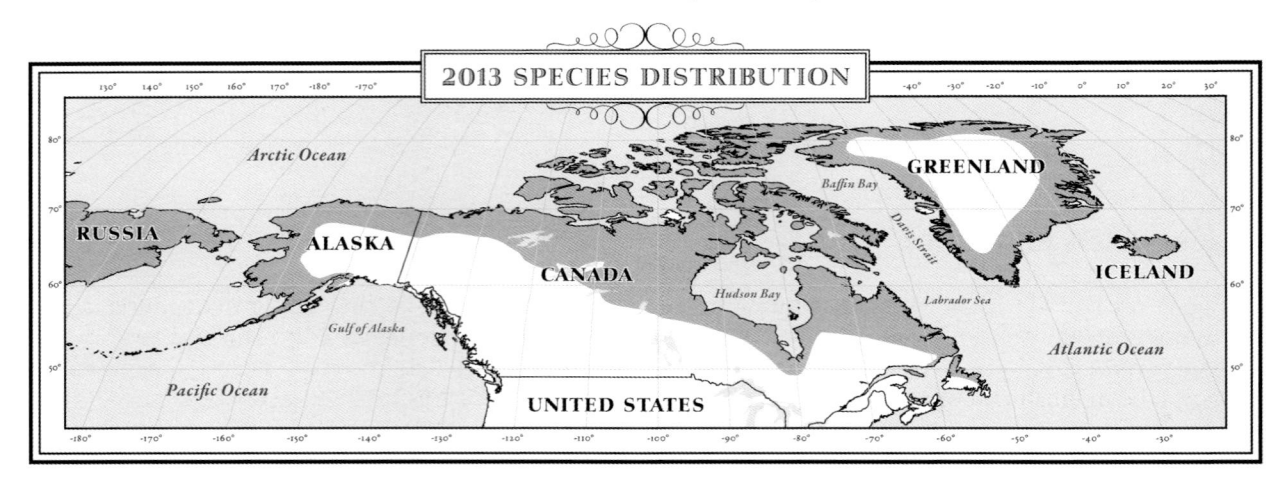

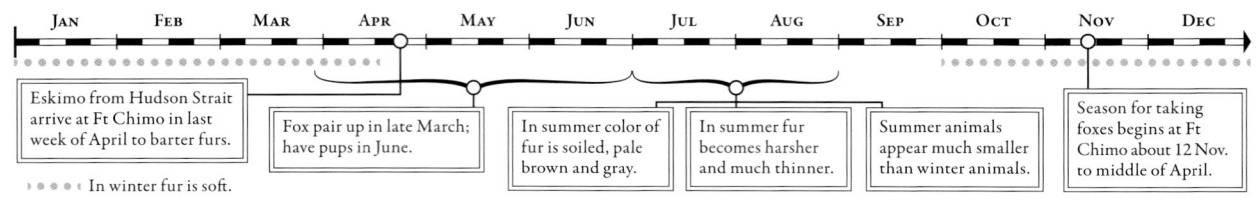

2013 SPECIES DISTRIBUTION

Jan — **Feb** — **Mar** — **Apr** — **May** — **Jun** — **Jul** — **Aug** — **Sep** — **Oct** — **Nov** — **Dec**

- Eskimo from Hudson Strait arrive at Ft Chimo in last week of April to barter furs.
- Fox pair up in late March; have pups in June.
- In summer color of fur is soiled, pale brown and gray.
- In summer fur becomes harsher and much thinner.
- Summer animals appear much smaller than winter animals.
- Season for taking foxes begins at Ft Chimo about 12 Nov. to middle of April.
- In winter fur is soft.

Chapter Two: Mammals of Ungava and Labrador

Descriptions by Turner

The White or Arctic Fox is very abundant throughout the region under discussion. Generally speaking it is the more plentiful in the vicinity of the coast and appears to be an inhabitant of the more open ground.

Along the coast they prefer the rolling or hilly country to the rugged, mountainous portions. Along the water courses of the tracts they are most common. The range of this animal is over all the territory north of latitude 50° and in the extreme northern parts of America they are as plentiful as along the middle region of their range. They undergo seasonal changes becoming white with a faint greenish tinge in the fresh subject, somewhat that of a sheep wool, but becoming a pure dead white after death. The tip of nose and edge of eyelids jet black. The tip of tail usually with many grayish hairs, rarely absent. Iris hazel-brown in life but darkening as [1504] as the animal expires.

In the winter the fur is long and soft but in summer it becomes somewhat harsher and much thinner. The color is quite different in summer, being of a soiled, pale brown and gray varying with each individual in age and sex although by the color alone the sex is not to be determined. The summer animal appearing much smaller than the winter animal and is due to the difference in the length of the fur. The size varies greatly. The height is about 14 inches. Tip of nose to root of tail 21 to 25 inches and the tail 11 to 14 inches. The weight ranges to slightly less than five pounds to rarely more than nine to seven pounds. Many examples freshly killed were weighed and none found to exceed the range given.

The food of this fox consists; verily, of anything. All is food that it is able to masticate. It is one of the greatest enemies of the [1505] Ptarmigan and a destroyer of all mice, young birds and eggs. In the summer time they wander along the coast picking up refuse from the beach, consisting of a cast up dead fish and not a few living fish are secured from among the tangle left bare by the receding tide. They are known to breed twice each year, bringing forth five to twelve young at a birth. They resort to the pockets of sandy or fine gravel abounding along the coast and excavate torturous galleries leading to widened chambers within which they make their nest of grass and leaves. These chambers often intercommunicate and the labor of digging them out is extremely fatiguing. They meet each other in late March and bring forth in June.

The rutting season is known by a peculiarly intolerable stench issuing from their dens.

The White Fox is very cunning and scarcely less so than the Red Fox. It is not an easy [1506] matter to entrap it if it becomes suspicious of the presence of a steel trap. With this trap many are caught each year. A hunter's rule is that each seventh year is a "Fox Year." They are so plentiful that as many as several thousand have been taken in the region belonging to a single trading post and a single person has been known to take as many as 800 animals of this kind during a single season. They are so eager for food during times of scarcity, that they have been known to sit but few yards from the person baiting a trap, and step into it as soon as the person retired.

When taken in a trap and yet alive, when the affair is visited, the trapper kills the fox by putting his snowshoe upon it until it is safe to put his foot upon its ribs and squeeze the life out of it. This done in order that the blood be not extravasated to the part where if a blow were given the wound would show a stain on [1507] the skin and deteriorate its quality as the skin of the animal is scarcely thicker than tissue paper. The skin is removed from the

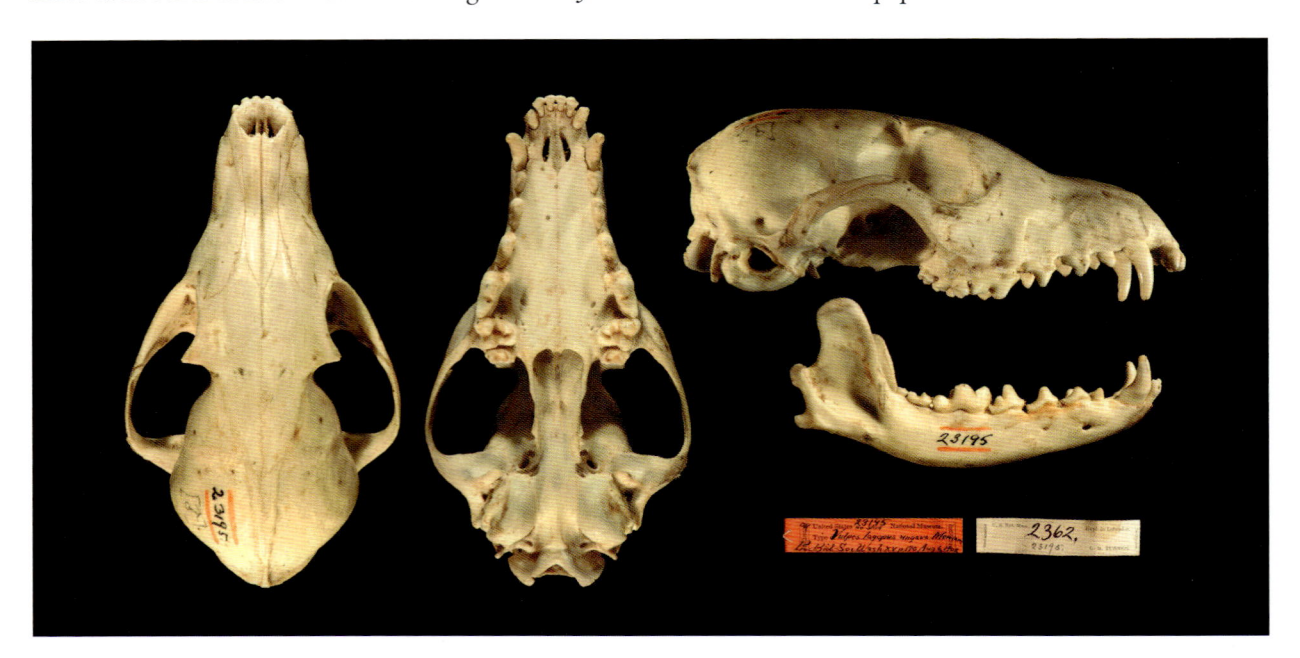

Arctic Fox Skull
This skull, collected by Turner at Ft. Chimo is the type specimen of Vulpes lagopus ungava, *a subspecies of arctic fox described by Merriam (1902). USNM A23195, Type Collection, Division of Mammals, NMNH. Photo by Angela Frost, 2011.*

Chapter Two: Mammals of Ungava and Labrador

body by an incision made between the hinder limbs and the skin stripped off. The tail is skinned from the caudal bones and if very fat is partly split to allow it to dry quickly. The longer the skin is exposed to dry the worse the character of the skin for the moisture should dissipate quickly as possible and to further this object the skin is carefully scraped to remove all traces of fat and ligament and then stretched upon a v-shaped board or frame. The skin may be much improved by exposing it to the severe cold which contracts the tissue and forces the oil from it, rendering the skin of a nearly white condition.

The season for taking foxes begins at Ft. Chimo about the twelfth of November and continues until the middle of April. The White Fox is quite erratic [1508] in disposition, often showing but little concern at approaching danger and if frightened will run with surprising velocity, apparently skimming the surface with the ease of a bird on the wing. It is said to be able to overtake the Polar Hare. This I doubt. It may be able to seize a rabbit but not a hare as anyone who has witnessed a hare run will be able to comprehend.

I was once, August 10th, along the coast, near the mouth of the Koksoak River, hunting for some reindeer which had been seen in that vicinity, but a few days before. Several persons were in the party and each eager to discover the deer and secure it. I was in front and above some of the others. Ahead of me I saw an object lying on the bare rocks which moved its tail in a wagging manner similar to that of a young dog witnessing the approach of its master. I suspected the proximity of a camp of Eskimo and the object to be one of [1509] their dogs. Something made me keep the gun at my shoulder and finally compelled me to shoot. I fired and the object never moved. I then felt sure I had killed a dog and as I had about ninety-yards to go I came up to it and found that the bullet had cut out the heart, lungs, and a good portion of the shoulder where the ball made its exit on the rock where the fox was lying. Everyone of the party supposed I had shot a deer. It was the first White Fox I had seen in the summer pellage and prized it accordingly [USNM Catalog No. 14160]. The Eskimo dwelling toward the western end of Hudson Strait obtain many of these animals and a party of those people determined upon visiting Ft. Chimo to trade annually make the trip. The party which comes this year may not return to trade for several years as they alternate with these trips, usually bringing some new person along with them; and of these people there are yet many [1510] who have not yet seen a white man. The journey is extremely arduous and undertaken as soon as the early, permanent snow sets in about the middle of November. The party travels by easy stages hunting along the route and slowly adding to their stock of furs. About the last week of April they arrive and barter their furs for ammunition and other articles, always securing tobacco and such things to supply their companions.

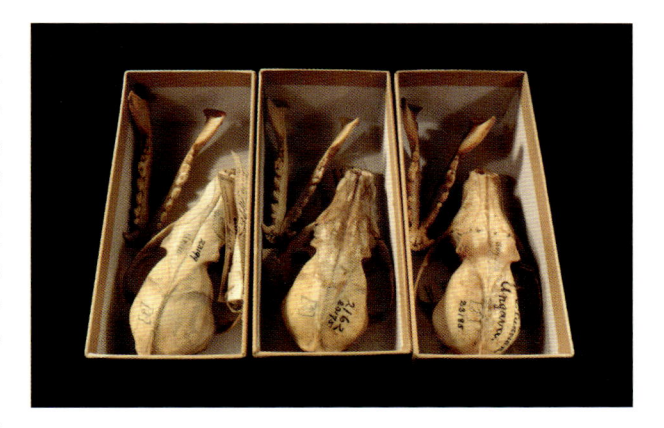

Arctic Fox Skulls in Boxes

These foxes were collected by Turner at Ft. Chimo in 1882. Turner wrote: "The food of this fox consists; verily, of anything... It is one of the greatest enemies of the Ptarmigan and a destroyer of all mice, young birds and eggs." From left to right USNM A23199, A23175, and A23185, Division of Mammals, NMNH. Photo by Angela Frost, 2011.

Some of the party are entrusted with the "catch" of other individuals who instruct the one to purchase certain articles. It often happens that some one of the party acts as a sort of middleman and procures articles of trade with which to secure the furs from his people who are unable to undertake the journey. The skins obtained from that locality are of best quality but are usually in soiled condition as the ultra Eskimo is not particular and doubtless has but meager facilities to keep the skins in clean condition. They are wrapped [1511] in soiled and often filthy covering which discolors the skin and advantage is taken of this fact by the trader to give a less price for them although the quality is not in the least impaired. In the course of a few days the new comers are satisfied with their visit and return. The warm days of spring are now near and before they go many day's travel they are impeded by the rapidly lessening snow so that by the time they get near their homes they have to complete their journey by walking. The coming of the "Northerners" is always an event at Ft. Chimo and serves to enliven an otherwise dull season of the year.

The "Northerners" use various kinds of deadfalls, of ice or stone, to entrap their foxes. The bait is a piece of meat of any kind. With steel traps the bait is cut into fine particles and scattered over the snow crust put upon the jaws of the trap.

[1512] The Blue or Stone Fox is considered as a race of the White Fox and is but only common in certain portions of the country. At Ft. Chimo it is quite rare, not more than four or five skins being annually obtained. I was unable to learn anything noteworthy of this animal. The Eskimo name is *aṅg zhûk*. They, however, apply the term *ka kók tak* to the White Fox and has bout the signification of "Whitey."

Arctic Fox Skins
Summer and winter coats of the arctic fox, collected by Turner at Koksoak River. Turner wrote: "The Blue or Stone Fox is considered as a race of the White Fox and is but only common in certain portions of the country. At Ft. Chimo it is quite rare, not more than four or five skins being annually obtained." Collected by Turner at Koksoak River, 10 August 1882. From top to bottom, USNM 14160 (collected 10 August 1882), 14785 (collected 11 December 1883), and 14784 (collected 19 November 1883). Photo by Angela Frost, 2011.

Foxes – Terms and Definitions

Syllabary	Modern Nunavimmiutitut Term and Definition	Term by Turner	Definition recorded by Turner
ᐊᑯᓐᓇᑐᖅ	***akunnatuq*** (cf. Schneider p.15 - *akunnatuq* = lit. "between two" … cross-bred fox […] thus named because they have a black cross on their beige fur)	*á ku nak tok; a ku nák tok*	Cross fox. A word having the particular meaning of, neither the one nor the other, but partaking of both; hybrid [p. 1502]; Cross fox. *Vulpes decussatus*. The word means that which is neither one nor the other, but of both. [p. 2166]
ᐊᖖᒐᓴᖅ; ᑎᕆᒐᓐᓂᐊᖅ	***anngasaq*** (cf. Schneider p. 33 - *anngasaq* = blue fox, white fox [*Vulpes lagopus*]). The "blue" variation of the white fox. Also *tiriganniaq* (Tivi Etok, Pers. Comm., 2010).	*añg zhúk*	White fox. The term Ka kók tok is applied to the white fox, too for its signification of "whitey". [p. 1503]
ᐊᖖᒐ ᓴᖅ	***anngasaq*** (see above)	*an'g zhuk*	Blue fox *Vulpes lagopus* [p. 2175]
ᑲᔪᕐᑐᖅ ; ᑲᔪᖅᑐᖅ	***kajurtuq***; *kajuqtuq* (cf. Schneider p.115 - *kajurtuq*; *kajuqtuq* = red fox [*Vulpes vulpes*]	*kai ók tok*	Red fox [p. 2262]
ᖃᑯᖅᑕᔪᖅ (?)	***qakuqtajuq*** (?) from *qakurtaq* ; *qakuqtaq* = white. (cf. Peck p.87 - kakkortak = "something white; a white fox"). Note in Schneider p.411 - *tiriganniaq* = fox in general, and specifically white fox [*Vulpes lagopus*]).	*ká kok tá zhuk*	White fox [p. 2264]
ᕿᕐᓂᑕᖅ	***qirnitaq*** (cf. Schneider p.310 - *qirnitaq* = that is black) *qiarngatuq* (also cf. Schneider p. 295 – silver fox)	*king ńik tok; kûng ńik tok*	Black fox (*Vulpes argentatus*) [p. 2305]; Black or silver fox (means black) [p. 1501]
ᓱᕐᕕᓂᖅ	***surviniq*** (cf. Schneider p.381 - *surviniq* = the odor of fox fur drying)	*su viñg nik*	The odor arising from a fox during the rutting season [p. 2791]
ᑎᕆᒐᓐᓂᐊᖅ	***tiriganniaq*** (cf. Schneider p.411 - *tiriganniaq* = fox in general, and specifically white fox [*Vulpes lagopus*])	*ti ghún yak*	Denotes all kinds of foxes, and is equivalent to English word for fox. To distinguish the species, the Innuit designate them specific names. [p. 1501]

Carnivores

ᐊᒪᕐᖁ

LABRADOR WOLF

amaruq

Canis lupus griseo-albus (Linné) Sabine.

[1513]

ORDER NAME USED BY TURNER	*Carnivora*
CURRENT ORDER NAME	*Carnivora*
FAMILY NAME USED BY TURNER	*Canidae*
CURRENT FAMILY NAME	*Canidae*
SCIENTIFIC NAME USED BY TURNER	*Canis lupus griseo-albus (Linné) Sabine.*
CURRENT SCIENTIFIC NAME	*Canis lupus* Linnaeus
SYLLABARY	ᐊᒪᕐᖁ
MODERN NUNAVIMMIUTITUT TERM	*amaruq (cf. Schneider p.22 - amaruq = grey wolf [Canis lupus])*
ESKIMO TERM BY TURNER	*a máu ghok*
DEFINITION OF ESKIMO TERM BY TURNER	*Wolf. Canis lupus, var. griseo-albus [gray wolf]. Woman's name [p. 2170; 695]; Wolf. Three types; 1) hunter. 2) chaser. 3) curly hair or shaggy head. The Innuit do not have more than one name for the wolf [p. 1519].*
INNU TERM BY TURNER	*a hé kan [p.708]; ma he kan [p. 1834]*

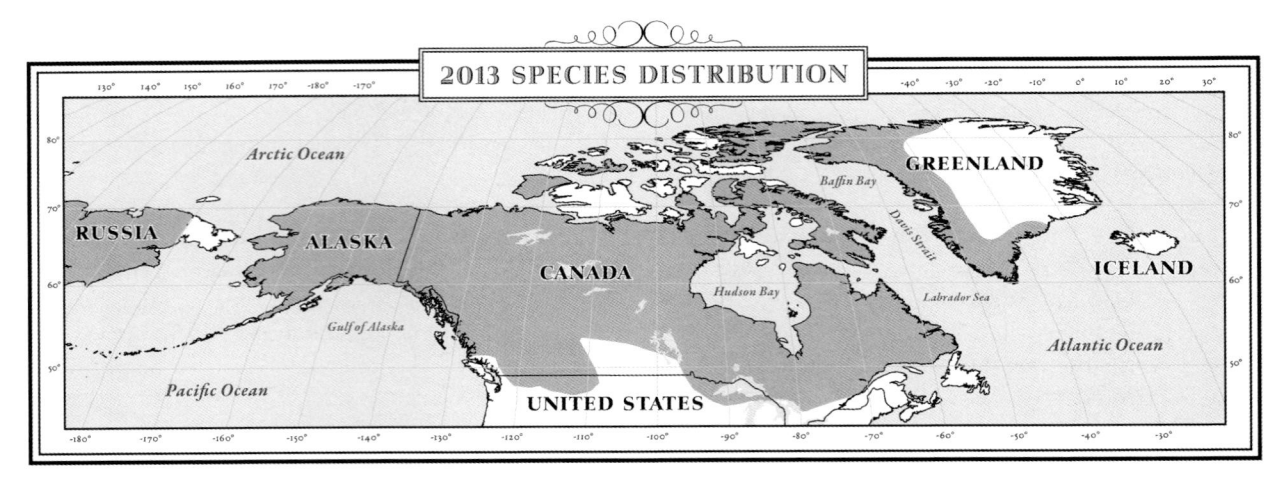

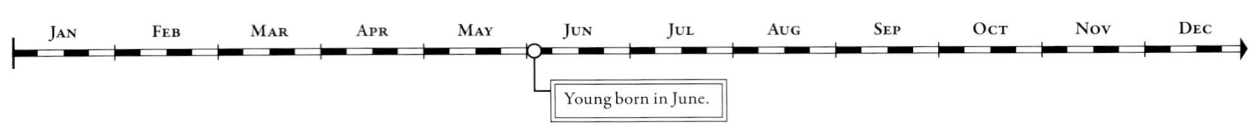

2013 SPECIES DISTRIBUTION

Jan — Feb — Mar — Apr — May — Jun — Jul — Aug — Sep — Oct — Nov — Dec

Young born in June.

CHAPTER TWO: MAMMALS OF UNGAVA AND LABRADOR

Descriptions by Turner

The Labrador wolf is plentiful throughout the entire region. More common in certain portions than others, due to their roving dispositions in search of food; the principal source of which is the reindeer, abounding throughout the region and wherever these animals are found in abundance not far off will be found a goodly number of wolves. In its habits the wolf, in this region, is often found solitary, fewer than four form a "gang" and in the vicinity of herds of reindeer they form "troops" frequently numbering fifty or more.

The females are somewhat larger than the males, having longer legs, thinner body, and larger tail. I was unable to observe any special sexual differences in the skins of those which I had an opportunity to examine. The variation in color is from a very dark, somewhat grizzly, gray to almost white. The skins of the white animals were, however, apparently smaller than the darker colored individual. The coloration of this [1514] wolf differs but little from that of the pure Eskimo dog. I have seen skins of wolves that could not be distinguished from those of dogs running about the place. In appearance, however, the wolf differs greatly from the Eskimo dog. The forefeet of the wolf are always larger, the imprint being not so circular as that of the dog. In the snow the mark of the claws of the wolf's tracks are always shown,

Wolf Skull

Collected by Turner at Ft. Chimo, 1882. Turner wrote: "I have not heard of any person in the region being killed by wolves. Two of the white men were, on separate occasions, chased by these brutes. One escaped by climbing a tree; the wolves lingered near, treating him to a serenade and left." USNM A23138, Division of Mammals, NMNH. Photo by Angela Frost, 2011.

while those of the dog are but rarely shown. Near the mouth of the Larch River, flowing into the Koksoak, I measured the track of a wolf to be six and a quarter inches across and six inches long. To judge, by the imprint, the claws were excessively long. There is great disparity between the tracks made by the fore and the hind feet, the latter being much less difference in the tracks of those members of the dog.

While the coating of the hair, may at times, of the wolf be nearly identical with that of the dog as to render the difference inappreciable upon the most critical examination, yet [1515] in the form of the body the two animals may be instantly distinguished. The head of the wolf is rarely elevated on a line, or above a line, with the back, mostly lowered and the snout turned aside. The barrel is narrow and deep in the wolf; a front view of the animal giving it quite a thin appearance. The tail is never elevated, but droops, held at less than forty-five degrees and the tip nearly touching the ground.

The wolf is cowardly and only when hard pressed for food will it advance upon man. A single wolf never, in this region, attacks a person, several may, however, combine and attack, although, of the numerous instances which I have heard related I think the wolves pursued because the person fled while if a bold stand had been made I believe they would have dispersed. I have not heard of any person in the region being killed by wolves. Two of the white men were, on separate occasions, chased by these brutes. One escaped by climbing a tree; the [1516] wolves lingered near, treating him to a serenade and left. The other person arrived within the sounds of the Post and the wolves ceased their pursuit. Neither of these men related that the wolves appeared very eager to close in on them.

That the wolf is cowardly none will deny. A wounded beast would doubtless give a serious resistance.

The people, white and natives, assert that there are three kinds of wolves in this region. The first are called "Hunters" which are doubtless the males. These are reported to hunt the game in company with another, a long-legged, kind, suspected to be the females which are known as "Chasers." The "Hunters" find the game and two or more of the "Chasers" pursue it. If the animal, being run after, is a reindeer separated from the remainder of the herd, the Chasers continue to run until they have gone over a distance of about ten miles. The reindeer when pursued by wolves nearly always runs in a circle [1517] having a diameter about six or eight miles. The "Chasers" pursue it for half of this circumference and their place taken at a point to which a relief of one or two more have gone across the country to either intercept it or else continue to drive it to where the remainder of the wolves are awaiting an opportunity to seize the exhausted animal as it draws near whence it started. One or two wolves rarely attempt to kill a reindeer unless it be wounded, for they are simply unable to catch it on account of the fleetness of the deer and the facility with which it makes its way through drifts of snow in which the wolves would simply flounder.

The food of the wolf is just what it can obtain in the way of flesh of any kind. The wounded deer and untouched carcass of those animals which the Indians and Eskimo so lavishly slaughtered in the spring and fall, that hundreds of these bodies lie along the shores of the streams where the temporary camps, of the natives were erected, [1518] were situated. Hares, Rabbits and Ptarmigan form a part of the food of the Wolves. These latter creatures are mostly obtained by stratagem.

I could not learn the time of coming together of the sexes. The young are born in June and do not attain their full size for two years. The number of young produced at a time varies from three in the younger females to five or six in the older females. The gravid female betakes herself to a retired spot where under the shelter of a rock or in a crevice she brings forth her young. When the young have attained sufficient strength to follow the dam they accompany her on the nightly excursions in search of food, consisting, at this time of mice, eggs and birds.

I have heard of a third kind of Wolf in this country. It is said to have curly hair and the head quite shaggy. I suspect it to be a dog run wild. A man informed me that he saw a soiled wild Grayhound near the mouth of George's River. It was doubtless one having escaped from [1519] a vessel along the coast. The man suspected that his own deerskin clothing frightened the animal and caused it to bound away. I have heard of no instances along the coast where the dog and wolf have crossed. There is, at times, at Ft. Chimo abundant opportunity for the wolves to do so as they are often quite near.

The Eskimo give the name of *a mán ghok* to the wolf. The Indians call it *ma hé kan*. Neither of these people have more than one name for the wolf. Quite a number of skulls of this wolf were obtained by me. I saw skins of young supposed to be about six or eight weeks old. The color of the back and upper sides was a fulvous-brown while the head, neck, legs and part of the tail light gray. The feet and tip of the tail somewhat darker.

Scarcely two of the adults are similar and none alike. The summer pellage is sparse and lighter colored.

How Children Became Wolves
An Inuit story, Turner (1887a: 301)

A poor woman had so many children she was unable to restrain their cries for food. They were changed into wolves and to this day their mother may be heard endeavouring to console her clamouring children with the hope she will find a nice, fat deer for them.

The Wolf and Her Otter Husband

An Innu story, Turner (1887b: 660-674)

An old wolf and his wife had a large family of children so young that it was with the greatest difficulty the parents could provide sufficient food and clothing to keep them alive. The eldest child was a daughter who helped her father and mother all that was in her power. They were so poor that the mother decided her daughter should marry in order that the husband should help them provide for their other children.

The Indians had left the region because there was so little game to be found that they too had nearly starved. Winter was drawing near and they had no resources left. The mother called her daughter to her and informed her that she must go and seek her lover. The daughter replied she had no lover. The mother answered, "You have a sweetheart. We will all starve if you do not find him and ask him to procure some food for us. Your little brothers have nothing to eat or to wear. The Otter is your lover; go and seek him." The daughter inquired where the Otter dwelt. The mother told her the Otter dwells in the waters of the "Narrows" of the lake. "Go there and you will find him. Let him not escape." The daughter replied that on the next day she would see her lover.

When the early morning came with its gray clouds hanging low, a deathlike stillness of the air that forbode a storm the daughter hesitatingly set out for the narrow portions of the lake and by noon had reached the top of a high hill near by and from it she soon saw an open hole in the ice; and in a few moments the Otter appeared from the opening; and, creeping on the snow which covered the ice began to roll and tumble in the soft and fluffy snow, plunging here and there, often hidden in its depths and as often appearing with the glittering crystals sparkling from his glossy black fur that made the Wolf admire his beauty and motions.

At an unguarded moment the wolf stealthily crept toward the place where the Otter gambolled in fancied security; as the wolf drew near the Otter observed her approach and with a dash he plunged toward the hole in the ice; and, leaping into the water was about to disappear when the Wolf cried out, "Do not dive. Come here; my mother says you are my lover." The Otter was astonished, and inquired, "How can I be your lover when I live on the water and you live on the land?" The Wolf replied, "You can live on the land as well as in the water." The Otter rejoined, "I will not live on the land." The Wolf replied, "You must come out on the land and live with me, I shall smother you in the water if you do not come out on the land."

The Otter informed her that she could not smother him in the water for he had a number of breathing-holes and you cannot watch all of them. He dove under the water and was lost to sight. The Wolf began to howl most dismally when the Otter disappeared. The clouds rolled over and the wind began to blow. In a few minutes the soft snow began to drift and with furious gusts swept blindly over the lake. The flying snow fell into the water of the Otter's breathing-holes in the ice; and, with the intense cold, which the wind had brought with it, the whole surface of the lake was rapidly freezing over. The holes were soon tightly frozen. The Wolf sat at the one where the Otter disappeared; and, with her paws, kept the water free from the slushy ice made by the drifting snow.

After awhile she heard the struggles of the Otter below the ice, striving to obtain a breath of air. She stood with open jaws ready to seize him the instant he should appear. Soon he came, nearly exhausted to the water and the instant his nose rose above the surface, the Wolf retreated a few steps. The Otter came out and began to flounder in the dry snow to free his fur from moisture. When he saw the Wolf he exclaimed, "I shall live with you. I shall live with you." The Wolf said, "Did I not tell you I would smother you if you did not come out of the water."

The Otter made no reply to this, but inquired if she had a piece of line to give him, adding that he would catch some fish for her supper. He got a piece of line from her and was then directed by him to go and prepare a tent for the night. The wind had abated, the snow fell to the ground and the sky rolled away its clouds while the bright sun slowly sank behind the hills beyond the lake.

The Otter went into the water and in a short time re-appeared with a quantity of fish which he had strung on the line the Wolf had given him. He fastened one end of the line to the edge of the ice and then rolled himself in the snow to dry his fur. In a few minutes he followed the tracks made by the Wolf, and presently found himself at the door of the tent she had prepared. He told the Wolf to go to the hole in the lake and bring the fish he had put on the string.

The Wolf gladly went to the spot and was so overjoyed to perceive such a fine string of fish which her lover had caught. From looking at them she began to devour them, eating so many that she could scarcely move. The Otter was such an expert fisherman that she could not eat all he had taken. She dragged the remainder to the tent. She found the Otter asleep when she returned; and, without disturbing him, prepared a fire; soon having the kettle of fish boiling for her lover's supper.

He aroused, after a time, but complained of being weary; and declined to eat the prepared fish. They retired and the next day a young Otter and a young Wolf had made their appearance. The Wolf prepared breakfast; and, after they both had eaten, the Wolf sat by the fireside with her head hung on her breast. The Otter observed her dejection and inquired what ailed her. She replied she was wondering whence she might obtain some young reindeer skins with which to prepare some garments for the newly-born children. She remarked that it would give her so much pleasure to dress them nicely in skin clothing. The Otter bade her open the door of the tent. She did so; whereupon he indicated a locality where he frequently killed as many deer as he desired. The next morning the Otter went away by the break of day. He soon found a herd of thirty deer, feeding upon the hillside, which he had pointed out to the Wolf the evening before. He had no gun with which to kill the reindeer but knowing he was able to jump so well he plunged at the deer, and by jumping through the body of the beast he soon had them all killed.

He now rolled himself in the snow to clear his fur; and then returned to the tent where he arrived a short time after sunset. The Wolf saw him coming and had a great pot of fish cooked for him by the time he got to the doorway of the tent.

After he had eaten his supper he told his wife he had killed thirty deer and that she should, on the morrow, go for them; she could follow his tracks and find the place where the deer lay dead. She soon had them hauled home by bringing four at a time. She laid them before the door of the tent and quietly slipped to bed. Before daylight the Otter awakened her, bidding her make the fire and prepare breakfast as she had to carry home the carcasses of the thirty deer.

The wife replied that she had already brought home the entire number of the deer which he had killed. The husband was amazed and inquired how it was possible for her to accomplish so much during the night. The only answer was, "If you do not believe it look out of the tent and you will see them." The Otter, still doubting, opened the door of the tent and a huge pile of dead reindeer was before his eyes. He then sat down, but made no remark to his wife, looking at her with a steadfast gaze until she could endure it no longer. She hastily asked, "Why do you look at me in that way?" The Otter replied that he was only wondering how she could bring home so much meat in so short a time during the dark night.

The Wolf then bade her husband to come and eat his breakfast, adding that he would have to help her skin the deer. After breakfast was finished they began to take the skins from the bodies; and, when that labor was nearly completed, the Wolf requested her husband to erect a staging of posts and poles, on which to place the meat. He did so; and, while she was removing the fleshy parts from the skins, he placed the meat upon the stage to dry. The skins of the deer were hung around the tent to dry, in order that they might soon be converted into clothing for the children and the parents. They then took their supper; and, when the meal was finished, the Wolf returned to the farther side of the tent, sitting there with bowed head.

The Otter finally inquired why she was so quiet. She replied she was thinking of her poor parents and suffering brothers and sisters. She added, "I suppose that are starving to death. My old father told me to ask you to place a mark in the middle of the lake, when you had killed some deer, so that they might visit me and thus get something to eat."

The Otter replied that he would go immediately and place the staff on the ice of the lake that the relatives of the Wolf might see it and come to her. The Wolf was delighted at the liberality of her husband. He went and soon had a stick erected which would be seen from a great distance. At the time this was being done the brothers of the Wolf were sitting on the top of the hill watching for the Otter to place the stake there. When they saw the Otter creeping through the snow they knew food would soon be had in abundance. They waited awhile and when the staff was placed upright the little wolves scampered back to the tent where their father and mother were shivering with cold, and gave the glad tidings that their sister had saved them from starvation.

In a short time they were all on their way to the staff on the lake. When they came to it they followed the trail of the Otter and soon arrived near the tent on the edge of the woods. They saw the deerskins hanging on the tent and the great stack of meat piled on the stage.

The little children of the Otter and Wolf were playing outside of the tent when the visitors approached, making so much uproar in their greedy haste to feed upon the stores of their new brother, the Otter, that the little children heard them and ran terrified into the tent where their father and mother were sitting.

The young otter threw himself into his father's arms, screaming with fright. The father asked him what had caused him to be so alarmed. The little thing clung closer and replied it had run from the hungry. The mother smiled and said perhaps it was her parents and brothers and sisters. The Otter told his wife to go out and see if it was her relations. She opened the door of the tent and there stood a row of gaunt wolves, scarcely more than skin and bones.

Without waiting they fell to and began to devour the pile of meat on the stage. The Otter's wife remonstrated with them for their greediness, saying, "My husband is not a stingy man. I take my meals when he is asleep, and pretend to not eat much in the day time." They satisfied their present want and went into the tent. The Otter had little to say and went to bed. When he was asleep they went to eating again and soon had all the flesh eaten.

The next morning the Otter appeared sullen; and, when his wife inquired what the matter was with him, he answered, "I think your brothers are going to make a fool of me." The Wolf asked him what cause he had to think so. He replied, "They look at me so hard I do not know where to turn my head."

After breakfast the Otter told his brothers, the Wolves, they must help him hunt some deer. They soon disappeared to search for them and quickly they came upon a great number quietly feeding on the side of hill. The Otter bade his brothers go and kill them. They ran toward the herd but were unable to overtake the frightened animals except one which broke from the heard and was quickly dispatched by the wolves.

The Otter witnessed the actions of the wolves and concluded the herd would escape them. He sprang after them and in a short time had killed all the remainder. He cleaned himself by rolling in the snow and then returned to the tent. He found that the wolves had arrived before him. They inquired if he had killed any deer and how many. He replied he had killed all that were in the herd, except the one they had killed, adding that they would have to go the next day after them. They signified their willingness to go after the carcasses of all the deer he had killed. When the morning came and the wolves were ready to start on their journey to obtain the meal of the deer slain by their brother, one of the wolves remarked to another, "Look at our brother, the Otter, he has a white mouth." The Otter turned to his wife and said, "Did I not tell you that your brothers would make a fool of me?" He grasped his child and with it dashed out of the tent door, telling her that she would have hereafter to live without him.

He plunged into the water of the lake and was never seen again.

Wolf Attack
An Inuit story, Johnny Sam Annanack (Heyes et al. 2003)

The story that my father would like to tell was about hunting. They used to go out hunting up the river, walking with the people from Killiniq. The more northern people would walk to another community. They had almost starved to death at one point. My father used to tell me what happened to them. He used to say that there were a bunch of men who were out hunting, and they came on a boat and walked back to their village. They almost starved to death, but they had gone hunting for caribou and they were blessed to have caught an animal. Our elders used to be happy to find what they had gone for, especially if they had caught it. Smaller game was especially fun to catch. There was this one particular man who had a vision, my father would tell me about this man.

On that day, when they didn't know what was going to happen, they had found an otter and all the men were enjoying trying to catch it. My father was a young man when this happened and he followed the hunters. One of the men was from Killiniq. He was called Sakaliasi, known as Noah Arnatuk's father.

One night without expecting anyone or anything because it was late in the night, just before they were asleep, they heard something. Some men were asleep by then. My father and his half brother had the same father with different mothers. When my father's half brother was younger he used to be scared all the time, (just the way he was), so he'd stay with my father all the time. While they were trying to sleep, something arrived. It was a wolf. It had entered the tent where my father and Jusipi were laying. It touched someone's leg and bit Jusipi's foot. My father doesn't even remember what happened after that, but I'm sure they did something about the wolf after he was bitten on the foot. The wolf now was in the tent, the one at the end of the tent fought with the wolf, he was the father of Noah Arnatuk.

The wolf and the man were fighting in the tent, they fought until they went out of the tent, and it was very dark, no one could shoot at it because he was fighting with the man. If it was more daylight, I'm sure they would've helped right away. The wolf was in a real fight, his arm was in the mouth of the wolf and thinking now that he was in control of the wolf he'd tell the men to shoot where he wanted them to. It was as if the wolf understood, he'd move and struggle whenever the man told the others to shoot.

Every person on earth always has a little bit of a bad side, and even if we think we're very nice, there's always a little bit of a bad side in us. He always thought it was a way of trying to make the man realize some thing that he was to understand.

If one day he was approached by the police, he'd only think of the wolf that once fought him, and made him more courageous. My grandfather couldn't kill the wolf because he couldn't while he was in a fight with a man. It was as though the wolf understood what the man was saying.

When morning came, the wolf was dead and the man was becoming sick from the fight. He had wanted the wolf to be put close to the door way and asked for his rifle. So he was given a rifle and said he wanted to shoot the wolf even if it was already dead. He wanted revenge and the other men let him do it. He also wanted the men to completely destroy the dead animal. Everyone did what he wanted. That's what happened, but after they destroyed the animal, the man thought he was dying and told the other hunters not to leave him if he died. He died. But he would tell the others that he would come back alive after dying for a short time. Although he had said that, my grandfather who was one of the well respected leaders of the camp, was very curious as to what was going to happen after he died and comes alive again. He was buried in a way that they used to, by covering up the body with rocks. They left the next morning leaving behind the buried man. They felt they had to leave at that moment.

After a year later, they went to see the grave to see if he had got up since he used to say that he would come back alive. He had actually got out of his grave. He always used to tell us that story. So ever since they left him like he didn't want, they always thought of him, so they went back a year later to see and the grave was empty. That's how my father used to tell us that story and that was the end, I'm just saying it like my father used to say it.

Wolf Skull
Collected by Turner at Ft. Chimo, 1882. USNM A23138, Division of Mammals, NMNH. Photo by Angela Frost, 2011.

Man Who Turns into a Wolf and Other Animals
An Inuit story, Tivi Etok (Heyes et al. 2003)

There was a man who would become any kind of animal such as a seal, wolf, caribou, walrus, you name it. People used to talk about him. Whichever kind of animal he turned into, he would be among the type of animals. I will tell you which animal he was at first. I think his first time, he became a caribou. They were the worst animals to be, because they did not stay in one place and were always aware of danger. They'd always be on the run and alert of predators. While he was a caribou, he'd rest a while but had to run away every so often, even in the night and in a blizzard. He could not stay in one place for long; he didn't like being a caribou.

During spring or winter, the caribou were always aware of danger, and they were hunted the most by their predators. In the winter when it was very cold, even if they had slept for a short time, they'd always suddenly run off trying to keep warm, so he was not so comfortable being a caribou.

And then he became a wolf. He liked being a wolf because if they had been eating well, they'd just relax and not move too much, not like being a caribou. But the wolves are always hunting for the caribou. The females are very good at hunting – better than the male wolves. The male usually follows behind, and when the wolf caught a caribou it would howl, and then the rest would come running, knowing that there was a catch.

When that happened the wolves ran off and the man that was a wolf could not keep up with the wolf pack, so he would often reach his destination only when the wolves had already almost finished the caribou. That's how the story goes, but one day he decided to ask another wolf "how can you run so fast?" The response was "when you start running, imagine that you are grabbing the ground with your feet, so that you can run faster."

So the next time the female wolf started howling, he started running trying to do what he was taught by the other wolves. The howling sounded like it was far, and he usually arrived when the catch was already finished, so doing what he was taught, he started arriving while there was still meat on the catch. So he now was like the other wolves and was able to keep up with the wolves after that. He liked being a wolf the most because they would just relax and take naps after a good meal and they did not worry about anything. They only moved when they were hungry and they didn't always have to be on the run.

And then he became a walrus, he was a human, but turned into an animal. Being a walrus was even better; they are not scared of anything. He liked being one because they were brave and strong. When the walruses started diving into the bottom of the ocean to feed on the clams, he would try and follow them, but even before reaching the bottom of the ocean, he'd turn right back up because he couldn't hold his breath long enough. He wanted some clams too and was getting hungry, but he had tried diving a few times and couldn't reach the bottom. So one day he decided to ask one of the walruses "how do you go down to the bottom of the ocean when you go for clams?" The walrus' reply was (whatever animal he became, he would understand them), "kick your flippers as if you're kicking to the North Pole, that's how we dive down to the bottom of the ocean." So when the walrus started diving down again, he kicked real hard and easily went down with the rest. When he learned to dive like the others, he became a real good diver just like the rest. That's the end of that story.

Recollections on Inuit Use of Wolves
Tivi Etok (Heyes 2010)

There are different kinds of wolves, just as there are different kinds of dogs. Some are slow. The male ones are generally not hunters. The females are generally the hunters. They run and hunt caribou. Females generally have thick fur and are faster runners than the males. When caribou are killed they make howling noises. The female howls to inform the males and her babies to come along out from their hiding places so that they can eat. The wolf would make their dens in hard sand, like foxes. After they make babies they would move to another place. Like humans, they would move around. Wolves are not like dogs. We used to have a dog and a wolf on a dogteam. We tried to train it like a dog, but they were not good at it; they couldn't have a rope around their head; they were not as strong. Dogs are stronger. Even when the *qamatik* (sled) is very heavy, a dog can pull it; even if an Inuit cannot move the sled.

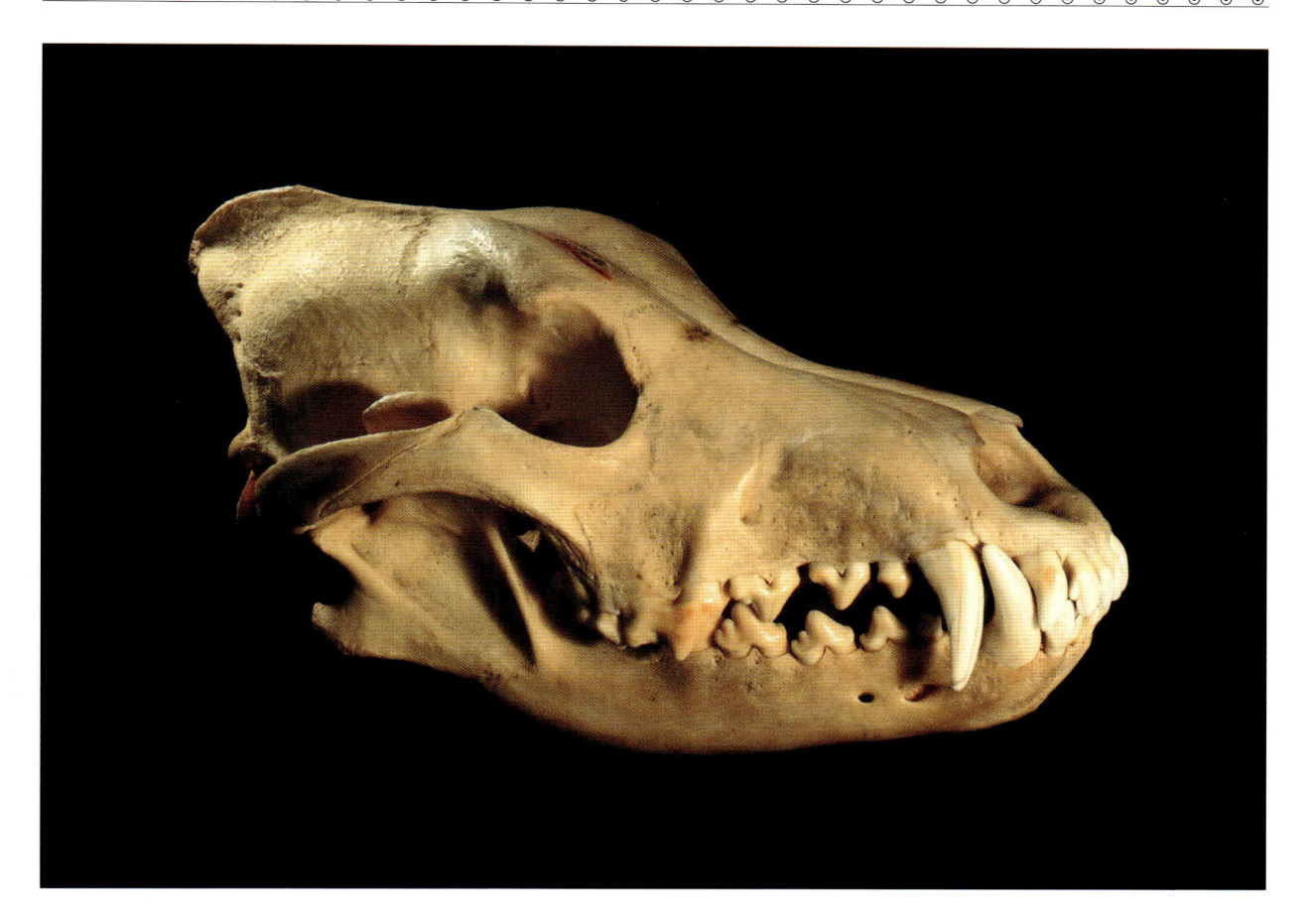

Female Wolf Skull, Type Specimen
This skull collected by Turner at Ft. Chimo in 1882 is the type specimen of Canis lupus labradorius, *the Labrador wolf, described as a new subspecies by Goldman (1937). USNM A23136, Type Collection, Division of Mammals, NMNH. Photo by Angela Frost, 2011.*

ᕿᒻᒥᖅ

Eskimo Dog

qimmiq

Canis familiaris (Linné) borealis?

[1520]

Order Name used by Turner	*Carnivora*
Current Order Name	*Carnivora*
Family Name used by Turner	*Canidae*
Current Family Name	*Canidae*
Scientific Name used by Turner	*Canis familiaris (Linné) borealis?*
Current Scientific Name	*Canis familiaris* Linnaeus
Syllabary	ᕿᒻᒥᖅ
Modern Nunavimmiutitut Term	*qimmiq (cf. Schneider p.304 - qimmiq = dog)*
Eskimo Term by Turner	*khiṅ mik*
Definition of Eskimo Term by Turner	*Dog [p. 2291; 1535]*
Innu Term by Turner	*a tum [p. 1830]*

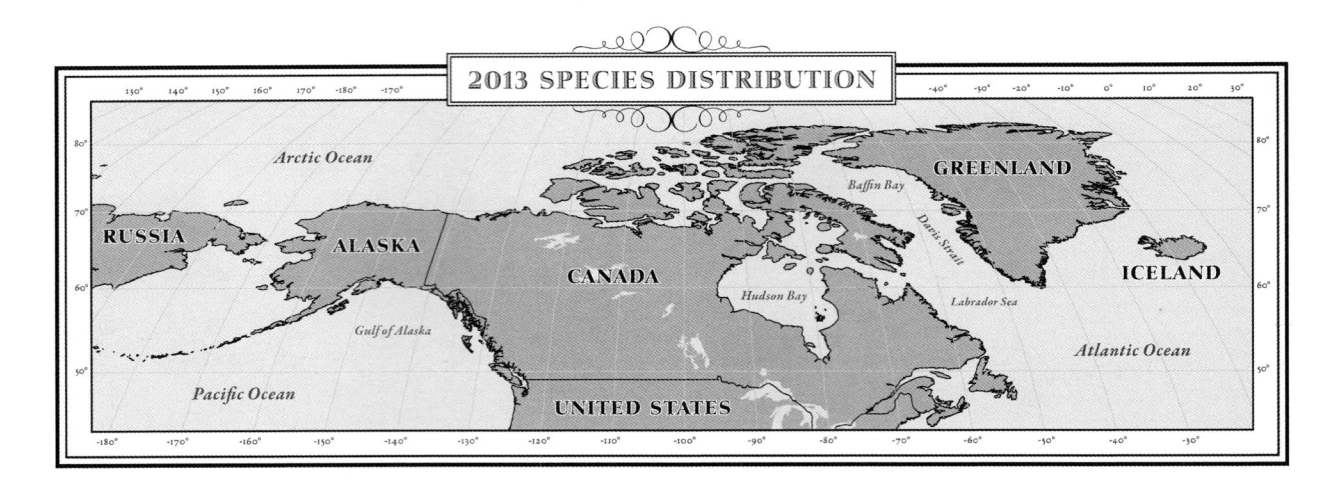

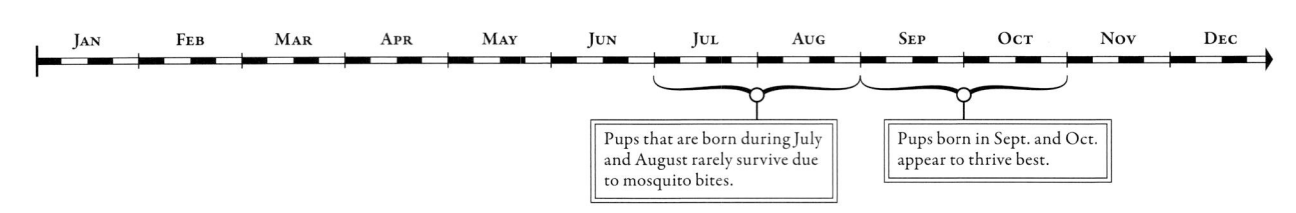

2013 SPECIES DISTRIBUTION

Pups that are born during July and August rarely survive due to mosquito bites.

Pups born in Sept. and Oct. appear to thrive best.

Chapter Two: Mammals of Ungava and Labrador

Descriptions by Turner

Without the dog the greater portion of the extended journeys made in the northern regions would never be performed. This brute serves all the purposes of draught which is necessary in the north. The dog is especially suitable, living upon flesh easily procured at certain seasons; and, during times of scarcity, able to withstand for many days the privation of food. There is great difference in these animals of various localities and even among the number comprising a team; the disparity of size, not altogether depending upon sex, is noticeable.

As a rule the females are smaller, yet certain males are sometimes no larger than large females. I think that the Eskimo dog of the Labrador and Hudson Strait coast is larger in every way than the dog of the Norton Sound coast of Alaska. The legs are longer, the body deeper and not as rounded. The head is longer, the muzzle noticeably longer, the tail is not nearly so profusely adorned with hair as the Alaskan Eskimo dog. In hauling [1521] a load I believe they are better but certainly not so swift.

A large Eskimo dog will weigh over a hundred pounds and is capable of hauling that number of pounds on a sled

"The winter coat of hair is so profuse that the skins of the Eskimo dogs are in great demand for rugs and other purposes."

for days at a time. The Labrador sled is four or five times as heavy as the Alaskan sled and is not, comparatively, so easy to drag. The number of dogs comprising a team varies from seven to twenty. The style of harnessing and sled used is described in another connection. The Eskimo driver is equipped with a huge, plaited whip composed of several stout thongs of sealskin at the rear end, tapering to a point or lash of a single thong. The whip may be as much as over thirty feet in length and when wielded by a lusty Eskimo, skillful in handling the awkward affair, it becomes a most formidable weapon, sufficient to urge the laziest brute to utmost exertion. The lash may be directed to touch any spot the driver desires and with a cringe and mingled moan and howl the stricken brute calls [1522] all its energy into action under the smarting blow delivered with such effect that the brute does not soon forget the whip behind is in hands willing to direct it.

The Eskimo dog is very quarrelsome and not good natured even to its master. The numbers of a team usually league together and pounce upon another dog that may be seen far or near. There is usually a superior dog in each team and he holds the remainder in subjection and in their marauding parties he usually takes the first nip at the strange dog. The remainder of the team are not slow to come to the rescue and often make the mistake to get hold of the wrong dog. They fight with savage ferocity. The head, neck and legs appear to be the favorite places for seizing the opponent. I have seen them so firmly locked upon each other's lower jaw that only by prying the jaws apart with a stick could they be separated. The wounds made by their strong jaws [1523] upon the legs of the opponent are often serious and result in either permanent or else long continued injury. Two dogs of equal strength sometimes, of a single team, struggle for mastery and the combat is terrific to behold.

Their freaks in fighting are very peculiar and often unknown how it will terminate as two dogs may at times engage in battle while the remainder are apathetic to the issue while at other times the least snarl at fancied intrusion will cause the entire pack of dogs to engage in a free and tumble fight.

Inuit Dog
An Inuit dog belonging to Kangiqsualujjuamiut hunter Daniel Annanack. Beliefs relating to dogs were recounted to Turner between 1882 and 1884. He recorded: "Dogs are not permitted to eat foetal reindeer lest that act displease the guardian spirit of the deer." (1887a: 65). Turner also wrote: "The shaman may decree that the person slit the ear of a favorite dog or cut off the tail of that brute..." (1887a: 70). In another account Turner (1886c: 692-693) described dogs catching fish: "The Eskimo dogs search for these fish [Pleuronectes americanus, winter flounder] and feel them with their paw. When one is discovered the dog remains motionless with his foot on the fish; then puts his head under the water and seizes it, carrying it to the shore to eat it. I saw an old dog teaching her pups to fish in this manner at Rigolet. She was known as an expert fisher and on one occasion I saw her obtain three fish each of which was seized by the pups while she held foot on it. She conveyed the intelligence of the capture of the fish to her pups simply by a look upon which the pup went galloping through the water to her." Photo by Scott Heyes 2008.

When several dogs pitch upon a single individual the beaten dog is quickly thrown upon his back, exposing not only the legs which will be each seized but pulled so as to lift the brute from the ground. In one instance I saw the victim actually torn to pieces. I have seen, also, individual dogs which, when endeavoring to seize another dog, always strive to grasp it [1524] by a certain part of the body, and castration results.

The sluts receive the males at any season of the year. The period of gestation is eight weeks and two days, sometimes four days, a period carefully noted by me. During the period of heat the sluts receive any male. The stronger dogs usually obtaining such favors simply by force of strength to drive off other suitors. It is not infrequent for the female to desert two dogs fighting for her and slip off with another weaker dog loitering at a respectful distance in anticipation of the affray. During the act other dogs never interfere.

The number of whelps varies from two or three for the first litter produced to twelve to fifteen for an adult female of four to seven years. The number of followers influencing the number of pups. The females are very savage, toward other dogs, while nursing her young and especially so to another female. It depends much upon the character [1525] of the female as to where she will bring forth her young. Those most petted and considered as favorites usually whelp about some of the buildings and are only too glad to have a bed prepared for them. Others, wild by nature, retire to some secluded spot and there rear their young until they are able to return to the posts. Those reared away from the sight of man are very wild and snap like wolves on being approached by a person. An instance occurred at Ft. Chimo where a slut occupied a wooden coffin containing the skeleton of a favorite Eskimo who had died several years previous to the box being tenanted by five young dogs and their mother. The box was placed alongside of a huge boulder and loose stones placed on it, and as the box crushed beneath the weight imposed a sort of shelter was afforded to the dogs. I passed by the locality, little suspecting it to be occupied by living creatures and thus discovered it. The pups born during July and August are generally so persecuted by [1526] mosquitoes that they rarely survive. Those born in September and October appear to thrive best and grow into sturdy brutes.

Sluts of only five or six months receive attention from the males and may become mothers at seven or eight months of age. Such animals are not only retarded in their growth and prove inferior draught animals but also their progeny appear to be less strong. Such results might be easily obviated by little more attention from those whose duty it is to look after the teams.

The dogs receive good attention during the periods they labor; and, during the summer season they find but a precarious living. Their food consists of all manner of flesh; and, the post teams, used for hauling wood, water and making more or less extended trips for meat or hunting parties; receive also an occasional mixture of damaged flour, meal, etc., with chopped meat of the deer, whale and seal. Such numbers of ptarmigan were plentiful that during [1527] the period I was at Ft. Chimo the dogs, in winter, consumed hundreds of these birds. The bodies of the birds were frozen hard as stones and two of these frozen bits were considered a daily ration for each dog. The winter coat of hair is so profuse that the skins of the Eskimo dogs are in great demand for rugs and other purposes. They form excellent lap-robes and for this purpose many skins are sent to friends in different parts of the civilized world where, doubtless, the robes are known as wolf robes. The Eskimo dogs attain an age of over fifteen years and then become so infirm as to be scarcely able to move at more than a walk. The teeth drop out and only tenderest care prolongs the life of these favored brutes. Mosquitoes, excessive cold and attacks from other dogs shorten their lives and many of the pups die from the first two causes. Among the dogs over four months of age a disease mysteriously breaks [1528] out. The first symptoms are a greater or less rigidity of the legs, followed by a desire to stand or lie in perfect quiet, disturbed only by spasmodic efforts to bark or whine. These efforts are apparently attended with a severe pain about the shoulders and throat. The eyes then become glassy, exhibiting an indifferent stare at all objects. The creature now becomes much worse and may pass through the stages described in less than thirty six hours and so rapid may be the progress of the disease that only was it recognized by the latter symptoms. An aversion to food and water is now strongly manifested, having increased in the aversion each day and in the course of two to five days the creature dies, apparently choked to death. The dogs developing the disease rapidly do not froth or foam at the mouth, only those do so which have been less rapidly attacked.

The dog endeavors, during the progress of the disease, to snap at imaginary enemies but as other dogs rarely [1529] molest them, and even avoid them, no such thing as communicating the disease by wound has ever come under my observation. I never knew a person bitten by them for in all the instances witnessed by me the poor creature appeared to expect some relief by human agency. The dogs under the circumstance exhibiting only such fear or courage as would have been done in a state of health; and if it desired to escape it was done only in order to prevent molestation by the person.

The disease is always fatal. I never knew a case of recovery and so destructive is the disease that it has been known to carry off nearly all the dogs of a district or particular region of coast. The season of the year has nothing to do with the disease; although, it appears more virulent during the summer but this may be accounted for by the discomfort of the heavily clad animal suffering from the heat which the more quickly causes the intolerable thirst apparently raging and unable to be satiated.

[1530] Never having seen a genuine case of rabies in the dog I must confess that the disease at times afflicting the Eskimo dog does not agree with the symptoms ascribed to rabies. That there are hydrophobic symptoms none will deny.

Along the northern Alaskan coasts it is the practice of the trading companies to castrate the greater number of their dogs destined for draught animals. These as well as the untouched animals are afflicted with the disease. Along the Labrador and Hudson Strait coast the male dogs are rarely, if ever, castrated. All ages, over four months, and sexes are liable to the disease. Mr. L. Kumlein, Naturalist of the Howgate Expedition to Cumberland Gulf, in Bulletin No. 15, U.S. National Museum 1879, endeavors to prove a theory which is not at all supported by facts observed in other places equally favorable for such a condition of affairs as at Cumberland Gulf.

Dogs which have never done a day's work or received a blow from [1531] a person or attacked by other dogs, have to my certain knowledge, contracted the disease and died from it.

The people who have anything to do with these dogs do not fear them and apply the terms rabies and hydrophobia so loosely that it is rarely to be doubted whether the person has a proper conception of the terms. That for a great portion of the life of these brutes they procure but insufficient food, and this often if such character as to afford but little nourishment, there can be no doubt. At other times they are permitted to gorge themselves to their utmost capacity. Their diet is also, at times, but a single kind of food for months and frequently of the leanest character while circumstances may, tomorrow, afford them an opportunity to revel in the strongest fats and oils. Half-bred dogs are liable to this disease but full blooded dogs of other breeds than the Eskimo dog are apparently exempt from [1532] its ravages.

That the dogs are badly treated by some of their masters there can be no dispute as anyone who has dwelt in the countries may well witness. The Eskimo and many of the white men are cruel masters and often vent their grievances upon a dog instead of their own kind. Blows from the heavy stock of the whip and the doubled, thick lash are applied from about the head or back with little regard to the future usefulness of the dog. Superstition causes the Eskimo to mutilate the ears, tail and other parts of the dog; so that, take it all in all, the dog has a sorrowful life before him. Many theories have been advanced, upon the cause of this disease, by the white men of the countries, but none of them appear to be substantiated by actual facts. I do not venture to assign any cause, for the simple reason that I am not satisfied that I know it.

> *"...the dog slinks from the sight of a wolf and the former certainly evinces fear if it passes only a fresh track of the wolf."*

The Eskimo dog is not at all [1533] cowardly. It will attack anything that has life, even man. The dog is a keen observer and detects the least sign of fear in the object it may, for a moment, have hesitated to attack. Several instances have occurred on the Labrador coast where the dogs have torn people to pieces, and in certain instances, devoured the remains. Small children are liable to be torn not specially from any desire on the part of the dog but because it cries when accidently pushed over and

Mushing
A young Inuit hunter learning how to mush near Kangiqsualujjuaq. Photo by Scott Heyes, 2008.

in its struggles seems to invite the attack. Any unusual object excites their curiosity and will be satisfied before the dog gives up the examination. A Roman Missionary was attacked at Nakvak where a local race of huge dogs are raised. The black robes of the person attracted a dog which came up and on the person, grasping his robe to frighten the animal, was seized and other dogs sprang in. They would have made short work of him if assistance had not come opportunely.

[1534] There are but two things I know of which an Eskimo dog fears. One is a whip, although, the merest motion to cast an object at a dog suspected of evil design will often frighten it away, until better means of defense may be procured. The mere presence of a whip is sufficient to secure the most respectful attention from one of these dogs. The sound of the lash causes all in the team to lie on their bellies and howl with terror at the expected blow, while those running free about the place quickly secrete themselves under any shelter.

The Eskimo dog is naturally afraid of a wolf and the stories of interbreeding with that animal are facts only when the wolf is able to drive off the male dogs and secure the female by force alone. I have little faith that the slut seeks at her own instance a favor from a male wolf.

It is a recognized fact that in those countries that the dog slinks from the sight of a wolf and the former certainly evinces fear if it passes only a fresh track of the wolf.

[1535] Mr. James Irvine of Ft. Chimo related to me that they were hauling wood from near the "Chapel" to the post. A trap set near that hill had caught a wolf which was dead when found. The wolf was dragged to the loaded sled and as soon as its scent was detected by the team they became uneasy and desired to run off. The frozen wolf was placed in a standing position on the load of wood and the team started. Irvine affirms, and doubtless correctly, that no team ever made quicker time than was done on that occasion.

The Eskimo of Labrador and Hudson Strait apply the name of *Khĭng mĭk* to the dog. This word is so near the word *Kĭng mĭk* meaning heel that it requires a correct ear to distinguish the difference.

Carnivores

Illustration of a Story about the Man and His Wife Who Had No Dogs
By Johnny George Annanack, 2 May 2005. Annanack described the story in a 2005 interview (Heyes 2007): "This drawing is about a man and wife who did not have dogs. They had to pull their sleds by themselves over great distances (as shown by the open harnesses at the bottom of the illustration). When they became tired and when it was starting to get dark, they built an igloo on the ice. The woman is using her ulu (cresent-moon shaped knife) to cover the gaps in snow-blocks of the igloo with fresh snow. The man is cutting out blocks of snow using his pana (snow knife). The drawing is based on a story that I once heard." Reduced from original, 11″ × 17″. Lead pencil on bond paper. Scott Heyes Private Collection.

Other races of dogs are occasionally introduced, by traders and from the vessels, into the country and, often with a desire to improve the breed, some misguided individual permits the crossing of the [1536] introduced race with the native dog and always with bad results. The Eskimo dog is specially adapted to withstand the rigors of the severest weather and undergo privations sufficient to render a common dog utterly worthless. The result of such unions is a cross characteristics of neither the one nor the other. The Eskimo dog does not bark; the half-breed dogs do bark and any attempt to improve upon the pure Eskimo dog is time thrown away. These dogs are addicted to howling in a most dismal manner, causing the utmost confusion for the time.

Notwithstanding the fact that dog-driving is not calculated to inspire a novice with a desire to perpetually indulge in the task where every object along the path must be inspected, every passing bird jumped at and the entanglement of the traces which is certain to result, upon attempted disengagement, in a fight, piling dog upon dog until the driver exhausts not only his strength [1537] but repertory of vocables in all the languages he has ever heard.

In this connection I will state that the Indians have a small breed of black dogs. These animals are smooth, short-haired, clean limbed and of vivacious nature. Their color is always black as a dog can be, some of them have only a small spot of white on the throat. They are called "Hunting" dogs but for no other apparent reason than their prowling nature would warrant. It is, however, stated that these dogs are trained to search for porcupines and that when one of them is discovered the dog utters a sharp bark to attract its master. The Indian dog is a vicious brute and quickly advances upon a white man.

These dogs lie around and in the tents and during severe weather are protected; otherwise, their short coat of hair would afford insufficient warmth. Other dogs were observed but the Eskimo dog and the Indian dogs are [1538] the only ones worthy of mention. I consider it necessary to give no description of the Eskimo dog so far as its color is concerned as this is so variable that I content myself by stating that it is comprised within the range of darkest brown (not black) to nearly pure white. The spotted dogs or gray ones are considered the best for general purposes. In this region I observed a greater number of dark colored dogs than were noticed in Alaska where I saw that dark colored dogs were objects certain to attract the special attention of the other dogs and lead to fighting. For this reason gray or whitish dogs were preferred in that country.

Certain dogs, having a spot above the eye are, in the eastern coasts, preferred to any others. They receive a special name, *Tǔ holik*, said to mean "double-eyed." I very much doubt the asserted origin of the word and claim that it refers to another part of the body, having no connection with the eye.

Eskimo Dog Ethnography
Turner

DOG WHIP. HUDSON STRAIT INNUIT. UNGAVZA DISTRICT. H.B. TERR. (1887A: 162-164)

The construction of the most important aid to dog-driving depends greatly upon the taste of the one who uses it. Some of the Innuit are specially fitted for the work of driving dogs while others will strive far more and accomplish only one-fourth as much.

The team readily discovers whether a whip is forgotten or whether the lash is short or long and above all who wields it. A good driver is always in demand for a trip and will deserve extra compensation for his efforts, hence he is usually prepared with that necessity for urging the brutes to do their utmost at the least expense of their strength.

The whip consists of a number of thongs knitted together so as to form a lash two inches wide and half an inch thick, gradually tapering to a single thong. The body of the whip may be as much as eight to fifteen feet while the lash may extend fifteen to twenty feet longer. The body or larger end is fitted to a short wooden handle of eight to ten inches, shaped like the handle of a sword to prevent slipping and additionally protected by a loop through which the hand passes and rests.

In order to use the whip effectively requires much practice but in the hand of one accustomed to throwing the lash it becomes an instrument that causes the offending dog to howl with pain from the stroke or cringe in terror as it passes hurtling over his head. The lash is allowed to trail behind the person when not in use and must be in that position when required to be thrown. It is a movement that calls forth a great strength in the wrist and shoulder muscles. By elevating the arm to the height of the shoulder and giving it a half sweep the lash is urged past the head and when checked the continuation of its length streams along to the spot where the tip of the lash is directed.

Any portion of a dog's body may be struck by one skilled in handling the whip. The ear, jaw, foot or ribs of a lazy dog may be made to tingle for the remainder of the day or even the skin torn from its body by the snap of the lash.

In the hands of an unskilled person the lash is certain to wind about his neck in a manner calculated to disturb his thoughts, or to curb additional aspiration to become a dog-driver.

The Eskimo dog possesses many virtues as a draft animal but his vices fully counterbalance his good, hence nothing but a whip of such generous proportion and scorching character will govern his wont to indulge in his vicious propensity to fight on every occasion and to exhibit those idiosyncrasies often observed in his more civilized kindred by attempting to stop at each crook, crevice or corner to indulge in a moments repetition of what occurred a minute before.

The intrusion of one dog on the pathway of another is often the course of a snap and snarl that will enlist the sympathy of the entire team and where but a moment before was a calm placidity of apparently fatigued beasts is now a writhing mass of infuriated beasts tearing each other as only Eskimo dogs are able to fight. The lines are being entangled and possibly choking some dog to death.

Here the driver acts well his part; retiring to a convenient distance the lash is sent amongst the dogs with a crash that causes them to separate as though a load of shot had been fired among them. A few more such strokes are distributed to the principals and quiet restored while the tracks are separated and a crack of the whip urges them on at the top of their speed.

Dog Whip
A dog whip made from sealskin that was collected by Turner at Ft. Chimo, probably in 1883. Turner described the challenges of using a whip: "In the hands of an unskilled person the lash is certain to wind about his neck in a manner calculated to disturb his thoughts, or to curb additional aspiration to become a dog-driver." USNM E74480, Museum Support Center, NMNH. Photo by Donald Hurlbert, 2013.

COMPLETE SET OF HARNESS FOR NINE ESKIMO DOGS: HUDSON STRAIT INNUIT (1887A: 125-130)

While there is but little diversity in the general plan of constructing harness for the Eskimo dog used as a draught animal in the locality here designated, there is some modification to be allowed when a few or a great number of those brutes are employed to drag a sledge. In general, the difference consists in the greater or less length given to the main-trace and to the single trace leading from the individual dog to the main trace. The collar and shoulder straps area seldom varied except in material.

The main-trace is usually prepared from the skin of a walrus or the hide from a fully adult male Square-flipper Seal (*E. barbatus*). The skin from either of these creatures will average nearly half an inch thick. The hide is cut into a continuous strip of five-eighths of an inch wide and thirty to forty feet in length. The angles are trimmed so as to form a somewhat flattish ovoid thong; all inequalities of surface are removed so as to render this trace smooth.

To the rear of this trace is attached a similar piece of thong which is braided to it in a peculiar manner; the main line has a slit made of sufficient size to permit the other strand to pass through it. A slit is now made in the piece and the main line run through that, by a continuation of the alternate slittings and passings through the two lines are knitted together, leaving free ends of two feet or less in length. Each of these ends has an ivory toggle or button of variable shape to fit into the loop on the outside of the sledge runner. These two ends are termed the yoke.

The yoke may be a different form when the other described arrangement is fitted to the sledge. This consists in the supplementary piece ending in a long loop which fastens over the toggle on the side of the sledge; the main trace also having a similar loop to fit on the opposite side.

If there be but a few, three to five, dogs to be hitched to a sledge the main trace may not be more than ten feet in length while if from seven to fifteen dogs are attached the main-trace may be so much as forty feet in length. At the forward end of the main trace is affixed an ivory toggle which is to be slipped into either a supplementary thong knitted to the main trace or it may be of itself an additional piece of thong. This forms a long loop which serves a purpose to be noted farther on.

The body harness is usually prepared from such material as may be sufficiently strong.

The skin of a two-year-old Harp Seal is considered the best as it does not become so stiff as to chafe the dog upon which it is placed. The manner of constructing the harness is as follows: A piece of hairy sealskin is cut double the width required and about forty inches in length. The flesh side of the skin is folded together and the edges sewed, leaving the hair out. The forward end is now turned over so as to have a loop fourteen or fifteen inches in length. The end is now stitched to the other part and leaves a single end of some ten or twelve inches. A similar piece is prepared and the two free ends are sewed to a stout piece of tanned seal skin which forms a loop of two or three inches in length. There are now two loops of hairy skin which must be joined by two pieces of similar skin, one of which is placed slightly back of the other or as to bind over the shoulder of the dog while the second strap is placed below so as to be across the breast or chest of the dog.

Behind this body harness is a long thong of stout sealskin from which the hair has been removed. This thong is of variable length, depending on the particular position to be taken in the team by the dog. The dogs nearest to the loop of the main trace may be distant from that loop only fifteen feet while the trace of the dog farthest from it may be as much as thirty-five feet from it and seventy-five feet from the forward end of the sledge. The different lengths

Inuit Dogs with Harness at Rigolet, Labrador (c. 1882)
Photo reads: "Regoulette, Labrador. Dogs in Sleigh Harness." Photo by J.R.H. Used with permission of McGill University Rare Books and Special Collections, 120608.

of the individual traces permit the dogs to spread out without interference until they begin to fatigue when they pass and re-pass under or over each others' lines and form a perfect rope of traces which shorten the line of the rear dogs until they have not the necessary freedom; necessitating a stoppage and disentanglement of the strands.

Inuit Dog
An Inuit dog belonging to Kangiqsualujjuamiut hunter Daniel Annanack. Photo by Scott Heyes, 2008.

The sledge is usually loaded first; then the harness attached to the sledge, or each dog is captured and successively brought to the sledge and his trace slipped into the loop of the main-trace. The placing of the harness on the dog is attended with little trouble for one accustomed to such work. The body harness must be held in the hand so that the shoulder strip will be above and through the aperture between the shoulder and breast straps the head if the dog is thrust. Each fore-leg of the dog must be lifted so that the underpart of the side-loop will pass under it. The dog will now have two straps between his forelegs, one across his breast, one over his shoulder while the forward end of the loop encircles the side of the neck. The rear ends of the harness pass along the side and are united near the small of the back of the dog; the trace continuing behind to the main trace.

The number of dogs usually hitched to a sled depends on the amount of load, condition of snow and the speed desired. It is safe to estimate each dog capable of pulling seventy-five pounds exclusive of sledge weight. Seven to nine dogs will draw five hundred pounds as much as forty miles over a good surface in ten hours time or nine dogs will accomplish the same in an hour less. Ten to twelve dogs will easily pull seven-hundred pounds the same distance and in the same time. I have travelled twenty miles in three hours over excellent surface with a weight of five hundred pounds and seven dogs. That was exceptional and may be considered as a test of speed rather than of draft.

The condition of the snow, the driver, the weight, the direction, whether toward home, or over a new region; and many other things tend to increase or decrease the rate of travel.

THE HABITS OF THE KIGŬKHTAGMYUT (1887A: 11)

The various species of waterfowl migrating, in the spring, to breed farther to the north afford a supply of food not obtainable at any other season. There are no reindeer on those islands [islands of the eastern waters of Hudson Bay], so that the people [*Kigu˘khtagmyut*] are compelled to clothe themselves with the skins of seals and the skins of the dog, which they raise for that purpose, as there is little journeying by sled on those islands. My informant knew of no tradition that explained how those islanders came to inhabit the islands distant one hundred to one hundred and fifty miles from the mainland. They have a speech said to be distinct from their mainland neighbors, who laughed as they spoke of their squeaky voices and chattering expression of words.

ORIGIN OF THE WHITE PEOPLE
An Inuit story, Turner (1887a: 283)

Many ages ago the Innuit placed two puppies, one on a large chip and the other in an old boot, and sent them adrift upon the water. The one which was on the chip became an Indian. The puppy in the boot returned after many years as a white man in a vessel.

This had its origin in this way. Some Innuit who were on the verge of starvation remembered that one of the shamans of their earlier times had predicted that sometime they would have people visit there that were not of their kind and that the newcomers would help them. At the time of starvation the white people came and relieved their wants. The old woman who told this to me rubbed her hands and said she was glad the Innuit sent for the white people to come to their land.

The Evil Dark Cloud
An Inuit story, Tivi Etok (Heyes 2007)

This story occurred when I was out on the sea ice with my father around Nachvak Fiord, Labrador. My father was out on the edge of the sea hunting seals. I was a boy at the time. I was on the dog sled while my father was hunting seal some distance away. Suddenly the sled started to be pulled very fast by the dogs. The dogs went very straight and fast. I yelled at them to stop but the dogs kept on running. Then I noticed a dark cloud appear as a shadow of the dogs. But it was a very bright and sunny day; there were no clouds to be seen anywhere else on the horizon. The dogs kept running, following the coastline. The dogs were running that fast that their front legs were in the air. As the dogs ran, the dark cloud hovered over them. The dogs continued to speed up. As the dogs reached the shoreline, evil voices then started to speak from all directions. The dark cloud then started to come towards me. Soon enough the cloud was over me, it engulfed me and blocked out the sun. Where the dogs stopped on the shore I was caught off guard. I had no weapons to defend myself. Then I saw footprints appearing in the snow coming towards me, but there was no person there. The dark cloud made the dogs dance. Voices were appearing out of nowhere and were spooking the dogs, yet I didn't understand what the voices were saying. At this time my father was a long way from me hunting seals. I was all alone. I picked up a couple of rocks in readiness to hit the spirits as they closed in on me. These were my only weapons. I thought that they were going to take me for good. The dark cloud was trying to collect souls.

Then I received a message, a thought from God. I was just sitting in the sled shaking rapidly. God gave me the words to fight the evil. The evil spirits shouted back to me that you are not safe from the words of your God. I yelled to them that "you guys are consumed by evil; I don't need to be bothered." At that point the dark cloud slowly moved away from me and then it disappeared into thin air. If we believe in good and pray, bad things will not overcome us. I have seen this dark cloud more than once since this day. I told my father what happened when he returned from hunting. I was afraid to tell him what I had experienced. My father told me that if weird things happen to you in life it is important to talk about it and let it out.

Illustration of a Story about the Evil Dark Cloud
By Kangiqsualujjuamiut elder Tivi Etok, 28 April 2005. Reduced from original, 11" × 17". Lead pencil on bond paper. Scott Heyes Private Collection.

The Miraculous River Crossing
An Inuit story, Tivi Etok (Heyes 2007)

It was late spring, there was no snow left on the mountains. The ice had broken up sooner than we thought it would on the Korac River. We were essentially stranded on the Northern side of the River; we would have had to travel very far inland to find a safe crossing point. My father and the rest of the family were on the other side of the river. Harry and I decided to cross the Korac at the narrowest point with our dogs. We waited about 10 days to do so. It spanned about 200-300 metres. We were wearing caribou parkas and we knew that they would get very wet. We tied up all our belongings fast to the sled. I tied up the lead dog and threw him in the water. The dogs then began to pull us along, with all our belongings affixed to the sled. The water was very cold. The current was strong and the rapids were running. I didn't even drain my clothes when I got to the other side. I just went to bed. My father didn't want us to cross. He yelled out across the river for us to make a raft from the nearby trees. But we did not listen to my father. My father was angry and disappointed that we crossed. My father knew that dogs were not trained for crossing rivers. My father and my older brother raced down to the water's edge when they saw that we were crossing. Harry was a very tall man and he was older than me. He listened to everything I told him to do. I was the leader.

Where we crossed it was very deep and the waves were high. I had never crossed a river before like this. But when I began the crossing, something from above gave me a message on the path to take across the river. The lead dog changed directions upon my commands. When we got to the other side I didn't remember anything about the crossing except for us entering the river; the journey across was just a blur. I truly believe that something or somebody like God helped us cross. All our belongings, which were wrapped in Caribou hides that we had acquired from the Hudson Bay Company were bone dry; it was as though our belongings had not even entered the water! It was very strange. When I went back to the tent I slept all day long. When I woke up the next morning I needed to relieve myself. My hand was completely frozen; I could not walk. Harry suffered the same fate too. I was stiff for five days. We had to postpone our walk across to Nachvak Fiord (to another camp) to allow me time to recover. I truly felt that "someone" helped us to cross the river that day.

Illustration of a Story about the Miraculous River Crossing
By Kangiqsualujjuamiut elder Tivi Etok, 28 April 2005. Reduced from original, 11" × 17". Lead pencil on bond paper. Scott Heyes Private Collection.

Dogs and the Inugagulliq

An Inuit story, Benjamin Jararuse (Heyes 2007)

The Inugagulliq (little persons: plural, Inugagulliit) is very small. For instance it is as large as my arm. An Inugagulliq had everything; it used to have a qajaq, a harpoon, an avataq. There was once this Inugagulliq who was approached by a normal man. This man had never seen a small human being; he even circled the little human to take a good look at it. The little person jumped on the big man's head, curled his legs and arms around the man's head, this is how the Inugagulliit killed, by suffocating their victims with their stomachs. As the big man was starting to suffocate, he started struggling to take the little person off of his face, and then managed to get it off of him. He then started asking his dogs to go over to him, shouting "Au, au, au," so the dogs started running to their owner. The dogs knew their owner was being attacked. The dogs bit the little person and started fighting it, and it died. The Inugagulliq had everything; they only hunted their game with their only hunting gears. We were saved by our elders who tried everything to make their generations, so we should teach our children our way of life. There are so many young people who do not know our traditional way of life, us older people who know these things should teach our younger generations in order to keep our traditional heritage alive because our generations before us worked so hard to survive and we mustn't lose that.

Anyway, back to the Inugagulliq, an Inugagulliq caught a bearded seal; he was out on the edge of the sea ice and cut up his catch. He even had a qajaq to fetch his bearded seal out in the sea. I have done this myself, although I didn't own my own qajaq. You go out into the sea water to fetch the seal that was just harpooned hooked onto an avataq, and drag it to the shore or the ice edge. That's how it goes, I have experienced it.

Eskimo Dog – Terms and Definitions

Syllabary	Modern Nunavimmiutitut Term and Definition	Term by Turner	Definition recorded by Turner
ᐊᓄ	*anu* (cf. Schneider p.35 - *anu* = harness)	*á nut*	Harness for dogs [p. 2177; 696]
ᐊᓄᔨᕗᖅ	*anujivuq* (cf. Schneider p.35 - *anujivuq* = harnesses it)	*a no ǵwak*	Putting the dog harness on [p. 2176]
ᐊᐅᒃ(?)	*auk; au* (?) (cf. Schneider p. 49 - *au* = command to halt [...] sled dogs)	*auk*	A word spoken to a team of dogs directing them to the right [p. 2204]
ᕼᐊᕋ	*ha'ra* (cf. Schneider p.55 - *ha'ra* = turn left! [to dogs]: *ha'ra-ha'ra*)	*ha ŕa*	A word spoken to a team of dogs, directing their course to the left. It is more than probable that this word has been sadly corrupted from *agha* meaning proceed, go ahead etc. The illiterate English-speaking white people put the letter *h* where it should not be placed; and through inability to pronounce the hard, guttural *gh* it was sounded as the letter *r* and finally adopted by the Innuit. [p. 2216]

Chapter Two: Mammals of Ungava and Labrador

Eskimo Dog – Terms and Definitions

Syllabary	Modern Nunavimmiutitut Term and Definition	Term by Turner	Definition recorded by Turner
ᐃᐱᕋᐅᑕᖅ	*ipirautaq*; *iparautaq* (cf. Schneider p.90 - *ipirautaq* = whip, long Inuit whip)	*í pi gháu tak*	Dog-whip [p. 2242; 697]
ᐃᐱᐅᑕᖅ	*ipiutaq* (cf. Schneider p.90 - *ipiutaq* = tie rope)	*í pi á tut*	Harness trace [p. 697]
ᐃᓯᕋᖅᑐᓂᒃᖅ	*isurartuniq*; *isuraqtunik* (cf. Schneider p.103 -*isuraatuniq* = the sled dog with the longest trace; the head dog, leader)	*i shú gak tú nik*	Leader of team of dogs [p. 2244]
ᐱᑐᒃ	*pituk* (cf. Schneider p.264 - *pituk* the "v" shaped part of the rope that joins the dog harness to the sled; includes buckle through which passes the trace *putusiut*)	*pí tuk*	Halter (of harness for dogs) [p. 2761]
ᕿᒻᒥᖅ	*qimmiq* (cf. Schneider p.304 - *qimmiq* = dog)	*khíng mik*	Dog [p. 2291; 1535]
ᕿᒧᒃᓯᖅᑐᖅ	*qimuksiqtuq* (cf. Schneider p.304 - *qimuksiqtuq* = he travels on a sled). Drives dogs.	*kí muk shúk tok*	Drives (dogs) [p. 2302]
ᕿᒧᑦᓯᖅ	*qimutsiq* (cf. Schneider p. 304 - *qimutsiq* = dog team)	*ki-mik-shit*	Dog team [p. 700]
ᕿᒧᑦᓯᖅ ᐳᔪᓕᒃ (?)	*qimutsiq pujulik* (?) (cf. Schneider p. 304 - *qimutsiq* = dog team, and p.269 - *pujuq* = water vapour)	*ki múk shit su yú lik*	A railway train. The word is composed of *kimukshit*, a team of dogs and sledge, and *puyuk*, steam. [p. 2302]
ᑕᖁᓕᒃ	*taqulik* (cf. Schneider p. 396 – *taqulik* = white mark over the eyes of some dogs)	*tú ho lik*	Eskimo dogs with a spot above the eye in the eastern coasts is preferred; term means double-eyed. [p. 1538]
ᐅᐃᑦ	*uit* (cf. Schneider p. 436 - *uit* = exclamation to make dogs leave a house or to excite them into pulling a sled faster)	*huit or hwit*	A word spoken to a team of dogs to urge them to greater speed. I have also hear [sic] the word *Tuk tú i*, repeated several times, used for the same purpose. The word means Reindeer; and, the mention of the name of that animal is supposed to excite the dogs in increased action. [p. 2216]
ᐅᓇᔪᖅ	*unajuq* (cf. Schneider p.448 - *unajuq* = [dog] shows its delight in [at the return of its master]; cf. Spalding 1998, p.185 - animal plays or gambols)	*u ná vok*	It is friendly, kindly disposed; said of a dog (or person) perceiving acquaintance [p. 2847]

Carnivores
355

ᐱᖅᑐᓯᕋᖅ

Canadian Lynx
pirtusiraq

Lynx borealis canadensis (Gray) Mivart.

[1539]

Order Name used by Turner	*Carnivora*
Current Order Name	*Carnivora*
Family Name used by Turner	*Felidae*
Current Family Name	*Felidae*
Scientific Name used by Turner	*Lynx borealis canadensis (Gray) Mivart.*
Current Scientific Name	*Lynx canadensis* Kerr
Syllabary	ᐱᖅᑐᓯᕋᖅ; ᐱᖅᑐᓯᕋᖅ
Modern Nunavimmiutitut Term	*Pirtusiraq* (cf. Schneider p.259 -piqtusiraq = lynx)
Eskimo Term by Turner	Not Recorded
Definition of Eskimo Term by Turner	*Lynx*

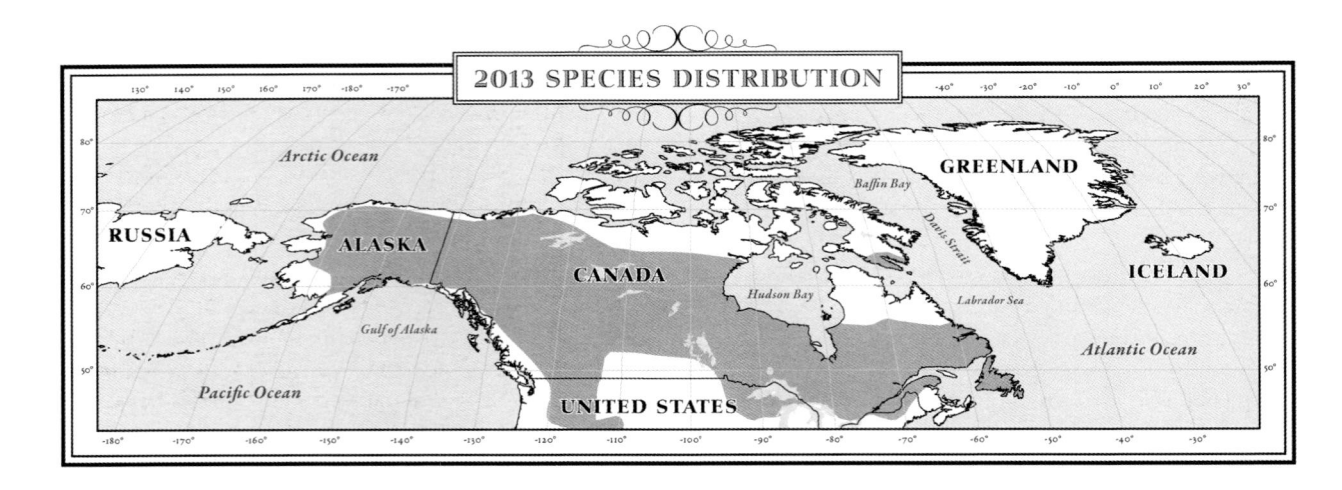

2013 SPECIES DISTRIBUTION

Chapter Two: Mammals of Ungava and Labrador

Descriptions by Turner

The Canada Lynx is quite rare in the Ungava district[28], probably not more than a score of these animals become tributary to the annual number of furs collected at Ft. Chimo. They rarely wander so far as the limit of trees, within a dozen miles north of Ft. Chimo. An individual was captured near the "Chapel," three miles north of Ft. Chimo, during the winter of 1881-2. Tracks of two others were seen in "Hunting Bay" in 1883. South of the "Heighth of Land" the Lynx is not common but becomes plentiful south of latitude 55 degrees where the timber is larger and affords not only a better protection to this animal but serves to furnish a greater supply of food adapted to its nature.

The principal food is the Rabbit, spruce partridge, and other birds. The animal lies in wait for an unlucky bird or rodent to come near when with [1540] a spring it leaps upon its victim which it seizes with its huge paw. The Lynx is taken in snares or steel traps. The former method is employed to greater advantage. A small sapling is bent down and the limbs removed. A stout thong is affixed to the upper portion and communicates with a figure 4 trigger which is surrounded with a circle of pegs around which the noose is arranged. The loop passes over the head or foot and the sapling or spring is of sufficient strength to lift the animal from the surface.

A Lynx is, by nature, a cowardly animal and flees with incredible bounds from the sight of man. If, however, it rests among the branches of a tree it endeavors to crouch out of sight and will make no attempt to disclose its presence. If taken by the leg in a snare it is an amusing sight to witness its antics to escape when the trapper makes his visit to the locality. The bounds and struggles of the animal [1541] to free itself are wonderful. It at such times exhibits a stubborn disposition also, and spits, strikes with its feet, reversing its hair, giving it a formidable appearance.

The breeding habits of this animal were not learned as the Indians themselves knew but little about it.

---

28. Inuit hunters from Kangiqsualujjuaq report that a lynx was sighted near the community at the end of winter, 2012. Many seasoned hunters in Kangiqsualujjuaq have never seen one, but have observed their tracks when passing through the wooded areas south of Kangiqsualujjuaq

General Mammal Terms and Definitions

Syllabary	Modern Nunavimmiutitut Term and Definition	Term by Turner	Definition recorded by Turner
ᐋᑦᑐᕇᖅᐳᖅ	*aatturiirpuq* (-riir- = "finishes to do it"); *cf.* Schneider p.3 *aatturpaa* - he skins an animal)	*ak tu ģhi pok*	Finishes skinning or flaying [p. 2168]
ᐋᑦᑐᑐᖅ	*aattutuq* (See above)	*ak tú tok*	He skins or flays [p. 2168]
ᐊᒃᓴᓯᔪᖅ	*Aksasijuq* - a river inlet immediately east of Kuujjuaq, called "False River" in English. Name also said to derive from *atsaq* ; *aksaq* - black bear (cf. Schneider p. 47).	*Ák sak sĭģ yok*	"False River" The word is derived from *aksaktok*, rolls, waves, undulates, sways. The hills in the vicinity are, at certain seasons of the year, literally covered with reindeer and when viewed, from a distance, appear to be in motion from the masses of these creatures. [p. 2167]
ᐊᑯᓪᓕᒑᖅᐳᖅ (?)	*akulligarpuq* (?) - derives from *akulliq* meaning "in between" (cf. Schneider p.14). Note: in the North Baffin Island area the term *Akullirut* refers to a "moon month" to approximately September, dividing summer from winter, when caribou hair begins to thicken. This is the period when the skins are considered best for clothing (cf. MacDonald p.197).	*a kú li gák pok*	Not yet complete; applied to the period or condition when the summer hair is not yet fallen off and the winter hair is well advanced. It specially refers to the tardy condition of the summer hair. [p. 2165]
ᐊᓗᑦᓴᖅ ; ᐊᓗᒃᓴᖅ	*alutsaq/ aluksaq* (cf. Peck p.27 – *alluklaek; aluklaik* = "the fat of the kidneys of seals and river horses"; also Schneider p.22 - *aluksaq* = caribou fat)	*a lú ki ek*	Fat from kidneys of marine mammals [p. 2139]
ᐊᒫᒪᒃ	*amaamak* (*cf.* Schneider p.22 - *amaamak* = mammillary, breast [of human] and udder [animals])	*a ṁa mŭk*	Mammal, breast of a woman. Teats of mammals in general [p. 2170]
ᐊᒥᖅ	*amiq* (cf. Schneider p.23 - *amiq* = skin with fur on it, pelt, particularly caribou hide […])	*á mik; á mik*	Skin, hide [p. 2171]. Bark; skin of boat; hide of mammal after removal [p. 695]
ᐊᑕᐅᑦᓯᑦ	*atautsit* (most probably) (cf. last syllable –chin, with initial ch- (not s) and final dental –n (not k) *atausiq*; *atautsit* (?) (cf. Schneider p.46 - *atausiq; atautsit* = one, pl. –t, [or *atautsit*, rare] with collective plurals])	*a taŭ chin*	Herd, flock [p. 2178]
ᐊᕘᓐᓂᑦ	*avunniit* (cf. Spalding 1998, p. 15 - *avunit* = month of March [when baby seals are born]; cf. MacDonald p. 196 - *avunniit* = a moon month corresponding to March when seal pups are aborted or born prematurely)	*a vú nĭk*	Prematurely born mammal [p. 2204]
ᐊᕘᓐᓂᒃ (?)	*avunnik* (?) See *avujuq* above. (cf. MacDonald p. 196 - *avunniit* [deriving from *avujuq* ?] refers to the "moon month" (March/April) when seal pups are aborted or born prematurely)	*a vú nĭk*	Prematurely born mammal [p. 2204]

Chapter Two: Mammals of Ungava and Labrador

General Mammal Terms and Definitions

Syllabary	Modern Nunavimmiutitut Term and Definition	Term by Turner	Definition recorded by Turner
ᐃᕙᒃᑐᖅ	*ivakkatuq* (cf. Schneider p.111 - *ivakkatuq* = to trot [dog, horse])	*i vû kák tok*	Trots [p. 2257]
ᐃᕙᓗ	*ivalu* (cf. Schneider p.111 - *ivalu* = thread made of a tendon [of a caribou or whale])	*i vá lu; i vû lu*	Sinew, thread; The name also of the strings for a violin [p. 2257; 697]
ᑲᑎᒪᔪᑦ	*katimajut* (cf. Schneider p.126 - *katimajut* = who are together, have joined company etc.)	*ka tí ma uk; ká ti m̓ai ut; k̓û ti m̓ai ut*	Herd, flock. A Labrador word. See word for assemble [p. 2272]; Herd, drove, flock [p. 698]; A Labrador word for herd or flock. See word for assemble, congregate. [p. 2324]
ᑲᐅᖕᓂᐅᑦ	*kaungniut* (cf. Schneider p.129 - *kaungnaqtut* -*kaungniut* = trap made in such a way as to make a stone fall on a fox). From *kautaq* = hammer.	*kaúg ni ut; kaung ni ut*	Deadfall trap [p. 699, p. 2278]
ᑲᐅᖕᓂᐅᖅᕕᒃ (?)	*kaunniuqarvik*(?) (See above)	*kaung ni uti há vik*	A place where a deadfall is set [p. 2278]
ᑭᓕᐅᑕᖅ	*kiliutaq* (cf. Schneider p.137 - *kiliutaq* = woman's instrument [...] for scraping and softening skins)	*kil yú tûk*	Stone skin-dresser [p. 700]
ᑰᐃᒃᑰᓯ	*kuukkuusi* (cf. Schneider p.153 – *kuukkuusi* = pig [...] loan word from Naskapi Amerindian). Also cf. Dorais p.25 – *kuingiingi* = Pig [Labrador] probably an onomatopoeia.	*kwing ingi a*	Pork, Hog (an onomatopoeic word, from sound made by pigs seen on the vessels) [p. 2325]
ᒪᐅᔭᖅ (?)	*maujaq* (?) (cf. Schneider p.167 - *maujaq* = any ground that gives under one's steps: mud, marsh, soft snow). The term is usually used for deep soft snow.	*maú yak*	Settings from oil or other liquid fat [p. 2331]
ᒥᑭᒋᐊᖅ	*mikigiaq* (cf. Schneider p.167 -*mikigiaq* = trap)	*mi kí ghi ak*	Trap [p. 2332]; steeltrap [p. 701]
ᒥᑭᒋᐊᕐᐸᕕᒃ	*mikigiarpavi*(?): *mikigiaq+pa(q)* ("big") + *vik* ("big") = "very big trap(?)"; (cf. Schneider p. 168 - *mikigitjiturvik* = place for laying out traps)	*mi kí ghi a p̓a vik*	A place where a trap is set [p. 2332]
ᒥᕐᕿᐊᔪᖅ (?)	*mirqiajuq* (cf. Schneider p.173 - *miqqiaq* = [animal] that has molted, changed fur)	*mí ki yái yuk*	Shedding hair (applied only to mammals) [p. 2332]
ᓇᐅᓚᖅ	*naulaq* (cf. Schneider p.199 - *naulaq* = point of sealing harpoon)	*náu lûk*	Spear for seals and other marine mammals [p. 2686]
ᓂᒐᖅ	*nigaq* (cf. Schneider p.200 - *nigaq* = snare for birds, mice etc.)	*ni gak; ní gak*	Snare (for mammals or birds). Also a species of spider. From the spiders web may have originated the snare [p. 2687]; Snare [p. 702]

Carnivores

General Mammal Terms and Definitions

Syllabary	Modern Nunavimmiutitut Term and Definition	Term by Turner	Definition recorded by Turner
ᓂᒃᑯ	*nikku* (cf. Schneider p.202 - *nikku* = dried meat). But not dried fish which is *pipsi* or *pipsik*.	nĭ´ku	Dried flesh [p. 702]
ᓂᕐᔪᑦ	*nirjut* (cf. Schneider p.212 - *nirjut* = general term for animal)	nŭg´ zhut	A term applied to the large land mammals in contradistinction to the smaller kinds. A mammal larger than a wolf [p. 2713]
ᓂᐅᒃ (?); ᓄᑭ (?)	*niuja* (?) Perhaps conflicted with *"niuk"* = leg; *nuki* (cf. Schneider p.220 - *nuki* = muscle; tendon; physical strength)	ni ú ya	Tendon [p. 2693]
ᓄᓕᐊᕐᐳᖅ	*nuliarpuq* - general term for copulation (cf. Schneider p. 221 - *nuliarniq* = the conjugal act)	nú li ák tok	Cohabits (for all other mammals) [p. 2720]
ᐸᒥᐅᒃ ; ᐸᒥᐅᖅ	*pamiuk*; *pamiuq* (cf. Schneider p.237 - *pamiaq* = tail of land animals [not fish or cetaceans])	pú mi uk	Tail (of mammal) [p. 2768]
ᐱᕈᔭᖅ	*pirujaq* (cf. Schneider p.259 - *pirujaq* = cache of caribou meat; also cf. Spalding 1998, p. 95). Note: meat usually cached in rocks in the fall. It ferments lightly and is collected in the winter months.	pú gú yak; pĭ gú yak	Caches (stone pile) for fresh meats [p. 2734]; cache for fresh meat [p. 702]
ᐳᐊᓗ ; ᐳᐊᓗᒃ	*pualu*; *pualuk* (cf. Schneider p.267 - *pualu* ; *pualuk* = mitten)	po á lu	Mitten (fur or skin) [p. 2766; 703]
ᕿᓯᒃ	*qisik* (cf. Schneider p.310 - *qisik* = skin in general except for caribou [*amiq*] or fish [*amiraq*])	khí sik	Skin, hide (mammal) [p. 2291]
ᓴᒡᒐᖅ	*saggaq* (cf. Schneider p.334 - *saggaq* = skin or living animal whose fur is short [...] any skin with short fur). Note: short fur in Arctic mammals is a summer phenomenon.	shá gak	Summer hair (of mammals) [p. 2775]
ᓴᑭᐊᖅ	*sakiaq* (cf. Schneider p.335 - *sakiaq* = bones of the thorax: breast, chest of an man and animal, rib cage)	chû ki ak; sá ki á gok	Brisket of mammal [p. 2212]; Brisket (of mammal) [p. 2774]
unknown	unknown	sí la lí ok	Sheds fur hair. Applied only to quadrupeds [p. 2779]
ᓯᐅᕆᐊᖅᑐᖅ	*siuriartuq*: combined infixes - *siur* + *riar* -carrying the meaning "starts to (-*riar*-) go seeking..." and by extension, "starts to go hunting."	si ú ghi ák tok	Added to animal names signifies to hunt, go in quest of those animals. The word is either specific or generic in application. [p. 2783]
ᓯᕗᖅᑕᑎ (?)	*sivukatakti* (?); *sivuqatauti* (?) (cf. Schneider p.369 -*sivulirtaittuq* = lead dog). From *sivulliq* "the one in front."	sí vu ka tŭk ti	Leader of a herd or flock of mammals or birds [p. 2784]

Chapter Two: Mammals of Ungava and Labrador

General Mammal Terms and Definitions

Syllabary	Modern Nunavimmiutitut Term and Definition	Term by Turner	Definition recorded by Turner
ᓱᓇᒥᐊᕐᐳᖅ	*sunamiarpuq* (cf. Peck p.266 -*sunamiarpok* = "the reindeer [caribou] has a brown tail")	*su ńa mi ák pok*	It (specially referring to a reindeer) has its hair in best condition, ready for use, to make garments of. The word implies a fitness, suitable condition of hair and does not refer to the pelt. [p. 2791]
unknown	*sunamiarut* (cf. Peck p.266 - *sunamiarut* = "the month of October because the tail of the reindeer [caribou] is full-grown"). Note: in North Baffin Island the equivalent period is *akullirut* when caribou skins are prime for clothing (MacDonald p.197).	*su ńa mi á gut*	The period when the hair of the reindeer is best for use in preparing garments. About the beginning of October [p. 2791]
ᑐᓗᕆᐊᖅ	*tuluriaq* (cf. Schneider p.418 -*tuluriaq* = canine tooth [of dogs, bear, man])	*tu lú ghi ak*	Fang (of mammal) [p. 2812]
ᑐᐱᖅ	*tupiq* (cf. Schneider p.422 - *tupiq* = tent)	*tú pik*	Skin tent [p. 705]
ᑐᕐᕈᔪᖅ	*turrujuq* (cf. Spalding 1998, p.172 - *turrujuq* = it is long haired or thick haired [as an animal – caribou, musk-ox, dog – in winter])	*to ǵo yuk*	Winter hair (of mammals) [p. 2806]
ᑐᑦᑐᐊᓗᔭᖅ (?) ᑐᑦᑐᕙᔭᖅ	*tuttualujaq* (?) lit. "the meat of a big caribou"; see Schneider p.430 -*tuttuvajaq* = canned beef; also *tuttuvak* = moose; domestic cattle	*túk tu a lú yak*	Beef (meat) [p. 2809]
ᑐᑦᑐᔫᑦ ; ᑐᒃᑐᔪᖅ	*Tuttujuit*; *Tuktujuq* (cf. Schneider p.430 - *Tuttujuit*; *Tuktujuq* = the Great Bear stars; also MacDonald p.79)	*tuk tú yuk*	Great Bear (constellation)
ᑐᑦᑐᕕᓂᖅ ; ᑐᒃᑐᕕᓂᖅ	*tuttuviniq*; *tuktuviniq* (cf. Schneider p.430 - *tuttuviniq* = caribou meat)	*túk tu vái ya*	Beef. This is the Labrador word. [p. 2810]
ᐅᑭᐅᓕᖅ	*ukiuliq* (cf. Schneider p. 439 - *ukiuliq* = animal in winter [white] fur)	*ó ki ú lik*	Any mammal in winter pelage [p. 2727]
ᐅᓚᐅᑦ ; ᐅᓚᐅᑎ	*uliut*; *uliuti* (cf. Schneider p.444 - *uliuti* = bundle of tendons of a filet [caribou, whale] when it is whole before being separated into threads). Note: in caribou often referred to as the "tenderloin" or "filet mignon" from which the sinew is removed for thread. (See *ivalu* above)	*ú li ut*	Sinew from the back of mammals [p. 705]
ᐅᒥᖕᒪᖅ; ᐅᒥᒻᒪᒃ	*umimmak* in modern Nunavimmiutitut. *umingmaq* (cf. Schneider p.447 -*umingmaq* = muskox [*Ovibos moschatus*])	*ú ming múk*	Musk ox. Applied also to domestic cattle (Onomatopoeic) [p. 2841]

Carnivores

361

Appendix

Turner Collections at the National Museum of Natural History

Type Specimens of Mammals Collected by Turner

Current Identification	USNM #	Date Collected	Location	Published Type Name(s)
Microtus pennsylvanicus	186495	15 Nov 1882	Fort Chimo [= Kuujjuaq]	*labradorius* Bailey, 1898 (holotype)
Synaptomys borealis	14838	-- Spring 1884	Fort Chimo [= Kuujjuaq]	*innuitus* True, 1894 (holotype)
Myodes gapperi	186492	12 May 1883	Fort Chimo [= Kuujjuaq]	*ungava* Bailey, 1897 (holotype)
Phenacomys ungava	186487	4 Feb 1883	Fort Chimo [= Kuujjuaq]	*latimanus* Merriam, 1889 (holotype)
	186488	-- Spring 1884	Fort Chimo [= Kuujjuaq]	*ungava* Merriam, 1889 (holotype)
Lepus arcticus	14149	28 Sep 1882	Fort Chimo [= Kuujjuaq]	*labradorius* Miller, 1899 (syntype)
	A23132	28 Sep 1882	Fort Chimo [= Kuujjuaq]	*labradorius* Miller, 1899 (syntype)
Canis lupus	A23136	-- --- 1882	Fort Chimo [= Kuujjuaq]	*labradorius* Goldman, 1937 (holotype)
Vulpes lagopus	A23195	-- --- ----	Fort Chimo [= Kuujjuaq]	*ungava* Merriam, 1902 (holotype)

Mammal Specimens Collected by Turner

Current Identification	USNM #	Date Collected	Location	Field #(s)	Other #(s)	Preparation(s)
Tamiasciurus hudsonicus	14170	Mar 1882	Northwest River	1176	A39020	Skin; Skull
	14783	-- --- ----		No Number		Skin
	A23212	-- --- ----		2336		Skull
	A23213	-- --- ----		2334		Skull
	A23350	-- --- ----		2311		Skull
	A23351	-- --- ----		2330		Skull
	A23352	-- --- ----		2331		Skull
	A23353	-- --- ----		2332		Skull
	A23354	-- --- ----		2333		Skull
	A23355	-- --- ----		2335		Skull
	A23356	-- --- ----		2337		Skull
	A23357	-- --- ----		2338		Skull
	A23358	-- --- ----		2339		Skull
	A23359	-- --- ----		2340		Skull
	14166	25 Nov 1882	Fort Chimo [= Kuujjuaq]	1042		Skin
	14167	29 Nov 1882	Fort Chimo [= Kuujjuaq]	1044		Skin
	14168	2 Dec 1882	Fort Chimo [= Kuujjuaq]	1130		Skin
	14169	2 Dec 1882	Fort Chimo [= Kuujjuaq]	1131		Skin
	14171	14 Dec 1882	"Forks"	1329		Skin
	14172	20 Dec 1882	"Forks"	1330		Skin
	14173	20 Dec 1882	"Forks"	1331		Skin
	14174	22 Dec 1882	"Forks"	1332/7		Skin
	14175	13 Dec 1882	"Forks"	1333		Skin
	14176	8 Dec 1882	"Forks"	1334		Skin
	14177	11 Dec 1882	"Forks"	1335		Skin
	14178	16 Dec 1882	"Forks"	1336		Skin
	14179	12 Dec 1882	"Forks"	1337		Skin
	14180	20 Dec 1882	"Forks"	1338		Skin
	14181	31 Jan 1882	Fort Chimo [= Kuujjuaq]	1374		Skin
	14182	19 Apr 1883	Fort Chimo [= Kuujjuaq]	1663		Skin
	14187	1 Mar 1883	Fort Chimo [= Kuujjuaq]	302		Fluid
	14232	11 Nov 1882	Fort Chimo [= Kuujjuaq]	831	A39046	Skin; Skull

Current Identification	USNM #	Date Collected	Location	Field #(s)	Other #(s)	Preparation(s)
Glaucomys sabrinus	14782	1 Nov 1883	Fort Chimo [= Kuujjuaq]	4333		Skin
	14162	28 May 1882	Rigolet, North West River, midway between	1177	A38384	Skin; Skull
Castor canadensis	A23538	-- --- ----		2378		Partial Skull: mandible
Microtus pennsylvanicus	14860	3 Jul 1884	Fort Chimo [= Kuujjuaq]	613		Fluid
	14865	19 Jul 1884	Fort Chimo [= Kuujjuaq]	619		Fluid
	14188	2 Jul 1882	Rigolet	94		Fluid
	14190	18 Jul 1882	Davis Inlet	180		Fluid
	14194	14 Oct 1882	Fort Chimo [= Kuujjuaq]	285		Fluid
	14195	14 Oct 1882	Fort Chimo [= Kuujjuaq]	286		Fluid
	14196	14 Oct 1882	Fort Chimo [= Kuujjuaq]	287		Fluid
	14203	3 May 1883	Fort Chimo [= Kuujjuaq]	310		Fluid
	14206	9 May 1883	Fort Chimo [= Kuujjuaq]	314		Fluid
	14210	20 May 1883	Fort Chimo [= Kuujjuaq]	318		Fluid
	14214	24 May 1883	Fort Chimo [= Kuujjuaq]	322		Fluid
	14215	24 May 1883	Fort Chimo [= Kuujjuaq]	323		Fluid
	14216	24 May 1883	Fort Chimo [= Kuujjuaq]	324		Fluid
	14218	27 May 1883	Fort Chimo [= Kuujjuaq]	326		Fluid
	14219	28 May 1883	Fort Chimo [= Kuujjuaq]	327		Fluid
	14220	28 Feb 1883	Fort Chimo [= Kuujjuaq]	328		Fluid
	14223	3 Jun 1883	Fort Chimo [= Kuujjuaq]	333		Fluid
	14224	10 Jun 1883	Fort Chimo [= Kuujjuaq]	335	A37425	Skull; Remainder in Fluid
	14225	11 Jun 1883	Fort Chimo [= Kuujjuaq]	336		Fluid
	14226	23 Jun 1883	Fort Chimo [= Kuujjuaq]	337		Fluid
	14818	-- --- ----	Fort Chimo [= Kuujjuaq]	402		Fluid
	14820	14 Oct 1883	Fort Chimo [= Kuujjuaq]	404		Fluid
	14821	30 Oct 1883	Fort Chimo [= Kuujjuaq]	405		Fluid
	14822	May-Jun 1884	Fort Chimo [= Kuujjuaq]	481	A37631	Skull; Remainder in Fluid
	14825	May-Jun 1884	Fort Chimo [= Kuujjuaq]	484	A37632	Skull; Remainder in Fluid

Current Identification	USNM #	Date Collected	Location	Field #(s)	Other #(s)	Preparation(s)
Microtus pennsylvanicus	14831	May-Jun 1884	Fort Chimo [= Kuujjuaq]	497		Skull; Remainder in Fluid
	14832	-- Spring 1884	Fort Chimo [= Kuujjuaq]	500		Fluid
	14833	-- Spring 1884	Fort Chimo [= Kuujjuaq]	501		Fluid
	14834	-- Spring 1884	Fort Chimo [= Kuujjuaq]	502		Fluid
	14836	-- Spring 1884	Fort Chimo [= Kuujjuaq]	504	A37633	Fluid
	14837	-- Spring 1884	Fort Chimo [= Kuujjuaq]	505		Fluid
	14839	-- Spring 1884	Fort Chimo [= Kuujjuaq]	493	A37426	Skull; Remainder in Fluid
	14840	-- Spring 1884	Fort Chimo [= Kuujjuaq]	496		Fluid
	14841	-- Spring 1884	Fort Chimo [= Kuujjuaq]	498		Fluid
	14842	-- Spring 1884	Fort Chimo [= Kuujjuaq]	499		Fluid
	14843	-- Jun 1884	Fort Chimo [= Kuujjuaq]	518		Fluid
	14844	-- Jun 1884	Fort Chimo [= Kuujjuaq]	519		Fluid
	14845	-- Jun 1884	Fort Chimo [= Kuujjuaq]	520	A37427	Skull; Remainder in Fluid
	14846	-- Jun 1884	Fort Chimo [= Kuujjuaq]	521		Fluid
	14847	-- Jun 1884	Fort Chimo [= Kuujjuaq]	522		Fluid
	14848	-- Jun 1884	Fort Chimo [= Kuujjuaq]	523		Fluid
	14849	-- Jun 1884	Fort Chimo [= Kuujjuaq]	524		Fluid
	14851	-- Jun 1884	Fort Chimo [= Kuujjuaq]	525		Fluid
	14852	-- Jun 1884	Fort Chimo [= Kuujjuaq]	585		Fluid
	14854	1 Jul 1884	Fort Chimo [= Kuujjuaq]	596		Fluid
	14855	1 Jul 1884	Fort Chimo [= Kuujjuaq]	597		Fluid
	14856	1 Jul 1884	Fort Chimo [= Kuujjuaq]	598		Fluid
	14857	3 Jul 1884	Fort Chimo [= Kuujjuaq]	610	A37430	Skull; Remainder in Fluid
	14858	3 Jul 1884	Fort Chimo [= Kuujjuaq]	611		Fluid
	14859	3 Jul 1884	Fort Chimo [= Kuujjuaq]	612		Fluid
	14861	10 Jul 1884	Fort Chimo [= Kuujjuaq]	614		Fluid
	14862	10 Jul 1884	Fort Chimo [= Kuujjuaq]	615		Fluid
	14863	10 Jul 1884	Fort Chimo [= Kuujjuaq]	616		Fluid

Current Identification	USNM #	Date Collected	Location	Field #(s)	Other #(s)	Preparation(s)
Microtus pennsylvanicus	14864	10 Jul 1884	Fort Chimo [= Kuujjuaq]	617		Fluid
	14868	30 Jul 1884	Fort Chimo [= Kuujjuaq]	620		Fluid
	15231	-- --- 1884	Fort Chimo [= Kuujjuaq]	486		Fluid
	15232	-- Jun 1884	Fort Chimo [= Kuujjuaq]	507		Fluid
	15235	-- Jun 1884	Fort Chimo [= Kuujjuaq]	510		Fluid
	15236	-- Jun 1884	Fort Chimo [= Kuujjuaq]	511		Fluid
	15238	-- Jun 1884	Fort Chimo [= Kuujjuaq]	513		Fluid
	15239	-- Jun 1884	Fort Chimo [= Kuujjuaq]	514		Fluid
	15240	-- Jun 1884	Fort Chimo [= Kuujjuaq]	515		Fluid
	15241	-- Jun 1884	Fort Chimo [= Kuujjuaq]	516		Fluid
	15242	-- Jun 1884	Fort Chimo [= Kuujjuaq]	517		Fluid
	186495	15 Nov 1882	Fort Chimo [= Kuujjuaq]	296; CHM 5481/6566		Skin; Skull
	190720	28 Mar 1883	Fort Chimo [= Kuujjuaq]	5463/6151		Skin; Skull
	190721	1 Mar 1883	Fort Chimo [= Kuujjuaq]	5464/6152		Skin; Skull
	190722	12 Apr 1883	Fort Chimo [= Kuujjuaq]	5470/6157		Skin; Skull
	190723	-- Spring 1884	Fort Chimo [= Kuujjuaq]	5474/6564		Skin; Skull
	190724	28 Mar 1883	Fort Chimo [= Kuujjuaq]	5467/6562		Skull; Remainder in Fluid
	190725	-- Spring 1884	Fort Chimo [= Kuujjuaq]	5476/6565		Skull; Remainder in Fluid
	190726	-- Spring 1884	Fort Chimo [= Kuujjuaq]	5482/6567		Skull; Remainder in Fluid
	190727	19 Oct 1883	Fort Chimo [= Kuujjuaq]	5483/6568		Skull; Remainder in Fluid
	190740	-- Spring 1884	Fort Chimo [= Kuujjuaq]	5465/6153		Skin; Skull
	190741	-- Spring 1884	Fort Chimo [= Kuujjuaq]	5466/6154		Skin; Skull
	190742	-- Spring 1884	Fort Chimo [= Kuujjuaq]	5472/6563		Skin; Skull
	190743	-- Spring 1884	Fort Chimo [= Kuujjuaq]	5473/6569		Skull; Remainder in Fluid
	190744	-- Spring 1884	Fort Chimo [= Kuujjuaq]	CHM 5475		Fluid
	190745	-- Spring 1884	Fort Chimo [= Kuujjuaq]	CHM 5477		Fluid
	190746	-- Spring 1884	Fort Chimo [= Kuujjuaq]	CHM 5478		Fluid

Current Identification	USNM #	Date Collected	Location	Field #(s)	Other #(s)	Preparation(s)
Microtus pennsylvanicus	190747	8 May 1883	Fort Chimo [= Kuujjuaq]	CHM 5479		Fluid
	190748	-- Spring 1884	Fort Chimo [= Kuujjuaq]	CHM 5480		Fluid
	A36939	22 May 1883	Fort Chimo [= Kuujjuaq]	321		Skull
Dicrostonyx hudsonius	14189	17 Jul 1882	Davis Inlet	179		Fluid
	14191	25 Aug 1883	Fort Chimo [= Kuujjuaq]	216		Fluid
	14192	27 Aug 1882	Fort Chimo [= Kuujjuaq]	217		Fluid
	14193	14 Oct 1882	Fort Chimo [= Kuujjuaq]	284	A38896	Skull; Remainder in Fluid
	14208	12 May 1883	Fort Chimo [= Kuujjuaq]	316		Fluid
	14221	29 May 1883	Fort Chimo [= Kuujjuaq]	329		Fluid
	14222	29 May 1883	Fort Chimo [= Kuujjuaq]	330	A36940	Skull; Remainder in Fluid
	14227	23 Jun 1883	Fort Chimo [= Kuujjuaq]	338		Fluid
	14835	-- Spring 1884	Fort Chimo [= Kuujjuaq]	505		Skull (broken); Remainder in Fluid
	190375	27 May 1883	Fort Chimo [= Kuujjuaq]	325		Skull; Remainder in Fluid
	190376	12 May 1883	Fort Chimo [= Kuujjuaq]	315		Skull; Remainder in Fluid
	190377	8 May 1883	Fort Chimo [= Kuujjuaq]	312		Skull; Remainder in Fluid
	190378	20 May 1883	Fort Chimo [= Kuujjuaq]	320		Skull; Remainder in Fluid
Synaptomys borealis	14838	-- Spring 1884	Fort Chimo [= Kuujjuaq]	506	A24729	Skull; Remainder in Fluid
Myodes gapperi	186492	12 May 1883	Fort Chimo [= Kuujjuaq]	317; CHM 5471/6158	14209	Skin; Skull
Ondatra zibethicus	14163	29 Nov 1882	Fort Chimo [= Kuujjuaq]	1045		Skin
	14164	29 Nov 1882	Fort Chimo [= Kuujjuaq]	1046		Skin
	14165	10 Oct 1882	"Forks"	1178		Skin

Current Identification	USNM #	Date Collected	Location	Field #(s)	Other #(s)	Preparation(s)
Phenacomys ungava	186487	4 Feb 1883	Fort Chimo [= Kuujjuaq]	300; CHM 5484/6159		Skin; Skull
	186488	-- Spring 1884	Fort Chimo [= Kuujjuaq]	525; CHM 5468/6155		Skin; Skull
Erethizon dorsatum	14157	5 May 1883	Fort Chimo [= Kuujjuaq]	1741/2314	192612	Skin; Skull
	14158	5 May 1883	Fort Chimo [= Kuujjuaq]	1742/2315	192613	Skin; Skull
Lepus americanus	A23125	-- --- ----		2301		Skull
	A23134	-- --- ----		2328		Partial Skull: cranium only
	14152	17 Dec 1882	"Forks"	1340		Skin
	14153	5 May 1883	Fort Chimo [= Kuujjuaq]	1740/2313	A23128	Skin; Skull
	14789	5 Aug 1883	Fort Chimo [= Kuujjuaq]	4148		Skin: foot only
	14790	28 Nov 1883	"Forks"	4425		Skin
	14791	28 Nov 1883	"Forks"	4426		Skin
	14792	19 Nov 1883	"Forks"	4427		Skin
	A23119	-- --- ----		2185		Skull
	A23122	-- --- ----		2188		Skull
	A23123	-- --- ----		2189		Skull
Lepus arcticus	14151	19 Jul 1882	Solomons Island, Near Davis Inlet	1182		Skin
	A23126	-- --- ----		2303		Skull
	A23129	-- --- ----		No Number		Skull
	A23130	-- --- ----		2324		Skull
	A23131	-- --- ----		2325		Skull
	A23133	-- --- ----		2327		Skull
	14149	28 Sep 1882	Fort Chimo [= Kuujjuaq]	1180	A37138	Skin; Skull
	14150	20 Oct 1882	Fort Chimo [= Kuujjuaq]	1181	A38460	Skin; Skull
	14788	10 Jun 1883	Fort Chimo [= Kuujjuaq]	4086		Skin: headless
	14793	6 Jul 1884	Fort Chimo [= Kuujjuaq]	5720		Skin
	14794	20 Jul 1884	Fort Chimo [= Kuujjuaq]	5938		Skin
	15246	5 Aug 1883	Fort Chimo [= Kuujjuaq]	4148		Skin
	A23120	-- --- ----		2186		Skull
	A23121	-- --- ----		2187		Skull
	A23124	-- --- ----		2300		Skull
	A23127	-- --- ----		2304		Skull
	A23132	28 Sep 1882	Fort Chimo [= Kuujjuaq]	2326		Skull

Current Identification	USNM #	Date Collected	Location	Field #(s)	Other #(s)	Preparation(s)
Sorex cinereus	14183	15 Nov 1882	Fort Chimo [= Kuujjuaq]	295		Fluid
	14184	9 Mar 1883	Fort Chimo [= Kuujjuaq]	303	A38768	Skull; Remainder in Fluid
	14185	13 Apr 1883	Fort Chimo [= Kuujjuaq]	307	A38756	Skull; Remainder in Fluid
	15245	May-Jun 1884	Fort Chimo [= Kuujjuaq]	489	A38757	Skull; Remainder in Fluid
	304692	-- Spring 1884 *	Fort Chimo [= Kuujjuaq]	15243		Skull
Sorex hoyi	14869	30 Jul 1884	Fort Chimo [= Kuujjuaq]	621	A38848	Skull; Remainder in Fluid
Canis lupus	A23136	-- --- 1882	Fort Chimo [= Kuujjuaq]	2190		Skull
	A23137	-- --- ----	Fort Chimo [= Kuujjuaq], probably vicinity of	2302		Skull
	A23138	-- --- 188-	Fort Chimo [= Kuujjuaq], probably vicinity of	2305		Skull
	A23139	-- --- ----	Fort Chimo [= Kuujjuaq], probably vicinity of	2316		Skull
	A23140	-- --- 188-	Fort Chimo [= Kuujjuaq], probably vicinity of	2317		Skull
Vulpes lagopus	14160	10 Aug 1882	Koksoak River	1183		Skin
	14784	19 Nov 1883	Fort Chimo [= Kuujjuaq]	4423		Skin
	14785	11 Dec 1883	Fort Chimo [= Kuujjuaq]	4424		Skin
	239664	-- --- ----		No Number		Partial Skull: mandible
	A23143	-- --- ----		2170		Skull
	A23145	-- --- ----		2172		Skull
	A23146	-- --- ----		2175		Skull
	A23161	-- --- ----		2348		Skull
	A23162	-- --- ----		2367		Skull
	A23163	-- --- ----		2368		Skull
	A23164	-- --- ----		2369		Skull
	A23172	-- --- ----		2159		Skull
	A23173	-- --- ----		2160		Skull
	A23174	-- --- ----		2161		Skull

Current Identification	USNM #	Date Collected	Location	Field #(s)	Other #(s)	Preparation(s)
Vulpes lagopus	A23175	-- --- ----		2162		Skull
	A23176	-- --- ----		2163		Skull
	A23177	-- --- ----		2164		Skull
	A23178	-- --- ----		2165		Skull
	A23179	-- --- ----		2166		Skull
	A23180	-- --- ----		2167		Skull
	A23181	-- --- ----		2168		Skull
	A23182	-- --- ----		2169		Skull
	A23183	-- --- ----		2173		Skull
	A23184	-- --- ----		2308		Skull
	A23185	-- --- ----		2309		Skull
	A23186	-- --- ----		2342		Skull
	A23187	-- --- ----		2344		Skull
	A23189	-- --- ----		2349		Skull
	A23190	-- --- ----		2350		Skull
	A23191	-- --- ----		2351		Skull
	A23192	-- --- ----		2352		Skull
	A23193	-- --- ----		2353		Skull
	A23194	-- --- ----		2361		Skull
	A23195	-- --- ----	Fort Chimo [= Kuujjuaq]	2362		Skull
	A23196	-- --- ----		2363		Skull
	A23197	-- --- ----		2365		Skull
	A23198	-- --- ----		2366		Skull
	A23199	-- --- ----		2364		Skull
	A23200	-- --- 1882		2954		Skull
	A23201	-- --- 1882		2955		Skull
	A23202	-- --- ----		2956		Skull
	A23203	-- --- ----		2957		Skull
	A23204	-- --- 1882		2958		Skull
	A23205	-- --- 1882		2960		Skull
	A23206	-- --- 1882		2312		Skull
	A23210	-- --- ----		No Number		Skull
Vulpes vulpes	A23157	-- --- ----		2310		Skull
	A23158	-- --- ----		2341		Skull
	A23159	-- --- ----		2343		Skull

Current Identification	USNM #	Date Collected	Location	Field #(s)	Other #(s)	Preparation(s)
Vulpes vulpes	A23160	-- --- ----		2347		Skull
	A23165	-- --- ----		2370		Skull
	A23169	-- --- ----		2374		Skull
	A23170	-- --- ----		2375		Skull
	A23171	-- --- ----		2376		Skull
	A23188	-- --- ----		2346		Skull
	A23209	-- --- ----		No Number		Skull: cranium damaged
	14786	10 Feb 1885	Fort Chimo [= Kuujjuaq]	4471		Skin
	A23144	-- --- ----		2171		Skull
	A23147	-- --- ----		2176		Skull
	A23148	-- --- ----		2177		Skull
	A23149	-- --- ----		2178		Skull
	A23150	-- --- ----		2179		Partial Skull: cranium
	A23151	-- --- ----		2180		Skull
	A23152	-- --- ----		2181		Skull
	A23153	-- --- ----		2182		Skull
	A23154	-- --- ----		2183		Skull
	A23155	-- --- ----		2174		Skull
	A23156	-- --- ----		2306		Skull
	A23166	-- --- ----		2371		Skull
	A23167	-- --- ----		2372		Skull
	A23168	-- --- ----		2373		Skull: cranium slightly damaged
Ursus maritimus	A23207	30 Jul 1882 to 1 Aug 1882	Gull/Beacon Island, mouth of George River	2381		Partial Skull: cranium
Odobenus rosmarus	A22014	Jun 1883	Ungava Bay	2383		Skull
Lontra canadensis	14156	7 Apr 1883	"Forks"	1616/2307	A23217	Skin; Skull; Partial Skeleton
Gulo gulo	14781	28 Sep 1883	Fort Chimo [= Kuujjuaq], Koksoak River	4314		Skin
	141911	-- --- ----	Fort Chimo [= Kuujjuaq], vicinity of	2192		Skull

Current Identification	USNM #	Date Collected	Location	Field #(s)	Other #(s)	Preparation(s)
Gulo gulo	A23142	-- --- ----	Fort Chimo [= Kuujjuaq], vicinity of	2319		Skull
	A23254	21 Feb 1889	Fort Chimo [= Kuujjuaq], vicinity of	2191		Skull
	A23256	-- --- ----	Fort Chimo [= Kuujjuaq], vicinity of	2320		Skull
Martes americana	A23214	-- --- ----		2377		Skull
	A23215	-- --- ----		2184		Skull
Mustela erminea	A23216	-- --- ----		2329		Skull
	14161	1 Dec 1882	"Forks"	1339		Skin
	14186	8 May 1883	Fort Chimo [= Kuujjuaq]	313		Fluid
	14866	11 May 1884	Fort Chimo [= Kuujjuaq]	490		Fluid
Mustela nivalis	A23135	-- --- ----		No Number		Skull
Rangifer tarandus	14777	1 Feb 1884	Fort Chimo [= Kuujjuaq]	4430	A37110	Skin: skin is head only; Skull, with antlers
	14778	1 Feb 1884	Fort Chimo [= Kuujjuaq]	4431	A37111	Skin; Skull
	14779	1 Feb 1884	Fort Chimo [= Kuujjuaq]	4432	A37112	Skin; Skull
	14780	1 Feb 1884	Fort Chimo [= Kuujjuaq]	4433	A36932	Skin; Skull
	A21631	-- --- ----		6017		Partial Skull: no mandible; Antlers
	A21632	-- --- ----		6018		Skull; Antler or Horn
	A21633	-- --- ----		6019		Partial Skull: no mandible
	A21634	-- --- ----		6020		Partial Skull: no mandible
	A21635	-- --- ----		6021		Partial Skull: no mandible
	A21637	-- --- ----		6023		Partial Skull: no mandible; Antlers
	A21638	-- --- ----		6024		Partial Skull: no mandible
	A21639	-- --- ----		6025		Partial Skull: no mandible

Current Identification	USNM #	Date Collected	Location	Field #(s)	Other #(s)	Preparation(s)
Rangifer tarandus	A21640	-- --- ----		6026		Partial Skull: no mandible
	A21641	-- --- ----		6027		Partial Skull: no mandible
	A21642	-- --- ----		6028		Partial Skull: no mandible
	A21643	-- --- ----		6029		Partial Skull: no mandible; Antlers
	A21644	-- --- ----		6030		Partial Skull: no mandible; Antlers
	A21645	-- --- ----		6031		Partial Skull: no mandible; Antlers
	A21647	-- --- ----		6033		Partial Skull: no mandible; Antlers
	A21648	-- --- ----		6034		Skull
	A21649	-- --- ----		6035		Partial Skull: no mandible
	A21650	-- --- ----		6036		Partial Skull: no mandible
	A21651	-- --- ----		6037		Partial Skull: no mandible
	A38207	7 Oct 1883	Fort Chimo [= Kuujjuaq]	4315	14774	Skull; Antlers
Delphinapterus leucas	A23208	1882 or 1883	Ungava Bay	2382		Skull
	14867	25 Jul 1884	Fort Chimo [= Kuujjuaq]	622		Fluid: Fetus

Glossary

AVATAQ — An inflated sealskin bladder used as a buoyant float and often attached to the tether of a harpoon.

AVATAQ CULTURAL INSTITUTE — An Inuit owned and operated Institute mandated under the 1975 James Bay Northern Quebec Agreement to uphold and record the traditional ways of the Inuit of Northern Quebec, Nunavik. The Institute has full-time archaeologists and social scientists on staff and houses a number of artifacts from the Dorset and Thule Inuit periods.

CARNIVORA — The order comprising the carnivores or flesh-eating mammals. Lucien Turner used this order to categorize seals, walrus, bears, weasels, dogs, and cats.

CETACEA — An order of marine mammals, including the whales, dolphins, and porpoises. Like ordinary mammals, they breathe by means of lungs and bring forth living young, which they suckle for some time. Lucien Turner used this order to describe the white whale (beluga), bowhead whale, and narwhal.

COLOR MORPH — A color variant within a single species (for example, the different fur color variations of the red fox).

DENTARY — Either of a pair of bones of the lower jaw of most vertebrates.

ESKIMO — The former name used to describe the Inuit people of Canada. The term Eskimo is now considered pejorative. In linguistics, Eskimo is used to refer to the Eskimo-Aleut family of languages that is spoken by the northern peoples of Canada, Russia, Greenland, and Alaska.

FORT CHIMO — Now known as the Inuit community of Kuujjuaq in Nunavik, Northern Quebec. Ft. Chimo was the village that Lucien Turner was based in from 1882-1884. Kuujjuaq is situated 50 km inland of Ungava Bay on the banks of the Koksoak River, and is regarded as the political nerve-center of, and gateway to, the Nunavik homeland. It is the most populated Inuit community in Nunavik, with approximately 2000 residents. Ft. Chimo was a trading post operated by the Hudson's Bay Company in the 1830s. The Inuit and the Innu (Montagnais and Naskapi) regularly travelled to Ft. Chimo to trade with HBC during the time that Turner was based there. According to Turner (1887a: 8; 1887b: 222), the place name "Ft. Chimo" originates from the Inuit salutation term, *cháimo* or *chamái*. Note that the *"Eskimo"* designation derives from this salutation. It is said that the Naskapi would be greeted by the Inuit with *"Ai-ish-chi-ma-yu."*

GEORGE RIVER — The name often used to describe the Inuit community of Kangiqsualujjuaq. It is also the river that flows beside the community, beginning near Schefferville in the South and opening into Ungava Bay, about 15 km west of the Village.

HBC — The Hudson's Bay Company. This was an English company chartered in 1670 to trade in all parts of North America drained by rivers flowing into Hudson Bay. The HBC was actively operating in Ungava when Lucien Turner was in the region from 1882-1884.

HEIGHTH OF LAND — Correctly: Height of Land. The geographical term used during the 1800s for describing a watershed by early explorers and geographers in relation to the Ungava District.

HUDSON'S BAY TERRITORY — Also known as Rupert's Land or Prince Rupert's Land. This historical region was a territory in British North America and was nominally "owned" for 200 years from 1670 to 1870 by the Hudson's Bay Company. The Territory encompassed what is now known as Northern Ontario and Northern Quebec, as well as southern Nunavut, Manitoba, and some parts of Saskatchewan, Alberta, Minnesota, North Dakota, South Dakota, and Montana.

INNU — A term used to describe the aboriginal peoples of Northern Quebec and Labrador. The Innu generally inhabit the more forested tracts of land south of the land occupied by the Inuit. The Innu are comprised of the Montagnais (who largely identify with the land along the North Shore of the St Lawrence River) and Naskapi people (who largely identify with the lands North of the Montagnais), as well as several distinctions that are based on regional affiliations and dialects of the Innu language. Lucien Turner interacted with both the Montagnais and Naskapi groups during his time in Ungava and Labrador from 1882 to 1884.

INSECTIVORES — Small, principally nocturnal mammals of the order Insectivora, characteristically feeding chiefly on insects. Lucien Turner used this order to categorize moles and shrews.

INUIT — Formerly spelled by Turner and other early explorers and scientists as Innuit. The aboriginal peoples that inhabit the Arctic regions of Canada, Greenland, the Unites States, and Russia. The Inuit of the Nunavik region of Northern Quebec are descendants of the Thule people, who migrated to the region over 1000 AD. The Dorset culture occupied the Arctic previous to the arrival of the Thule. In Nunavik, the mother tongue of the Inuit is Inuktitut. The Inuit were formerly known as the Eskimo people.

INUKSUIT — Navigational marker, a cairn; made by Inuit to find their way. Inuksuk is singular.

INUKTITUT — The language of the Inuit of Nunavik and Labrador. It is sometimes referred to as Inuttitut.

JBNQA — The 1975 James Bay Northern Quebec Agreement, a land-claim treaty between Nunavik Inuit, James Bay Cree and the Governments of Canada and Quebec.

KANGIQSUALUJJUAMIUT — The Inuit people who live in Kangiqsualujjuaq. Local identity is maintained today by Inuit applying the suffix "miut" to the name of the village from which they originate. For example, a resident of Kangiqsualujjuaq is therefore a Kangiqsualujjuamiut – the last letter "q" is dropped for pronunciation of the morpheme.

KANGIQSUALUJJUAQ — An Inuit village on the banks of the George River estuary in Nunavik (Northern Quebec), Canada. Kangiqsualujjuaq (ᑲᖏᖅᓱᐊᓗᔾᔪᐊᖅ) is sometimes spelled as Kangirsualujjuaq (ᑲᖏᕐᓱᐊᓗᔾᔪᐊᖅ). The placename means "very large bay" in Inuktitut (Qumaq 1992: 121). Pronunciation guide: Kang-ik-swal-loo-jew-ak. "Kang" as in kangaroo; "iq" as in "ick" in tick; "su" as in "swa" in swallow; "al" as in "ull" in gull; "lu" as in loo; "jju" as in jew; and "aq" as in "ack" in pack. Kangiqsualujjuaq has also been historically known as Fort Severight, Fort George River, George River, and Port-Nouveau-Québec. The Hudson's Bay Company operated a post south of the community of Kangiqsualujjuaq called Illutaliviniq. The post operated during the periods of 1838-42, 1876-1915 and 1923-32. Lucien Turner visited the post on at least one occasion during his time in Ungava and Labrador. The community of Kangiqsualujjuaq, constructed in 1964, is positioned above a small cove of the George River called Akilasakallak, about 16 km downstream from the site of the now defunct Hudson's Bay Company trading post.

KHAIAK — See QAJAQ

LABRADOR COAST — An historical region. Regarded as the stretch of coastline between Cape Chidley and Hamilton Inuit. This region forms part of the Nunatsiavut homeland, an Inuit self-governing region that was created in 2000.

MANDIBLE — The lower jaw in vertebrates that is hinged to the open mouth.

MONTAGNAIS — See Innu.

NASKAPI — See Innu.

NMNH — National Museum of Natural History.

NUNAVIK — The territory of Nunavik, formerly Rupert's Land, and known through different epochs as

Novae Franciae, North East Territory (Low 1895), Arctic Quebec, the District of Ungava, Kativik (since 1975), Nord-du-Québec and Nouveau Québec (predominately by French-Canadians since 1912), was incorporated within the boundaries of Canada in 1867. The 1912 Boundaries Extension Act transferred federal (Dominion) jurisdiction of the District of Ungava (Nunavik) to the Province of Quebec on condition that outstanding indigenous rights to the District be settled. Although Nunavik is a culturally-bound territory to Quebec Inuit in its own right, it remains under the political umbrella, jurisdiction and legal framework of the Government of Quebec. Nunavik is the Inuit homeland to some 11,000 Inuit people of Northern Quebec who live within 14 communities. The territory of Nunavik comprises a land area of 443,684.71 km² above the 55th parallel. The Nunavik landscape is characterized by innumerable lakes, rivers, tributaries and fiords that abound with fish and wildlife. Four large rivers empty into the seas of Nunavik: the Great Whale River; Leaf River; Koksoak River and George River. The average daily temperature in the Southern Ungava Bay region of Nunavik is minus 6°C; the temperature in winter averages minus 30°C and in summer the temperature averages 12°C. Southern Ungava Bay receives about 280 mm of rainfall and 260 cm of snow per annum. Winters are generally long and cold and summers tend to be short, cool and moist. The sea-ice in the estuaries, inlets and near-shore environment breaks-up between May and June and forms again in December, but is often not safe to travel on by snowmobile until January. Bands of continuous and discontinuous permafrost keep the crust of the land frozen throughout the year.

NUNAVIMMIUT — The Inuit people of the Nunavik homeland.

NUNAVIMMIUTITUT — The Nunavimmiut dialect of Inuktitut in syllabics.

PELLAGE — The fur of a mammal (now spelled pelage).

PLANTERS — Term used to describe the Labrador/Newfoundland settlers of English origin. Many of the Planters were cod-fisherman or sealers.

QAJAQ or QAYAQ — Inuit word for kayak; sometimes spelled as Kaiak or Khaiak by Turner.

QALLUNAAT — Inuit term for describing a Caucasian, a Euro-Canadian, a Southerner, regardless of race, language, or ethnicity.

QAMUTIK — An Inuit sledge that is affixed to a snowmobile or dog team. Often made from marine plywood and/or fiberglass.

QUEBEC-LABRADOR PENINSULA — The north-eastern portion of Quebec and the northernmost tip of Labrador that is bordered by Ungava Bay, Hudson Strait and the Labrador Sea.

RODENTS — Mammals of the order Rodentia which are characterized by large incisors adapted for gnawing or nibbling. Turner used this order to categorize squirrels, beaver, muskrat, lemmings, voles, mice, and porcupines, as well as hares and rabbits (the latter now considered Lagomorphs).

SI — Smithsonian Institution.

SYLLABICS — An Inuit writing system used in Nunavik (Northern Quebec) and Nunavut. The system was first introduced by Anglican Missionary, Edmund Peck in the 1870s, and was adapted from Cree scripts.

TORNGAK — Legendary, spiritual being that inhabits the land and sea of the Quebec-Labrador Peninsula, and is particularly associated with cliffs and caves of the Torngat Mountains around the Abloviak and Nackvak Fiords. It is said that the Torgak can change their color, form and remain invisible. Also spelled as Turngait, Tuurnngait, and Torngat.

TYPE SPECIMEN — The museum specimen on which a scientific name (for example a new species or subspecies) is based.

TUNDRA — A geographical region in the Arctic where large trees are relatively absent and where the soil or subsoil is permanently frozen. In the Northern reaches of Quebec and Labrador the tundra consists largely of worn and weathered rock with patches of dwarf bushes, mosses, sedges, and grasses. The zone at which the boreal forest and the tundra meet in Northern Quebec and Labrador is known as the tree-line.

UMIAQ — A small sealskin boat, usually made from the skins of a bearded seal.

UNGAVA — An historical district sometimes referred to as the District of Ungava. From 1895-1920 it was part of Canada's Northwest Territories. In 1912, as part of the Boundaries Extension Act, the jurisdiction of the District of Ungava was transferred to the Province of Quebec on the condition that outstanding indigenous rights to the District be settled. The Ungava region encompassed the northern portion of what is now known as Quebec, as well as the interior of Labrador and the surrounding islands off the coast of Quebec and Labrador. The offshore islands now form part of the territory of Nunavut.

UNGAVA BAY — A large horseshoe-shaped bay that is situated between Hudson Strait to the north, the Quebec-Labrador Peninsula to the east, and the Ungava Plateau to the west. It drains several large rivers, including the Payne, Leaf, Koksoak, George, and Korac Rivers. Inuit communities are sparsely situated along its fringes, which includes: Quaqtaq, Kangirsuk, Aupaluk, Tasiujaq, Kuujjuaq, and Kangiqsualujjuaq. Akpatok Island is a large feature to the northwest of Ungava Bay, and is regularly frequented by walrus and polar bears. Ungava Bay experiences some of the largest tidal ranges in the world of up to 16 metres, and does not reach depths beyond 150 metres.

UNGULATE — An animal having hooves. Lucien Turner used this description to classify caribou and moose.

USNM — United States National Museum. From 1911 it became known as the Smithsonian Institution's National Museum of Natural History.

Bibliography

Allen, G.M. 1942. *Extinct and Vanishing Mammals of the Western Hemisphere with the Marine Species of All the Oceans.* American Committee for International Wild Life Protection, Cambridge, Massachusetts.

Anderson, R.M. 1934. "Mammals of the Eastern Arctic and Hudson Bay." *Canada's Eastern Arctic: its History, Resources, Population and Administration.* Ed. W.C. Bethune. Department of Interior, Ottawa: 67-108.

Anderson, R.M. 1948. "A Survey of Canadian Mammals of the North." *Annual Report of the Province of Quebec Association for the Protection of Fish and Game* 1948: 9-17.

Bailey, V. 1897. "Revision of the American Voles of the Genus *Evotomys*." *Proceedings of the Biological Society of Washington* 11: 113-138.

Bailey, V. 1898. "Descriptions of Eleven New Species and Subspecies of Voles." *Proceedings of the Biological Society of Washington* 12: 85-90.

Bailey, V. 1898. "Preliminary Description of *Microtus pennsylvanicus labradorius*." *Proceedings of the Biological Society of Washington* 12: 85-90.

Banfield, A.W.F. 1961. "A Revision of the Reindeer and Caribou Genus *Rangifer*." *Bulletin of the National Museum of Canada* 177: 1-137.

Bangs, O. 1898. "A List of Mammals of Labrador." *The American Naturalist* 32(379): 491.

Bangs, O. 1912. "List of the Mammals of Labrador." *Labrador, the Country and the People.* Ed. W.T. Grenfell. Macmillan, New York: 458-468.

Bell, R. 1884. *Observations on the Geology, Zoology, and Botany of Hudson's Strait and Bay [made in 1884].* Geological and Natural History Survey of Canada. Dawson Brothers, Montreal.

Bell, R. 1885. *Observations on the Geology, Zoology, and Botany of Hudson's Strait and Bay, Made in 1885.* Geological and Natural History Survey of Canada. Dawson Brothers, Montreal.

Bergerud, A.T., Luttich, S.N., & Camps, L. 2008. *The Return of the Caribou to Ungava.* McGill-Queens University Press, Montreal.

Bourquin, T. 1894 (1966). *Grammar of the Eskimo Language.* Trans. W.W. Perrett. Happy Valley, Labrador.

Burnham, D.K. 1992. *To Please the Caribou: Painted Caribou-Skin Coats Worn by the Naskapi, Montagnais, and Cree Hunters of the Quebec-Labrador Peninsula.* Royal Ontario Museum, Toronto.

Cabot, W.B. 1912. *In Northern Labrador.* Richard Badger, Boston.

COSEWIC. 2003. *COSEWIC Assessment and Update Status Report on the Wolverine* Gulo gulo *in Canada.* Committee on the Status of Endangered Wildlife in Canada, Ottawa.

Cronin, M.A., Macneil, M.D., & Patton, J.C. 2005. "Variation in Mitochondrial DNA and Microsatellite DNA in Caribou (*Rangifer tarandus*) in North America." *Journal of Mammalogy* 86: 495-505.

The Daily Republican. West Chester, Pennsylvania. Monday 15 June, 1885.

D'Anglure, B. Saladin. 1964. *L'organisation sociale traditionnelle de Esquimaux de Kangirsujuaq (Noveau Quebec)*. PhD Dissertation, University of Montreal.

D'Anglure, B. Saladin. 1984. "Inuit of Quebec." *Handbook of North American Indians: Arctic. Volume Five*. Ed. D. Damas. Smithsonian Institution, Washington, DC: 477-478.

Dorais, L-J. 1983. *Uqausigusiqtaat: An Analytical Lexicon of Modern Inuktitut in Quebec-Labrador.* Les Presses de l'Université Laval, Quebec.

Dorais, L-J. 2010. *The Language of the Inuit: Syntax, Semantics, and Society in the Arctic.* McGill Queen's University Press, Montreal.

Doutt, J.K. 1942. "A Review of the Genus *Phoca*." *Annals of the Carnegie Museum* 29(4): 61-125.

Doutt, J.K. 1954. "Observations on Mammals along the East Coast of Hudson Bay and the Interior of Ungava." *Annals of the Carnegie Museum* 33(14): 235-249.

Dunbar, M.J. 1952. "The Ungava Bay Problem." *Arctic* 5(4): 4-16.

Dunbar, M.J. 1954. "A Note on Climatic Change in the Sea." *Arctic* 7(1): 27-30.

Dunbar, M.J. 1983. "A Unique International Polar Year Contribution: Lucien Turner, Capelin, and Climate Change." *Arctic* 36(2): 204-205.

Elton, C. 1931. "Epidemics among Sledge Dogs in the Canadian Arctic and Their Relation to Disease in the Arctic Fox." *Canadian Journal of Research* 5: 673-692.

Elton, C.S. 1942. *Voles, Mice, and Lemmings: Problems in Population Dynamics*. Oxford University Press, Oxford.

Elton, C.S. 1954. "Further Evidence about the Barren-Ground Grizzly Bear in Northeast Labrador and Quebec." *Journal of Mammalogy* 35: 345-357.

Fabricus, O. 1780. *Fauna Groenlandica: systematice sistens animalia Groenlandiae occidentalis hactenus indagata, quoad nomen specificum, triviale, vernaculumque: synonyma auctorum plurium, descriptionem, locum, victum, generationem, mores, usum, capturamque singuli, prout detegendi occasio fuit : maximaque parte secundum proprias observationes.* Impensis Ioannis Gottlob Rothe, Copenhagen.

Ft. Chimo Journal. 1883. *Hudson Bay Company Journal Records on Microfilm,* Reel IM998. Library and Archives of Canada.

Goldman, E.A. 1937. "The Wolves of North America." *Journal of Mammalogy* 18: 37-45.

Hall, E.R. 1981. *The Mammals of North America*. Second Edition, two volumes. John Wiley and Sons, New York.

Harper, F. 1961. "Land and Fresh-Water Mammals of the Ungava Peninsula." *Miscellaneous Publications of the Museum of Natural History, University of Kansas* 27: 1-178.

Harper, F. 1964a. "The Friendly Montagnais and Their Neighbors in the Ungava Peninsula." *Miscellaneous Publications of the Museum of Natural History, University of Kansas* 37: 1-120.

Harper, F. 1964b. "Plant and Animal Associations in the Interior of the Ungava Peninsula." *Miscellaneous Publications of the Museum of Natural History, University of Kansas* 38: 1-58.

Hearne, S. 1795. *A Journey from Prince of Wales's Fort, in Hudson's Bay, to the Northern Ocean in the Years 1769, 1770, 1771 & 1772*. A. Strahan and T. Cadell, London.

Heyes, S.A. 2007. *Inuit Knowledge and Perceptions of the Land-Water Interface.* PhD Thesis. Department of Geography, McGill University, Montreal.

Heyes, S.A. 2010. *Interviews with Kangiqsualujjuamiut Elders on Mammals.* Trans. Lucina Annanack. Transcripts held on file by the author.

Heyes, S.A., Labond, C., & Annanack, T. 2003. *The Social History of Kangiqsualujjuaq, Nunavik.* Avataq Cultural Institute, Montreal, Canada.

Higdon, J.W., & Ferguson, S.H. 2011. "Reports of Humpback and Minke Whales in the Hudson Bay Region, Eastern Canadian Arctic." *Northeastern Naturalist* 18: 370–377.

Howell, A.B. 1926. "Voles of the Genus *Phenacomys*." *North American Fauna* 48: 1-66.

Hoyt, E. 1984. *The Whales of Canada*. Camden House Publishing, Camden East, Ontario.

Hutton, S.K. 1912. *Among the Eskimos of Labrador*. J.B. Lippincott, Philadelphia.

Jeddore, R. 1976. *Labrador Inuit Uqausingit*. St. John's Department of Education, St. John's, Newfoundland.

Kays, R.W., & Wilson, D.E. 2009. *Mammals of North America*. Second Edition. Princeton University Press, Princeton, New Jersey.

Loring, S., & Spiess, A. 2007. "Further Documentation Supporting the Former Existence of Grizzly Bear *(Ursus arctos)* in Northern Quebec-Labrador." *Arctic* 60: 7-16.

Low, A.P. 1896. "Report on Explorations in the Labrador Peninsula along the East Main, Koksoak, Hamilton, Manicuagan and Portions of Other Rivers, 1892-93-94-95." *Geological Survey of Canada Annual Report* (n.s.) 8: 1L-387L.

MacDonald, J. 2000. *The Arctic Sky: Inuit Astronomy, Star Lore, and Legend*. Royal Ontario Museum, Nunavut Research Institute, Toronto.

Makivik Corporation. 1984. *Kangiqsualujjuaq Land-Use Interviews*. Makivik Corporation, Montreal.

Mason, O.T. 1902. "Aboriginal American Harpoons: A Study of Ethnic Distribution and Invention." *Report of the United States National Museum for 1900*. Government Printing Office, Washington, DC, 189-304.

McLean, J. 1849. *Notes of a Twenty-Five Years' Service in the Hudson's Bay Territory*. Two volumes. Richard Bentley, London.

Merriam, C.H. 1889. "Description of Fourteen New Species and One New Genus of North American Mammals." *North American Fauna* 2: 1-52.

Merriam, C.H. 1902. "Four New Arctic Foxes." *Proceedings of the Biological Society of Washington* 15: 167-172.

Miller, G.S., Jr. 1899. "A New Polar Hare from Labrador." *Proceedings of the Biological Society of Washington* 13: 39-40.

Mitchell, E.D., & Reeves, R.R. 1988. "Records of Killer Whales in the Western North Atlantic, with Emphasis on Eastern Canadian Waters." *Rit Fiskideildar* 11: 161-193.

Mørk, T., & Prestrud, P. 2004. "Arctic Rabies – A Review." *Acta Veterinaria Scandinavica* 45 (1-2): 1-9.

National Anthropological Archives (NAA). Negative 3208. Photograph in L.M. Turner Collection. SPC E Canada Naskapi NM No ACC # Cat 175484 00297600. Smithsonian Institution, Washington, DC.

National Museum of Natural History. Ledger Book, entry 14159 in the "Skins" *Ledger Book,* Vol. 4, Division of Mammals Library, Washington, DC.

Payne, F.F. 1887-88. "Eskimo of Hudson's Strait." *Proceedings of the Canadian Institute* (7): 213-23.

Peacock, F.W. 1974. *Eskimo-English Dictionary*. Memorial University of Newfoundland, Saint-John.

Peck, E.J. 1925. *Eskimo-English Dictionary*. (Compiled from Erdman's Eskimo-German Edition 1864.) Ed. Walton, W.G. Church of the Ascension Thank Offering Mission Fund, Under the Direction of the General Synod of the Church of England in Canada, Hamilton.

Periodical Accounts (PA) "Relating to the Missions of the Church of the United Brethren Established Among the Heathen, 1790–1960." Extracts from the *Diary of Hebron and Extracts from the Diaries of Congregations in Labrador*. Moravian Records, London, UK.

Reeves, R.R., Finley, K.J., Mitchell, E., & MacDonald, J. 1986. "Strandings of Sperm Whales, *Physeter catodon*, in Ungava Bay, Northern Québec." *Canadian Field Naturalist* 100: 174-179.

Reeves, R.R., & Mitchell, E.D. 1988a. "Killer Whale Sightings and Takes by American Pelagic Whalers in the North Atlantic." *Rit Fiskideildar* 11: 7-23.

Reeves, R.R, & Mitchell, E.D. 1988b. "Distribution and Seasonality of Killer Whales in the Eastern Canadian Arctic." *Rit Fiskideildar* 11: 136-160.

Roth, J.D., Marshall, J.D., Murray, D.L., Nickerson, D.M., & Steury, T.D. 2007. "Geographical Gradients in Diet Affect Population Dynamics of Canada Lynx." *Ecology* 88: 2736-2743.

Schneider, L. 1970. *Dictionnaire Esquimau-Français du Parler de L'Ungava.* Les Presses de l'Université Laval, Quebec.

Schneider, L. 1985. *Ulirnaisigutiit: An Inuktitut-English Dictionary of Northern Quebec, Labrador and Eastern Arctic Dialects.* Trans. Dermot Ronan F. Collis. Les Presses de l'Université Laval, Quebec.

Schultz-Lorentzen, C.W. 1926 (re-issue 1967). *Dictionary of the West Greenland Eskimo Language.* Meddelelser om Grønland, Bind LXIX, C.A. Reitzels Forlag, Copenhagen.

Slough, B.G. 2007. "Status of the Wolverine in Canada." *Wildlife Biology* 13 (suppl. 2):76-82

Smith, R.J. 1999. *The Lacs des Loups Marins Harbour Seal,* Phoca vitulina mellonae *Doutt 1942: Ecology of an Isolated Population.* PhD Thesis. University of Guelph, Guelph, Canada.

Smith, R.J., Hobson, K.A., Koopman, H.N., & Lavigne, D.M. 1996. "Distinguishing between Populations of Fresh- and Salt-Water Harbour Seals (*Phoca vitulina*) Using Stable-Isotope Ratios and Fatty Acid Profiles." *Canadian Journal of Fisheries and Aquatic Sciences* 53: 272-279.

Smith, R.J., Lavigne, D.M., & Leonard, W.R. 1994. "Subspecific Status of the Freshwater Harbor Seal (*Phoca vitulina mellonae*): A Re-Assessment." *Marine Mammal Science* 10: 105-110.

Smithsonian Institution Archives. Record Unit 305, Accession 15388. List of supplies Turner sent from Ft. Chimo to Washington, DC, 1884.

Smithsonian Institution Archives. Record Unit 305, Accession 15388. Turner's shipping manifest of material sent from Ft. Chimo to Washington, DC, 1884.

Smithsonian Institution Archives. Record Unit 305, Accession 13922. Memo from Lucien M. Turner to Spencer F. Baird.

Spalding, A. 1998. *Inuktitut: A Multi-Dialectal Outline Dictionary.* Nunavut Arctic College, Iqaluit.

Spiess, A. 1976. "Labrador Grizzly (*Ursus arctos* L.): First Skeletal Evidence." *Journal of Mammalogy* 57: 787-790.

Spiess, A., & Cox, S. 1976. "Discovery of the Skull of a Grizzly Bear in Labrador." *Arctic* 29: 194-200.

Stirling, E. 1884. "The Grizzly Bear in Labrador." *Forest and Stream* 22: 324.

Strong, W.D. 1930. "Notes on Mammals of the Labrador Interior." *Journal of Mammalogy* 11: 1-10.

True, F.W. 1885. "Report upon the Department of Mammals of the U.S. National Museum for 1884." *Annual Report of the Board of Regents of the Smithsonian Institution for the year 1884.* Government Printing Office, Washington, DC, 130.

True, F.W. 1894. "Diagnoses of New North American Mammals." *Proceedings of the United States National Museum* 17: 241-243.

Turner, L.M. 1884. *Language of the "Koksoagmyut" Eskimo at Ft. Chimo, Ungava, Labrador Peninsula (1882-1884).* Three volumes, unpublished manuscript, National Anthropological Archives, BAE 2505-a. Museum Support Center, Smithsonian Institution, Suitland, Maryland.

Turner, L.M. 1885. "List of the Birds of Labrador, Including Ungava, East Main, Moose, and Gulf Districts of the Hudson Bay Company, together with the Island of Anticosti." *Proceedings of the U.S. National Museum* 8: 233-254.

Turner, L.M. 1886. *Contributions to the Natural History of Alaska: Results of Investigations Made Chiefly in the Yukon District and the Aleutian Islands; Conducted under the Auspices of the Signal Service, United States Army, Extending from May, 1874, to August, 1881.* Government Printing Office, Washington, DC.

Turner, L.M. 1886a. *Notes on the Mammals Ascertained to Occur in the Labrador, Ungava, East Main, Moose, and Gulf Region.* Unpublished manuscript. Smithsonian Institution Archives, Record Unit 7192, Box 2 of 2, Turner, Lucien M, 1848-1909, Lucien M. Turner Papers, Washington, DC.

Turner, L.M. 1886b. *Contributions to the Natural History of Labrador and Ungava, Hudson Bay Territory.* Unpublished manuscript. Smithsonian Institution Archives, Record Unit 7192, Box 2 of 2, Turner, Lucien M, 1848-1909, Lucien M. Turner Papers, Washington, DC.

Turner, L.M. 1886c. *Manuscripts on the Turner Natural History Collections from Labrador Including Works on Algae, by William Gilson Farlow; Botany, by Turner; Arachnids, by Turner and George Marx; Moths, by Charles Henry Fernald; Lepidoptera, by William Henry Edwards; Mollusks, by William H. Dall; Crustacea, by Turner and Sidney Irving Smith; and Fishes, Apparently by Turner.* Unpublished manuscript. Smithsonian Institution Archives, Record Unit 7192, Box 2 of 2 (2 folders), Turner, Lucien M, 1848-1909, Lucien M. Turner Papers, Washington, DC.

Turner, L.M. 1887a. *Descriptive Catalogue of Ethnological Collections Made by Lucien M. Turner in Ungava and Labrador, Hudson Bay Territory, June 24, 1882 to October 1884, Part 1 on Innuit.* Unpublished manuscript. Smithsonian Institution Archives, Record Unit 7192, Box 1 of 2 (3 folders), Turner, Lucien M, 1848-1909, Lucien M. Turner Papers, Washington, DC.

Turner, L.M. 1887b. *Descriptive Catalogue of Ethnological Collections Made by Lucien M. Turner in Ungava and Labrador, Hudson Bay Territory, June 24, 1882 to October 1884, Part 2 on Naskapi Specimens.* Unpublished manuscript. Smithsonian Institution Archives, Record Unit 7192, Box 1 of 2 (3 folders), Turner, Lucien M, 1848-1909, Lucien M. Turner Papers, Washington, DC.

Turner, L.M. 2001 (1894). *Ethnology of the Ungava District: Hudson Bay Territory.* Introduction by S. Loring, McGill-Queen's University Press, Montreal. Co-published by the Smithsonian Institution Press. Originally published in 1894 as part of the *Eleventh Annual Report of the Bureau of Ethnology (BAE), Smithsonian Institution, 1889-1890.*

Turner, L.M. 2010. *An Aleutian Ethnography.* Ed. Raymond L. Hudson. University of Alaska Press, Fairbanks.

Turner, L.W. 2008. *Lucien McShan Turner: Genealogy and History 1847 to 1909.* Milton-Freewater, Oregon.

Wilson, D.E., & Reeder, D.M. (Editors). 2005. *Mammal Species of the World: A Taxonomic and Geographic Reference.* Third Edition. Johns Hopkins University Press, Baltimore, Maryland.

Wright, M.O. 1898. *Four-Footed Americans and Their Kin.* Ed. F. M. Chapman. MacMillian Company, New York.

Letters from Lucien M. Turner

Letter from Lucien M. Turner to HBC Factor Samuel K. Parson of Montreal. Dated 6 Aug 1882. HBC Archives, A.10/110. Document consulted at the Smithsonian Archives, NAA, Museum Support Centre, Vertical File "T," Lucien M. Turner.

Letter from Lucien M. Turner to Major J. W. Powell. Dated 6 March 1886. Source: Box 89 (Thomas-Tylor). Series 1, Correspondence, Letters Received 1879-1887. Records of the Bureau of American Ethnology. National Anthropological Archives, Smithsonian Institution.

Letter from Lucien M. Turner to Robert Ridgway. Dated 2 June 1882, Ft. Chimo. *Robert Ridway Collection.* McGill University Rare Books and Special Collections, Montreal.

Letter from Lucien M. Turner to Robert Ridgway. Dated 1 Sept 1882, Ft. Chimo. *Robert Ridway Collection.* McGill University Rare Books and Special Collections, Montreal.

Letter from Lucien M. Turner to Robert Ridgway. Dated 28 Aug 1883, Ft. Chimo. *Robert Ridway Collection.* McGill University Rare Books and Special Collections, Montreal.

Letter from Lucien M. Turner to Robert Ridgway. Dated 3 Sept 1884, Ft. Chimo. SI Archives, Record Unit 305, Accession 15388.

Letter from Lucien M. Turner to Spencer F. Baird. Dated 10 Oct 1884, St John's Newfoundland. SI Archives, Record Unit 305, Accession 15388.

Letter from Lucien M. Turner to Spencer F. Baird. Dated 5 Dec 1881. SI Archives, Record Unit 305, Accession 15388.

Letter from Spencer F. Baird to William Armit, Secretary, HBC, London. Dated Washington, 17 March 1882. HBC Archives, A.10/110. Document consulted at the SI Archives, NAA, Museum Support Center, Vertical File "T," Lucien M. Turner.

About the Authors

Lucien McShan Turner (1847-1909) was an American naturalist and Arctic ethnographer with an extraordinary knowledge of birds, fish, and mammals. Working for the US Army Signal Service under the auspices of the Smithsonian Institution, he spent seven years taking meteorological observations and recording Aboriginal knowledge in Alaska from 1874 to 1881, as well as in Northern Quebec and Labrador for two years from 1882 to 1884. Two highly regarded books on his Arctic observations were published by the Smithsonian Institution: *Contributions to the Natural History of Alaska* (1886) and *Ethnology of the Ungava District, Hudson Bay Territory* (1894). His unpublished fieldnotes on mammals, recorded over 130 years ago, appear in this book for the first time.

Scott A. Heyes is a Research Associate at the Smithsonian Institution's Arctic Studies Center in the Department of Anthropology, National Museum of Natural History. He is also Assistant Professor in the Faculty of Arts and Design at the University of Canberra, Australia and is a Research Associate at the Frost Centre for Canadian and Indigenous Studies at Trent University, Canada. Scott has been conducting research on Inuit and Australian Aboriginal knowledge and perceptions of landscapes for more than a decade. He carried out his PhD research in the same region of Ungava Bay where Lucien Turner was based. He was the 2010-2011 Roberta Bondar Fellow in Canadian and Indigenous Studies at the Frost Center at Trent University. He holds a PhD in human geography from McGill University, Canada and holds graduate degrees in landscape architecture from the University of Adelaide, Australia. He is an avid outdoorsman, naturalist, and storyteller.

Kristofer M. Helgen is a Research Zoologist at the Smithsonian Institution and Curator-in-Charge of the Division of Mammals at the National Museum of Natural History in Washington, DC. He has undertaken research on mammals in 50 countries and in nearly 100 museums around the world. He earned a B.A. from Harvard University and a PhD from the University of Adelaide, Australia, and holds honorary positions at the American Museum of Natural History (New York), Bernice P. Bishop Museum (Honolulu, Hawaii), George Mason University (Fairfax, Virginia), and the National Geographic Society (Washington, DC). His interests include documenting mammalian biodiversity, preventing species extinctions, use and stewardship of natural history museum collections, and indigenous knowledge of mammals.